Lipases and Phospholipases in Drug Development

Edited by

Günter Müller and Stefan Petry

Further Titles of Interest

J. Östman, M. Britton, E. Jonsson (Eds.)

Treating and Preventing Obesity

2004. ISBN 3-527-30818-0

T. Dingermann, D. Steinhilber, G. Folkers (Eds.)

Molecular Biology in Medicinal Chemistry

2004. ISBN 3-527-30431-2

A. K. Duttaroy, F. Spener (Eds.)

Cellular Proteins and Their Fatty Acids in Health and Disease

2003. ISBN 3-527-30437-1

H. Buschmann et al. (Eds.)

Analgesics –
From Chemistry and Pharmacology to Clinical Application

2002. ISBN 3-527-30403-7

G. Molema, D. K. F. Meijer (Eds.)

Drug Targeting

2001. ISBN 3-527-29989-0

Lipases and Phospholipases in Drug Development

From Biochemistry
to Molecular Pharmacology

Edited by
Günter Müller and Stefan Petry

WILEY-VCH Verlag GmbH & Co. KGaA

Dr. Günter Müller
Dr. Stefan Petry
Aventis Pharma Germany
Industrial Park Höchst
65926 Frankfurt am Main
Germany
guenter.mueller@aventis.com
stefan.petry@aventis.com

Library of Congress Card No.: applied for

British Library Cataloguing-in-Publication Data
A catalogue record for this book is available from the British Library.

Bibliographic information published by Die Deutsche Bibliothek
Die Deutsche Bibliothek lists this publication in the Deutsche Nationalbibliografie; detailed bibliographic data is available in the Internet at <http://dnb.ddb.de>

Printed in the Federal Republic of Germany
Printed on acid-free paper

Composition K+V Fotosatz GmbH, Beerfelden
Printing Strauss Offsetdruck GmbH, Mörlenbach
Bookbinding Litges & Dopf Buchbinderei GmbH, Heppenheim

ISBN 3-527-30677-3

Contents

Preface *XIII*

List of Contributors *XV*

1 **Purification of Lipase** *1*
Palligarnai T. Vasudevan
1.1 Introduction *2*
1.2 Pre-purification Steps *2*
1.3 Chromatographic Steps *3*
1.4 Unique Purification Strategies *7*
1.5 Theoretical Modeling *9*
1.5.1 Model Formulation *9*
1.5.1.1 Mobile Phase *9*
1.5.1.2 Stationary Phase *10*
1.5.1.3 Boundary Conditions *10*
1.5.2 Solution *11*
1.5.3 Method of Moments *13*
1.5.4 Model Evaluation *15*
1.5.5 Simulation Results *16*
1.5.5.1 Effect of Feed Angle *16*
1.5.5.2 Effect of Flow Rate *17*
1.5.5.3 Effect of Rotation Rate *17*
1.5.5.4 Effect of Column Height *19*
1.6 Conclusions *19*
1.7 Acknowledgements *20*
1.8 References *20*

2 **Phospholipase A_1 Structures, Physiological and Patho-physiological Roles in Mammals** *23*
Keizo Inoue, Hiroyuki Arai, and Junken Aoki
2.1 Introduction *23*
2.2 Phosphatidylserine-specific Phospholipase A_1 (PS-PLA_1) *27*
2.2.1 Historical Aspects *27*

2.2.2 Biochemical Characterization and Tissue Distribution *27*
2.2.3 Structural Characteristics *29*
2.2.4 Substrate Specificity *29*
2.2.5 Possible Functions *30*
2.3 Membrane-associated Phosphatidic Acid-selective Phospholipase A_1s ($mPA-PLA_1\alpha$ and $mPA-PLA_1\beta$) *32*
2.3.1 Historical Aspects *32*
2.3.2 Characterization and Distribution *33*
2.3.3 Structural Characteristics *34*
2.3.4 Function *34*
2.4 Phosphatidic Acid-preferring Phospholipase A_1 ($PA-PLA_1$) *35*
2.4.1 Historical Aspects *35*
2.4.2 Characterization and Distribution *36*
2.4.3 Substrate Specificity *36*
2.4.4 Function *37*
2.5 KIAA0725P, a Novel PLA_1 with Sequence Homology to a Mammalian Sec23p-interacting Protein, p125 *37*
2.5.1 Historical Aspects *37*
2.5.2 Characterization and Distribution *37*
2.6 References *38*

3 Rational Design of a Liposomal Drug Delivery System Based on Biophysical Studies of Phospholipase A_2 Activity on Model Lipid Membranes *41*
Kent Jørgensen, Jesper Davidsen, Thomas L. Andresen, and Ole G. Mouritsen
3.1 Introduction *41*
3.2 Role for Secretory Phospholipase A_2 ($sPLA_2$) in Liposomal Drug Delivery *43*
3.3 Lateral Microstructure of Lipid Bilayers and its Influence on $sPLA_2$ *43*
3.4 $sPLA_2$ Degradation of Drug-delivery Liposomes: A New Drug-delivery Principle *46*
3.4.1 Liposomes Protected by Polymer Coating *46*
3.4.2 Biophysical Model Drug-delivery System to Study $sPLA_2$ Activity *47*
3.4.3 Effect of Lipid Composition on $sPLA_2$-triggered Drug Release and Absorption *48*
3.4.4 Effect of Temperature on Liposomal Drug Release and Absorption by $sPLA_2$ *49*
3.4.5 Liposomal Drug Release as a Function of $sPLA_2$ Concentration *50*
3.5 Conclusion *51*
3.6 Acknowledgments *51*
3.7 References *52*

4 **Phospholipase D** *55*
John H. Exton
4.1 Introduction *55*
4.2 Structure and Catalytic Mechanism of Mammalian Phospholipase D *56*
4.3 Cellular Locations of PLD1 and PLD2 *58*
4.4 Post-translational Modification of PLD *59*
4.5 Regulation of PLD1 and PLD2 *60*
4.5.1 Role of PIP_2 *60*
4.5.2 Role of PKC *61*
4.6 Role of Rho Family GTPases *64*
4.7 Role of Arf Family GTPases *65*
4.8 Role of Tyrosine Kinase *66*
4.9 Role of Ral *66*
4.10 Cellular Functions of PLD *66*
4.11 Role of PLD in Growth and Differentiation *67*
4.12 Role of PLD in Vesicle Trafficking in Golgi *68*
4.13 Role of PLD in Exocytosis and Endocytosis *68*
4.14 Role of PLD in Superoxide Formation *69*
4.15 Role in Actin Cytoskeleton Rearrangements *70*
4.16 Role in Lysophosphatidic Acid Formation *71*
4.17 Role of PA in Other Cellular Systems *71*
4.18 References *72*

5 **Sphingomyelinases and Their Interaction with Membrane Lipids** *79*
Félix M. Goñi and Alicia Alonso
5.1 Introduction and Scope
5.2 Sphingomyelinases *80*
5.2.1 Types of Sphingomyelinases *80*
5.2.1.1 Acid Sphingomyelinase (aSMase) *80*
5.2.1.2 Secretory Sphingomyelinase (sSMase) *81*
5.2.1.3 Neutral, Mg^{2+}-dependent Sphingomyelinases (nSMase) *81*
5.2.1.4 Mg^{2+}-independent Neutral Sphingomyelinases *84*
5.2.1.5 Alkaline Sphingomyelinase from the Intestinal Tract *85*
5.2.1.6 Bacterial Sphingomyelinase-phospholipase C *85*
5.2.2 Sphingomyelinase Mechanism *85*
5.2.2.1 Binding of Magnesium Ions *85*
5.2.2.2 Binding of Substrate *85*
5.2.2.3 Mechanism of Catalysis *86*
5.2.3 Sphingomyelinase Assay *88*
5.2.4 Sphingomyelinase Inhibitors *89*
5.3 Sphingomyelinase–Membrane Interactions *89*
5.3.1 Lipid Effects on Sphingomyelinase Activity *90*
5.3.2 Effects of Sphingomyelinase Activity on Membrane Properties *91*
5.3.2.1 Effects on Membrane Lateral Organization *91*

5.3.2.2 Effects on Membrane Permeability *93*
5.3.2.3 Effects on Membrane Aggregation and Fusion *94*
5.4 Acknowledgments *96*
5.5 References *96*

6 Glycosyl-phosphatidylinositol Cleavage Products in Signal Transduction *101*
Yolanda León and Isabel Varela-Nieto
6.1 Introduction *101*
6.2 GPI Structure and Hydrolysis by Specific Phospholipases *102*
6.3 Diffusible Factors and the Regulation of GPI Levels *104*
6.4 IPG Structure and Biological Activities *106*
6.5 GPI/IPG Pathway and the Intracellular Signaling Circuit *109*
6.6 Acknowledgments *112*
6.7 References *113*

7 High-throughput Screening of Hormone-sensitive Lipase and Subsequent Computer-assisted Compound Optimization *121*
Stefan Petry, Karl-Heinz Baringhaus, Karl Schoenafinger, Christian Jung, Horst Kleine, and Günter Müller
7.1 Introduction *121*
7.1.1 Lipases in Metabolism *121*
7.2 Lipases Show Unique Differences in Comparison to Other Drug Targets *122*
7.3 Lipase Assays *123*
7.4 Hormone-sensitive Lipase (HSL) as a Drug Target in Diabetes *125*
7.4.1 Biological Role of HSL *125*
7.4.2 Characteristics of HSL *126*
7.4.3 Inhibitors of HSL *128*
7.5 Perspective *134*
7.6 References *134*

8 Endothelial Lipase: A Novel Drug Target for HDL and Atherosclerosis? *139*
Karen Badellino, Weijun Jin, and Daniel J. Rader
8.1 Introduction *139*
8.2 Structure of Endothelial Lipase *140*
8.3 Tissue Expression of Endothelial Lipase and Its Implications *141*
8.4 Enzymatic Activity and Effects on Cellular Lipid Metabolism of Endothelial Lipase *142*
8.5 Regulation of Endothelial Lipase Expression *145*
8.6 Physiology of Endothelial Lipase *146*
8.7 Variation in the Human Endothelial Lipase Gene *149*
8.8 Endothelial Lipase as a Potential Pharmacologic Target *151*
8.9 References *151*

9 **Digestive Lipases Inhibition: an In vitro Study** *155*
Ali Tiss, Nabil Miled, Robert Verger, Youssef Gargouri, and Abdelkarim Abousalham
9.1 Introduction *155*
9.1.1 3-D Structure of Human Pancreatic Lipase *156*
9.1.2 3-D Structure of Human Gastric Lipase *158*
9.2 Methods for Lipase Inhibition *159*
9.2.1 Method A: Lipase/Inhibitor Pre-incubation *162*
9.2.2 Method B: Inhibition During Lipolysis *162*
9.2.3 "Pre-poisoned" Interfaces *163*
9.2.3.1 Method C *163*
9.2.3.2 Method D *163*
9.3 Inhibition of Lipases by E_{600} and Various Phosphonates *164*
9.3.1 Inhibition of PPL, HGL and RGL by Radiolabeled E_{600} *165*
9.3.2 Interfacial Binding to Tributyrin Emulsion of Native and Chemically Modified Digestive Lipases *167*
9.3.3 Inhibition of Lipases by Phosphonates and the 3-D Structures of Lipase-inhibitor Complexes *167*
9.3.3.1 Synthesis of New Chiral Organophosphorus Compounds Analogous to TAG *167*
9.3.3.2 The 2.46 Å Resolution Structure of the Pancreatic/Procolipase Complex Inhibited by a C_{11} Alkylphosphonate *170*
9.3.3.3 Crystal Structure of the Open Form of DGL in Complex with a Phosphonate Inhibitor *173*
9.4 Inhibition of Digestive Lipases by Orlistat *174*
9.4.1 Introduction *174*
9.4.2 Inhibition of Digestive Lipases by Pre-incubation with Orlistat (Method A) *175*
9.4.2.1 Inhibition of Gastric Lipases *175*
9.4.2.2 Inhibition of Pancreatic Lipases *176*
9.4.2.3 Kinetic Model Illustrating the Covalent Inhibition of HPL in the Aqueous Phase *180*
9.4.3 Inhibition of Digestive Lipases During Lipolysis (Method B) *181*
9.4.4 Inhibition of Digestive Lipases on Oil Emulsions "Poisoned" with Orlistat (Method C) *181*
9.4.5 Inhibition of Digestive Lipases on Oil Substrate "Poisoned" with Orlistat (Method D) *184*
9.4.5.1 Inhibition of Pancreatic Lipase on Emulsion "Pre-poisoned" with Orlistat *184*
9.4.5.2 Inhibition of Gastric and Pancreatic Lipases on Mixed Films Containing Orlistat *185*
9.4.5.3 Inhibition of Pancreatic Lipase on Oil Drop "Pre-poisoned" with Orlistat *185*
9.5 References *187*

10 **Physiology of Gastrointestinal Lipolysis and Therapeutical Use of Lipases and Digestive Lipase Inhibitors** *195*
Hans Lengsfeld, Gabrielle Beaumier-Gallon, Henri Chahinian, Alain De Caro, Robert Verger, René Laugier, and Frédéric Carrière
10.1 Introduction *195*
10.2 Tissular and Cellular Origins of HGL and HPL *196*
10.3 Hydrolysis of Acylglycerols by HGL and HPL *199*
10.3.1 Substrate Specificity *199*
10.3.2 Specific Activities of HGL and HPL *200*
10.3.3 Lipase Activity as a Function of pH *202*
10.3.4 Effects of Bile Salts on the Activity of HGL and HPL *202*
10.4 Gastrointestinal Lipolysis of Test Meals in Healthy Human Volunteers *204*
10.4.1 Test Meals *205*
10.4.2 Experimental Device for Collecting Samples *in vivo* *207*
10.4.3 Gastric and Duodenal pH Variations *207*
10.4.4 Lipase Concentrations and Outputs *207*
10.4.5 Lipolysis Levels *211*
10.5 HGL and HPL Stability *213*
10.6 Potential Use of Gastric Lipase in the Treatment of Pancreatic Insufficiency *215*
10.7 Inhibition of Gastrointestinal Lipolysis by Orlistat for Obesity Treatment *216*
10.7.1 The Lipase Inhibitor Orlistat *216*
10.7.2 Design of Clinical Studies for Quantification of Lipase and Lipolysis Inhibition *217*
10.7.3 HGL Inhibition by Orlistat *218*
10.7.4 HPL Inhibition by Orlistat *219*
10.7.5 Effects of Orlistat on Gastric Lipolysis *220*
10.7.6 Effects of Orlistat on Duodenal Lipolysis *221*
10.7.7 Effects of Orlistat on Overall Lipolysis *221*
10.7.8 Effects of Orlistat on Fat Excretion *221*
10.7.9 Weight Management by Orlistat in Obese Patients *222*
10.7.10 Conclusions *224*
10.8 References *224*

11 **Physiological and Pharmacological Regulation of Triacylglycerol Storage and Mobilization** *231*
Günter Müller
11.1 Metabolic Role of Triacylglycerol *231*
11.1.1 Triacylglycerol and Energy Storage *231*
11.1.2 Lipolysis and Re-esterification *234*
11.1.3 TAG Storage/Mobilization and Disease *236*
11.1.3.1 Diabetes Mellitus and Metabolic Syndrome *236*
11.1.3.2 Lipotoxicity *238*

11.1.3.2.1 β-Cells *238*
11.1.3.2.2 Cardiac Myocytes *239*
11.1.3.2.3 Molecular Mechanisms *240*
11.1.3.3 Inborn Errors of TAG Storage and Metabolism *241*
11.2 Components for TAG Storage and Mobilization *242*
11.2.1 TAG in Lipoproteins *242*
11.2.2 TAG in Adipose Cells *243*
11.2.2.1 Enzymes of TAG Synthesis *244*
11.2.2.2 Lipid Droplets *246*
11.2.2.2.1 Morphology and Lipid Composition *246*
11.2.2.2.2 Protein Composition *248*
11.2.2.2.3 Biogenesis *252*
11.3 Mechanism and Regulation of TAG Mobilization *259*
11.3.1 cAMP *259*
11.3.2 Phosphorylation of HSL *260*
11.3.3 Dephosphorylation of HSL *263*
11.3.4 Intrinsic HSL Activity *263*
11.3.5 Translocation of HSL *264*
11.3.5.1 Mechanism *264*
11.3.5.2 Involvement of Perilipins *266*
11.3.5.3 Involvement of Lipotransin *268*
11.3.6 Intrinsic Activity of HSL *270*
11.3.6.1 Feedback Inhibition *270*
11.3.6.2 Adipocyte Lipid-binding Protein *272*
11.3.7 Expression of HSL *274*
11.3.8 Release of Lipolytic Products *275*
11.3.8.1 FA Transport *275*
11.3.8.2 Glycerol Transport *276*
11.3.8.3 Cholesterol Transport *277*
11.4 Physiological, Pharmacological and Genetic Modulation of TAG Mobilization *278*
11.4.1 Muscle Contraction *278*
11.4.2 Nutritional State *279*
11.4.3 Hormones and Cytokines *279*
11.4.3.1 Insulin *279*
11.4.3.1.1 Molecular Mechanisms *279*
11.4.3.1.2 Desensitization *281*
11.4.3.2 Leptin *282*
11.4.3.3 Growth Hormone *283*
11.4.3.4 Glucose-dependent Insulinotropic Polypeptide *283*
11.4.3.5 TNF-α *283*
11.4.4 ASP *285*
11.4.5 Acipimox and Nicotinic Acid *286*
11.4.5.1 Mode of Action *287*
11.4.5.2 Molecular Mechanism *288*

11.4.5.3 Desensitization *289*
11.4.6 Glimepiride and Phosphoinositolglycans *290*
11.4.7 Differences in Regulation of TAG Storage and Mobilization between Visceral and Subcutaneous Adipocytes *292*
11.4.8 Up-/Down-regulation of Components of TAG Storage and Mobilization *294*
11.4.8.1 HSL *294*
11.4.8.2 ALBP *296*
11.4.8.3 Perilipin *297*
11.4.8.4 PKA *299*
11.4.8.5 ASP *300*
11.4.8.6 Caveolin *301*
11.5 Concluding Remarks *302*
11.6 References *303*

Subject Index *333*

Preface

Within the last decade the interest in lipases has increased dramatically, and no doubt, this interest will be maintained in the future. Novel and powerful tools of molecular biology, crystallography, NMR technology and molecular modeling will continue to reveal new amino acid sequences and three-dimensional structures of lipases. In addition, transgenic and knockout animal models as well as the use of specific inhibitors will add knowledge about their mode of interaction with lipid substrates, cleavage mechanism and physiological roles in humans.

After having clarified many basic aspects of lipolysis (i.e. structure, function and regulation of lipases and their catalytic mechanism) the focus will shift to the pathophysiological role of lipases in metabolic diseases and will result in strategies for pharmacological intervention.

The common interest of academic and industrial pharmaceutical research is based on the search for selective small molecule modulators of lipase activity for the development of drugs suited to the therapy of diabetes, atherosclerosis, obesity and for in depth analysis of the underlying molecular defects in nutritional signaling by lipolytic cleavage products.

The discovery of appropriate starting points for the development of potent and specific inhibitors is still a challenge. In addition to the usual problems of low abundance and purity that enzymologists and structural biologists are generally faced with, lipases present a unique additional difficulty. Unlike other hydrolytic enzymes, e.g. esterases or proteases, the substrates hydrolyzed by lipases are insoluble in water and therefore must be efficiently presented to the enzyme in a separate lipidic phase. The presence of a suitable second phase apparently leads to increased lipase activity and may effect subtle but critical changes to the enzyme's three-dimensional structure.

This biphasic substrate recognition represents something like the unifying theme present in all the contributions.

With the help of an international authorship, we have attempted to cover major aspects of lipase research, from genomics to drug discovery via validation of targets, structural biology, rational drug design, and drug–lipase interaction.

We are convinced that this book will be a valuable compendium for researchers already engaged in this fascinating field and will motivate talented young scientists to enter it.

Frankfurt, November 2003

Günter Müller
Stefan Petry

List of Contributors

Abdelkarim Abousalham
Laboratoire de Lipolyse Enzymatique
UPR 9025/CNRS
31 Chemin Joseph Aiguier
13402 Marseille Cedex 20
France

Alicia Alonso
Dpto. de Bioquímica
Fac. de Ciencias
Universidad del País Vasco
Apdo 644
48080 Bilbao
Spain

Thomas L. Andresen
LiPlasome Pharma A/S
Technical University of Denmark
Building 207
2800 Lyngby
Denmark

Junken Aoki
Faculty of Pharmaceutical Sciences
The University of Tokyo
7-3-1 Hongo, Bunkyo-ku
Tokyo 113
Japan

Hiroyuki Arai
Faculty of Pharmaceutical Sciences
The University of Tokyo
7-3-1 Hongo, Bunkyo-ku
Tokyo 113
Japan

Karen Badellino
Preventive Cardiovascular Medicine
and Lipid Research
University of Pennsylania School
of Medicine
421 Curie Blvd.
Philadelphia, PA 19104
USA

Karl-Heinz Baringhaus
Aventis Pharma Germany
Industrial Park Höchst, Bldg. G 878
65926 Frankfurt am Main
Germany

Gabrielle Beaumier-Gallon
Laboratoire de Lipolyse Enzymatique
UPR 9025/CNRS
31 Chemin Joseph Aiguier
13402 Marseille Cedex 20
France

Frédéric Carrière
Laboratoire de Lipolyse Enzymatique
UPR 9025/CNRS
31 Chemin Joseph Aiguier
13402 Marseille Cedex 20
France

Henri Chahinian
Laboratoire de Lipolyse Enzymatique
UPR 9025/CNRS
31 Chemin Joseph Aiguier
13402 Marseille Cedex 20
France

Jepser Davidsen
LiPlasome Pharma A/S
Technical University of Denmark
Building 207
2800 Lyngby
Denmark

Alain De Caro
Laboratoire de Lipolyse Enzymatique
UPR 9025/CNRS
31 Chemin Joseph Aiguier
13402 Marseille Cedex 20
France

John H. Exton
Dept. of Molecular Physiology
& Biophysics
Vanderbilt University School
of Medicine
831 Light Hall
Nashville, TN 37232-0295
USA

Youssef Gargouri
Unité de Lipolyse Enzymatique
ENIS – Ecole Nationale d'Ingénieurs
de Sfax
BPW
3038 Sfax
Tunisia

Félix Goñi
Dpto. de Bioquímica
Fac. de Ciencias
Universidad del País Vasco
Apdo 644
48080 Bilbao
Spain

Keizo Inoue
Faculty of Pharmaceutical Sciences
Teikyo University
Sagamiko, Tsukui
Kanagawa 199-0195
Japan

Weijun Jin
Preventive Cardiovascular Medicine
and Lipid Research
University of Pennsylvania School
of Medicine
421 Curie Blvd.
Philadelphia, PA 19104
USA

Kent Jørgensen
LiPlasome Pharma A/S
Technical University of Denmark
Building 207
2800 Lyngby
Denmark

Christian Jung
Aventis Pharma Germany
Industrial Park Höchst, Bldg. G 878
65926 Frankfurt am Main
Germany

Horst Kleine
Aventis Pharma Germany
Industrial Park Höchst, Bldg. G 878
65926 Frankfurt am Main
Germany

René Laugier
Laboratoire de Lipolyse Enzymatique
UPR 9025/CNRS
31 Chemin Joseph Aiguier
13402 Marseille Cedex 20
France

Hans Lengsfeld
F. Hoffmann-La Roche Ltd.
Pharmaceuticals Division
CH-4070 Basel
Switzerland

Yolanda León
Departamento de Biología
Universidad Autónoma de Madrid
Carretera de Colmenar Km 15
Cantoblanco
28049 Madrid
Spain

NABIL MILED
Laboratoire de Lipolyse Enzymatique
UPR 9025/CNRS
31 Chemin Joseph Aiguier
13402 Marseille Cedex 20
France

OLE G. MOURITSEN
Department of Physics
University of Southern Denmark
Campusvej 55
5230 Odense M
Denmark

GÜNTER MÜLLER
Aventis Pharma Germany
DG Metabolic Diseases
Industrial Park Höchst, Bldg. H 825
65926 Frankfurt am Main
Germany

STEFAN PETRY
Aventis Pharma Germany
Industrial Park Höchst, Bldg. G 878
65926 Frankfurt am Main
Germany

DANIEL J. RADER
Preventive Cardiovascular Medicine
and Lipid Research
University of Pennsylvania School
of Medicine
654 BRB II/III, 421 Curie Blvd.
Philadelphia, PA 19104
USA

KARL SCHOENAFINGER
Aventis Pharma Germany
Industrial Park Höchst, Bldg. G 878
65926 Frankfurt am Main
Germany

ALI TISS
Laboratoire de Lipolyse Enzymatique
UPR 9025/CNRS
31 Chemin Joseph Aiguier
13402 Marseille Cedex 20
France

ISABEL VARIELA-NIETO
Inst. de Investigaciones Biomedicas
Alberto Sols
CSIC-Universidad Autónoma
de Madrid
c/Arturo Duperier, 4
28029 Madrid
Spain

PALLIGARNAI T. VASUDEVAN
Department of Chemical Engineering
University of New Hampshire
Durham, NH 03824
USA

ROBERT VERGER
Laboratoire de Lipolyse Enzymatique
UPR 9025/CNRS
31 Chemin Joseph Aiguier
13402 Marseille Cedex 20
France

1
Purification of Lipase

Palligarnai T. Vasudevan

Abbreviations

a	hydrodynamic radius of the solute (m)
C_m	mobile phase concentration (mol l^{-1})
C_f	feed concentration (mol l^{-1})
C_s	stationary phase concentration (mol l^{-1})
C'_m	normalized mobile phase concentration
C'_s	normalized stationary phase concentration
$\overline{C}'_m$	normalized mobile phase concentration in the Laplace domain
$\overline{C}'_s$	normalized stationary phase concentration in the Laplace domain
C_D	drag coefficient
D_m	dispersion coefficient, $m^2\ s^{-1}$
D_s	diffusion coefficient in the stationary phase, $m^2\ s^{-1}$
D_∞	bulk diffusivity, $m^2\ s^{-1}$
k	mass transfer coefficient, m s^{-1}
K_{eq}	equilibrium partition coefficient
L	length of the column, m
m_n	n^{th} moment about the origin
N	Avogadro number
Nu_s	Nusselt number in the stationary phase
Nu_m	Nusselt number in the mobile phase
Pe	Peclet number
R	particle size, m
r_0	pore radius, m
Re	Reynolds number
S	Particle surface area per unit column volume, m^{-1}
Sh	Sherwood number
Sc	Schmidt number

Greek letters

β	internal porosity
ε	external porosity
ν	kinematic viscosity, $m^2\ s^{-1}$

Lipases and Phospholipases in Drug Development
Edited by G. Müller and S. Petry

ISBN: 3-527-30677-3

θ_f feed angle (deg)
ω rotation rate (deg h^{-1})
μ viscosity of the solvent, kg m^{-1} s^{-1}
μ'_n n^{th} normalized moment
μ_n n^{th} moment about the mean

1.1 Introduction

Lipases (EC 3.1.1.3) find wide use, ranging from the food industry, e.g. in cheese making, to manufacturing, where they can act as catalysts of high specificity and selectivity. Research has focused on the purification and the characterization of lipases from sources as diverse as the unicellular *Pseudomonas putida* to cod (*Gadus morhua*). Aires-Barros et al. [1] have reviewed the isolation and purification of lipases, mainly from microbial and mammalian sources, while different purification techniques have been reviewed by Palekar et al. [2].

1.2 Pre-purification Steps

Lipases obtained from different sources are usually subjected to certain pre-purification steps before they are purified further. Typically, this is a one-step procedure involving precipitation by saturation with an ammonium sulfate [$(NH_4)_2SO_4$] solution. The lipase is thus separated from the extract solution. It can then be subjected to more specific purification steps. In some cases [3–9], the solution is concentrated by ultrafiltration to reduce the volume of the solution, and is then subjected to ammonium sulfate precipitation. The increase in lipase activity depends on the concentration of the ammonium sulfate solution used. Pabai et al. [10] demonstrated that the maximum increase in lipase activity occurred between 20–40% of saturation, with a 19-fold increase in purification level. Ammonium sulfate precipitation can be combined with other purification steps such as acid precipitation.

Other pre-purification steps are summarized in Tab. 1.1 (note that the purification factors are typical values, and where the range cannot be established, end-values are reported). Despite the wide range of lipase sources used, the purification levels obtained from any one pre-purification protocol (for example, ammonium sulfate precipitation) remain within a certain range.

Tab. 1.1 Summary of the pre-purification steps used and the purification levels typically attained.

No	Technique used	X-fold increase	Reference	Source
1	Ammonium sulfate precipitation	2 (range: 2–4)	11	Human Pancreatic Lipase in V79 Chinese hamster lung cells
2	Ultrafiltration	1.1 (range: 1–2)	5	*Pseudomonas putida* 3SK
3	Ultrafiltration and ammonium sulfate precipitation	12	8	*Serratia marcescens* Sr41 8000
4	PEG suspension	2.3	12	*Manduca sexta*
5	Hexanol extraction and Methanol precipitation	2.3	13	*Bacillus thermocatenulatus* (DSM 730)
6	Triton X-100 extraction	35	14	Human Hepatic Lysosomal Acid Lipase
7	Acrinol treatment	50	15	*Bacillus* sp.
8	Isoelectric focusing	7	16	*Pseudomonas* nov. sp. 109
9	Alcohol treatment	1.1	17	*Aspergillus niger*
		10	18	*Penicillium camembertii* U-150
10	Acid treatment	1.8	17	*Aspergillus niger*
		6.2	10	*Pseudomonas fragi* CRDA 323
11	Acetone precipitation	2.1	19	*Pseudomonas* sp. KWI-56
		7.1	20	Rape (*Brassica napus*) seedling

1.3 Chromatographic Steps

Most biological materials constitute themselves into a clear or a nearly clear solution for direct application to chromatographic columns after centrifugation or filtration. Almost all purification protocols use chromatographic steps after the pre-purification steps. Normally, a single chromatographic step is not sufficient to obtain the required level of purity. Hence, a combination of chromatographic steps is required. As a rule of thumb, to get a lipase of high purity, as measured by the level of purification (purification fold) and loss of activity (specific activity), at least four chromatographic steps are needed.

One might expect the order in which chromatographic steps are applied in a lipase purification protocol to be of minor importance. However, the actual situation is far from ideal. For example, a gel filtration step might be optimized to give extremely high levels of purity for the selected protein but only at the cost of time and sample volume. Likewise the selection of affinity chromatography as the first step would result in an extremely high purification factor. However, the cost of the adsorbent makes the use of smaller columns and repeated injections mandatory. Hence, the required process time and the possibility of product loss with/without structural modifications increase. Consequently, there are several practical rather than theoretical reasons why one should choose certain chromatographic techniques for the early steps and others for the final steps of a protein purification process. The choice is primarily governed by (1) the sample volume, (2) the protein concentration and viscosity of the sample, (3) the degree of purity of the protein product, (4) the presence of nucleic acids, pyrogens and proteolytic enzymes in the sample, and (5) the ease with which different types of adsorbents can be washed free from adsorbed contaminants and denatured protein. The last parameter governs the life of the adsorbent and, together with its purchasing price, the material cost of the particular purification step [21].

The logical sequence of chromatographic steps would be to start with the more robust techniques that combine a concentration effect with high chemical and physical resistance and low material cost. The obvious candidates are ion-exchange chromatography and, to some extent, hydrophobic-interaction chromatography. As the latter often requires the addition of salt for adequate protein binding, it is preferably applied after salt precipitation or after salt displacement from ion-exchange chromatography, thereby excluding the need for a desalting step. Thereafter, the protein fractions can be applied to a more specific and more expensive adsorbent. The protocol is often finished with a gel filtration step.

It is advisable to design the sequence of chromatographic steps such that buffer changes and concentration steps are avoided. The peaks eluted from an ion exchanger can, regardless of the ionic strength, be applied to a gel filtration column. This step also functions as the desalting procedure which means that the buffer used for the gel filtration should be chosen so as to allow the direct application of the eluted peaks to the next chromatographic step. The different chromatographic techniques have widely different capacities, even though several of the methods can be applied on a larger scale. However, in the initial stages of a purification scheme, it is most convenient to start with the methods that allow the application of large volumes and which have the highest capacities. Ion-exchange chromatography and hydrophobic-interaction chromatography belong to this category, but any adsorption chromatographic method can be used to concentrate larger volumes, especially in batch operations [21].

The final step aims to remove possible aggregates or degradation products and to condition the purified protein for its use or storage. The procedure will thus be different depending on the fate of the lipase. Aggregates and degradation products are preferably removed by gel filtration and if the protein is to be lyophilized, this step is also used for transferring the protein to a volatile buffer. This can sometimes be done by ion-exchange chromatography, but other forms of chromatography can, rarely, do this. If the protein solution is to be frozen, stored as a solution or used immediately the requirements for specific buffer salts might be less stringent. Several of the adsorption chromatographic techniques might be adapted to give peaks of reasonably high protein concentration. This is an advantage when gel filtration is chosen as the final step. Gel filtration will always dilute the sample and is often followed by a concentration step. The impact of an ion exchange step after the gel filtration step is well illustrated by analyzing data in the purification of lipase from *Penicillium camembertii* [18]. The impact of the latter step is minimized due to the use of a more specific purification step before it. Conversely, an ion exchange step before the gel filtration step is remarkably efficient.

Of crucial importance to the purification process is knowledge of the stability of the lipase in solution, and of the presence of any interfering activities and proteins in solution. All these contribute to an increased level of difficulty in handling the protein source. As may be expected, traditional animal and microbial sources can be replaced by genetically engineered microorganisms or cultured eukaryotic cells. By proper selection of a secreting strain, a considerable degree of purification is achieved at this early stage. This is more than ably demonstrated by the number of published instances of lipase expression through cloning [6, 22–25]. In fact, Simons et al. [24] conclude that for the same number of steps the cloning route leads to a 50-fold higher purification than that obtained when the species is not cloned. An analysis of the purification data for higher plants and animals and for microbial sources for cloned species confirms this observation.

Generally, lipases purified from higher plant and animal sources go through many purification steps (including numerous chromatographic steps) to attain the same level of purification as those obtained from microbial sources. The specific activities of the lipases from the former sources are also relatively lower than those of the latter. Studies dealing with the purification of Human Hepatic Lysosomal Acid Lipase [14], purification of Human Gastric Lipase [26], and the purification of a mono-acylglycerol associated with human erythrocytes [27] start out with the same range of specific activity, 0.2–0.6 mU mg^{-1}, and, except for the one dealing with human gastric lipase, they require a many steps to attain greater purification levels. Lipases obtained from higher plants and animals, however, seem to be more stable (activity loss during the actual purification steps) than those obtained from microbial sources. In fact, there are few reported instances of instability in a protein from a higher plant or animal source [12, 28].

Ncube et al. [20] have reported the purification lipase from rape (*Brassica napus*) seedlings by a scheme involving homogenization, centrifugation, chromatography on DEAE-Sephadex, Octyl-Sepharose, and finally Sephacryl S-300 (Tab. 1.2). The extremely high level of purification obtained by hydrophobic-interaction chromatography clearly dominates the purification protocol followed. A 20-fold purification is obtained over the previous step, i.e., ion-exchange chromatography using DEAE-Sephadex, which is not matched by the other purification steps. This suggests the use of hydrophobic-interaction chromatography as one of the finishing steps.

Tab. 1.2 Purification data for rape (*Brassica napus*) seedling lipase – Ncube et al. [20].

Purification step	*Nature of chromatography*	*Specific activity (U mg^{-1})*	*% Activity recovery*	*Purification fold*
100 000 g supernatant	–	6.7	100	1
Acetone precipitate	–	47.4	80	7.1
DEAE-Sephadex	Ion exchange	94.8	78	14.2
Octyl-Sepharose	Hydrophobic interaction	1900	31	283
Sephacryl S–300	Gel filtration	2140	16	321

Hydrophobic-interaction chromatography (HIC) and affinity chromatography (AFC) [29–31] have been used in lipase purification. The ligands are very specific and each ligand is only applicable for the separation of the lipase from a certain source. Polypropylene glycol is reported to be a suitable ligand for the fractionation of *Chromobacterium viscosum* lipase [29].

The molecular rationale for using hydrophobic chromatography is seen in the importance of the hydrophobic surface regions on the lipase protein. The hydrophobic nature of the surface (pocket or lid) of the lipase is due to alkyl groups, which are the source of the different enzymatic activities of heterogeneous lipase. The catalytic properties of lipases (selectivity, stereospecificity) can be easily modulated by the reaction conditions. This dependence of the enzyme properties on their environment may be a consequence of their complex mechanism of action (interfacial activation) that involves significant conformational changes of the enzyme structure [32]. This suggests that lipases may be very susceptible to alternations of the interactions between their hydrophobic surface and the medium, which can modify the equilibrium between closed and open structures and, perhaps, the exact shape of the active center [33].

Lipases are interphase-active enzymes with hydrophobic domains. The hydrophobic surface (loop) on lipase is thought to enable lipophilic interfacial binding with substrate molecules that actually induces the conformational changes in lipases. The open conformation will provide substrate with access to the active site, and vice versa. In certain types of lipases, the movements of a short α-helical hydrophobic loop in the lipase structure cause a conformational change that exposes the active sites to the substrate. This movement also increases the nonpolarity of the surface surrounding the catalytic site [30, 32, 34, 35]. Obviously, the hydrophobic surface plays an important role in the activity of lipase as an enzyme.

Investigators have examined the purification of lamb pre-gastric lipase [36], of dog gastric lipase [37], of diacylglycerol lipase from bovine aorta [38], of intestinal acid lipase from rat [39], purification and characterization of bovine pancreatic lipase [40], purification and characterization of rat phospholipase [41], and the purification of lipoprotein lipase from different rat tissues [42]. An important conclusion is that lipases from different tissues subjected to identical purification steps are purified to the same extent. Cod (*Gadus morhua*) lipases have been treated in two investigations [43, 44], which have dealt with two lipases from the same parent source but from different tissue. An identical number of purification steps gave the same extent of purification, lending further credence to the assumption that lipases from the same source show similar extents of purification.

Thermophilic bacteria, and their possible practical use, have attracted much interest, as is reflected in the research available for the purification of thermally stable lipases from such bacteria [9, 13, 15, 19, 45, 46]. The microbial sources give extremely low purification levels after the precipitation steps. Sugihara et al. [15] attribute this to a viscous material excreting into the culture fluid that made the salting out of the enzyme incomplete. However, ion-exchange chromatography followed by gel filtration chromatography results in about a 200-fold purification of lipase.

1.4 Unique Purification Strategies

While most purification strategies used chromatography as their mainstay, a few investigators have used entirely different strategies, not always involving the use of chromatography. These efforts [47–50] have concentrated on trying to make the process of lipase purification continuous and/or one that enables easy scale-up. Making the process continuous should make the use of lipases as industrial catalysts more common. These methods are discussed below.

Bompensieri et al. [51] have described the rapid purification of lipase from *Acinetobacter calcoaceticus* in temperature-induced aqueous two-phase systems. The influence of ion strength and Triton X-114 concentration on the extractive purification of the lipase was examined. The rapid procedure allows direct extraction from culture broths in three successive steps with no additional operations. This technique enabled them to extract 68% of the enzyme and achieve a purification factor of 41. Bompensieri et al. [47] have also compared the performance of a hydrophobic-interaction chromatography column to that of various aqueous two-phase systems (formed by either polyethylene glycol and dextran, or those based on detergents) for the purification of lipase from *Acinetobacter calcoaceticus* AAC323-I. Triton X-114 gave the best performance and although the yield obtained (81%) is greater than that with hydrophobic-interaction chromatography (42%), the purification factor of 68 obtained in the former case is low compared to the 140 obtained from chromatographic separation. The investigators claim ease of scale-up owing to the simplicity of the method.

Sztajer and Bryjak [48] have taken an entirely different approach to the purification of the lipase from *Pseudomonas fluorescens* by investigating the use of ultrafiltration capillary membranes. A two-step procedure involving continuous fractionation of the protein on polyacrylonitrile membrane followed by concentration on polysulfone membranes is suggested for continuous lipase recovery. They observe that the permeate fluxes through both membranes are similar, thereby suggesting that changing the production scales should not be difficult.

Terstappen et al. [49], like Bompensieri et al. [47], have investigated the use of detergent-based aqueous two-phase systems to purify a lipase from *Pseudomonas cepacia* DSM 50181. An investigation of the partitioning of a variety of extracellular lipases revealed that all procaryotic lipases showed a preference for the detergent-based coacervate phase. The eucaryotic lipases were significantly excluded from this phase, which is attributed to the glycosylation of the protein. A single purification step gave a four-fold concentration of the lipase and a purification factor of 24.

Genest et al. [50] have investigated the purification of crude porcine lipase using a continuous rotating annular chromatograph. The purification of the lipase was carried out using single-step size-exclusion chromatography with Sephadex and Sephacryl packing materials. They found that the batch runs showed comparable purification fold and activity recovery values to those of continuous runs. The major advantage of using such a purification technique, according to Genest et al., is the higher rate of lipase purification and the ease of scale-up. Vasudevan et al. [52] optimized the purification process in a rotating annular size-exclusion chromatography unit. The parameters that were optimized for efficient functioning of the column included mobile phase flow rate, lipase flow rate, introduction of multiple feeds to maximize throughput of the column and rotation rate. Continuous annular chromatography can be used to recover and purify products from multicomponent mixtures at high productivities. Both size-exclusion packing and ion-exchange packing can be used in the annulus. The chromatogram of a continuous separation run using Sephadex with two feeds is shown in Fig. 1.1. The parameters for the chromatogram are presented in Tab. 1.3. As can be seen, the purification fold factors are 11.9 and 11.5 for the two lipase fractions recovered, and the activity recoveries are 53.4 and 49.8%, respectively. The productivities of the two peaks are 3.5 and 3.2 mg lipase (mg gel-h)$^{-1}$, respectively. The introduction of

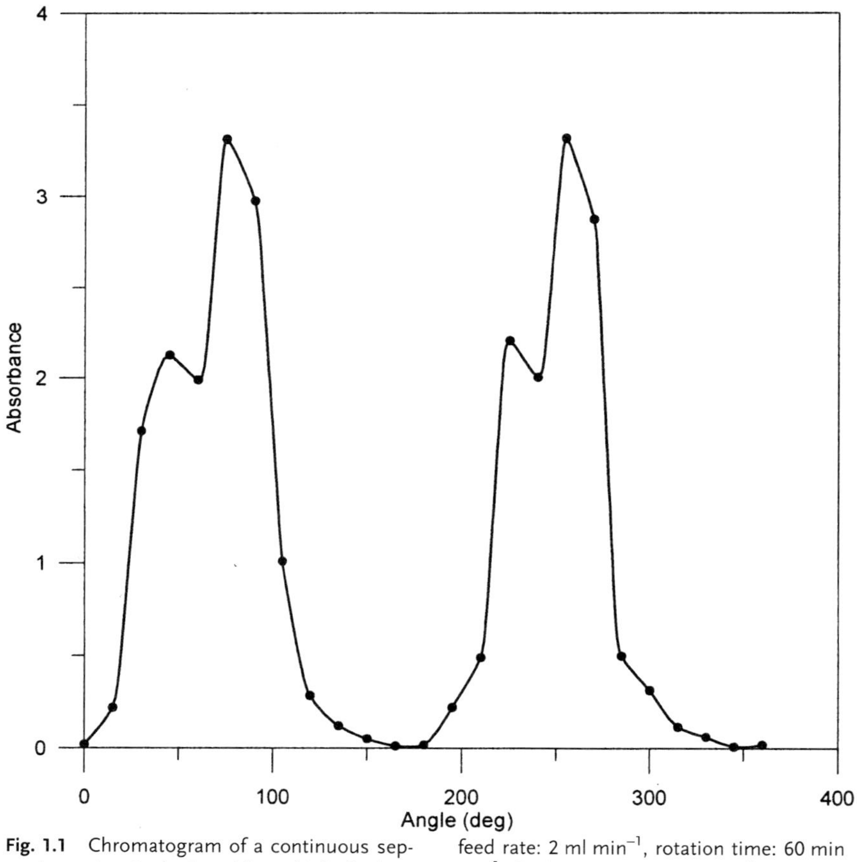

Fig. 1.1 Chromatogram of a continuous separation using Sephadex with multiple feeds. Packing height: 0.14 m, flow rate: 45 ml min^{-1}, feed rate: 2 ml min^{-1}, rotation time: 60 min rev^{-1}, feed concentration: 40 g l^{-1}.

Tab. 1.3 Parameters for chromatogram in Fig. 1.1.

Sample	*TPC*[a] *(mg dl*$^{-1}$*)*	*Activity units (ml)*	*SPA*[b] *units (mg)*	*Purification fold value*	*% Activity recovered*	*Productivity [mg lipase (mg gel-h)*$^{-1}$*]*
Crude	133.8	103.5	77	1.0	100.0	
Eluted (1)	6.4	59.0	919	11.9	53.4	3.5
Eluted (2)	6.2	55.0	887	11.5	49.8	3.2

a) Total protein content.
b) Specific protein activity.

multiple feeds thus results in better utilization of the column and consequently higher throughputs.

Theoretical modeling of a continuous size-exclusion unit is presented below.

1.5 Theoretical Modeling

A mathematical model is set up using the rate theory approach and mass balance equation for a solute over a differential volume in each of the two phases. The differential equations forming the model are to be solved using the Laplace transformation [53, 54].

1.5.1 Model Formulation

The model for the column is formulated using the rate theory. Using this model both the elution profile and the effect of some parameters on the elution profile and resolution can be obtained. The assumptions made in this model are

- The stationary phase is made up of porous, rigid spherical particles.
- Particles are of uniform pore size.
- The mobile phase is treated as a continuous phase.
- Equilibrium exists at the particle interface.
- Solute adsorption in the stationary phase is neglected.
- There is no velocity in the radial direction in the column.

1.5.1.1 Mobile Phase

The mobile phase is the interstitial space in the packed column. The mass balance for a solute in the mobile phase can be obtained from the continuity equation. This is the steady-state equation for an observer in the laboratory reference frame.

$$\omega\frac{\partial C_{\mathrm{m}}}{\partial\theta}+v\frac{\partial C_{\mathrm{m}}}{\partial z}=D_{\mathrm{m}}\frac{\partial^2 C_{\mathrm{m}}}{\partial z^2}+\frac{D_\theta}{r^2}\frac{\partial^2 C_{\mathrm{m}}}{\partial\theta^2}+\frac{kS\varepsilon}{\beta(1-\varepsilon)}\left[C_{\mathrm{s}}|_{r=R}-C_{\mathrm{m}}K_{\mathrm{eq}}\right] \tag{1}$$

The various symbols appearing in the equation are defined in the nomenclature. The last term of the equation accounts for solute transfer between mobile and stationary phase. The driving force is the difference between the stationary phase concentration at equilibrium and the concentration at the interface. The rest of the terms arise from the standard continuity equation. The coefficient of the first term on the right-hand side is the dispersion coefficient in the mobile phase D_{m}. This model corresponds to the Fickian analogy where the dispersion coefficient is treated as a diffusion coefficient.

1.5.1.2 Stationary Phase

A similar mass balance is obtained for the solute in the stationary phase.

$$\omega\frac{\partial C_{\mathrm{s}}}{\partial\theta}=D_{\mathrm{s}}\left[\frac{1}{r^2}\left(\frac{\partial}{\partial r}r^2\frac{\partial C_{\mathrm{s}}}{\partial r}\right)\right] \tag{2}$$

These stationary and mobile phase equations form the model for a continuous chromatograph.

1.5.1.3 Boundary Conditions

For any model a set of boundary conditions is derived from the physical aspects of the model and these boundary conditions are then transformed into mathematical equations to solve the model. The boundary conditions for this model are

$$\begin{array}{l}\text{Mobile phase}\\ z=0 \quad \theta=0 \quad C_{\mathrm{m}}=C_{\mathrm{s}}=0\\ z=0 \quad 0<\theta\leq\theta_{\mathrm{f}} \quad C_{\mathrm{m}}=C_{\mathrm{f}}\\ z=0 \quad \theta_{\mathrm{f}}<\theta<\infty \quad C_{\mathrm{m}}=0\\ z=\infty \quad \text{all } \theta \quad C_{\mathrm{m}}=0\end{array} \tag{3}$$

These boundary conditions come from the physics of the column. Here z is the axial coordinate along the column axis, while θ is the azimuthal angle. At the inlet of the column ($z=0$), the feed is introduced over an angle, θ_{f}. From the beginning of the feed introduction to the feed angle we have the mobile phase concentration as the feed concentration. At $z=\infty$ there is no solute in the mobile phase as all the solute has eluted at the column length L. The stationary phase boundary conditions are

Stationary phase

$$r = 0 \quad \frac{\partial C_s}{\partial r} = 0$$

$$r = R \quad \frac{-D_s}{r}\frac{\partial C_s}{\partial r} = kS\left[C_s|_{r=R} - C_m K_{eq}\right] \tag{4}$$

These boundary conditions are in spherical coordinate for the spherical particles. The first condition corresponds to symmetry at the center of the pore. The second condition corresponds to the mass transfer at the interface between mobile and stationary phase.

1.5.2 Solution

This section deals with the solution of Eqs. (1) and (2) subject to the boundary conditions given in Eqs. (3) and (4).

A good way to solve these differential equations is to use dimensionalization techniques.

The dimensionless variables and parameters are defined as:

$$C'_m = C_m/C_f \quad z' = z/L \tag{5}$$

$$C'_s = C_s/C_f \quad r' = r/R \tag{6}$$

$$W_m = \omega L^2/D_m \quad Pe = \upsilon L/D_m \tag{7}$$

$$W_s = \omega R^2/D_s \quad Nu_s = kSR^2/D_s \tag{8}$$

$$Nu_m = \varepsilon kSL^2/\beta(1-\varepsilon)D_m \tag{9}$$

Using these variables and parameters, the mobile and stationary phase equations as well as the boundary conditions are made dimensionless. In the mobile phase, the angular dispersion term is neglected owing to the difficulty in estimating, and the low order of magnitude of, D_θ. Substituting the dimensionless parameters we obtain

Mobile phase

$$W_m \frac{\partial C'_m}{\partial \theta} + Pe\frac{\partial C'_m}{\partial z'} = \frac{\partial^2 C'_m}{\partial z'^2} + Nu_m\left[C'_s|_{r'=1} - C'_m K_{eq}\right] \tag{10}$$

Stationary phase

$$W_s \frac{\partial C'_s}{\partial \theta} = \frac{1}{r'^2}\left(\frac{\partial}{\partial r'} r'^2 \frac{\partial C'_s}{\partial r'}\right) \tag{11}$$

Boundary conditions

The dimensionless boundary conditions are

$$\begin{aligned}
&z' = 0 \quad \theta = 0 \quad C'_m = C'_s = 0\\
&z' = 0 \quad 0 < \theta \le \theta_f \quad C'_m = 1\\
&z' = 0 \quad \theta_f < \theta < \infty \quad C'_m = 0\\
&z' = \infty \quad \text{all } \theta \quad C'_m = 0\\
&r' = 0 \quad \frac{\partial C'_s}{\partial r'} = 0\\
&r' = 1 \quad \frac{\partial C'_s}{\partial r'} = -N_{us}\left[C'_s|_{r=1} - C'_m K_{eq}\right]
\end{aligned} \tag{12}$$

The dimensionless Eqs. (10) and (11) are solved with the Laplace transformation.

The Laplace transformation of Eq. (11) reduces it to an ordinary differential equation in r'. The following solution is obtained, using the transformed boundary conditions.

$$\overline{C}'_s = \frac{A}{r'} \sinh \sqrt{(pW_s)} r'$$

where

$$A = \frac{Nu_s \overline{C}'_m K_{eq}}{\sqrt{pW_s} \cosh \sqrt{pW_s} + (Nu_s - 1) \sinh \sqrt{pW_s}} \tag{13}$$

The Laplace transformation of Eq. (10) is a second-order ordinary differential equation in z'. Using the transformed mobile phase boundary conditions, the mobile phase solution in the Laplace domain is obtained.

$$\overline{C}'_m = \left(\frac{1 - \exp(-p\theta_f)}{p}\right) \exp\left(\frac{Pe}{2} - \sqrt{\left(\frac{Pe}{2}\right)^2 - \phi}\right) z'$$

where

$$\phi = A \sinh(\sqrt{pW_s}) Nu_m - Nu_m K_{eq} - pW_m \tag{14}$$

The inversion of $\overline{C}'_{\mathrm{m}}$ to the θ domain is quite difficult. Statistical methods (the method of moments) are therefore used to obtain an expression for the elution profile (the mobile phase concentration distribution).

1.5.3 Method of Moments

Every statistical distribution can be described by its moments. If the distribution is defined by a polynomial expansion, then the coefficients of the polynomial are related to the moments. The peak-like form of the concentration profile suggests that we can define it by its moments. The zeroth moment measures the area under the curve, the first moment gives the mean residence angle of the solute sample and the second moment gives the variance of the peak. The higher moments gives the skewness and flatness. If concentration is denoted by $C_{\mathrm{m}}(\theta, z)$ then the moments about the origin of θ are defined by

$$m_0 = m_0(z) = \int_0^\infty \theta^0 C(z, \theta)\mathrm{d}\theta \tag{15}$$

$$m_1 = m_1(z) = \int_0^\infty \theta^1 C(z, \theta)\mathrm{d}\theta \tag{16}$$

$$m_2 = m_2(z) = \int_0^\infty \theta^2 C(z, \theta)\mathrm{d}\theta \tag{17}$$

$$m_n = m_n(z) = \int_0^\infty \theta^n C(z, \theta)\mathrm{d}\theta \tag{18}$$

If all the moments m_n exist for a function $\overline{C}'_{\mathrm{m}}(\theta)$, then

$$\overline{C}'_{\mathrm{m}}(p) = \int_0^\infty \mathrm{e}^{-p\theta} C'_{\mathrm{m}}(\theta)\mathrm{d}\theta = m_0 - m_1 p + \ldots + \frac{(-1)^n}{n!} m_n p^n \tag{19}$$

The Laplace transformation may therefore be regarded as a moment-generating function. The coefficients in this expansion can be evaluated by

$$m_n = \lim_{p \to 0} (-1)^n \overline{C}'^{(n)}_{\mathrm{m}}(p) \tag{20}$$

where $\overline{C}'^{(n)}_{\mathrm{m}}(p)$ is the nth derivative of $\overline{C}_{\mathrm{m}}(p)$.

The moments of Eq. (14) are evaluated at the outlet of the column $z' = 1$. The first four moments of the mobile phase concentration distribution are given below.

$$m_0 = \theta_\mathrm{f} \tag{21}$$

$$m_1 = \frac{\theta_\mathrm{f}^2}{2} + \frac{\theta_\mathrm{f}}{Pe}\left(\frac{W_\mathrm{m}}{3K_\mathrm{eq}} + W_\mathrm{m}\right) \tag{22}$$

$$m_2 = \frac{\theta_\mathrm{f}^3}{3} + \frac{\theta_\mathrm{f}^2}{2Pe}\cdot temp + \left[\frac{\theta_\mathrm{f}^2}{2} + \frac{\theta_\mathrm{f}}{Pe}\cdot temp\right]\cdot\frac{temp}{Pe} + \frac{2\theta_\mathrm{f}}{Pe^3}\cdot temp^2 + \frac{\theta_\mathrm{f}}{Pe}\cdot temp2 \tag{23}$$

$$\begin{aligned} m_3 = {} & \frac{\theta_\mathrm{f}^4}{4} + \frac{3\theta_\mathrm{f}^3 temp}{Pe} + 2\theta_\mathrm{f}^2\left(\frac{2temp^2}{Pe^3} + \frac{temp2}{Pe}\right) + 1.5\theta_\mathrm{f}^2\left(\frac{temp}{Pe}\right)^2 \\ & + \frac{\theta_\mathrm{f}\, temp^3}{Pe^3} + \theta_\mathrm{f}\left(\frac{4temp^3}{Pe^4} + \frac{2temp\, temp2}{Pe^2}\right) + \frac{\theta_\mathrm{f}^2 temp^2}{Pe^3} \\ & + \frac{2\theta_\mathrm{f}\, temp^3}{Pe^4} + \frac{4\theta_\mathrm{f}\, temp\, temp2}{Pe^3} + \frac{96\theta_\mathrm{f}\, temp^3}{8Pe^5} + \frac{\theta_\mathrm{f}^2 temp2}{2Pe} \\ & + \frac{\theta_\mathrm{f}\, temp\, temp2}{Pe^2} - \frac{\theta_\mathrm{f} W_\mathrm{s}^3 Nu_\mathrm{m}\, temp1}{PeK_\mathrm{eq}Nu_\mathrm{s}^3} + \frac{2temp\, 2temp}{Pe^3} \end{aligned} \tag{24}$$

where

$$\begin{aligned} temp &= \frac{W_\mathrm{m}}{3K_\mathrm{eq}} + W_\mathrm{m} \\ temp1 &= \left(\frac{-630}{2835} - \frac{252Nu_\mathrm{s}}{2835} - \frac{36Nu_\mathrm{s}^2}{2835}\right) \\ temp2 &= \frac{2Nu_\mathrm{s}Nu_\mathrm{m}W_\mathrm{s}^2(5 + Nu_\mathrm{s})}{45K_\mathrm{eq}Nu_\mathrm{s}^3} \end{aligned} \tag{25}$$

These moments describe the properties of the mobile phase concentration distribution at the unit outlet. To obtain the elution profile, and to study the effect of parameters, an analytical expression is required. The mobile phase concentration distribution is represented as an expansion using Hermite polynomials. The Hermite polynomials are used for the series expansion because the z variable has a domain of $[-\infty, \infty]$. The Hermite polynomials cover this domain and the zeros of the polynomial can be obtained over the entire domain. The Hermite polynomial expansion is given as

$$c(\theta) = \sum_{n=0}^{\infty} a_n H_n\left[\frac{-(\theta - \mu_1')}{\sqrt{2\mu_2}}\right]\exp\left[\frac{-(\theta - \mu_1')^2}{\sqrt{2\mu_2}}\right] \tag{26}$$

where the Hermite polynomial is given by

$$H(\tau) = \sum_{k=0}^{\frac{n}{2}} \frac{(-1)^k n!}{(k!(n-2k)!)}(2\tau)^{n-2k} \tag{27}$$

The coefficient of the Hermite polynomial a_n can be found using the orthogonality property of the polynomial. Thus

$$a_n = \frac{1}{2^n n! \sqrt{\pi}} \int_0^\infty C(t) H_n \left(\frac{t - \mu_1'}{\sqrt{2\mu_2}} \right) \frac{dt}{\sqrt{2\mu_2}} \tag{28}$$

These coefficients can be related to the moments about the origin and the mean.

1.5.4
Model Evaluation

The effect of four parameters on the elution profile was studied

- Feed angle
- Eluent flow rate
- Rotation rate
- Column height

The bulk diffusion coefficient of lipase was estimated [55]. The dispersion coefficient is used to characterize the axial dispersion in a packed bed. This parameter accounts for the dispersion due to molecular diffusion as well as eddy diffusion due to velocity differences around the particles. A correlation used to estimate the dispersion coefficient D_m in fixed beds was developed by Chung and Wen [56].

$$\frac{D_m \rho}{\mu} = \frac{Re}{0.20 + 0.011 Re^{0.48}} \tag{29}$$

A correlation by Ohashi et al. [57] is used to estimate the mass transfer coefficient

$$Sh = \frac{2kR_p}{D_\infty} = 2.0 + 0.51 \left[\frac{\kappa^{1/3} (2R_p)^{4/3}}{\mu} \right]^{0.6} Sc^{1/3} \tag{30}$$

$$\kappa = \frac{50(1 - \varepsilon)\varepsilon^2 C_D v^3}{2R_p} \tag{31}$$

To evaluate the solute diffusion coefficient in the stationary phase, D_s and the solute partition coefficient, K_{eq}, a model for the pore is required. A simple model where the pore is considered as an infinitely long cylinder and the solute is a rigid sphere adequately describes the elution process [58]. Using this model, D_s, the solute diffusivity within the porous particles, can be estimated from the hydrodynamic theory of hindered diffusion [59]:

$$\frac{D_s}{D_\infty} = \left[1 - 2.104 \frac{a}{r_0} + 2.089 \left(\frac{a}{r_0} \right)^3 - 0.948 \left(\frac{a}{r_0} \right)^5 \right] \tag{32}$$

The hydrodynamic radius of the solute a is evaluated using the Stokes-Einstein equation [55]:

$$a = \frac{RT}{6\pi\mu D_\infty N} \tag{33}$$

The equilibrium partition coefficient, K_{eq}, depends on the pore size in the porous particle as well as the size and conformation of the solute molecule. Giddings et al. [60] have shown that the characteristic size parameter for all shapes of molecules is the mean length of the molecular projection. For a rigid spherical solute this is simply the radius a. By considering the ratio of the area available for the solute molecule inside the cylindrical pore to the actual pore area, and neglecting any interaction between the solute and pore wall, we get

$$\begin{aligned} K_{eq} &= \left(1 - \frac{a}{r_0}\right)^2 \qquad a < r_0 \\ K_{eq} &= 0 \qquad\qquad\quad\; a \geq r_0 \end{aligned} \tag{34}$$

The equations were programmed in C++, and solved numerically. The results of the numerical analysis are discussed below.

1.5.5 Simulation Results

The parameters used in this simulation are presented in Tab. 1.4.

1.5.5.1 Effect of Feed Angle

The effect of changing feed angle on the elution profile is shown in Fig. 1.2 (feed angles are specified in the figure). The peaks shift to a greater elution angle as

Tab. 1.4 Parameters and their values used in the model.

Parameters	***Value***
Eluent flow rate	$1.5–4\ l\ h^{-1}$
Rotation rate	0.5–4 rph
Feed angle	$1–40^\circ$
Column length	0.14 m
Particle size	40 μm
Pore size	40 Å
Internal porosity	0.4
External porosity	0.4
Viscosity of eluent	$0.001\ kg\ m^{-1}\ s^{-1}$
Eluent density	$1000\ kg\ m^{-3}$
Outer diameter of annulus	0.1395 m
Inner diameter of annulus	0.1268 m
Diffusivity of lipase (estimated)	$5\times10^{-11}\ m^2\ s^{-1}$

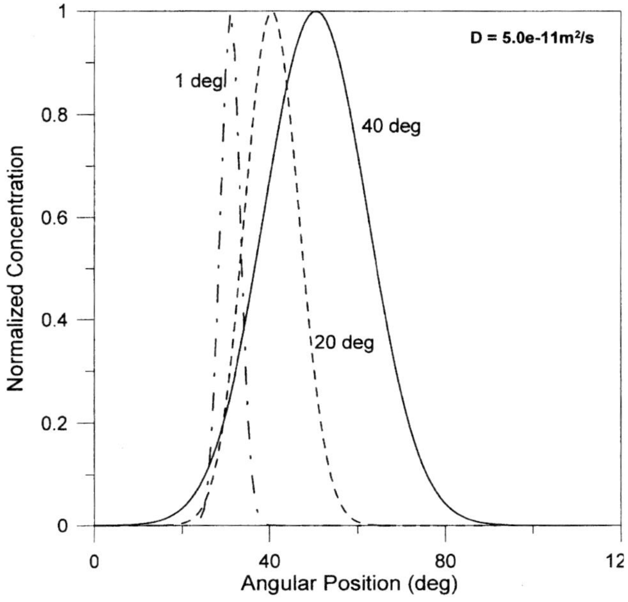

Fig. 1.2 Effect of feed angle on elution profile; rotation rate 2 rph, flow rate 3.5 l h^{-1}.

the feed angle increases. This is because more material is being introduced. The increase in peak width with increasing feed angle may be attributed to an increase in the initial band width. An increase in the feed angle increases the first and second moments, which are related to the elution angle and the variance of the peak, respectively.

1.5.5.2 **Effect of Flow Rate**

As the flow rate increases, the peak shifts to a lower elution angle (Fig. 1.3), due to a decrease in residence time. Mathematically, as the velocity increases, the first moment decreases, resulting in a lower elution angle. A decrease in flow rate leads to an increase in the solute band width due to eddy diffusion. This has been observed in practice [52].

1.5.5.3 **Effect of Rotation Rate**

As the rotation rate increases, the peaks shift to a greater elution angle (Fig. 1.4). The rotation rate also affects the peak width, with an increase in the rate resulting in band broadening. Overlapping of peaks in the purification of lipase has been reported by Genest et al. [50].

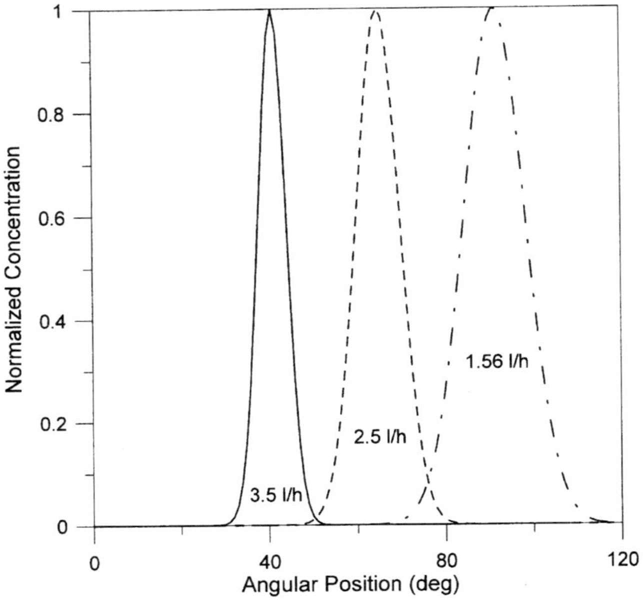

Fig. 1.3 Effect of flow rate on elution profile; rotation rate 2 rph, feed angle 1°.

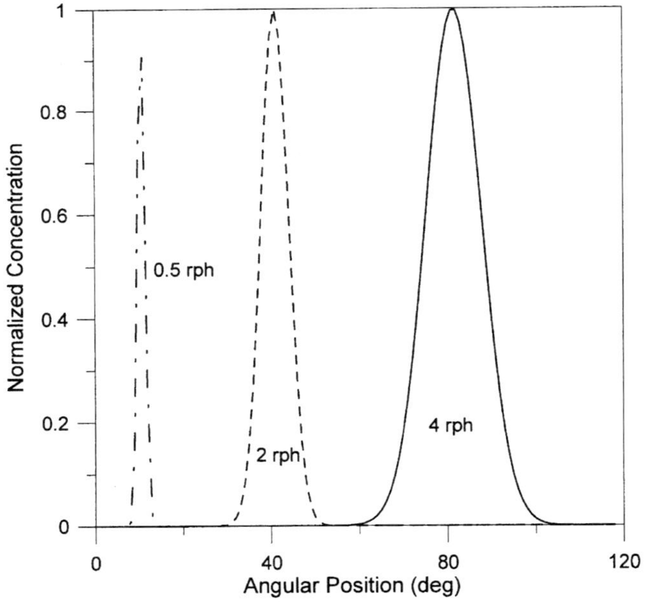

Fig. 1.4 Effect of rotation rate on elution profile; feed angle 1°, flow rate 3.5 l h^{-1}.

1.5.5.4 **Effect of Column Height**

With increasing column height (more packing), the peak shifts to a greater elution angle (Fig. 1.5). An increase in the column height also results in a slight increase in peak width.

Clearly, size-exclusion chromatography is a valuable technique for the separation of protein from non-protein components as well as in the separation of various proteins components present in a mixture. The model has to be improved to take into account angular dispersion, interaction of components and also adsorption on the porous media.

1.6 Conclusions

This chapter examines in some detail the different techniques available to an investigator in the field of lipase purification. Lipases tend to follow very specific patterns of purification. The techniques used to purify the lipase need to be optimized to obtain maximum efficiency. Purification strategies need to be arranged in the proper order to maximize the level of lipase purification. Clearly, more detailed work is needed to establish what factors enable the selection of one strategy over the other. Last, but not the least, theoretical models of continuous purifica-

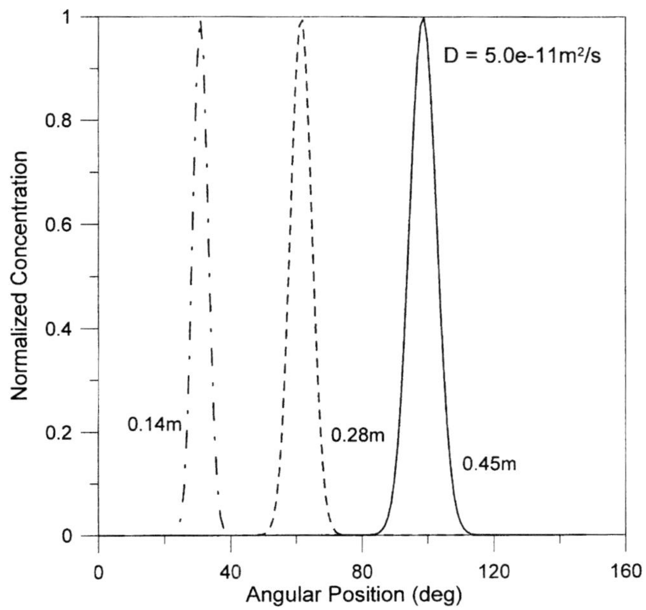

Fig. 1.5 Effect of column height on elution.

tion of proteins need to be improved to make this process efficient and economical. The few studies that have been carried out certainly reveal the potential for such processes.

1.7 Acknowledgements

The author wishes to thank his former graduate students, Paul Genest, Ashish Palekar and Shuo Yan. The author also acknowledges Taylor and Francis (www.tandf.co.uk) for granting permission to use material from two articles published by the author [2, 52].

1.8 References

1 Aires-Barros, M.R., Taipa, M.A., and Cabral, J.M.S. *Lipases* (Woolley, P., Petersen, S.B., eds.) Cambridge University Press, Cambridge, **1994**, 243–270.

2 Palekar, A.A., Vasudevan, P.T., and Yan, S., *Biocatal. Biotransform.*, **2000**, *18*, 177–200.

3 Chartrain, M., Katz, L., Marcin, C., Thien, M., Smith, S., Fisher, E., Goklen, K., Salmon, P., Brix, T., Price, K., and Greasham, R., *Enzyme Microb. Technol.*, **1993**, *15*, 575–580.

4 Kojima, Y., Yokoe, M., and Mase, T., *Biosci., Biotechnol. Biochem.*, **1994**, *58*, 1564–1568.

5 Lee, S.Y. and Rhee, J.S., *Enzyme Microbial Technol.*, **1993**, *15*, 617–621.

6 Li, X., Tetling, S., Winkler, U.K., Jaeger, K., and Benedik, M., *Appl. Environ. Microbiol.*, **1995**, *61*, 2674–2680.

7 Mase, T., Matsumiya, Y., and Akiba, T., *Biosci., Biotechn., Biochem.*, **1995**, *59*, 1771–1772.

8 Matsumae, H. and Shibatani, T., *J. Ferment. Bioeng.*, **1994**, *77*, 152–158.

9 Sharma, R., Soni, S.K., Vohra, R.M., Gupta, L.K., and Gupta, J. K., *Process Biochem.*, **2002**, *37*, 1075–1084.

10 Pabai, F., Kermasha, S., and Morin, A., *Appl. Microbiol. Biotechn.*, **1995**, *43*, 42–51.

11 Canalias, F., Visvikis, A., Thioudellet, C., and Siest, G., *Clin. Chem.*, **1994**, *40*, 1251–1257.

12 Arrese, E.L., and Wells, M.A., *J. Lipid Res.*, **1994**, *35*, 1652–1660.

13 Schmidt-Dannert, C., Sztajer, H., Stöcklein, W., Menge, U., and Schmid, R., *Biochim. Biophys. Acta*, **1994**, *1214*, 43–53.

14 Ameis, D., Merkel, M., Eckerskorn, C. and Greten, H., *Eur. J. Biochem.*, **1994**, *219*, 905–914.

15 Sugihara, A., Tani, T., and Tominaga, Y., *J. Biochem.*, **1991**, *109*, 211–216.

16 Ihara, F., Kageyama, Y., Hirata, M., Nihira, T., and Yamada, Y., *J. Biol. Chem.*, **1991**, *266*, 18135–18140.

17 Torossian, K. and Bell, A.W., *Biotechnol. Appl. Biochem.*, **1991**, *13*, 205–211.

18 Yamaguchi, S. and Mase, T., *Appl. Microbiol. Biotechnol.*, **1991**, *34*, 720–725.

19 Iizumi, T., Nakamura, K., and Fusake, T., *Agric. Biol. Chem.*, **1990**, *54*, 1253–1258.

20 Ncube, I., Adlercreutz, P., Read, J., and Matiasson, B., *Biotechnol. Appl. Biochem.*, **1993**, *17*, 327–336.

21 Ersson, B., Rydén, L., and Janson, J.-C., *Protein Purification* (Janson, J.-C., Rydén, L., eds.) VCH, New York, **1989**.

22 Lee, Y.P., Chung, G.H., and Rhee, J.S., *Biochim. Biophys. Acta*, **1993**, *1169*, 156–164.

23 Thirstrup, K., Carrière, F., Hjorth, S., Rasmussen, P.B., Woldike, H., Nielsen, P.F., and Thim, L., *FEBS Lett.*, **1993**, *327*, 79–84.

24 Simons, J.-W.F.A., Boots, J.-W.P., Slotboom, A.J., and Verheij, H.M., *J. Mol. Catal., B: Enzym.*, **1997**, *3*, 13–23.

25 Kim, M.H., Kim, H.K., Lee, J.K., Park, S.Y., and Oh, T.K., *Biosci. Biotechnol. Biochem.*, **2000**, *64*, 280–286.

26 Aoubala, M., Douchet, I., Laugier, R., Hirn, M., Verger, R., and De Caro, A., *Biochim. Biophys. Acta*, **1993**, *1169*, 183–188.

27 Somma-Delpéro, C., Valette, A., Lepetit-Thévenin, J., Nobili, O., Boyer, J., and Vérine, A., *Biochem. J.*, **1995**, *312*, 519–525.

28 Fuchs, C., Vine, N., and Hills, M.J., *J. Plant Physiol.*, **1996**, *149*, 23–29.

29 Diogo, M.M., Silva, S., Carbral, J.M.S., and Queiroz, J.A., *J. Chromatogr. A*, **1999**, *849*, 413–419.

30 Sabuquillo, P. and Reina, J., *Biochim. Biophys. Acta*, **1998**, *1388*, 337–348.

31 Queiroz, J.A., Garcia, F.A.P., and Carbral, J.M.S., *J. Chromatogr. A*, **1996**, *734*, 213–219.

32 Brzozowski, A.M. and Derewenda, U., *Nature*, **1991**, *351*, 491–494.

33 Lowrier, A., Drtina, G.J.J., and Klibanov, A.M., *Biotechnol. Bioeng.*, **1996**, *50*, 1–5.

34 Reetz, M.T., *Adv. Mater.*, **1997**, *9*, 943–954.

35 Bezzine, S., Carrière, F., and De Caro, A., *Biochemistry*, **1998**, *37*, 11846–11855.

36 De Caro, J., Ferrato, F., Verger, R., and De Caro, A., *Biochim. Biophys. Acta*, **1995**, *1252*, 321–329.

37 Carrière, F., Moreau, H., Raphel, V., Laugier, R., Benicourt, C., Junien, J.-L., and Verger, R., *Eur. J. Biochem.*, **1991**, *202*, 75–83.

38 Lee, M.W. and Severson, D.L., *Biochem. J.*, **1994**, *298*, 213–219.

39 Rao, H.R. and Mansbach, C.M., II, *Biochim. Biophys. Acta*, **1990**, *1046*, 19–26.

40 Tanaka, H., Mierau, I., and Ito, F., *J. Biochem.*, **1999**, *125*(5), 883–890.

41 Tojo, H., Ichida, T., and Okamoto, M., *J. Biol. Chem.*, **1998**, *273*(4), 2214–2221.

42 Soteriou, A. and Cryer, A., *Int. J. Biochem.*, **1993**, *25*, 1483–1490.

43 Gjellesvik, D.G., Lombardo, D., and Walther, B.T., *Biochim. Biophys. Acta*, **1992**, *1124*, 123–134.

44 Gjellesvik, D.G., Raae, A.J., and Walther, B.T., *Aquaculture*, **1989**, *79*, 177–184.

45 Sugihara, A., Ueshima, M., Shimada, Y., Tsunasawa, S., and Tominaga, Y., *J. Biochemistry*, **1992**, *112*, 598–603.

46 Nawani, N. and Kaur, J., *Mol. Cell. Biochem.*, **2000**, *206*, 91–96.

47 Bompensieri, S., Gonzalez, R., Kok, R., Miranda, M.V., Nutgeren-Roodzant, I., Hellingwerf, K.J., Cascone, O., and Nudel, B.C., *Biotechnol. Appl. Biochem.*, **1996**, *23*, 77–81.

48 Sztajer, H. and Bryjak, M., *Bioprocess Eng.*, **1989**, *4*(6), 257–259.

49 Terstappen, G.C., Geerts, A.J., and Kula, M.-R., *Biotechnol. Appl. Biochem.*, **1992**, *16*, 228–235.

50 Genest, P.W., Field, T.G., Vasudevan, P.T., and Palekar, A.A., *Appl. Biochem. Biotechnol.*, **1998**, *73*, 215–230.

51 Bompensieri, S., Mahler, G.F., and Nudel, B.C., *Biotechnol. Techniq.*, **1998**, *12*(8), 611–613.

52 Vasudevan, P.T., Palekar, A.A., and Yan, S., *Biocatal. Biotransform.*, **2002**, *20*(3), 189–199.

53 Dalvie, S., *Theoretical Modeling of a Rotating Annular Size Exclusion Chromatograph*, MS thesis, Clarkson University, **1988**.

54 Yan, S., *theoretical Modeling of a Rotating Annular Size Exclusion Chromatography Unit and Purification of Porcine Lipase by HIC*, MS thesis, University of New Hampshire, **2000**.

55 Young, M.E. and Carroad, P.A., *Biotechnol. Bioeng.* **1980**, *22*, 947–955.

56 Chung, S.F. and Wen, C.Y., *A.I.Ch.E.J.*, **1968**, *14*, 857–866.

57 Ohashi, H., Sugawara, T., Kikuchi, K. and Konno, H., *J. Chem. Eng. Jpn.*, **1981**, *14*, 433–438.

58 Knox, J.H. and Scott, H.P., *J. Chromatogr.*, **1984**, *316*, 311–332.

59 Anderson, J.L. and Quinn, J.A., *Biophys. J.*, **1974**, *14*, 130–150.

60 Giddings, J.C., Kucera, E., Russell, C.P., and Myers, M.N., *J. Phys. Chem.*, **1968**, *72*, 4397–4408.

2
Phospholipase A_1 Structures, Physiological and Patho-physiological Roles in Mammals

Keizo Inoue, Hiroyuki Arai, and Junken Aoki

2.1
Introduction

Phospholipase A_1s (PLA_1s) hydrolyze the *sn*-1 fatty acid residue from phospholipids and produce mostly saturated or mono-unsaturated fatty acids and 2-acyl-lysophospholipids. PLA_1s have not attracted as much attention as mammalian acylhydrolases, PLA_2s, which are involved in the production of bio-active lipids such as prostanoids and platelet-activating factor (PAF) [1, 2].

Membrane phospholipids have a high turnover rate [3], for reasons that are unclear. PLA_1 appears to be one of the enzymes responsible for the rapid turnover of cellular phospholipids. In addition to such a general role, some PLA_1s play a specific role through production of functional 2-acyl-1-lysophospholipids. One type of PLA_1, membrane-associated phosphatidic acid-selective PLA_1 (mPA-$PLA_1\alpha$ in Tab. 2.1), for example, produces 2-acyl-1-lysophosphatidic acid (2-acyl-lysoPA) with an unsaturated fatty acid residue [4]. 2-Acyl-lysoPA specifically activates LPA_3/EDG7, a receptor preferentially interacting with 2-acyl-lysoPA [5, 6]. Knock out of the gene coding phosphatidylserine-specific PLA_1 (PS-PLA_1 in Tab. 2.1) in mice affected their patho-physiological conditions (unpublished observation), which further supports the idea that PLA_1s play important specific roles in mammals.

Although PLA_1 activity has been detected in many mammalian tissues and cells [7–14], only a few PLA_1s have been purified and cloned. Some lipases that hydrolyze triacylglycerol such as hepatic lipase and lipoprotein lipase also show PLA_1 activity [15]. However, these lipases will not be considered here.

Group designations of PLA_1s are now required as new PLA_1s have been discovered and known enzymes have been better characterized. Previously, two phosphatidic acid (PA)-preferring and PA-selective PLA_1s have been identified and characterized [4, 16]. However, this nomenclature is no longer appropriate, because the observed substrate specificities do not reflect the structural aspects of the enzymes, and there are at least two "phosphatidic acid-preferring PLA_1s" with completely different structures [4, 16].

Two types of PLA_1s, type I and type II, have been described, with different amino acid sequences (Tab. 2.1). Most PLA_1s of both types contain a lipase consensus sequence (GXSXG). An exception is PA-PLA_1 (type II), which contains SXSXG.

Lipases and Phospholipases in Drug Development
Edited by G. Müller and S. Petry

ISBN: 3-527-30677-3

Tab. 2.1

Group	Proposed name (proposed abbreviation)	Location	Cellular localization	Size (kDa) (a.a.)	Substrate specificity	Molecular characteristics	References
I Extracellular type	Phosphatidylserine-specific phospholipase A_1 (PS-PLA_1)	Platelets (rat) Various tissues	Secreted	55 kDa (456 a.a.)	PS, lysoPS	Catalytic triad Heparin affinity lysoPS-producing enzyme	17 23 29 30
	Phosphatidylserine-specific phospholipase A_1-short form (PS-$PLA_1\Delta C$)	Platelets (rat) Various tissues	Secreted	40 kDa (376 a.a.)	lysoPS	Catalytic triad Heparin affinity Splicing variant	23
	Membrane-associated phosphatidic acid-selective phospholipase A1α (mPA-PLA1α)	Colon, intestine, skin, lung Prostate, ovary, testis	Secreted Membrane-associated	58 kDa (451 a.a.)	PA	Catalytic triad Heparin affinity lysoPA-producing enzyme	29
	Membrane-associated phosphatidic acid selective phospholipase $A_1\beta$ (mPA-$PLA_1\beta$)	Testis	Secreted Membrane-associated	58 kDa (460 a.a.)	PA	Catalytic triad Heparin affinity lysoPA-producing enzyme	4 51
II Intracellular type	Phosphatidic acid-preferring phospholipase A_1 (PA-PLA_1)	Testis, brain (bovine)	Intracellular	98 kDa (875 a.a.)	PA, PE	Catalytic center (SXSXG) Highly phosphorylated	16
	KIAA0725 protein	Ubiquitous	Intracellular	81 kDa (711 a.a.)	PA, PE, PS	Catalytic center (GXSXG)	18
	p125	Ubiquitous	Intracellular	111 kDa (1,000 a.a.)	?	sec23p-interacting protein	19

Enzymes that exhibit high similarity to conventional lipase family members (pancreatic lipase, hepatic lipase, and lipoprotein lipase) are designated as type I (also called extra-cellular type enzymes) in this chapter. The amino acid sequence of one member of the type I PLA_1 family ($PS\text{-}PLA_1$ in Tab. 2.1) shares about 30% homology with the sequence of conventional lipases (Fig. 2.1) [17]. Type I enzymes are secreted or localized extra-cellularly on surface membranes as ecto-type enzymes. The observed distribution of the enzymes agrees with a putative signal sequence that they carry. Type I enzymes conserve the active serine residue, two other residues (Asp and His) that with serine form the catalytic triad (asterisk in Fig. 2.1), and the residues surrounding each of these three residues. They also conserve six cysteine residues (Fig. 2.1) which play an important role in the maintenance of tertiary structures and the formation of "lid" structure in lipase family. Most members of the lipase family have surface loops called "lid" and "β9", which cover the active site and are implicated in substrate recognition. Both phosphatidylserine-specific PLA_1 ($PS\text{-}PLA_1$) [17] and membrane-associated phosphatidic acid-selective PLA_1 ($mPA\text{-}PLA_1\alpha$ [4] and $mPA\text{-}PLA_1\beta$ [51]) contain such loops, though the loops conserved in the phospholipases A_1 are shorter or deleted as compared with those of conventional lipases (Fig. 2.1). Shorter "lid" and shorter (deleted) "β9" structures observed in these phospholipase A_1 might be responsible for their inability to hydrolyze triacylglycerol. All type I enzymes strictly recognize the head group structure of phospholipids; $PS\text{-}PLA_1$ recognizes a glycero-phospho-L-serine structure [17] and $mPA\text{-}PLA_1\alpha$ recognizes a glycero-phosphate structure [4]. The amino acid sequence of human $mPA\text{-}PLA_1\alpha$ has 34.0% identity with that of human $PS\text{-}PLA_1$ (Fig. 2.1). A phylogenetic tree of type I PLA_1s shows that $mPA\text{-}PLA_1$s and $PS\text{-}PLA_1$ form separate clusters (Fig. 2.2).

The amino acid sequence of the other group of the PLA_1s, intracellular PLA_1s, shows little if any homology to that of type I enzymes (Fig. 2.3). Three intracellular PLA_1s (PA-preferring PLA_1, p125, and KIAA0725) have been identified in mammals. These enzymes are now classified into either type II PLA_1 (or intracellular PLA_1 since they localize intra-cellularly). It is not known whether type II enzymes also contain a typical catalytic triad composed of Ser, Asp and His. Neither the "lid" nor the "β9" structure should be present in the intracellular enzymes, because stable intra-molecular S–S bonds between cysteine residues in proteins cannot be kept intracellularly. A newly discovered enzyme (KIAA0725) shows 22.4% homology with $PA\text{-}PLA_1$ [18]. KIAA0725 was discovered as a homologous protein of P125, which has amino acid sequence homology to $PA\text{-}PLA_1$ but lacks PLA_1 activity [19].

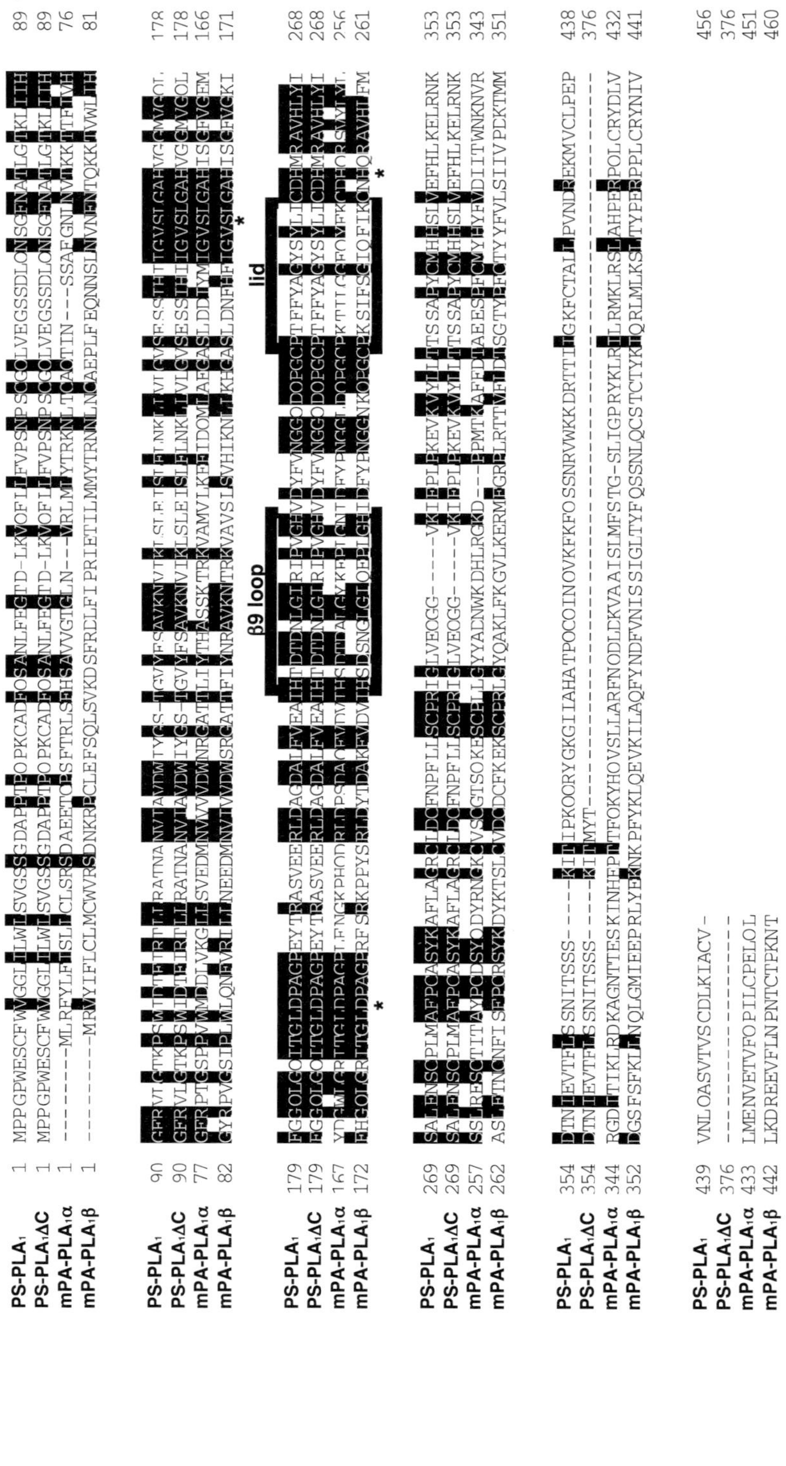

Fig. 2.1 Comparison of the amino acid sequences of human type I PLA$_1$s (PS-PLA$_1$, PS-PLA$_1$ΔC, mPA-PLA$_1α$, and mPA-PLA$_1β$). The amino acid residues are numbered at both sides. The asterisks indicate the active serine, aspartic acid and histidine residues, which may make up a catalytic triad. The lid domains and β9 loops are indicated by boxes. Highly conserved residues (residues that are same in three out of the four proteins) are represented by black boxes.

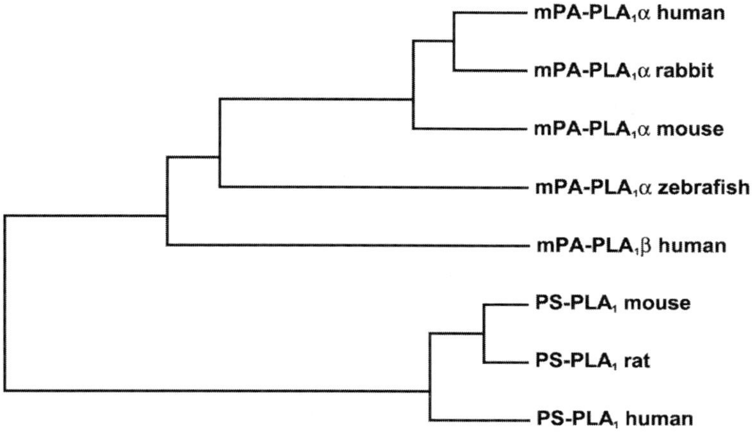

Fig. 2.2 Phylogenetic relationship of type I PLA_1s. The phylogenetic tree was generated from ClustalW alignment data using the GENETYX-MAC v10.1.6 (Software Development Co., Ltd., Tokyo, Japan).

2.2 Phosphatidylserine-specific Phospholipase A_1 (PS-PLA_1)

2.2.1 Historical Aspects

The enzyme was found originally as a lysophosphatidylserine-selective lysophospholipase (lysoPS-lysophospholipase) that was secreted from activated rat platelets together with type II PLA_2 (recently often referred to as sPLA_2-IIA) [20, 21]. It was later found that the lysoPS-lysophospholipase activity was always recovered from the same fraction as PS-PLA_1 activity on several types of column chromatography, which suggests that both activities are derived from the same protein. This was finally confirmed by the findings that the recombinant protein showed both PS-PLA_1 and lysoPS-lysophospholipase activities [17].

2.2.2 Biochemical Characterization and Tissue Distribution

Activated rat platelets secreted PS-PLA_1 as well as sPLA_2-IIA, both of which are stored in α-granules [20]. The PS-PLA_1 was purified from culture medium of thrombin-activated platelets prepared from 1000 rats [17]. In rats, PS-PLA_1 expressed not only in platelets but also in heart and lung [22]. The enhanced expression was detected in many tissues when rats were treated with bacterial lipo-polysaccharide *in vivo* (unpublished observation). The expression of PS-PLA_1 differed between species. For example, the mRNA encoding the enzyme was poorly expressed in human platelets but highly expressed in rat platelets.

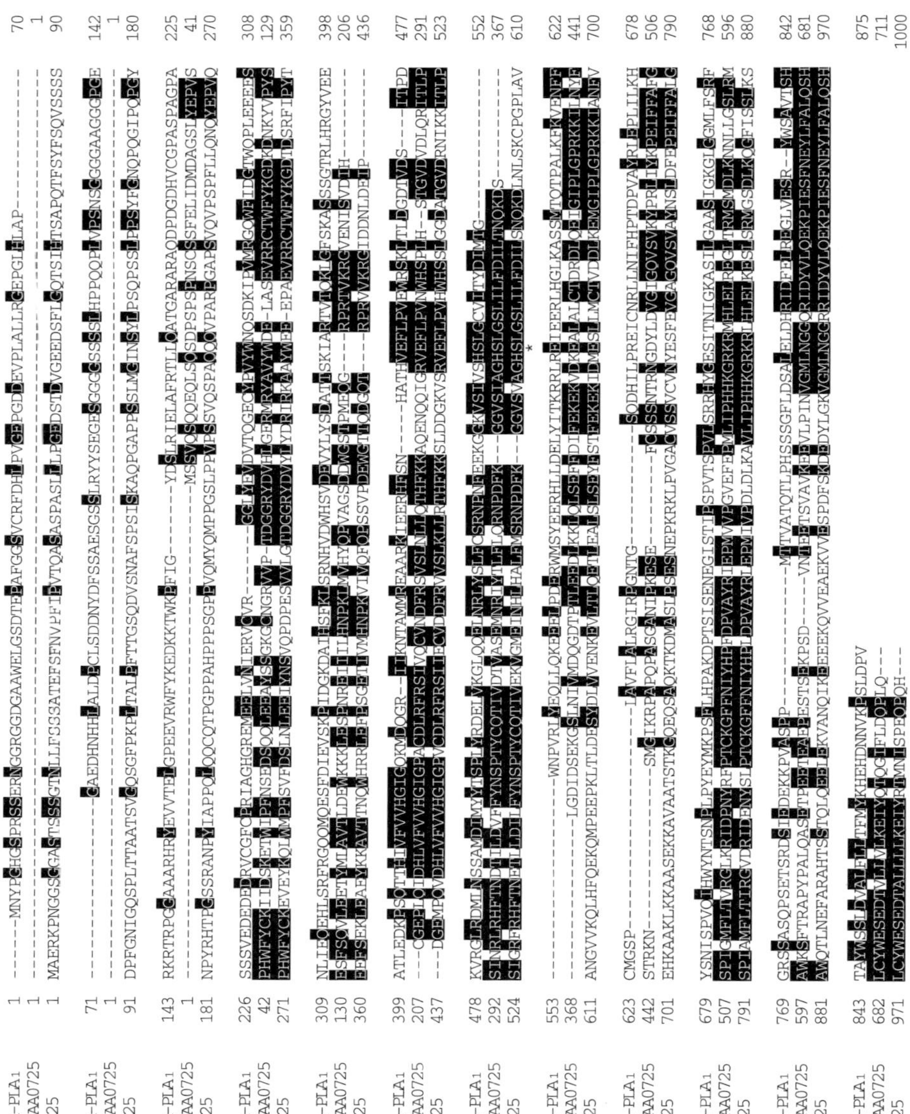

Fig. 2.3 Comparison of the amino acid sequences of human type II PLA_1s (PA-PLA_1, KIAA0725, and p125). The amino acid residues are numbered at both sides. Asterisks indicate the putative active serine residues. Highly conserved residues (residues that are same in three out of the four proteins) are represented by black boxes.

Rat platelet PS-PLA_1 was purified as a glycoprotein. The deduced amino acid sequence of the enzyme has three potential N-glycosylation sites. The enzyme binds strongly to lectins such as concanavalin A, WGA, and RCA. The difference between the molecular mass of the enzyme calculated from the cDNA nucleotide sequence (47 kDa) and that of the purified protein (55 kDa) reflects the addition of an asparagine-linked sugar moiety to the protein, because treating the protein with N-glycanase reduced the molecular mass considerably. The recombinant protein produced in a baculovirus system (50 kDa) showed exactly the same activity as the native PS-PLA_1 from rat platelets. This suggests that the larger carbohydrate structure detected in rat platelet PS-PLA_1 is not required for enzyme activity, because the carbohydrate structure of the recombinant protein is probably different from that of the platelet-derived enzyme.

2.2.3
Structural Characteristics

Fig. 2.1 shows the deduced amino acid sequences of human type I PLA_1 including PS-PLA_1. The sequence homology of PS-PLA_1 is about 80.0% between human [23] and rat [17]. The deduced amino acid sequence contains a catalytic triad composed of active Ser, Asp and His residues, and "lid" surface loops (Fig. 2.1). Interestingly, conventional lipases have long "lid" (22–23 residues) and "$\beta 9$" loops (18–19 residues), whereas PS-PLA_1 has a shorter "lid" (12 residues) and a shorter (deleted) "$\beta 9$" loop (13 residues) [17, 23]. Both loops, which may be supported by S–S bonds of 14 cysteine residues, have been implicated in substrate recognition. These findings are compatible with the inhibition of the enzyme by diisopropylfluorophosphate and dithiothreitol [21].

While isolating human PS-PLA_1 cDNA we found a similar cDNA, designated PS-$PLA_1\Delta C$, with four extra bases (GTAC) inserted at the boundary of the exon–intron junction (nt 1122) [23]. The insertion of these four bases caused a frame shift, resulting in the loss of about 80 amino acid residues at the C-terminus (Fig. 2.1). The mRNA having this sequence was found in various human tissues and cell lines [23]. Although it is not clear whether the short form of the enzyme is actually expressed at the protein level, experiments using the recombinant proteins prepared with the baculovirus system revealed the functional difference of the two isoforms (see below) [23].

2.2.4
Substrate Specificity

PS-PLA_1 reacts specifically with PS and 1-acyl-lysoPS [20, 24]. It can not appreciably hydrolyze any other phospholipids including phosphatidyl-D-serine. The enzyme is not a type of phospholipase B because it exclusively hydrolyzes an acyl residue bound at *sn*-1 position of either PS or 1-acyl-lysoPS. Incubation of the enzyme with 1-acyl-2-radioactive acyl-PS produced radioactive lysoPS but only a small amount of radioactive-free fatty acid, even after long-term incubation [17]. Thus, PS-PLA$_1$ recognizes the 1-acyl-glycerophospho-L-serine structure and hydrolyzes an acyl residue at the *sn*-1 position. PS-PLA$_1$ may first interact with the surface of the bilayer structure to gain access to PS molecules buried in the bilayer. A kinetics study revealed that the enzyme could directly interact with monomeric lysoPS molecules. PS-PLA$_1$ΔC showed hydrolytic activity to lysoPS, but not to PS [23], indicating that the domain responsible for interaction with the surface of bilayer is in the C-terminal 80 amino acid residues.

2.2.5 Possible Functions

Activation of rat platelets with a calcium ionophore resulted in a marked accumulation of lysoPS having unsaturated fatty acids as well as lysoPS with saturated fatty acids [24]. The accumulation of lysoPS with unsaturated fatty acids suggests that PS-PLA$_1$ participates in regulating the functions of platelets. For example, it might regulate blood clotting through the elimination of PS, which is implicated in the blood-clotting reaction.

LysoPS has been shown to stimulate histamine release from rat peritoneal mast cells triggered by cross-linking of FcεRI, a high affinity receptor for IgE [25–28]. Cross-linking of the receptors in the absence of lysoPS did not result in the release of appreciable amounts of histamine. The lysoPS appears to come from an exogenous source because its levels in pure mast cells are very low. It is not clear why rat mast cells require exogenous lysoPS. One possibility is that some molecules on the mast cell membrane interact with lysoPS, causing it to undergo a conformational change and become activated (Fig. 2.4). This may lead to the transmission of a signal within the cell, which in combination with other signals triggered by cross-linking of the Fc receptor, induces degranulation (i.e., release of serotonin and histamine).

Incubation of peritoneal mast cells with cross-linked Fc receptors with exogenous PS-PLA$_1$ (but no addition of lysoPS) also caused histamine release [29]. This indicates that PS-PLA$_1$ interacted with PS in the mast cell plasma membrane to produce lysoPS. However, the efficiency of lysoPS production from the mast cell membrane is very low, which results in a poor degranulation reaction, because the degranulation was greatly stimulated when mast cells were supplemented with apoptotic cells [29]. This clearly indicates that lysoPS does not come from mast cell membranes but from membranes of cells that surround the mast cells. PS molecules are located in the inner leaflet of the plasma membrane in unstimulated cells and are exposed in the outer leaflet when cells are in either apoptotic or activated conditions. The PS on the plasma membrane in mast cells is not easily accessible to PS-PLA$_1$, and thus may not be an important source of lysoPS. However, there may be other cells on which PS is more accessible, such as apoptotic or stimulated cells (upper right of Fig. 2.4). These PS-donor cells may be the

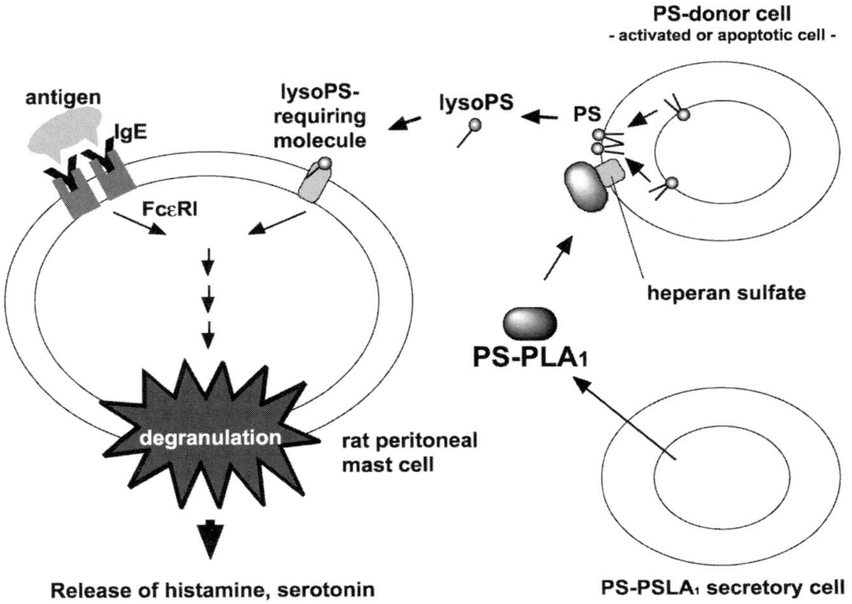

Fig. 2.4 A model of activation of rat peritoneal mast cells by lysoPS produced by PS-PLA_1. LysoPS (both 1-acyl and 2-acyl-lysoPS) stimulates degranulation of rat peritoneal mast cells (RPMC) triggered by FcεRI (high affinity receptor for IgE) cross-linking through a "putative lysoPS receptor". PS-PLA_1 produces and supplies the lysoPS by efficiently hydrolyzing PS exposed on activated or apoptotic cells.

main source of lysoPS. In support of the PS donor cell model, a crude preparation of rat peritoneal mast cells (containing other cell types such as mononuclear leukocytes and neutrophils) released appreciable amounts of histamine in the presence of PS-PLA_1 and concanavalin A, a Fc receptor cross-linker [29]. Pure mast cells under the same conditions would not do this because of a lack of a source of lysoPS.

Where does the PS-PLA_1 come from? Northern and western blot analyses indicate the protein is expressed by platelets, skin, lung, and heart in rats. We also found that an inflammation stimulus significantly increased PS-PLA_1 at both the mRNA and protein levels in various tissues of rat (unpublished observations). These PS-PLA_1 protein may supply lysoPS to mast cells. Indeed, stimulants such as nerve growth factor, in combination with activated platelets, activate mast cells [30], possibly due to an accumulation of lysoPS on the activated platelets that secrete PS-PLA_1.

The induced PS-PLA_1 might attack PS-sensitized cells in either an autocrine or paracrine manner. Heparin completely blocked degranulation of a crude mast cell preparation induced by Fc receptor-cross linkers and PS-PLA_1 [29]. This indicates that, under normal (heparin-free) conditions, PS-PLA_1 first anchors to surface heparan-sulfate proteoglycans on the membranes of sensitized cells, and then displays its hydrolytic activity against PS on target cells.

PS-PLA_1 produces 2-acyl-1-lyso-PS that has not been well characterized. 2-Acyl-1-lyso-PS has the same activity at inducing degranulation as 1-acyl-lyso-PS [29]. Although the acyl chain at the *sn*-2 position of 2-acyl-1-lysophospholipids can easily migrate to the *sn*-1 position, resulting in the formation of 1-acyl-2-lysophospholipids, this does not appear to explain the equal activities because the degranulation of mast cells occurs within 5 min, while the migration of a fatty acid residue on the glycerol backbone takes several hours [31].

2.3 Membrane-associated Phosphatidic Acid-selective Phospholipase A_1s (mPA-$PLA_1\alpha$ and mPA-$PLA_1\beta$)

2.3.1 Historical Aspects

Lysophosphatidic acid (lysoPA) is a lipid mediator with multiple biological functions, including induction of platelet aggregation, smooth muscle contraction, stimulation of cell proliferation, formation of actin stress fibers in fibroblast, and inhibition of neurite outgrowth in neuronal cells. Recent studies have identified a new family of receptor genes for lysoPA. Three G-protein-coupled receptors (GPCR) belonging to the EDG (endothelial differentiation gene) family, LPA_1/EDG2, LPA_2/EDG4 and LPA_3/EDG7 have been identified [6, 32, 33]. The latter reacts preferentially with lysoPA, which has an unsaturated fatty acid residue at the *sn*-2 position of the glycerol backbone [5, 6], and it also recognizes specific structures of the lysoPA, including the cis double bond at the $\Delta 9$ position of octadecanyl residues, whereas LPA_1/EDG2 and LPA_2/EDG4 have much broader ligand specificities. Both receptors interact with lysoPA, despite the difference in the fatty acyl chains. Hayashi et al. found that lysoPA with an unsaturated fatty acid residue stimulated dedifferentiation of rat smooth muscle cells isolated from blood vessels [34]. LysoPA with a saturated fatty acid residue did not show such activity. The rat smooth muscle cells did express LPA_3/EDG7, indicating the involvement of LPA_3/EDG7 in the phenomena.

In contrast to the molecular mechanisms of lysoPA receptors, the molecular mechanisms of lysoPA production are poorly understood. LysoPA is thought to be produced from PA by the action of PLA_2 [35–37], while PA is generated as a result of phospholipase D activation (PLD). However, lysoPA with polyunsaturated fatty acid species has been detected in several biological systems [38], suggesting that PLA_1 is involved in lysoPA production. "Phosphatidic acid-preferring phospholipase A_1" (PA-PLA_1 in Tab. 2.1), a type II enzyme, was first isolated from brain [39] and cloned from bovine testis [16]. This enzyme is localized within cells. It is not known whether it is involved in production of lysoPA as a ligand for GPCRs under physiological conditions. Two homologues of PS-PLA_1 recently been found in GenBank™ are both membrane-associated PA-selective PLA_1s: mPA-$PLA_1\alpha$ [4] and mPA-$PLA_1\beta$ [51].

2.3.2 Characterization and Distribution

Both mPA-$PLA_1\alpha$ and mPA-$PLA_1\beta$, like many phosphate-recognizing enzymes such as phosphatase and ATPase [40], are inhibited by vanadate [51]. When expressed in insect Sf9 cells, neither mPA-$PLA_1\alpha$ nor mPA-$PLA_1\beta$ were secreted. Both were recovered from the Triton X-100-insoluble fraction of the cell membrane. An immunofluorescence analysis confirmed that these proteins are localized exclusively to the plasma membrane. Both mPA-$PLA_1\alpha$ and -β may localize in the detergent-resistant membrane domain, referred to as raft structures, and function as lysoPA-producers under certain conditions [51].

The membrane fraction showed no hydrolytic activity against exogenously added phospholipids including PA, unless subjected to a freeze–thaw cycle and treated with cholic acid. LysoPA production was detected in Sf9 cells expressing mPA-$PLA_1\alpha$, when the cells were treated with phorbol ester or bacterial phospholipase D [4]. Both endogenous and exogenous PLD may produce PA in the membrane and supply substrates to PA-PLA_1. These findings suggest that mPA-PLA_1s expressed in membranes of Sf9 cells may primarily hydrolyze substrates integrated in the same membrane, as was previously found with the PLA_2 of sheep red blood cell membranes [41].

In a culture of mPA-$PLA_1\alpha$-expressing Sf9 cells treated with bacterial PLD to generate PA, the only lysoPA species to increase in the culture medium were 16:1- and 18:1-lysoPA [4], with no appreciable change in the level of any other lyso-phospholipids [4]. This indicates that mPA-$PLA_1\alpha$ has a high substrate specificity for PA and that it has PLA_1 activity. Thus, the enzyme selectively hydrolyses PA molecules located on the same membrane.

Human mPA-$PLA_1\alpha$ is most abundantly expressed in prostate, testis, ovary, colon, pancreas, kidney, lung, platelets and is expressed at lower levels in spleen, brain, and heart [4]. Interestingly, the expression pattern of mPA-$PLA_1\alpha$ is similar to that of LPA_3/EDG7, a receptor for lysoPA having an unsaturated fatty acid residue.

A closely homologous enzyme, mPA-$PLA_1\beta$, was found from the gene bank. mPA-$PLA_1\beta$ has the same characteristics as mPA-$PLA_1\alpha$, except that it was exclusively detected in human testes [51]. mPA-$PLA_1\beta$ was detected in sperm, but only a small amount was detected in seminal fluids [51].

2.3.3
Structural Characteristics

The deduced amino acid sequence of mPA-$PLA_1\alpha$ had 34.0% identity with that of human PS-PLA_1 [4], and the first half of the molecule, which corresponded to the N-terminal catalytic domain of PS-PLA_1, had an identity of about 40%. Three of the amino acid residues in the sequence of mPA-$PLA_1\alpha$, Ser-154, Asp-178, and His-248, are completely conserved in the whole lipase family (Fig. 2.1). The deduced amino acid sequences of mPA-$PLA_1\alpha$ and mPA-$PLA_1\beta$ are 45.9% identical [51]. The homologous regions were again most prominent in the first half of the molecule [51].

2.3.4
Function

The tissue distribution of mPA-$PLA_1\alpha$ is similar to that of LPA_3/EDG7. Thus, PLA_1-expressing cells are expected to be functionally coupled to LPA_3/EDG7-expressing cells (Fig. 2.5). Co-culturing with mPA-$PLA_1\alpha$-expressing Sf9 cells significantly stimulated the cells expressing LPA_3/EDG7 [4, 6]. Conditioned medium

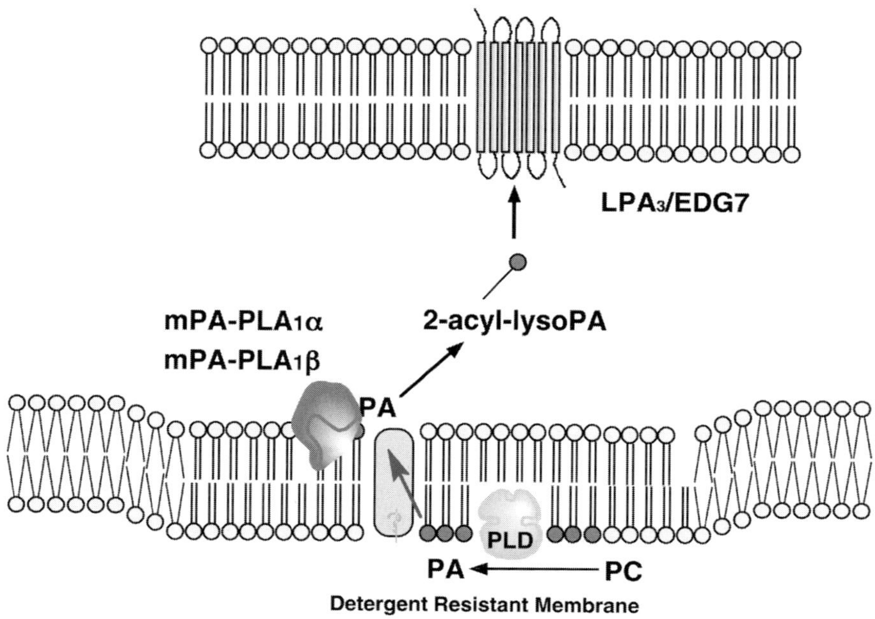

Fig. 2.5 A model of production of 2-acyl-lysoPA by mPA-PLA_1s and consequent activation of LPA_3/EDG7 by 2-acyl-lysoPA. mPA-PLA1α and -β hydrolyze PA generated as a result of PLD activation and produce 2-acyl-lysoPA in detergent-resistant membrane. LPA_3/EDG7 is specifically activated by 2-acyl-1-lysoPA.

prepared from Sf9 cells expressing mPA-$PLA_1\alpha$ did stimulate the cells expressing LPA_3/EDG7, indicating that a small amount of lysoPA was released to the medium.

Treatment of mPA-$PLA_1\alpha$-expressing Sf9 cells with PMA stimulated lysoPA formation [4]. This suggests that intrinsic PLD is involved in the production of lysoPA in infected Sf9 cells, because PMA is known to activate PLD through the activation of protein kinase C. PMA-stimulated production of lysoPA was suppressed by a PLD inhibitor, butan-1-ol [4], which supports the idea that endogenous PLD collaborate with mPA-$PLA_1\alpha$ to produce lysoPA in the present system (Fig. 2.5).

In mammalian plasma or serum, lysoPA is also produced by lysophospholipase D (lysoPLD) from lysophosphatidylcholine (lysoPC), which is present at the several hundred μM level [42]. It is noteworthy that lysoPLD is identical to autotaxin (ATX) [43], a tumor cell motility-stimulating factor, originally isolated from melanoma cell culture supernatants [44]. Thus, ATX might mediate its effects through lysoPA production. Because lysoPLD/ATX acts on both 1-acyl-2-lysoPC and 2-acyl-1-lysoPC, PLA_1 isozymes may be involved in the production of lysoPA by supplying 2-acyl-1-lysoPC to lysoPLD/ATX. Interestingly, lysoPA with unsaturated fatty acid residues (18:1 or 18:2) but not with saturated fatty acids (14:0, 16:0, or 18:0) specifically induces phenotypic modulation of vascular smooth muscle cells [34], which indicates that such PLA_1 isozymes may have a role in atherosclerosis.

2.4 Phosphatidic Acid-preferring Phospholipase A_1 (PA-PLA_1)

2.4.1 Historical Aspects

In addition to "membrane-bound" PLA_1s, mammals have cytosolic PLA_1s. Cytosolic PLA_1 activities have been studied in various tissues including heart, brain and testis. PA-PLA_1 has been purified from brain [45] and testes [39, 46]. Like other lipolytic enzymes, PLA_1 is affected by the assay conditions. Using a mixed micelle system Glomset and his colleagues [46] found that a 110 kDa enzyme from testes preferentially hydrolyzed phosphatidic acid. They cloned the PA-PLA_1 from bovine [16]. This PLA_1 lacks sequence similarity to type I PLA_1, lysophospholipase, LCAT, and triacylglycerol lipases.

Of two new members of the phosphatidic acid-preferring PLA_1 (type 2 PLA_1) family (Fig. 2.3), one, p125, was identified as a protein that interacts with mammalian Sec23p via its N-terminal proline-rich region [19]. Sec23p is a component of the COPII coat that functions in the production of vesicles from the ER [47]. p125 is localized in the ERGIC and/or cis-Golgi, and its overexpression causes the dispersion of these membrane compartments, suggesting its involvement in the early secretory pathway. Surprisingly, the enzyme did not have appreciable PLA_1 activity. Database searches using the amino acid sequence of p125 identified an homologous protein that is encoded by the human expressed sequence tag clone KIAA0725 [18]. This protein, named KIAA0725p, possessed PLA_1 activity preferentially for phosphatidic acid in the assay system containing Triton X-100.

2.4.2
Characterization and Distribution

The highest PA-PLA_1 activity was observed in high speed supernatant fractions from homogenates of bovine testes and brain, with little or no activity in corresponding fractions from newborn testes, liver, spleen, heart, and blood [46]. A northern blot analysis of mRNA from human tissues agreed with the observations with bovine tissues [16].

The molecular mass of the expressed PA-PLA_1 was estimated to be 110 kDa by SDS-PAGE, and 97.6 kDa by matrix-associated laser desorption/ionization [48]. A value of 97.6 kDa was also obtained from the deduced amino acid sequence of the open reading frame (ORF) [48]. The enzyme purified from bovine testes had an apparent molecular mass of 440 kDa, as determined by size exclusion chromatography [48], suggesting that PA-PLA_1 exists as a homotetramer of 98 kDa subunits in solution. The molecular mass of bovine brain PA-PLA_1, estimated by the same method, was 200 kDa, suggesting that it exists as a homodimer in the brain [48].

PA-PLA_1 was inhibited by methyl arachidonyl fluorophosphonate (MAFP), an inhibitor of certain PLA_2s, but not by DFP [39], which is a potent inhibitor of mPA-PLA_1. The activity of PA-PLA_1 was Ca^{2+}-independent at neutral pH but Ca^{2+}-dependent at alkaline pH [39].

2.4.3
Substrate Specificity

In a Triton X-100-mixed micelle system, the activity of PA-PLA_1 against PA was 4- to 10-fold greater than the activities of PA-PLA_1 against PI, PS, PE, and PC [39]. The enzyme hydrolyzed diunsaturated molecular species of PA (dioleoyl PA and diarachidonoyl PA) in preference to 1-palmitoyl-2-oleoyl PA and 1-stearoyl-2-arachidonoyl PA, in the same system. The substrate specificity of this enzyme varied according to the assay condition [39], as was also found by Nakajima et al. and Uchiyama et al. [18, 49]. In the absence of Triton X-100, PE was the best substrate for PA-PLA_1 [49]. Interestingly, hepatic lipase showed a substrate preference for PA and PS in the presence of Triton X-100 [50]. Further studies are needed to identify the real substrates within cells and *in vivo*.

2.4.4
Function

$PA\text{-}PLA_1$, which is expressed in mature bovine testes but not in newborn calf testes, may contribute to the formation or function of sperm. $PA\text{-}PLA_1$ may also be involved in the remodeling of phospholipids by hydrolyzing fatty acids at the *sn*-1 position and by subsequent acylation that occurs at this position.

2.5
KIAA0725P, a Novel PLA_1 with Sequence Homology to a Mammalian Sec23p-interacting Protein, p125

2.5.1
Historical Aspects

Vesicles coated with coat protein complex II (COPII) are involved in protein transport from the endoplasmic reticulum (ER) to the Golgi apparatus [47]. Sec23p is a component of CopII. A novel protein, named p125, was found to interact with Sec23p [19]; p125 has domains in the central and C-terminal regions that are significantly homologous to $PA\text{-}PLA_1$ and has the Gly-X-Ser-X-Gly consensus sequence that is considered to be the active site of lipases, but it did not show appreciable PLA_1 activity [19]. A database search revealed a protein encoded by the human expressed sequence tag clone KIAA0725 that exhibits strong sequence similarity (52.6%) to p125 throughout the entire sequence determined [18]. KIAA0725p has 711 amino acids with a calculated molecular mass of 81 003 Da, and like p125, has the Gly-X-Ser-X-Gly consensus sequence. Surprisingly, KIAA0025p showed PLA_1 activity, whereas p125 did not [18].

2.5.2
Characterization and Distribution

KIAA0725p does not have a signal sequence or transmembrane domain, suggesting that it is a cytosolic protein. The subcellular localization of KIAA0725 was confirmed by cell fractionation experiments. Both a northern blot analysis and an immunoblot analysis revealed KIAA0725 is expressed in all tissues tested [18]. KIAA0725p might play a fundamental role in cells, whereas $PA\text{-}PLA_1$ appears to have a fundamental role in brain and testes.

The postnuclear supernatant of cells transfected with KIAA0725p cDNA showed high hydrolytic activities against PE and PA and low activities against PS and PC in the absence of Triton X-100, while in the presence of Triton X-100 it had activity against PA but only weak activity against PE [18]. Interestingly, overexpression of KIAA0725p caused morphological change of organelles, such as dispersion of the ER-GIC and Golgi apparatus [18]. Further studies are needed to identify the function of the protein at the molecular level.

2.6 References

1 E.A. Dennis, *Trends Biochem. Sci.*, **1997**, *22*, 1–2.

2 M. Murakami, Y. Nakatani, G. Atsumi, K. Inoue and I. Kudo, *Crit. Rev. Immunol.*, **1997**, *17*, 225–283.

3 A. Lykidis and S. Jackowski, *Prog. Nucl. Acid Res. Mol. Biol.*, **2001**, *65*, 361–393.

4 H. Sonoda, J. Aoki, T. Hiramatsu, M. Ishida, K. Bandoh, Y. Nagai, R. Taguchi, K. Inoue and H. Arai, *J. Biol. Chem.*, **2002**, *277*, 34254–34263.

5 K. Bandoh, J. Aoki, M. Tsujimoto, H. Arai and K. Inoue, *FEBS Lett.*, **2000**, *478*, 159–165.

6 K. Bandoh, J. Aoki, H. Hosono, S. Kobayashi, T. Kobayashi, M.K. Murakami, M. Tsujimoto, H. Arai and K. Inoue, *J. Biol. Chem.*, **1999**, *274*, 27776–27785.

7 M. Robinson and M. Waite, *J. Biol. Chem.*, **1983**, *258*, 14371–14378.

8 G.L. Kucera, C. Miller, P.J. Sisson, R.W. Wilcox, Z. Wiemer and M. Waite, *J. Biol. Chem.*, **1988**, *263*, 12964–12969.

9 H. Ueda, T. Kobayashi, M. Kishimoto, T. Tsutsumi and H. Okuyama, *J. Neurochem.*, **1993**, *61*, 1874–1881.

10 H. Ueda, T. Kobayashi, M. Kishimoto, T. Tsutsumi, S. Watanabe and H. Okuyama, *Biochem. Biophys. Res. Commun.*, **1993**, *195*, 1272–1279.

11 H. Ueda, T. Kobayashi, M. Kishimoto, T. Tsutsumi and H. Okuyama, *Life Sci.*, **1993**, *53*, 629–634.

12 K.Y. Hostetler, M. F. Gardner and J. R. Giordano, *Biochemistry*, **1986**, *25*, 6456–6461.

13 M. Waite, *The Phospholipases* (M. Waite, ed.) Plenum, New York **1987**, pp. 79–110.

14 H. van den Bosch, *Phospholipases*, J.N. Hawthorne, ed., Elsevier, Amsterdam, **1982**, pp. 313–351.

15 F. Carrière, M.C. Withers, T.H. van, A. Roussel, C. Cambillau and R. Verger, *Biochim. Biophys. Acta*, **1998**, *1376*, 417–432.

16 H.N. Higgs, M.H. Han, G.E. Johnson and J.A. Glomset, *J. Biol. Chem.*, **1998**, *273*, 5468–5477.

17 T. Sato, J. Aoki, Y. Nagai, N. Dohmae, K. Takio, T. Doi, H. Arai and K. Inoue, *J. Biol. Chem.*, **1997**, *272*, 2192–2198.

18 K. Nakajima, H. Sonoda, T. Mizoguchi, J. Aoki, H. Arai, M. Nagahama, M. Tagaya and K. Tani, *J. Biol. Chem.*, **2002**, *277*, 11329–11335.

19 K. Tani, T. Mizoguchi, A. Iwamatsu, K. Hatsuzawa and M. Tagaya, *J. Biol. Chem.*, **1999**, *274*, 20505–20512.

20 K. Horigome, M. Hayakawa, K. Inoue and S. Nojima, *J. Biochem.*, **1987**, *101*, 53–61.

21 S. Higashi, T. Kobayashi, I. Kudo and K. Inoue, *J. Biochem.*, **1988**, *103*, 442–447.

22 J. Aoki, Y. Nagai, H. Hosono, K. Inoue and H. Arai, *Biochim. Biophys. Acta*, **2002**, *1582*, 26–32.

23 Y. Nagai, J. Aoki, T. Sato, K. Amano, Y. Matsuda, H. Arai and K. Inoue, *J. Biol. Chem.*, **1999**, *274*, 11053–11059.

24 K. Yokoyama, I. Kudo and K. Inoue, *J. Biochem.*, **1995**, *117*, 1280–1287.

25 T.W. Martin and D. Lagunoff, *Nature*, **1979**, *279*, 250–252.

26 N.Y. Tamori, K. Horigome, K. Inoue and S. Nojima, *J. Biochem. (Tokyo)*, **1986**, *100*, 581–590.

27 K. Horigome, N.Y. Tamori, K. Inoue and S. Nojima, *J. Biochem. (Tokyo)*, **1986**, *100*, 571–579.

28 A. Bruni, G. Monastra, F. Bellini and G. Toffano, *Prog. Clin. Biol. Res.*, **1988**, *282*, 165–179.

29 H. Hosono, J. Aoki, Y. Nagai, K. Bandoh, M. Ishida, R. Taguchi, H. Arai and K. Inoue, *J. Biol. Chem.*, **2001**, *276*, 29664–29670.

30 K. Kawamoto, J. Aoki, H. Hosono, A. Tanaka, A. Itakura, H. Arai, Y. Kiso and H. Matsuda, *J. Immunol.*, **2002**, *168*, 6412–6419.

31 A. Pluckthun and E.A. Dennis, *Biochemistry*, **1982**, *21*, 1743–1750.

32 J.H. Hecht, J.A. Weiner, S.R. Post and J. Chun, *J. Cell Biol.*, **1996**, *135*, 1071–1083.

33 S. An, T. Bleu, O.G. Hallmark and E.J. Goetzl, *J. Biol. Chem.*, **1998**, *273*, 7906–7910.

34 K. Hayashi, M. Takahashi, W. Nishida, K. Yoshida, Y. Ohkawa, A. Kitabatake, J. Aoki, H. Arai and K. Sobue, *Circ. Res.*, **2001**, *89*, 251–258.

35 F. Gaits, O. Fourcade, B.F. Le, G. Gueguen, B. Gaige, D. A. Gassama, J. Fauvel, J.P. Salles, G. Mauco, M.F. Simon and H. Chap, *FEBS Lett.*, **1997**, *410*, 54–58.

36 O. Fourcade, B.F. Le, J. Fauvel, M.F. Simon and H. Chap, *Adv. Enzyme Regul.*, **1998**, *38*, 99–107.

37 F. le Balle, M.F. Simon, S. Meijer, O. Fourcade and H. Chap, *Adv. Enzyme Regul.*, **1999**, *39*, 275–284.

38 J.M. Gerrard and P. Robinson, *Biochim. Biophys. Acta*, **1989**, *1001*, 282–285.

39 H.N. Higgs and J.A. Glomset, *J. Biol. Chem.*, **1996**, *271*, 10874–10883.

40 P.J. Stankiewicz, A.S. Tracey and D.C. Crans, *Metal Ions Biol. Syst.*, **1995**, *31*, 287–324.

41 Y. Tanaka, V.E. Riomin, Y. Kawamura, K. Inoue and S. Nojima, *J. Biochem.*, **1983**, *94*, 841–847.

42 J. Aoki, A. Taira, Y. Takanezawa, Y. Kishi, K. Hama, T. Kishimoto, K. Mizuno, K. Saku, R. Taguchi and H. Arai, *J. Biol. Chem.*, **2002**, *277*, 39696–39702.

43 G.M. Umezu, Y. Kishi, A. Taira, K. Hama, N. Dohmae, K. Takio, T. Yamori, G. B. Mills, K. Inoue, J. Aoki and H. Arai, *J. Cell Biol.*, **2002**, *158*, 227–233.

44 M.L. Stracke, T. Clair and L.A. Liotta, *Adv. Enzyme Regul.*, **1997**, *37*, 135–144.

45 D.S. Min, S.K. Park and J.H. Exton, *J. Biol. Chem.*, **1998**, *273*, 7044–7051.

46 H.N. Higgs and J.A. Glomset, *Proc. Natl. Acad. Sci. USA*, **1994**, *91*, 9574–9578.

47 C. Barlowe and R. Schekman, *Nature*, **1993**, *365*, 347–349.

48 M.H. Han, D.K. Han, R.H. Aebersold and J.A. Glomset, *J. Biol. Chem.*, **2001**, *276*, 27698–27708.

49 S. Uchiyama, Y. Miyazaki, Y. Amakasu, H. Kuwata, Y. Nakatani, G. Atsumi, M. Murakami and I. Kudo, *J. Biochem.*, **1999**, *125*, 1001–1010.

50 T. Thuren, P. Sisson and M. Waite, *Biochim. Biophys. Acta*, **1990**, *1046*, 178–184.

51 T. Hiramatsu, H. Sonoda, Y. Takanezawa, R. Morikawa, M. Ishida, K. Kasahara, Y. Sanai, R. Taguchi, J. Aoki and H. Arai, *J. Biol. Chem.*, **2003**, *278*, 49438–49447.

3
Rational Design of a Liposomal Drug Delivery System Based on Biophysical Studies of Phospholipase A_2 Activity on Model Lipid Membranes

Kent Jørgensen, Jesper Davidsen, Thomas L. Andresen, and Ole G. Mouritsen

Abstract

Secretory phospholipase A_2 (sPLA_2) belongs to a family of small interfacially active enzymes that catalyze the hydrolysis of phospholipids, yielding lysolipids and fatty acids. sPLA_2 is found in snake and bee venom and also in inflammatory and cancerous tissue. It is mainly active on organized lipid substrates such as micelles and lipid membranes, and its hydrolytic activity is moreover strongly influenced by the physical properties of the supramolecular lipid substrate. Systematic model studies of the activity of sPLA_2 on lipid-membrane substrates of different compositions have provided insight into the biophysical mechanisms of sPLA_2 activation. In particular, combined theoretical and experimental model studies have revealed an intimate relationship between the activity of sPLA_2 and the formation of small-scale lipid domains in the lipid membrane substrate. This has been used to design and develop a principle for liposomal drug targeting, release, and absorption that is based on the presence of high levels of sPLA_2 activity in certain diseased target tissue.

3.1
Introduction

Liposomes are self-assembled lipid membrane systems that are spontaneously formed when amphiphilic lipid molecules are mixed with water. Both the functional and structural behavior of liposomes such as the transmembrane permeability and the morphological stability are therefore largely controlled by non-specific physical interactions of the many-particle lipid system. Insight into the structural organization of liposomes on different length and time scales is crucial to manipulating and designing the functional behavior of the liposomes for specific drug delivery purposes [1, 2]. On the one hand, to be effective as drug delivery systems, liposomes have to be stable in the blood stream, i.e. they have to be able to retain the encapsulated drugs and circulate in the blood stream for sufficiently long time to passively target the diseased tissue. On the other hand, this stability has to be effectively balanced with an ability to release the encapsulated drugs at the diseased target site within an appropriate time window [3, 4].

Lipases and Phospholipases in Drug Development
Edited by G. Müller and S. Petry

ISBN: 3-527-30677-3

At present, one of the most successful ways of improving the *in vivo* stability of drug-carrying liposomes is to incorporate large amounts of cholesterol and/or small amounts of synthetic poly(ethylene glycol) (PEG)-lipids into the liposomal membranes [5]. Such highly stable liposomal drug delivery systems containing the anthracycline anticancer drug, doxorubicin, have been approved and marketed as Caelyx and Doxil in Europe and in the US, respectively [6]. However, the local bioavailability of the anticancer drug in the diseased tissue is low for these highly stable drug delivery liposomes that have been specifically designed to minimize systemic drug exposure and to maximize liposome accumulation in tumor tissue [3]. The efficacy of the liposomal drug delivery systems is thus currently not as high as desired.

The control of a rapid drug release close to the diseased target cells would significantly improve such liposomal drug-delivery systems. Different strategies developed to trigger rapid liposomal drug release in the target tissue include drug release triggered by local intervention techniques such as hyperthermia [7, 8], release by programmable fusogenic liposomes [9], and release triggered by the use of environmentally sensitive liposomes that respond to changes in pH [10, 11]. Furthermore, specific peptide-linked lipids incorporated into the drug-carrying liposomes might be turned into fusogenic agents by the action of proteases in tumors, thereby promoting intracellular drug transport [12].

A major advance in drug-delivery could be anticipated if the drug-carrying liposomes were to possess a dual triggering mechanism, involving enhanced drug release at the target with simultaneously enhanced transport of the released drug into the diseased target cells. The present chapter describes a novel principle along these lines for targeted drug release and absorption in the diseased tissue. The principle is built on the increased activity of secretory $sPLA_2$ in the diseased target site (Fig. 3.1), e.g. in cancerous and inflammatory tissue [13, 14].

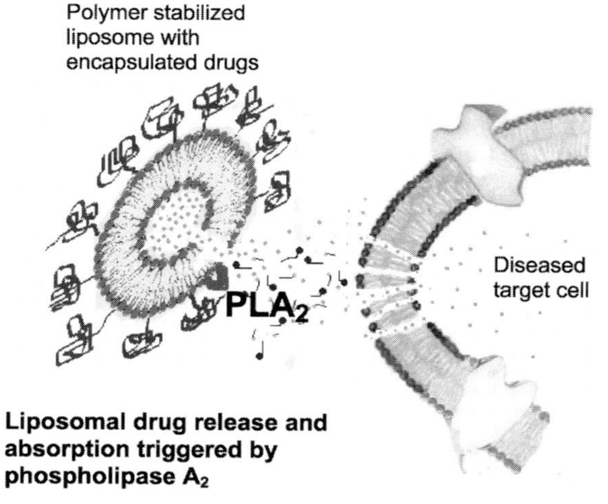

Fig. 3.1 Schematic illustration of a targeted drug-delivery principle that involves accumulation of polymer-covered liposomal drug carriers in porous diseased tissue where secretory $sPLA_2$ subsequently acts as a local trigger for liposomal drug release and drug transport across the target cell membrane.

3.2
Role for Secretory Phospholipase A_2 ($sPLA_2$) in Liposomal Drug Delivery

Secretory $sPLA_2$ belongs to a family of small (14 kDa) interfacially active enzymes that catalyze the hydrolysis of the ester-linkage in the *sn*-2 position of glycerophospholipids, producing free fatty acids and lysophospholipids [15, 16]. $sPLA_2$ is only weakly active on monomeric lipid substrates but is very active on organized lipid substrates such as micelles and liposomes. The enzyme's activity and its mode of action are largely controlled by physical properties, in particular the microstructure of liposomal lipid bilayers [17–19]. Since there is a substantial structural similarity between different secretory $sPLA_2$s from, for example, snake venom and mammals these small membrane-active enzymes probably share a common catalytic mechanism of action. This has been utilized to develop a principle for liposomal drug targeting, release and absorption that is based on elevated concentrations of $sPLA_2$ in the diseased target tissue.

This liposomal principle involves long-circulating polymer-covered liposomes that accumulate in the diseased tissue characterized by leaky blood vessels and high $sPLA_2$ activity. Subsequently, when the phospholipids constituting the liposomal membrane are converted into lysolipids and fatty acids, the drug-carrying liposomes start to disintegrate, resulting in the release of the encapsulated drugs specifically at the diseased target site [20–22]. Furthermore, the $sPLA_2$ generated lysolipids and fatty acids will subsequently act as locally generated permeability enhancers that promote the absorption of the released drugs across the cellular membrane into putative intracellular target sites of the diseased cell. The degradation of the liposomal drug carrier and the release rate of the encapsulated drugs will be determined by the lability of the liposomal membranes towards $sPLA_2$ hydrolysis as well as on the local concentration of $sPLA_2$ in the diseased tissue, e.g. in tumor or inflammatory tissue [23–26].

A fundamental understanding of the relationship between the physicochemical properties of the liposomes and the structural behavior of the lipid membrane, specifically the lateral organization and the formation of small-scale lipid domains, forms the basis for a more rational improvement of the performance of liposomes as drug delivery systems. In particular, this understanding could be used to optimize the fine balance between the liposomal lability required for rapid $sPLA_2$-catalyzed hydrolysis at the target site and the long-term *in vivo* stability in the blood stream needed for accumulation of the liposomes in the porous tumor tissue with high levels of secretory $sPLA_2$.

3.3
Lateral Microstructure of Lipid Bilayers and Its Influence on $sPLA_2$

Determination of the lateral organization of liposomal membranes on scales ranging from 10 to 1000 nm has proved elusive [27–30]. However, indirect and direct evidence from recent studies of model membranes as well as biological membranes has suggested that membranes are laterally heterogeneous and structured in terms of domains and so-called rafts [31]. The length and time scales of these domains remain controversial, and domain diameters from a few nanometers to many microns have been reported.

Lipid domains and rafts in biological membranes are stabilized by several different interactions, including membrane-cytoskeleton, lipid–protein, and lipid–lipid interactions, and the organization can be both equilibrium and non-equilibrium in nature [27]. Lipid-domain formation caused by cooperative phenomena in the lipid bilayers is particularly important for the activation of $sPLA_2$ [32–35]. The cooperative phenomena in lipid bilayers are caused by the fundamental interactions between the lipid molecules and are a consequence of the many-particle character of the supramolecular aggregate. The cooperativity leads to phase transitions and phase equilibria. The key cooperative event in many liposomal membranes is the so-called main phase transition, which takes the bilayer from a solid (gel) phase with conformationally ordered acyl chains to a fluid phase with conformationally disordered chains. The main transition in lipid bilayers is often accompanied by strong lateral density and compositional fluctuations. These fluctuations are manifested as dynamic lipid domains characterized by certain time and length scales that are determined by the thermodynamic conditions and the actual lipid species in question.

The top panel of Fig. 3.2 presents a selection of results for the lateral organization of some simple one- and two-component lipid bilayers that have been investigated by computer-simulation calculations and atomic force microscopy. A pronounced degree of lateral heterogeneity in terms of lipid domains is found. The lipid domains are either solid lipid patches in fluid bilayers or fluid lipid patches in solid bilayers. In lipid mixtures, the domains may reflect incomplete phase separation. The sizes, the morphology, and the topology of the lipid domain patterns depend on the lipid composition and the thermodynamic conditions. The domains can be enhanced or suppressed by adding further lipid components or solutes [36].

Similar lipid domain patterns have been found in other lipid bilayer systems using various experimental techniques, including fluorescence spectroscopy [37], fluorescence microscopy [38], single-particle tracking techniques [39], scattering techniques [40], and atomic-force microscopy [41, 42]. In particular, direct imaging techniques like atomic force microscopy applied to solid-supported bilayers and fluorescence microscopy applied to solid-supported lipid bilayers or giant unilamellar vesicles have revealed lipid domains in the range of tens of nanometers to microns, depending on the lipid membrane in question.

On being presented to lipid bilayer substrates that are laterally heterogeneous (top panel of Fig. 3.2) $sPLA_2$ can bind to and hydrolyze lipid bilayers in very different states: the domains, the background phase, and the domain interfaces. Importantly, the $sPLA_2$ enzyme only senses structure and structural variations over length scales that are smaller or comparable to the scale over which the enzyme diffuses within the time lapse of an actual hydrolytic event. Hence, it can be hy-

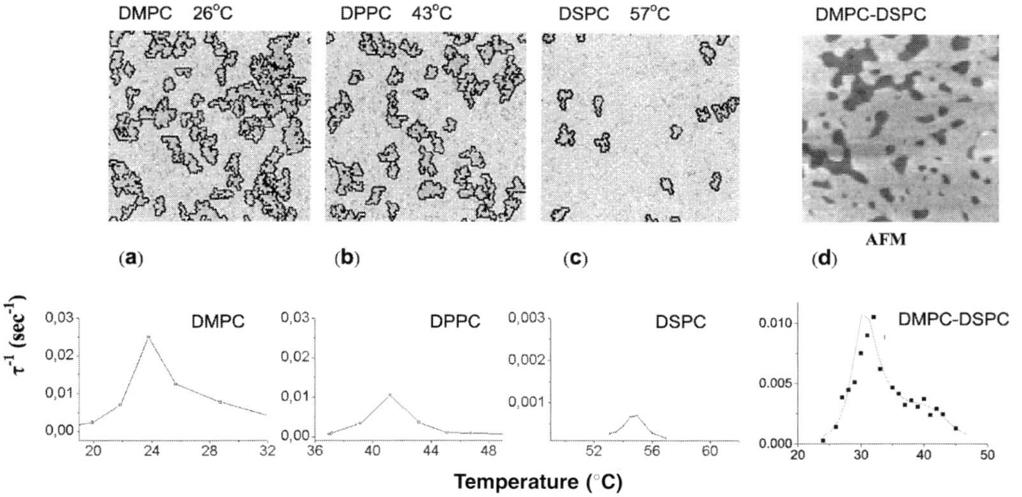

Fig. 3.2 Small-scale structure and domain formation in lipid bilayers (top panel) and the corresponding temperature dependence of sPLA$_2$ activity (bottom panel) measured by the inverse lag time, τ^{-1}. (a) DMPC; (b) DPPC; (c) DSPC; (d) 1:1 DMPC-DSPC (Courtesy of Pernille Høyrup). The domain structures for (a), (b), and (c) were obtained from computer-simulation calculations (frames are 100×100 nm), and for (d) by atomic force microscopy of a solid-supported bilayer (frame is 1×1 μm) (Courtesy of Thomas Kaasgaard). The lag-time data were obtained for snake venom sPLA (*Agkistridon piscivorus piscivorus*) by a combination of fluorescence spectroscopy, light-scattering, real-time HPLC analysis, and pH-stat titration.

pothesized that the enzyme responds to the small-scale structure of the substrate rather than its large-scale thermodynamic behavior.

This hypothesis is supported by data (bottom panel of Fig. 3.2) for the activation of sPLA$_2$ as a function of temperature for the same four lipid substrates as in the top panel. The rate of activation is measured here by the inverse of the so-called lag time, which is the time that describes the lag-burst kinetics that is characteristic for many sPLA$_2$s [17]; the smaller the lag time, the more rapid the activation of the enzyme. For the one-component lipid bilayers, we see a dramatic increase in activity at the phase transition temperature [17], and for the binary mixtures there is enhanced activity at the phase boundaries and in the phase-coexistence region [35, 43]. These observations suggest that the activity of sPLA$_2$ is modulated by the small-scale lateral structure of the lipid bilayer substrate. The relationship between sPLA$_2$ activity and lateral lipid bilayer heterogeneity can be made semi-quantitative via comparison with the results from computer-simulation calculations shown in the top panel of Fig. 3.2 [17, 32, 44]. This comparison suggests that sPLA$_2$ may be particularly active at domain interfaces that are sites of structural defects and hence good points of attack for the enzyme. Similar conclusions have been drawn from early work on lipid monolayers [45].

This general picture of $sPLA_2$ activation is supported by detailed analyses of atomic force microscopy images of solid-supported bilayers in the presence of $sPLA_2$ in the various stages before and after the lag phase [30, 33, 34]. These analyses strongly suggest that $sPLA_2$ attacks at areas with structural defects, such as domain boundaries, at regions with accumulated hydrolysis products, and at the edges of pre-existing holes in the bilayer. This is also in line with findings suggesting that $sPLA_2$ activity is enhanced on liposomes of decreasing size, i.e. increasing curvature [46], and therefore larger curvature tension. For bilayers made of lipids with different chain lengths, although $sPLA_2$ seems to bind uniformly across solid and fluid domains enriched in the long- and short-chain lipid species, respectively, the enzyme prefers the short-chain lipid species, which is hydrolyzed first [30, 35, 47]. The preferred hydrolysis of the short-chain lipids, which are in the fluid phase, might be due to the lower surface area density of the lipid head groups in this phase. This would accord with the enhanced $sPLA_2$ activity seen on giant unilamellar liposomes that are subject to an osmotic stress that lowers the surface density [48].

3.4 $sPLA_2$ Degradation of Drug-delivery Liposomes: A New Drug-delivery Principle

3.4.1 Liposomes Protected by Polymer Coating

Bare unilamellar liposomes are poor candidates for drug-delivery systems because they are quickly recognized and degraded by macrophages from the immune system. However, by covering the surface of the liposomes with polymers they become invisible to the macrophages and therefore enjoy increased circulation times in the blood stream. These so-called Stealth liposomes [1, 2] contain a certain fraction of lipopolymers, often of the type poly(ethylene glycol) (PEG) that is covalently attached to the lipid head group. The PEG moiety is highly soluble in water.

Although the polymer coat might be expected to inhibit the access of enzymes such as $sPLA_2$ to the lipid membrane surface, and hence suppress the enzymatic degradation of the polymer-covered liposomes, it actually decreases the lag time of $sPLA_2$ and therefore enhances the degradation of the PEG liposomes [20]. Systematic studies of the dependence of the lag time on the lipopolymer content and the polymer length have shown that the enhanced $sPLA_2$ activity is an electrostatic effect caused by the attraction between $sPLA_2$ and the negative charge of the phosphate group of the lipopolymers [21]. This was further underlined by results demonstrating that the activity of $sPLA_2$ was reduced when a methyl group was covalently attached to the phosphate group, thereby rendering the head group of the PEG-lipids uncharged [22]. The electrostatic character of the binding of $sPLA_2$ to liposome surfaces can obviously be exploited to optimize enzyme activity.

3.4.2
Biophysical Model Drug-delivery System to Study $sPLA_2$ Activity

Fig. 3.3 illustrates schematically a biophysical model of a drug-delivery system consisting of polymer-coated drug-carrying liposomes and liposomal target cell membranes. This system has been used to develop and investigate the potentials of a principle for liposomal drug targeting and absorption that involve triggering by $sPLA_2$ [14]. The drug-carrying liposomes are unilamellar liposomes made of DPPC lipids. The target membranes consist of liposomes made of 1,2-O-stearoyl PC (di-O-SPC) lipids whose stearoyl chains are ether-linked to the glycerol backbone, thereby rendering the di-O-SPC lipids inert towards $sPLA_2$-catalyzed hydrolysis. This simulates a situation where the target membranes may be inert to the endogeneous lipases. These target liposomal membranes are arranged to encapsulate a self-quenching fluorescent molecule, calcein. By monitoring the efflux of calcein, this type of experimental assay allows for a detailed investigation of the $sPLA_2$-catalyzed hydrolysis of the carrier liposomes while also monitoring the per-

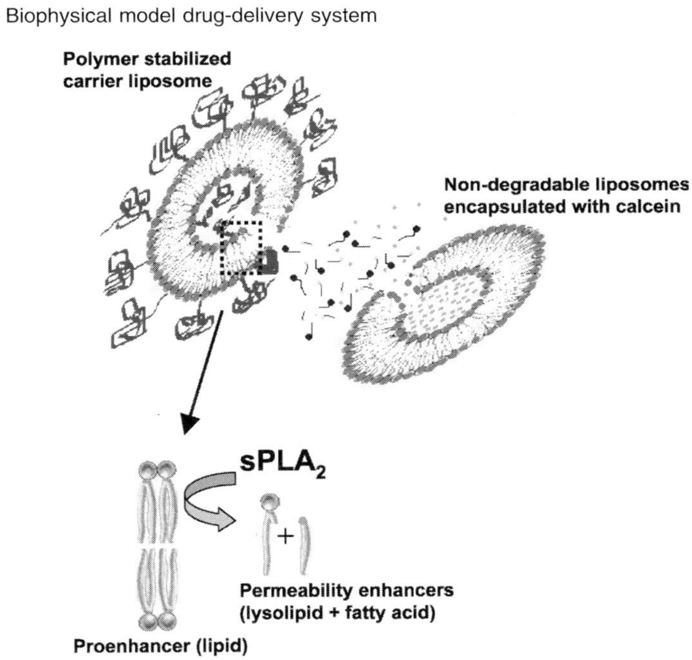

Fig. 3.3 Illustration of a supramolecular-based biophysical model drug-delivery system where the phospholipids of the carrier liposome, via $sPLA_2$-catalyzed hydrolysis, act as pro-destabilisers at the site of the carrier and as pro-enhancers at the site of the target membrane.

meability-enhancing effect of the generated lysopalmitoyl PC (lysoPPC) and palmitic acid (PA) on the inert di-O-SPC target membranes [49].

3.4.3
Effect of Lipid Composition on sPLA$_2$-triggered Drug Release and Absorption

When sPLA$_2$ is added to the liposome suspension, a slow release of calcein is observed from the target liposomes (Fig. 3.4). This period of slow release, which is clearly observed for the bare DPPC liposomes, corresponds to the well-known lag time of sPLA$_2$. A pronounced decrease in the lag time and a concomitant enhancement of the release of calcein across the target membranes is observed when 2.5 mol% DPPE-PEG$_{2000}$ is incorporated into the carrier liposomes. When short-chain di-capryl PC lipids (DCPC), that activate sPLA$_2$ [50], are incorporated into the carrier PEG-liposomes, the permeation of calcein across the target membranes increases almost instantaneously.

The release profiles suggest that the lysoPPC and PA hydrolysis products, which are formed in a 1:1 molar ratio by sPLA$_2$, are incorporated into the target membranes [51], leading to the increase in the permeability of the target liposomal membranes. These hydrolysis products, due to their non-bilayer forming molecular shapes, induce a curvature stress [52, 53] and/or form small-scale lipid domains [33, 36], which lead to membrane defects and consequently increased membrane

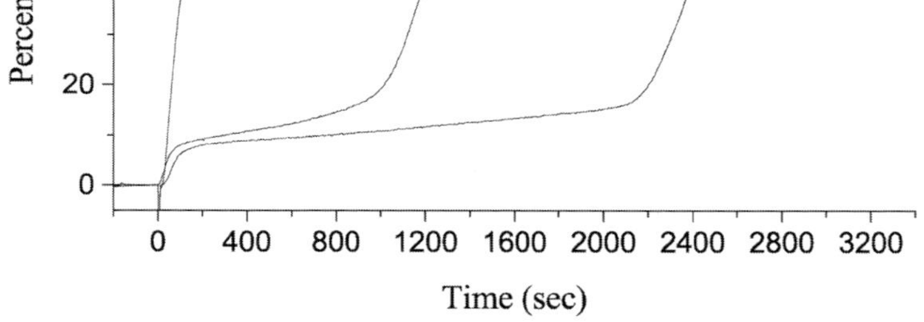

Fig. 3.4 sPLA$_2$ (*A. piscivorus piscivorus*) triggered permeation of fluorescent calcein across di-O-SPC target membranes as a function of time for different compositions of the carrier liposomes at 37 °C. In comparison with bare DPPC carriers, the rate of release of calcein across the target membranes is dramatically enhanced for the polymer-coated carriers. A further increase in the calcein release is obtained if the carrier also contains a short-chain phospholipid, DCPC, which acts as a local activator of sPLA$_2$.

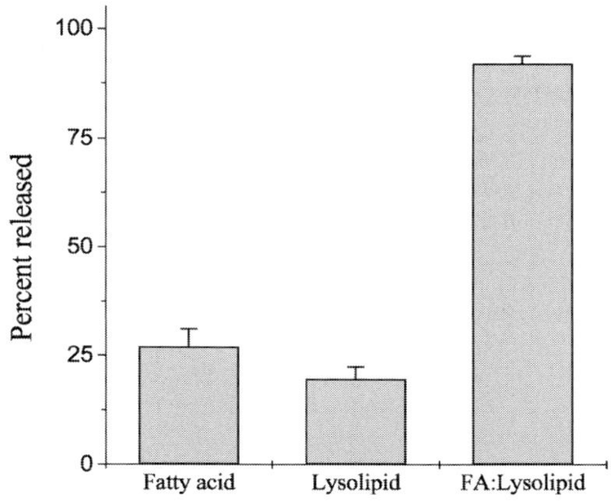

Fig. 3.5 Total calcein release from 1,2-O-SPC target liposomes 20 min after addition of 10 μm PA and lyso-PPC separately or simultaneously in an equimolar mixture. Concentration of the target liposomes was 20 μm in a HEPES buffer (pH = 7.5), temperature = 39 °C.

permeability. The unique ability of sPLA$_2$ to produce permeability enhancers is further underlined by the data in Fig. 3.5, showing that calcein release is dramatically enhanced only when lysoPPC and PA are added simultaneously from a 1:1 mixture (in the absence of sPLA$_2$). This effect suggests that the generated hydrolysis products act in a synergistic fashion as a pair of permeability enhancers [49].

Furthermore, when large amounts of cholesterol (>20 mol%) are incorporated into the carrier PEG-liposomes, sPLA$_2$ becomes inactivated. This is of particular interest in relation to the blood circulation times of PEG-liposomes, which are almost the same without cholesterol as with large amounts of cholesterol [54]. In fact, these results suggest that drug-carrying PEG-liposomes could be designed without cholesterol and still circulate in the blood stream for sufficiently long to accumulate in the diseased tissue where they, advantageously, will experience an increased susceptibility towards sPLA$_2$ degradation.

3.4.4
Effect of Temperature on Liposomal Drug Release and Absorption by sPLA$_2$

Temperature has a dramatic and highly non-trivial effect on sPLA$_2$ activation in the region of the main phase transition of saturated phospholipid bilayers [17, 19] (Fig. 3.2). As noted above, this is caused by dramatic lateral structural changes in the lipid bilayer [28]. It is possible to take advantage of this physical effect as a thermally activated release trigger mechanism in the biophysical drug-delivery model system, as illustrated by the data displayed in Fig. 3.6. As the temperature approaches the main phase transition temperature at 41 °C of the DPPC lipid bilayer, the rate of calcein release is dramatically enhanced as quantified by the time of 50% calcein release (insert in Fig. 3.6).

It has been suggested that hyperthermia could be used to enhance drug release, and that local heating at predefined tumor areas could be used to locally destabi-

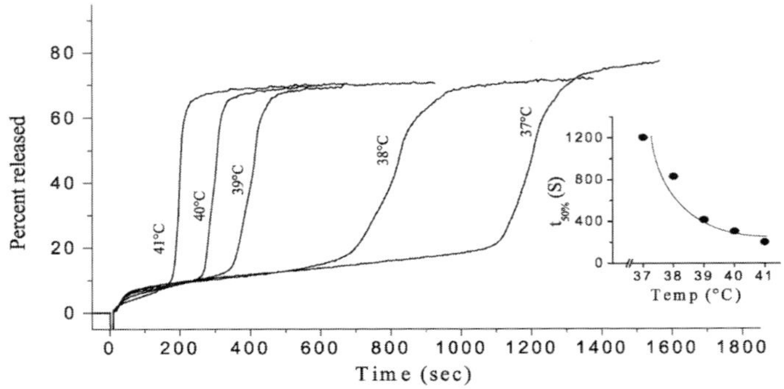

Fig. 3.6 Temperature-dependent $sPLA_2$ (*A. piscivorus piscivorus*) controlled permeation of calcein across 1,2-di-O-SPC target liposomes as a function of time at different temperatures. As the temperature is raised, calcein release is enhanced due to an increased activity of the enzyme induced by structural changes in the lipid membrane substrate of the carrier liposome consisting of DPPC + 2.5 mol% DPPE-PEG_{2000}. In the present assay, a maximum calcein release of about 70% was achieved in all cases. Insert: time of 50% calcein release, $t_{50\%}$, as a function of temperature.

lize drug-carrying liposomes, by exploiting the enhanced leakiness of liposomes at their phase transition [8]. In a similar way, local heating of the diseased tissue could enhance liposomal drug release and absorption via an enhancement of the $sPLA_2$-catalyzed degradation of the drug-carrying liposomes.

3.4.5
Liposomal Drug Release as a Function of $sPLA_2$ Concentration

Fig. 3.7 shows the $sPLA_2$-triggered release of calcein entrapped in liposomes composed of 1-O-DPPC (the ester linkage in the *sn*-1 position has been replaced by an ether linkage) incorporated with 10 mol% 1-O-DPPE-PEG_{350} [14]. These results demonstrate that rapid release, within minutes, can be obtained even at 1 nM PLA_2 ($\sim$14 ng ml^{-1}). Interestingly, the $sPLA_2$ concentrations found in effusions from patients with various cancer types ranged from 9 to 188 ng ml^{-1} [15]. Furthermore, $sPLA_2$ can hydrolyze the 1-O-DPPC and 1-O-DPPE-PEG_{350} mono-ether lipids, leading to the release of lysolipids that are stable against hydrolysis by, for example, lysophospholipases. The generated ether lysolipids are expected to display increased *in vivo* stability, thereby having the potential to induce a more pronounced permeability-enhancing effect in the diseased tissue.

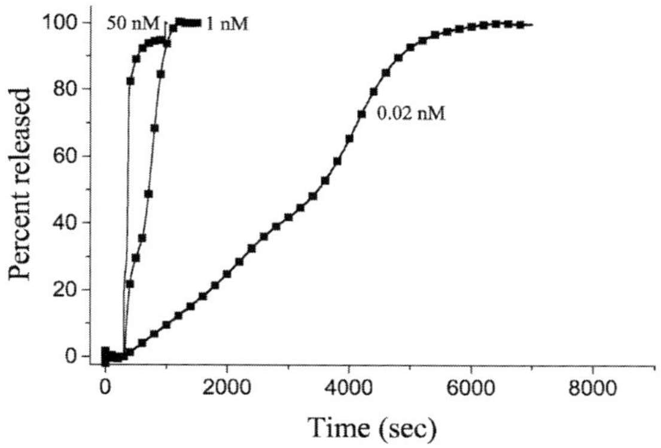

Fig. 3.7 Concentration-dependent $sPLA_2$-triggered release of calcein from 25 μm polymer-covered liposomes composed of 1-O-DPPC and 1-O-DPPE-PEG_{350} (9:1) suspended in a 10 nM HEPES-buffer (pH=7.5). $sPLA_2$ (*A. piscivorus piscivorus*) was added after 300 s; temperature=35.5 °C.

3.5 Conclusion

The results described here suggest that polymer-grafted liposomes, which accumulate in diseased tissue with leaky capillaries and increased levels of $sPLA_2$, can also be rationally designed and optimized to be susceptible to $sPLA_2$-catalyzed degradation. In this way it is possible to achieve a remote and site-specific release and absorption of the encapsulated drugs specifically at the diseased target site. This principle for targeted liposomal drug delivery to diseased tissue implies that the $sPLA_2$-controlled release will be enhanced where the $sPLA_2$ concentration is elevated and the polymer-covered liposomes accumulate due to leaky blood vessels that often characterize solid tumors or inflamed tissue. In particular, this principle for targeted drug delivery is independent of the size of the diseased region and does not require a preceding localization of the diseased tissue. Furthermore, by designing the carrier liposome to include special lipids whose degradation products, after exposure to $sPLA_2$, are converted into permeability enhancers, enzymatic activators, or certain drugs such as anticancer lysoetherlipids [55], a possible realization of Dr Erlich's 'magic bullet' can be imagined.

3.6 Acknowledgments

This work was supported by the Danish National Research Foundation via a grant to the MEMPHYS-Center for Biomembrane Physics, the Danish Medical Research Council via a grant to the Center for Drug Delivery and Transport, the Danish Natural Science Research Council, and the Danish Technical Research Academy.

3.7 References

1 D. D. Lasic, *Liposomes: From Physics To Application*, Elsevier Science B.V., Amsterdam, **1993**.

2 D. D. Lasic and D. Needham. The stealth liposome: a prototypical biomaterial. *Chem. Rev.*, **1995**, *95*, 2601–2627.

3 D. Needham, Materials engineering of lipid bilayers for drug carrier performance. *MRS Bull.*, **1999**, *24*, 32–40.

4 D. Needham and M. W. Dewhirst. The development and testing of a new temperature-sensitive drug delivery system for the treatment of solid tumors. *Adv. Drug Delivery Rev.*, **2001**, *53*, 285–305.

5 T. M. Allen and C. Hansen. Pharmacokinetics of stealth versus conventional liposomes, effect of dose. *Biochim. Biophys. Acta*, **1991**, *1068*, 133–141.

6 A. A. Gabizon, D. Goren, R. Cohen, and Y. Barenholz. Development of liposomal anthracyclines: from basics to clinical applications. *J. Control. Release*, **1998**, *53*, 275–279.

7 D. Needham, G. Anyarambhatla, G. Kong, and M. W. Dewhirst. A new temperature-sensitive liposome for use with mild hyperthermia: characterization and testing in a human tumor xenograft model. *Cancer Res.*, **2000**, *60*, 1197–1201.

8 M. B. Yatvin, J. N. Weinstein, W. H. Dennis, and R. Blumenthal. Design of liposomes for enhanced local release of drugs by hyperthermia. *Science*, **1978**, *202*, 1290–1293.

9 G. Adlakha-Hutcheon, M. B. Bally, C. R. Hew, and T. D. Madden. Controlled destabilization of liposomal drug delivery system enhances mitoxantrone antitumor activity. *Nat. Biotechnol.*, **1999**, *17*, 775–779.

10 J. Conner, M. B. Yatvin, and L. Huang. pH-sensitive liposomes: acid-induced liposome fusion. *Proc. Natl. Acad. Sci. USA*, **1984**, *81*, 1715–1718.

11 M. S. Hong, S. J. Lim, Y. K. Oh, and C. K. Kim. pH-sensitive, serum-stable and long-circulating liposomes as a new drug delivery system. *J. Pharm. Pharmacol.*, **2002**, *54*, 51–58.

12 P. Meers, Enzyme-activated targeting of liposomes. *Adv. Drug Delivery Rev.*, **2001**, *53*, 265–272.

13 K. Jørgensen, J. Davidsen, and O. G. Mouritsen. Biophysical mechanisms of phospholipase A_2 activation and their use in liposome-based drug delivery. *FEBS Lett.*, **2002**, *531*, 23–27.

14 J. Davidsen, K. Jørgensen, T. L. Andresen, and O. G. Mouritsen. Secreted phospholipase A_2 as a new enzymatic trigger mechanism for localised liposomal drug release and absorption in diseased tissue. *Biochim. Biophys. Acta*, **2003**, *1609*, 95–101.

15 O. G. Berg, M. H. Gelb, M.-D. Tsai, and M. K. Jain. Interfacial enzymology: the secreted phospholipase A_2-paradigm. *Chem. Rev.*, **2001**, *101*, 2613–2653.

16 Six, D. A. and E. A. Dennis. The expanding superfamily of phospholipase A_2 enzymes: classification and characterization. *Biochim. Biophys. Acta*, **2000**, *1488*, 1–19.

17 T. Hønger, K. Jørgensen, R. L. Biltonen, and O. G. Mouritsen. Systematic relationship between phospholipase A_2 activity and dynamic lipid bilayer microheterogeneity. *Biochemistry*, **1996**, *35*, 9003–9006.

18 J. D. Bell and R. L. Biltonen. Activation of phospholipase A_2 on lipid bilayers. *Methods Enzymol.*, **1991**, *197*, 249–258.

19 T. Hønger, K. Jørgensen, D. Stokes, R. L. Biltonen, and O. G. Mouritsen. Phospholipase A_2 activity and physical properties of lipid-bilayer substrates. *Methods Enzymol.*, **1997**, *286*, 168–190.

20 C. Vermehren, T. Keibler, I. Hylander, T. Hønger Callisen, and K. Jørgensen. Increase in phospholipase A_2 activity towards lipopolymer containing liposomes. *Biochim. Biophys. Acta*, **1998**, *1373*, 27–36.

21 K. Jørgensen, C. Vermehren, and O. G. Mouritsen. Enhancement of phospholipase A_2 catalyzed degradation of polymer grafted PEG-liposomes: effects of lipopolymer concentration and chain length. *Pharm. Res.*, **1999**, *16*, 1493–1495.

22 T.L. Andresen, O.G. Mouritsen, M. Begtrup, and K. Jørgensen. Phospholipase A_2 activity: dependence on liposome surface charge and polymer coverage. *Biophys. J.*, **2002**, *82*, 148a.

23 T. Abe, K. Sakamoto, H. Kamohara, Y. Hirano, N. Kuwahara, and M. Ogawa. Group II phospholipase A_2 is increased in peritoneal and pleural effusions in patients with various types of cancer. *Int. J. Cancer*, **1997**, *74*, 245–250.

24 J. Jiang, B.L. Neubauer, J.R. Graff, M. Chedid, J.E. Thomas, N.W. Roehm, S. Zhang, G. J. Eckert, M.O. Koch, J.N. Eble, and L. Cheng. Expression of group IIA secretory phospholipase A_2 is elevated in prostatic intraepithelial neoplasia and adenocarcinoma. *Am. J. Pathol.*, **2002**, *160*, 667–671.

25 S. Yamashita, M. Ogawa, K. Sakamoto, T. Abe, H. Arakawa, and J. Yamashita. Elevation of serum group II phospholipase A_2 levels in patients with advanced cancer. *Clin. Chem. Acta*, **1994**, *228*, 91–99.

26 E. Kaiser, Phospholipase A_2: its usefulness in laboratory diagnostics. *Crit. Rev. Clin. Lab. Sci.*, **1999**, *36*, 65–163.

27 O.G. Mouritsen and K. Jørgensen. Small-scale lipid-membrane structure: simulation versus experiment. *Curr. Opin. Struct. Biol.*, **1997**, *7*, 518–527.

28 L.K. Nielsen, T. Bjørnholm, and O.G. Mouritsen. Fluctuations caught in the act. *Nature*, **2000**, *404*, 352.

29 L.K. Nielsen, A. Vishnyakov, K. Jørgensen, T. Bjørnholm, and O.G. Mouritsen. Nano-meter-scale structure of fluid lipid membranes. *J. Phys.: Condens. Matter*, **2000**, *12*, 309–314.

30 T. Kaasgaard, C. Leidy, J.H. Ipsen, O.G. Mouritsen, and K. Jørgensen. *In situ* atomic force microscope imaging of supported bilayers. *Single Mol.*, **2001**, *2*, 105–108.

31 K. Jacobson and C. Dietrich. Looking at lipid rafts? *Trends Cell. Biol.*, **1999**, *9*, 87–91.

32 P. Høyrup, K. Jørgensen, and O.G. Mouritsen. Phospholipase A_2 – an enzyme that is sensitive to the physics of its substrate. *Europhys. Lett.*, **2002**, *57*, 464–470.

33 L.K. Nielsen, J. Risbo, T.H. Callisen, and T. Bjørholm. Lag-burst kinetics in phospholipase A_2 hydrolysis of DPPC bilayers visualized by atomic force microscopy. *Biochim. Biophys. Acta*, **1999**, *1420*, 266–271.

34 T. Kaasgaard, J.H. Ipsen, O.G. Mouritsen, and K. Jørgensen. *In situ* atomic force microscope imaging of phospholipase A_2 lipid bilayer hydrolysis. *J. Probe Microsc.*, **2001**, *2*, 169–175.

35 P. Høyrup, O.G. Mouritsen, and K. Jørgensen. Phospholipase A_2 activity towards vesicles of DPPC and mixtures of DMPC and DSPC with small amounts of SMPC. *Biochim. Biophys. Acta*, **2001**, *1515*, 133–143.

36 O.G. Mouritsen and K. Jørgensen. A new look at lipid-membrane structure in relation to drug research. *Pharm. Res.*, **1998**, *15*, 1507–1519.

37 C. Leidy, W.F. Wolkers, O.G. Mouritsen, K. Jørgensen, and J.H. Crowe. Lateral organization and domain formation in a two-component lipid membrane system. *Biophys.J.*, **2001**, *80*, 1819–1828.

38 L.A. Bagatolli and E. Gratton. Two-photon fluorescence microscopy of coexisting lipid domains in giant unilamellar vesicles of binary lipid mixtures. *Biophys. J.*, **2000**, *78*, 290–305.

39 C. Dietrich, B. Yang, T. Fujiwara, A. Kusumi, and K. Jacobson. Relationship of lipid rafts to transient confinement zones detected by single particle tracking. *Biophys. J.*, **2002**, *82*, 274–284.

40 C. Gliss, H. Clausen-Schaumann, R. Günther, S. Odenbach, O. Randl, and T.M. Bayler. Direct detection of domains in phospholipid bilayers by grazing incidence diffraction of neutrons and atomic force microscopy. *Biophys. J.*, **1998**, *74*, 2443–2450.

41 H.A. Rinia and B. de Kruijff. Imaging domains in model membranes with atomic force microscopy. *FEBS Lett.*, **2001**, *504*, 194–199.

42 T. Kaasgaard, O.G. Mouritsen, and K. Jørgensen. Lipid domain formation and ligand–receptor distribution in lipid bilayer membranes investigated by atomic force microscopy. *FEBS Lett.*, **2002**, *515*, 29–34.

43 W. R. Burack, A. R. Dibble, M. M. Allietta, and R. L. Biltonen. Changes in vesicle morphology induced by lateral phase separation modulate phospholipase A_2 activity. *Biophys. J.*, **1997**, *36*, 10551–10557.

44 P. Høyrup, K. Jørgensen, and O. G. Mouritsen. Nano-scale structure in membranes in relation to enzyme action – computer simulation vs. experiment. *Comp. Phys. Commun.*, **2002**, *147*, 313–320.

45 D. W. Grainger, A. Reichert, H. Ringsdorf, and C. Salesse. An enzyme caught in action: direct imaging of hydrolytic function and domain formation of phospholipase A_2 in phosphatidylcholine monolayers. *FEBS Lett.*, **1989**, *252*, 72–82.

46 W. R. Burack and R. L. Biltonen. Lipid bilayer heterogeneities and modulation of phospholipase A_2. *Chem. Phys. Lipids*, **1994**, *73*, 209–222.

47 S. A. Sanchez, L. A. Bagatolli, E. Gratton, and T. L. Hazlett. A two-photon view of an enzyme at work: *Crotalus atrox* venom PLA_2 interaction with single-lipid and mixed-lipid giant unilamellar vesicles. *Biophys. J.*, **2002**, *82*, 2232–2243.

46 J. Y. A. Lehtonen and P. K. J. Kinnunen. Phospholipase A_2 as a mechanosensor. *Biophys. J.*, **1995**, *68*, 1888–1894.

49 J. Davidsen, O. G. Mouritsen, and K. Jørgensen. Synergistic permeability enhancing effect of lysophospholipids and fatty acids on lipid membranes. *Biochim. Biophys. Acta*, **2002**, *1564*, 256–262.

50 N. E. Gabriel, N. V. Agman, and M. F. Roberts. Enzymatic-hydrolysis of short-chain lecithin long-chain phospholipid unilamellar vesicles: sensitivity of phospholipases to matrix phase state. *Biochemistry*, **1987**, *26*, 7409–7418.

51 D. Needham and D. V. Zhelev. Lysolipid exchange with lipid vesicle membranes. *Ann. Biomed. Eng.*, **1995**, *23*, 287–298.

52 R. B. Cornell and R. S. Arnold. Modulation of the activities of enzymes of membrane lipid metabolism by non-bilayer-forming lipids. *Chem. Phys. Lipids*, **1996**, *81*, 215–227.

53 A. Schmidt, M. Wolde, C. Thiele, W. Fest, H. Kratzin, A. V. Podtelejnikov, W. Witke, W. B. Huttner, and H. D. Soling. Endophilin I mediates synaptic vesicle formation by transfer of arachidonate to lysophosphatidic acid. *Nature*, **1999**, *401*, 133–141.

54 M. C. Woodle and D. D. Lasic. Sterically stabilized liposomes. *Biochim. Biophys. Acta*, **1992**, *1113*, 171–199.

55 M. Lohmeyer and P. Workman. Growth arrest vs direct cytotoxicity and the importance of molecular-structure for the in-vitro antitumor-activity of ether lipids. *Br. J. Cancer*, **1995**, *72*, 277–286.

4
Phospholipase D

John H. Exton

4.1
Introduction

Phospholipase D (PLD) is an enzyme that hydrolyzes phospholipids, principally phosphatidylcholine (PC), to phosphatidic acid (PA) and the free head group. It is widely distributed and has been purified or cloned from many mammalian, plant, protozoan and bacterial species and also yeast, *Drosophila*, *Anopheles*, and *C. elegans*. In mammals it occurs as splice variants of two gene products, and in plants there are multiple genes and isozymes.

All PLD isozymes share two conserved $HXKX_4DX_6GG/S$ sequences [1, 2], designated HKD motifs, which are required for catalytic activity. These motifs are also present in bacterial phospholipid synthases, bacterial endonucleases, pox virus envelope proteins, a murine toxin from *Yersinia* and tyrosyl-DNA phosphodiesterase [1, 3], although in this latter case the motifs lack the aspartate residues. Proteins containing HKD motifs are said to comprise a PLD superfamily.

Members of the PLD superfamily that exhibit PLD activity contain four conserved sequences (I–IV) of which sequences I and IV contain the HKD motifs [4] (Fig. 4.1). The functions of sequences I and III have not been defined, but it is assumed they participate in catalysis. Two adjacent domains, PX (Phox homology) and PH (pleckstrin homology), are located in the N-terminal sequences of mammalian, yeast and nematode PLDs and in the plant PLD ζ isozymes, but are absent from bacterial and most plant enzymes [5, 6]. The plant PLDs lacking the PX/PH tandem domain contain a Ca^{2+}/phospholipid binding domain, designated C2, at the N-terminus and require Ca^{2+} for activity [6].

Mammalian, yeast, *Tetrahymena* and certain plant PLDs depend on phosphatidylinositol 4,5-bisphosphate (PIP_2) for activity [6–10]. Although the PH domain can bind PIP_2 [9], there is a site between conserved sequences II and III which also binds PIP_2 and is required for activity [8]. Phosphatidylinositol 3,4,5-trisphosphate (PIP_3) can also stimulate activity, but other phospholipids have little or no effect.

As indicated above, there are multiple plant PLDs (at least 12 genes), which can be grouped into five types (α, β, γ, δ and ζ) [6]. PLDα, which is designated conventional, is active at millimolar concentrations of Ca^{2+}, whereas the β and γ iso-

Lipases and Phospholipases in Drug Development
Edited by G. Müller and S. Petry

ISBN: 3-527-30677-3

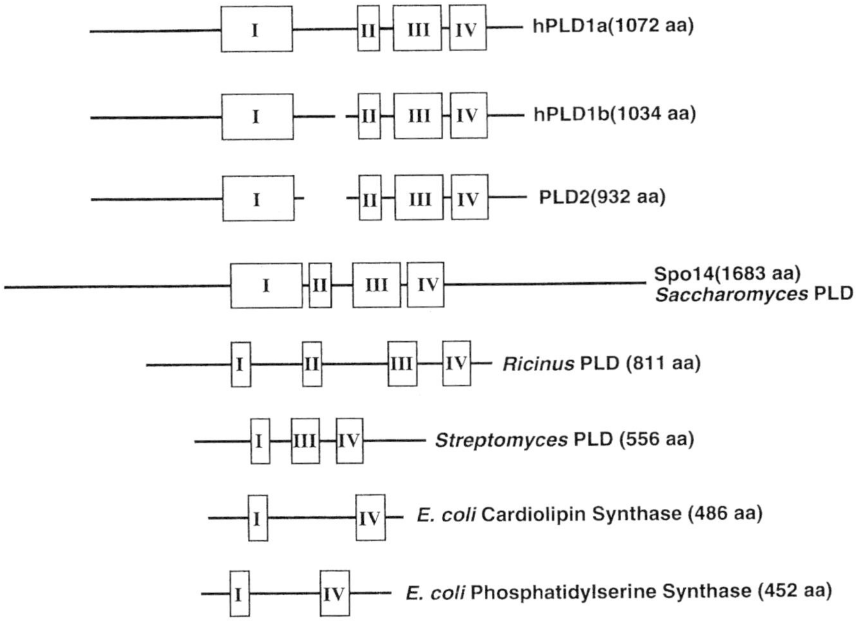

Fig. 4.1 Schematic depiction of members of the PLD superfamily. Boxes show conserved sequences I–IV; hPLD1a and b are the alternatively spliced versions of human PLD1. PLD2 is the mouse isozyme.

zymes depend on PIP_2 and are maximally active at micromolar Ca^{2+} [6]. The δ isozyme is activated by oleic acid, and the ζ isozyme contains PX and PH domains and is Ca^{2+}-independent and selective for PC [6]. In contrast, there are only two mammalian PLD genes whose protein products (PLD1 and PLD2) are alternatively spliced and show approximately 50% amino acid sequence identity [11, 12]. PLD1 is regulated *in vivo* and *in vitro* by protein kinase C (PKC) and small G proteins of the Rho and Arf families. In contrast, PLD2 has a high basal activity and is not affected by PKC or Rho or Arf proteins *in vitro*. The focus of this chapter is mainly on the mammalian PLD isozymes.

4.2
Structure and Catalytic Mechanism of Mammalian Phospholipase D

No mammalian PLD has been crystallized and its three-dimensional structure determined. However, the structure of a PLD from *Streptomyces* has been determined and that of the Nuc endonuclease from *E. coli*, which is a low molecular mass (16 kDa) member of the PLD superfamily [13, 14]. The crystal structure of another member of the PLD superfamily, tyrosyl-DNA phosphodiesterase, has been determined [15]. This enzyme catalyzes the hydrolysis of the phosphodiester bond between a tyrosine residue and a DNA 3′ phosphate. The Nuc protein crystallizes as a dimer, with the conserved HKD sequences adjacent to each other to produce a single active site [14]. Each monomer consists of an 8-strand β-sheet flanked by five α-helices. The active site residues are held together by a network of H bonds that involve not only the conserved His, Ser and Asn residues, but also adjacent Glu residues. Substrate binding involves the His, Lys and Asn residues of the HKD motif, with the Lys and Asn neutralizing the negative charge on the substrate phosphate and also the His residues during catalysis [14].

The *Streptomyces* PLD consists of two highly interacting components of similar topology [13]. Overall, the structure is very similar to the Nuc dimer. Each component consists of a β-sheet of 8–9 strands surrounded by nine α-helices. The catalytic center is also very similar to that of the Nuc dimer. Crystallization with a phosphate buffer indicates that the phosphate head group of the substrate lies in contact with His, Lys and Asn residues contributed from both HKD domains [13]. The tyrosyl-DNA phosphodiesterase is also composed of two similar domains related by a 2-fold axis of symmetry [15]. Conserved His, Lys and Asn residues are contributed by each domain to form the active site. Thus the structure and catalytic mechanism of this enzyme are very similar to those of other PLD superfamily members. A derived structure for the catalytic core of mammalian PLD1 is shown

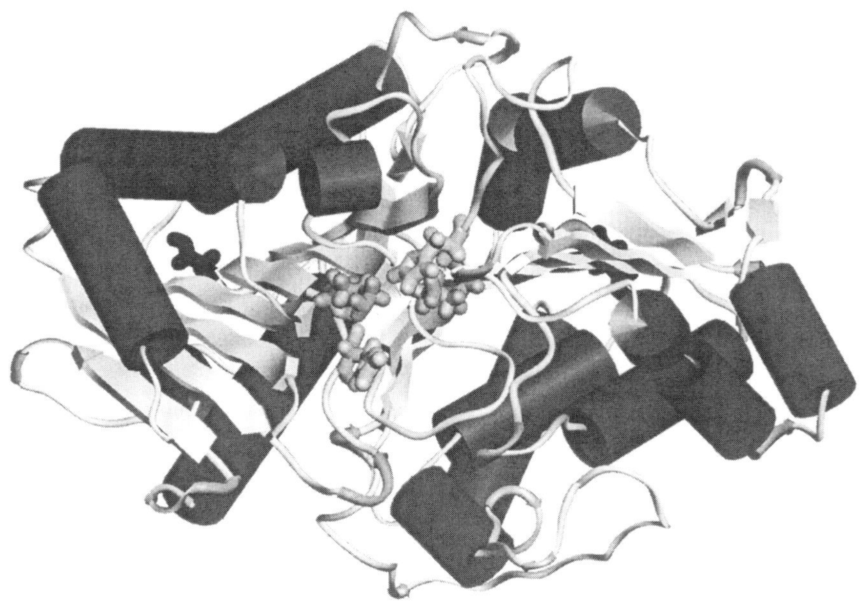

Fig. 4.2 Deduced catalytic core of PLD1. The structure was derived by molecular modeling of *Streptomyces* PLD. Structures in center are His and Lys residues contributed by the two HKD motifs.

in Fig. 4.2. This was derived by molecular modeling based on the *Streptomyces* enzyme [13].

The catalytic mechanism involves a two-step (ping-pong) reaction with the formation of a covalent phospho-enzyme intermediate, with one of the His residues acting as the nucleophile. The second His then acts as a general acid to donate a H^+ to the O of the leaving group [14]. This mechanism is supported by mutational studies [4], experiments with hydroxylamine, and also by the labeling of His with $^{32}P_i$ in phosphate–water exchange experiments [16].

The structure of the catalytic center of mammalian PLDs and also the catalytic mechanism are probably very similar to those reported above. In fact, there is evidence that the HDK motifs in conserved sequences II and IV must dimerize to form the catalytic center [17, 18] (Fig. 4.2). However, mutational studies indicate additional roles for conserved sequence III and also the extreme C-terminus [19, 20]. What roles these sequences play in catalysis are presently unknown. Interestingly, deletion of the PH and/or PX domains does not reduce the catalytic activity of mammalian PLDs [17, 21–23].

4.3 Cellular Locations of PLD1 and PLD2

The cellular locations of the mammalian PLD isozymes remain controversial. This is because immunological detection of the endogenous isozymes in cells has largely been unsuccessful, and most studies have utilized overexpression of epitope- or GFP-tagged PLDs. Consequently, the possibility of mislocalization of the overexpressed enzymes is a key issue in these studies. Subcellular fractionation studies have indicated that both PLD1 and PLD2 are membrane bound, except when the expression level is high [17, 24, 25]. Phosphorylation and palmitoylation of the enzymes can alter their membrane association [17, 24–26]. These post-translational modifications will be considered in the next section.

Studies of PLD activity in different subcellular fractions identified the enzyme in plasma membranes and intracellular membranes such as Golgi, nuclear membranes and secretory vesicles [27, 28]. However, these studies did not distinguish between PLD1 and PLD2. There has been one study of native PLD1 utilizing immunofluorescence with very sensitive antibodies, immunogold electron microscopy and cell fractionation [29]. This localized PLD1 predominantly to the Golgi apparatus, but there was some localization to lysosomes and late endosomes. In general, overexpressed tagged PLD1 exhibits a perinuclear and punctate cellular distribution. It has been reported to be present in Golgi, endoplasmic reticulum, early endosomes, lysosomes, secretory granules and the plasma membrane [12, 26, 28, 30–33]. However, there are large discrepancies between the localizations reported by different groups. Whether the differences are due to the extent of overexpression of PLD1 or result from the use of different methodologies and cell types remains to be resolved. Overexpressed PLD2 appears to be localized to the plasma membrane [12, 33–35], but there is evidence that the endogenous enzyme is associated with the Golgi [36]. PLD activity, including PLD1 and PLD2, has been identified in caveolae [37–42] which probably accounts, in part, for their presence at the plasma membrane. Co-localization of PLD1 and PLD2 with the actin cytoskeleton has also been reported [43].

4.4 Post-translational Modification of PLD

Several studies have shown that PLD1 and yeast PLD (Spo14) are phosphorylated under basal conditions. The phosphorylation involves Ser and Thr residues and results in relocalization of the enzymes and the appearance of slower migrating enzyme species on SDS polyacrylamide-gel electrophoresis [18, 24, 43, 44] (Fig. 4.3). The phosphorylation of PLD1 occurs predominantly in the N-terminal half of the enzyme and appears to have little effect on the intrinsic catalytic activity of the enzyme [18]. However, cell fractionation studies have shown that the phosphorylated form of overexpressed PLD1 is not present in the cytosol, in contrast to the non-phosphorylated form, indicating that phosphorylation increases membrane association of the enzyme [18]. Treatment of cells with phorbol ester or overexpression of PKCα results in further phosphorylation of PLD1, but this is associated with a loss of activity [45]. Basal Ser/Thr phosphorylation of PLD2 is low compared with that of PLD1. However, addition of phorbol ester or expression of PKCα greatly increases this.

Another post-translational modification of PLD1 and PLD2 is palmitoylation [25, 46]. This occurs on two specific Cys residues (C240, C241 in PLD1 and C223, C224 in PLD2) [24–26]. The requirements for palmitoylation have been studied in detail for PLD1. The modification requires the presence of the N-terminal 168

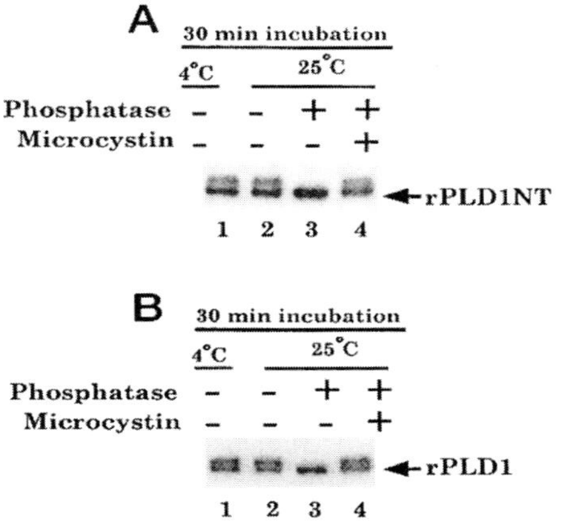

Fig. 4.3 Phosphorylated and non-phosphorylated forms of PLD1 shown by gel electrophoresis and phosphatase treatment. Panel (A) The N-terminal half of rat PLD1 (rPLD1NT) was expressed in COS7 cells and immunoprecipitated. The immunoprecipitates were treated with a Ser/Thr phosphatase in the absence and presence of a phosphatase inhibitor (microcystin) and then subjected to gel electrophoresis and Western blotting. The figure shows that the phosphatase eliminated the upper (phosphorylated) band, whereas microcystin restored it. (B) The same experiment utilizing full-length rat PLD1 (rPLD1).

amino acids of the enzyme and also the PH domain, but does not depend on the catalytic activity of the enzyme [24]. Mutation of C240 and C241 does not alter the catalytic activity of the enzyme *in vitro*, but causes some reduction *in vivo*. The mutation does, however, reduce the association of PLD1 with membranes and decreases its modification by Ser/Thr phosphorylation [24]. Similar changes in *in vivo* enzyme activity and membrane association are observed when C223 and C224 are mutated in PLD2 [25].

The determinants of membrane association of PLD1 and PLD2 are complex. As indicated above, phosphorylation and palmitoylation play a role. There is also evidence for involvement of the PH domain [9]. Surprisingly, mutation of a PIP_2 binding site located between conserved sequences II and III did not alter membrane association of PLD2 [8].

4.5 Regulation of PLD1 and PLD2

4.5.1 Role of PIP_2

The regulation of PLD is complex, but both mammalian isozymes depend on PIP_2 for activity, and no other phospholipid, except PIP_3, can substitute [8, 11, 12, 46, 47]. The PH domain can bind PIP_2 [9], but the relationship of this binding to the activity of the enzyme is unclear since deletion of the domain has no effect on the ability of PIP_2 to activate PLD2 [8]. Conversely, the PIP_2 binding site located between conserved sequences II and III of PLD2 contains critical residues that are required to stimulate the enzyme by PIP_2 [8]. These residues are in a highly conserved sequence of 21 amino acids, and are predominantly basic. Although mutations of these residues cause loss of activity, the mutant enzymes are still membrane-associated and show unchanged cellular localization [8]. As noted above, certain plant PLDs (β- and γ-isozymes) also depend upon PIP_2 [6]. Certain other phosphorylated inositol phospholipids are also active, but PIP_2 is the most efficacious [48]. The PIP_2 binding site has been identified in PLDβ [48] and corresponds to the site identified in mammalian PLD2 [8]. It contains several of the same conserved Arg and Lys residues, and mutation of these leads to a loss of activation by PIP_2. Interestingly, PIP_2 causes a conformational change that promotes the binding of PC to the catalytic domain of PLDβ [48]. This binding is reduced in enzymes with deletions or mutations of conserved Lys residues in the binding site.

Studies utilizing intact mammalian cells have shown that neomycin and *C. difficile* Toxin B, which reduce PIP_2 levels, inhibit PLD activity [49–51]. Conversely, co-expression of PLD1 or PLD2 and PI 4-P 5-kinase, which synthesizes PIP_2, results in increased PLD activity [52]. When a kinase-dead mutant of PI 4-P 5-kinase is transfected in place of the wild-type enzyme, the increase in PLD activity is much less [52].

4.5.2
Role of PKC

PKC was recognized early as an important regulator of PLD. Thus, addition of phorbol esters to intact cells *in vivo* and of PKC to membranes or purified preparations of PLD caused a marked increase in PLD activity [53]. A surprising observation emerging from *in vitro* studies was that PKC could activate PLD in the absence of ATP and without evident phosphorylation [11, 47, 54–57]. When the different isozymes of PLD were cloned and characterized, PLD1 was shown to be activated by PKCα *in vitro* whereas PLD2 was unresponsive [11, 12, 47]. Another unexpected finding was that the *in vitro* activation of PLD1 was only observed with the α- and β-isozymes of PKC [11, 47]. Surprisingly, addition of ATP did not enhance the activation of PLD1 by PKC, but caused inhibition [11, 47]. Consistent with a direct interaction between PKCα and PLD1, *in vitro* binding of the two proteins was demonstrated by several techniques and was enhanced by PMA [58, 59].

The interaction between PKCα and PLD1 predominantly involves the N-terminus of the phospholipase, although there is evidence for some interaction with the C-terminus [21, 23, 60]. The involvement of the N-terminus is shown by the fact that its deletion leads to a loss of activation of PLD1 by PKCα *in vitro* or by phorbol ester *in vivo* [17, 21, 23]. Measurements of binding between PKCα and PLD1 also implicated the N-terminus of the phospholipase [23, 59], although another site(s) might be involved [23]. The sites of interaction for PLD1 on PKCα are also multiple since neither the regulatory nor catalytic domain alone is sufficient to activate PLD1, and it appears that the holoenzyme is required [45]. Currently, the precise sequences in either PLD1 or PKCα that are involved in the interaction are unknown. As alluded to above, PLD1 can be phosphorylated by PKCα *in vivo* and *in vitro* [41, 45, 47, 51] and this leads to inhibition of its catalytic activity [45, 47]. The phosphorylation sites are multiple [61] and it is unclear which sites are associated with inhibition of activity. In one study, three phosphorylation sites (Ser2, Thr147 and Ser561) were identified in partially purified PLD1 phosphorylated by PKCα *in vitro* by analysis of P-peptides by mass spectrometry [61]. Single or triple mutations of these residues caused a partial loss of PLD1 activation by PMA in COS7 cells. Although these data indicate that phosphorylation of these residues is associated with some activation of PLD1 by PKC *in vivo*, since PMA (PKC) induced the phosphorylation of more and different residues *in vivo* than *in vitro*, it seems that the phosphorylation of other residues may be involved in the inhibition of PLD1 *in vivo*.

Both PLD1 and PLD2 can be activated by PKC in intact cells [62, 63], whereas only PLD1 can be activated *in vitro* [11, 12, 47]. However, this difference may reflect the *in vitro* assay conditions rather than an intrinsic difference between the two isozymes. The ability of phorbol esters such as PMA to activate PLD *in vivo* has been demonstrated in many cell types [62–64]. The role of PKC isozymes in this effect has been shown by the inhibitory effects of relatively selective PKC inhibitors such as Ro-31-8220, bisindolylmaleimide I, Gö6076, calphostin C, bryostatin and chelerythrine [53]. Most of these inhibitors target the ATP binding site of

PKC and block the phosphorylation of PLD [45]. However, as described above, *in vitro* and *in vivo* studies of the phosphorylation of PLD1 by PKCα indicate that phosphorylation causes inhibition rather than stimulation of the catalytic activity of this isozyme. Thus the decrease of PLD activity induced by the inhibitors *in vivo* cannot be explained in terms of reduced phosphorylation. It probably occurs because the inhibitors also block the interaction between PKCα and PLD1.

Another approach to exploring the role of phosphorylation in the activation of PLD by PKC *in vivo* has entailed the use of kinase-dead mutants of PKC isozymes. As expected, such mutants have greatly impaired ability to phosphorylate PLD1 or PLD2 *in vitro* or *in vivo*. In one study, the kinase-dead D481E PKCα mutant could still activate PLD1 in COS-7 cells, but the subsequent decline in PLD activity observed after the peak of activation was much less than seen with wild-type PKCα [45]. These data indicate that phosphorylation is not required for the initial activation of PLD by PKC, but causes a slow inhibition of PLD activity. In another report utilizing wild-type and kinase-dead PKCα in human melanoma cell lines, evidence was obtained for PKC regulation of PLD1 by both protein–protein interaction and phosphorylation mechanisms [65]. However, differences in the expression of PLD1 induced by wild-type and kinase-dead PKCα were not controlled for. In a study of the effects of wild-type and kinase-dead PKCδ on PLD2 activity in PC12 cells, the mutant kinase suppressed endogenous and PLD2 activity well below control levels [66], indicating that it acted mainly as a dominant negative to block the action of endogenous PKC isozymes. Although several PKC inhibitors decreased PLD2 phosphorylation and activity, there was not a good correlation between these changes. As noted above, these inhibitors can block the interaction of PKC with PLD. In summary, the role of phosphorylation in the activation of PLD1 and PLD2 by PKC remains open to serious questions.

Irrespective of the mechanism by which PKC activates PLD *in vivo*, there is much evidence that it plays a major role in the activation of the phospholipase by many agonists (Fig. 4.4). Thus PKC inhibitors often block the effects of agonists on PLD, although the extent of the inhibition may vary depending on the agonist and cell type [67]. Down-regulation of PKC by prolonged treatment with phorbol esters also causes partial or complete abrogation of the ability of growth factors and agonists coupled to G protein-linked receptors to activate PLD [67]. Mutant growth factor receptors that are unable to stimulate phosphoinositide phospholipase C (PLC) and thus cannot activate PKC are also incapable of activating PLD [68]. Likewise, in fibroblasts lacking PLCγ, the PLD response to platelet-derived growth factor (PDGF) is impaired [69]. In contrast, in fibroblasts overexpressing PLCγ the response is enhanced [70]. Overexpression of PKCα or PKCβ also increases the response of PLD to PMA, PDGF and other agonists [71–73], while depletion of PKCα by antisense methods decreases the activation [74]. In cells expressing a peptide that binds to RACK1, a receptor for PKC, there is inhibition of the actions of conventional PKC isozymes and a loss of PMA-stimulated PLD activity [75].

Since both PLD1 and PLD2 are membrane-associated, translocation of PKC isozymes to membrane loci is an essential component of the mechanism by which

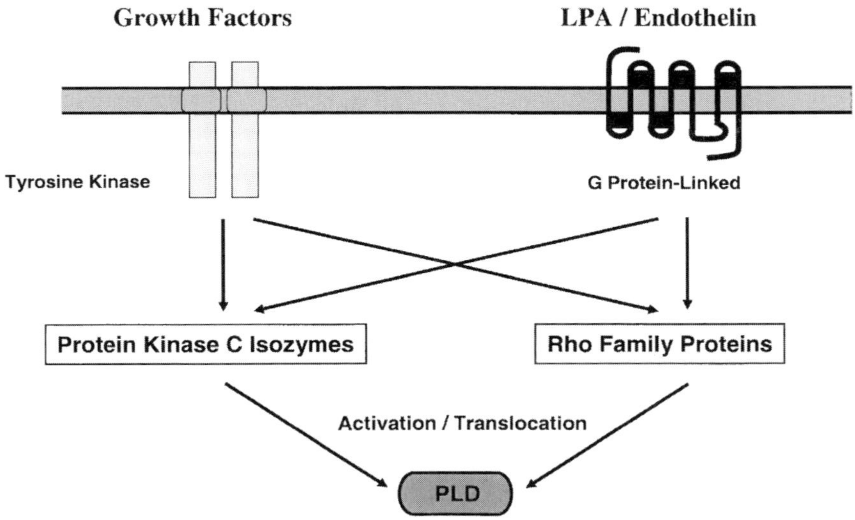

Fig. 4.4 Roles of PKC isozymes and Rho family proteins in the activation of PLD by growth factors and G protein-coupled agonists. The activation of PKC isozymes and Rho proteins by agonists activating either tyrosine kinases or G proteins is depicted. The activated PKC isozymes and Rho proteins translocate to membranes to induce activation of PLD there.

they activate the PLD isozymes. Numerous studies have demonstrated that phorbol esters and agonists that activate PLC induce the membrane translocation of conventional (classical) and novel PKC isozymes [76]. The translocation is rapid and occurs on a time scale consistent with the activation of PLD [31, 45]. There is also a rapid association between PKCα and PLD1 in cells treated with phorbol ester [31, 59]. The association of PKCα with the membrane fraction and with PLD1 occurs more rapidly than the phosphorylation of PLD1 [45], which is consistent with protein–protein interaction as the mechanism for the initial activation of the phospholipase.

In vitro observations have demonstrated a marked synergism between PKC and Rho or Arf in the activation of PLD1 [11, 56]. The molecular basis of this synergism is unknown, but binding of these regulators to the phospholipase must induce conformational changes that interact to produce enhanced activation. As indicated elsewhere in this chapter, PKC interacts with the N-terminal region of PLD1, but probably has an additional site(s). Rho binds to a sequence in the C-terminus, but the site of interaction of Arf is unknown.

4.6
Role of Rho Family GTPases

Early studies showed that purified or membrane-associated PLD could be stimulated *in vitro* by Rho and Arf in the presence of GTPγS [64]. Subsequently, the PLD1 isozyme, but not the PLD2 isozyme, was shown to respond to these GTPases [11, 12, 22, 46, 47, 77]. PLD1 was further shown to be stimulated by all the major members of the Rho family (RhoA, RhoB, Rac1, Rac2 and Cdc42Hs). However, Rho was more effective than Rac or Cdc42 [11, 78, 79]. Geranylgeranylation of these GTPases greatly enhanced their effects [78].

Expression of constitutively active V^{14}RhoA or V^{12}Rac1 enhances the activity of the endogenous PLD of COS cells or fibroblasts [60, 77, 80, 82], and expression of the dominant negative forms of these GTPases in these cells can suppress the activation of PLD by epidermal growth factor (EGF), PMA and $G\alpha_{13}$ [82–84]. In further support for a role of Rho proteins in the activation of PLD *in vivo*, many studies have shown that RhoA and Rac1 are activated by many agonists and translocated to cell membranes [53] (Fig. 4.4).

Clostridial toxins (C3 exoenzyme, Toxin B) that inactivate Rho proteins attenuate agonist activation of PLD in several cell lines [82–89]. The toxins also inhibit the activation of the enzyme by GTPγS *in vitro* [87, 89, 90]. These data implicate Rho in the regulation of PLD *in vivo*. This may involve direct activation of the enzyme by Rho [11, 47, 78, 79] or indirect regulation of its activity through changes in the level of PIP_2 [51]. The indirect mechanism arises because Rho activates PI 4-P 5-kinase, the enzyme that synthesizes PIP_2 [91–93].

Several studies have localized the interaction site for Rho proteins on PLD1 to a sequence in the C-terminus [4, 80, 81, 94]. These studies have utilized the yeast 2-hybrid system, co-immunoprecipitation and pull-down experiments with PLD1 fragments, and examinations of the effects of PLD1 mutations on RhoA binding and activation. Concerning the residues in RhoA involved in the activation of PLD1, these are located in the activation loop of this GTPase, as expected [78]. However, other residues determine the greater efficacy of RhoA compared with Rac and Cdc42. A detailed study of the interaction of Cdc42 with PLD1 showed that a critical residue (Ser124) in the insert helix of the GTPase is involved in the activation mechanism [79]. Interestingly, different mechanisms are involved in the activation of PLD1 by different Rho family members, accounting for their different efficacies [79]. RhoA, Rac1 and Cdc42 bind initially to PLD1 through their Switch 1 regions. Interaction between the phospholipase and the insert helix of RhoA, but not Rac1, results in further activation. Stimulation of PLD1 by Cdc42 only occurs through interaction with the insert helix, with Ser124 playing a critical role [79]. All Rho proteins show a strong synergistic action with PKCα and Arf to activate PLD1 *in vitro* [11, 56, 95]. The molecular basis for this synergism is unknown, but probably involves interactions between the conformational changes induced by these proteins.

4.7 Role of Arf Family GTPases

Arf was the first small GTPase shown to activate PLD [7, 96]. Subsequently it was found that all mammalian Arf isoforms (Arf 1–6) could stimulate the enzyme, although with differences in efficacy [97–99]. In line with other studies, the potency of Arf to stimulate PLD was greatly increased by myristoylation of a residue at the N-terminus (Gly2) [97–100]. PLD1 shows a large response to Arf *in vitro*, whereas PLD2 shows little or no response [11, 12, 22, 46, 47, 60, 101, 102]. As noted above, Arf and Rho proteins interact synergistically to activate PLD1, and this is true for Arf and PKCα [11, 56, 95]. The interaction site on Arf for PLD1 has been localized to the N-terminus [103, 104], and differs from that for cholera toxin activation or coatamer binding. Conversely, the site on PLD1 at which Arf interacts is undefined, although it is not at the N-terminus [23].

The role of Arf in the *in vivo* activation of PLD remains unclear for Arfs 1–5, but there is more evidence for Arf6. Whereas Arfs 1–3 are associated with the Golgi, the subcellular distribution of Arfs 4 and 5 is unclear. In contrast, Arf6 is predominantly associated with the plasma membrane where it cycles between this membrane and the cytosol and a vesicular compartment as a function of its GTP/GDP binding [105–109]. Stimulation of chromaffin cells induces translocation of Arf6 to the plasma membrane and an associated increase in PLD activity [109]. In addition, treatment of the cells with a myristoylated peptide corresponding to the N-terminal sequence of Arf6 inhibits PLD activation [109]. Similar results have been obtained with this peptide in myometrium [110].

In permeabilized HEK293 cells expressing the M3 muscarinic receptor, ARF1 restored the response of PLD to GTPγS [111] as found for HL60 cells [96]. The stimulatory effect of carbachol was also inhibited by brefeldin A, an inhibitor of some guanine exchange factors (GEFs) for Arf [111]. Other studies have reported inhibitory effects of brefeldin A on agonist-induced PLD activation [112–115], but this has not been uniformly observed [84, 116] perhaps due to the involvement of different GEFs.

There have been other approaches to defining a role for Arf in agonist activation of PLD. For example, PMA, GTPγS and certain agonists cause membrane translocation of Arf to sites of PLD activity [109, 117–120]. In two studies, expression of catalytically inactive mutants of Arf1 and Arf6 blocked agonist activation of PLD [112, 113], but there have been no other reports of the effects of these dominant negative forms of Arfs. Involvement of GEFs of the ARNO family in PLD activation has been indicated by the effects of overexpression of mSec7/ARNO on PLD activation [121]. ARP, an ARF-related protein that binds to ARNO, inhibited the stimulation of PLD induced by stimulation of the M3 muscarinic receptor [121]. Another approach to defining a role for Arf in PLD activation has involved the use of the Arf inhibitor Arfaptin 1 [122]. This agent impairs the activation of PLD *in vitro* and *in vivo* [99, 123, 124]. Like RhoA, Arf1 and Arf6 can activate PI 4-P 5-kinase [125, 126] and may therefore also regulate PLD through changes in PIP_2.

4.8
Role of Tyrosine Kinase

Growth factor receptors and receptors linked to the activation of soluble tyrosine kinases transduce signals leading to the activation of PLD [127]. However, the mechanism appears to be mainly indirect, i.e. it involves the PKC and Rho pathways. In many cells, inhibitors of tyrosine kinases decrease the activation of PLD by many agonists, but the specific kinases involved and the mechanisms by which they activate PLD are unclear [67, 128]. Vanadate, an inhibitor of protein tyrosine phosphatases, stimulates PLD activity in several cell lines either alone or in combination with H_2O_2 [67]. Although there is evidence that H_2O_2 and vanadate induce phosphorylation of PLD on Tyr residues, there is scant evidence that this occurs with agonists [67, 129, 130]. Furthermore, it is unclear that the phosphorylation seen with H_2O_2 or vanadate is responsible for the activation [129]. PLD2 is constitutively associated with the EGF receptor in transfected HEK293 cells and is activated by EGF in these cells [35]. The enzyme is phosphorylated on a specific Tyr residue (Tyr11), but mutation of this residue does not alter the activation of PLD2 by EGF [35].

4.9
Role of Ral

Recent evidence has implicated Ral, a member of the Ras subfamily of GTPases, in the activation of PLD by v-Src and v-Ras *in vivo* [131]. There is also a functional association between RalA and PLD1 in cell lysates, although this does not lead to activation of the phospholipase [131]. Later work showed that Arf could associate with Ral-PLD1 complexes [132, 133] and that RalA could enhance the effect of Arf1 on PLD1 activity [133]. Certain clostridial toxins that inactivate members of the Ras subfamily [88, 134] inhibit the activation of PLD induced by PMA *in vivo* or GTPγS *in vitro* [88, 135]. Ral, but not other GTPases, could restore the activation by PMA [88], and variants of one toxin further supported a role for Ral in the GTPγS effect [135]. Expression of dominant negative or constitutively active forms of RalA in 3Y1 cells also implicated this GTPase in the regulation of PLD [136, 137].

4.10
Cellular Functions of PLD

PLD has been implicated in many functions in mammalian and plant systems. In yeast, it is required for sporulation because of its role in completion of meiotic divisions and formation of the prospore membrane [138, 139]. It may also be involved in the lipid changes required for membrane trafficking [139]. In plants,

PLD plays important roles in the response to wounding and other stresses through mediating the actions and production of stress hormones [140]. PLDα is specifically involved in the actions of abscisic acid and ethylene in plant senescence and in the control of water loss [140, 141]. Different stresses in plants lead to different expression patterns for PLD isozymes.

In mammalian systems, PLD has been suggested to play roles in growth and differentiation, secretion, vesicle trafficking and transport, cytoskeletal rearrangements, superoxide production in phagocytes, protein translation and the production of lysophosphatidic acid (LPA).

4.11 Role of PLD in Growth and Differentiation

Early work implicating a role for PLD in growth regulation involved the addition of exogenous PA or PLD to cells. Although most PA preparations were contaminated with LPA, a known mitogen, some studies indicated that PA itself was active [67]. Some experiments with exogenous PLD likewise suffered from the possibility that LPA was generated, but in others the effects of LPA were controlled for [142]. Some PLD inhibitors (2,3-diphosphoglycerate and carbobenzyloxy-leucine-tyrosine-chloromethyl ketone) have also been used to define a role for PLD [143], but these are rather nonspecific. In contrast, primary alcohols, e.g. butan-1-ol, are frequently used to explore the role of PLD in the cellular actions of agonists (Fig. 4.5). Whereas primary alcohols reduce the formation of PA because they induce the formation of phosphatidyl alcohols through the transphosphatidylation reaction, secondary or tertiary alcohols, which are poor substrates, have little effect

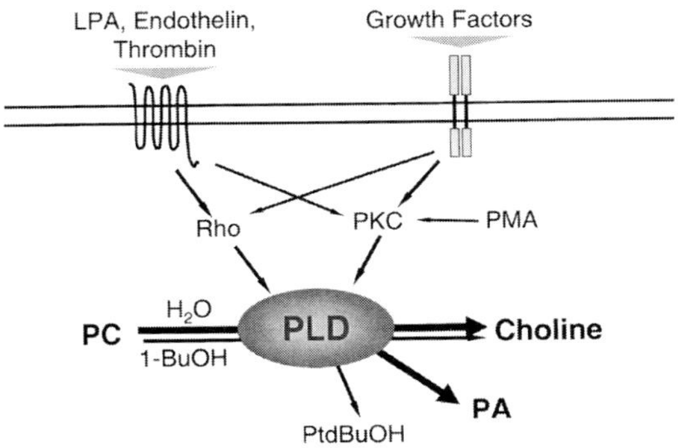

Fig. 4.5 Mechanisms of agonist activation of PLD and inhibition of PA formation by primary butanol. The roles of PKC isozymes and Rho proteins in agonist activation of PLD are described in the legend to Fig. 4.4. Phosphatidylcholine (PC) is the principal substrate of PLD. Primary butanol (1-BuOH) competes with H_2O as nucleophile in the enzyme reaction and phosphatidylbutanol (Ptd BuOH) is formed at the expense of PA.

on PA generation and can be used as controls. Thus, mitogenesis induced by angiotensin II in vascular smooth muscle cells is blocked by butan-1-ol, which inhibits PA production, and not by butan-3-ol, which does not alter PA levels [144]. Stable overexpression of PLD1 in fibroblasts results in enhanced MAP kinase activity in response to LPA, and this is decreased by butan-1-ol [145].

4.12
Role of PLD in Vesicle Trafficking in Golgi

Although there is much evidence that PLD plays a role in vesicle trafficking in the Golgi, there are some contrary findings. The function of Class I Arfs in the recruitment of coatamer (COP) proteins and adaptor proteins to Golgi membranes is well established. Golgi preparations possess an Arf-stimulated PLD activity [27, 28, 146, 147] and some studies have localized endogenous PLD1 and PLD2 to the Golgi [29, 36]. In support of a role for PLD in vesicle trafficking in this organelle, cells with high constitutive PLD activity have been observed not to require Arf to form coatamer-coated vesicles [148]. Furthermore, ethanol inhibited vesicle formation in Golgi incubated with Arf. Addition of PLD to Golgi membranes also induced coatamer binding and vesicle formation in the absence of Arf [148]. The addition of primary alcohols inhibited transport from the ER to the Golgi, an effect that was reversed by PA [149].

The release of secretory vesicles from the trans-Golgi network in permeabilized pituitary cells was stimulated by PLD1 and inhibited by butan-1-ol [147]. Vesicle budding was also increased when PA was generated by the combined action of PLC and DAG kinase [150]. In contrast, one group reported that PA levels in Golgi membranes declined during vesicle budding [151]. Another group made mutations in Arf3 and found a poor correlation between the effects of these mutations on PLD activity and coatamer recruitment [152]. Likewise, an N-terminal deletion mutant of ARF1 was observed to prevent the effect of Arf1 on coatamer binding to Golgi membranes, but have no effect on PLD activation [104]. Thus the role of PLD in Golgi trafficking remains controversial.

4.13
Role of PLD in Exocytosis and Endocytosis

Secretion in a number of cell types has been reported to be inhibited by primary alcohols [67]. Further support for a role of PLD comes from several studies. These include the stimulation of matrix metalloproteinase 9 secretion by PA [153], the identification of PLD in secretory granules, and the translocation of PLD-containing vesicles to the plasma membrane in response to f-Met-Leu-Phe (FMLP) or activation of the IgE receptor [119, 154]. In chromaffin cells, stimulation leads to translocation of Arf6 to the plasma membrane and concomitant activation of PLD [109]. In addition, a myristoylated peptide corresponding to the N-terminus of Arf6 blocked both PLD activity and catecholamine secretion [155]. Antibodies to ARNO also inhibited both of these actions. Overexpression of PLD1 in PC12 or chromaffin cells increased catecholamine secretion, whereas a catalytically inactive form of the enzyme inhibited this [33]. Microinjection of a catalytically dead mutant of PLD1 (K898R) into *Aplysia* neurons also inhibited acetylcholine release, most likely by controlling the number of functional release sites at nerve terminals [156].

Concerning endocytosis, the internalization and degradation of EGF receptors is inhibited by primary, but not secondary alcohols [157]. Overexpression of PLD1 and PLD2 reduces surface levels of the EGF receptor, but this is not observed with catalytically inactive mutants [157]. The activation of MAP kinase by EGF depends on the internalization of the EGF receptor [158], and there is evidence that this is inhibited by butan-1-ol, but not butan-2-ol [157]. The fusion of early endosomes may also involve PLD since this is stimulated by exogenous PLD and blocked by butan-1-ol, but not butan-2-ol [159]. A role for the phospholipase in the assembly of clathrin coats on lysosomes has also been reported [160].

PLD may play a role in the trafficking of the GLUT4 glucose transporter. Using Myc-tagged PLD1, the enzyme was shown by immunofluorescence to partly co-localize with the GLUT4 glucose transporter, which is involved in insulin-stimulated glucose transport [161]. Microinjection of PLD into adipocytes potentiated the effect of insulin on GLUT4 translocation [161]. Furthermore, butan-1-ol was observed to inhibit glucose transport in these cells.

4.14 Role of PLD in Superoxide Formation

Superoxide (O_2^-) production by neutrophils and macrophages is part of the host defenses against microorganisms. PLD has been implicated in this response because agents that elicit O_2^- formation also activate PLD [162–164]. In addition, *in vitro* experiments showed that PA elicited NADPH-dependent O_2^- formation in neutrophil extracts or intact neutrophils [165, 166]. Some studies also showed a high correlation between O_2^- production and PA accumulation in neutrophils stimulated with FMLP [167]. Furthermore, treatment of these cells with ethanol decreased both PA formation and O_2^- generation.

In vitro studies also showed a synergism between PA and diacylglycerol (DAG) in the activation of NADPH oxidase [168]. Evidence was presented that the effect was mediated by a PA-stimulated protein kinase [169] that phosphorylated the $p47^{phox}$ and $p22^{phox}$ subunits of the oxidase complex [170, 171]. Conventional PKC isozymes could also phosphorylate the $p22^{phox}$ subunit, implying a PA-independent mechanism [171]. Some studies have also shown direct regulation of the system by PA and DAG.

4.15
Role in Actin Cytoskeleton Rearrangements

The actin cytoskeleton is important in cell shape changes, polarization, motility, chemotaxis, metastasis and cell division, and Rho family proteins play critical roles in these effects. The involvement of PLD in changes in the actin cytoskeleton was first suggested by the effects of exogenous PLD and PA, which induced actin stress fiber formation and also increased the F-actin content of fibroblasts [172, 173]. Another study utilizing endothelial cells also showed that PA that was free of LPA induced actin stress fiber formation [174]. Furthermore, the effects of LPA to induce this change in the actin cytoskeleton and also PA formation were blocked by butan-1-ol, but not butan-2-ol [174] (Fig. 4.6).

More recently, Rat2 cell lines that stably express a dominant negative form of PLD1 have been generated. These show reduced PLD responses to PMA and LPA and fail to show normal actin stress fiber formation in response to LPA [32]. Measurements of RhoA activation and Rho kinase activity indicate that PLD is not required for these changes [32], and the role of PLD1 appears to be downstream of Rho kinase. The precise function of PLD1 in the actin cytoskeletal changes has not been defined, but may involve the maintenance of PIP_2 levels since PA activates type 1 PI 4-P 5-kinase [92, 175, 176] and PIP_2 affects the function of many actin-associated proteins. A recent study has indicated a role for PLD-generated PA in the actin cytoskeletal changes involved in membrane ruffling in antigen-stimulated mast cells [177]. The ruffling response is blocked by butan-1-ol and restored by removal of the alcohol. PLD2 was found to be localized to membrane ruffles together with Arf6 [177]. Antigen-stimulated PIP_2 formation depends on both Arf6 and PLD-derived PA.

The association of PLD with the actin cytoskeleton is indicated by other studies [42, 178]. Thus β-actin has been shown to bind directly to PLD1 and PLD2 [42]. Immunocytochemistry has also shown an association between both PLD isozymes with the actin cytoskeleton in intact cells.

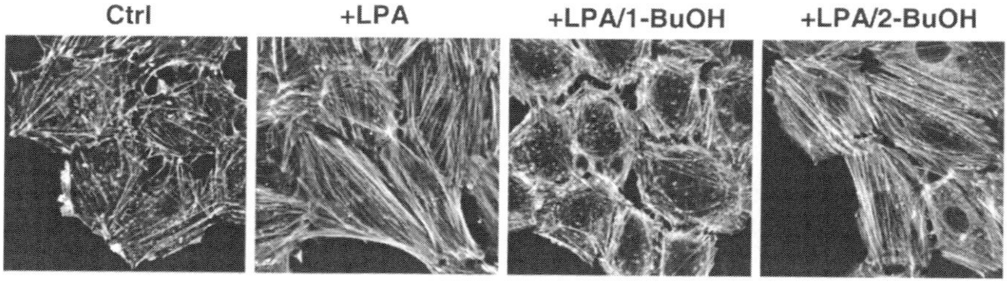

Fig. 4.6 Inhibition of LPA-induced stress fiber formation in fibroblasts by primary butanol. Rat 1 fibroblasts were treated with LPA and actin stress fibers were visualized with Texas red X-phalloidin. Addition of primary butanol (1-BuOH) inhibited stress fiber formation, whereas secondary butanol (2-BuOH) had no effect.

4.16
Role in Lysophosphatidic Acid Formation

LPA is an important extracellular signal that is released from platelets and other cells. It produces many cellular responses through interaction with several G protein-coupled receptor subtypes, formerly called Edg receptors and now termed LPA receptors [179, 180]. LPA can be generated by cells through several pathways. One is from PA through the action of specific phospholipases of the A_1 and A_2 subtypes [181–183]. These generate 2-acyl-LPA and 1-acyl-LPA, respectively. Another pathway is through the action of a lysophospholipase D (autotaxin) on cell surface lyso PC [184, 185]. Although LPA is an intermediate in the synthesis of triacylglycerol and most phospholipids in the endoplasmic reticulum, this form does not seem to be the source of released LPA.

4.17
Role of PA in Other Cellular Systems

It is generally agreed that PA is the major signaling molecule produced by PLD action. Numerous *in vitro* effects of PA have been reported. As indicated above, it has been shown to affect mitogenesis, the actin cytoskeleton, O_2^- production and vesicle trafficking in Golgi and other organelles. It also can activate PI 4-P 5-kinase [92, 175, 176], bind to Raf1 kinase [186], stimulate PTP1C tyrosine phosphatase [187] and PDE4A5 and PDE4D3 cAMP phosphodiesterases [188, 189], PP1 Ser/Thr phosphatase [190], and alter the activity of proteins that regulate Ras and Rac [191–193]. A novel microdomain (TAPAS-1) has been identified at which PA binds to PDE4A1 cAMP phosphodiesterase and anchors it to membranes [194]. Less defined effects are those on tyrosine phosphorylation, PLC isozymes, other Ser/Thr protein kinases and PKC isozymes [53]. Recently a role for PA in the regulation of protein translation has been indicated by the observation that PA can bind to mTOR, leading to phosphorylation and activation of the S6K1 protein kinase and the phosphorylation of 4E-BP1, which is an inhibitory binding protein for the eIF-4E initiation factor [195]. The phosphorylation of the S6 ribosomal protein and activation of eIF-4E promotes the initiation of translation of 5′ TOP and 5′ capped mRNAs.

PA derived from PC can be converted into DAG through the action of phosphatidate phosphatase. It remains controversial whether DAG derived from this source can activate PKC. Initial studies indicated that treatment of fibroblasts with thrombin induced a rapid, transient phase of membrane translocation of PKCα that was correlated with transient rises in DAG and Ca^{2+} due to PIP_2 breakdown [196]. In contrast, the membrane translocation of PKCε induced by thrombin persisted for at least an hour and was associated with a second prolonged phase of DAG accumulation [196]. Platelet-derived growth factor (PDGF) did not induce PKCα translocation, but caused a slow membrane association of PKCε which was correlated with the second phase of DAG accumulation [196]. This DAG was presumed to be derived from PC through PLD and phosphatidate phosphatase action, since it was not associated with IP_3 production or an increase in Ca^{2+}.

Other studies have indicated that PC and PLD are a major source of PA in cells stimulated with various agonists [197–200] and several studies have confirmed biphasic formation of DAG, with the first phase derived from PIP_2 and the second from PC via PA [201–205]. Analyses of the specific fatty acid composition of PA and PtdBut derived from cells treated with butanol and stimulated by bombesin or LPA also showed a predominance of saturated and mono-unsaturated fatty acids in the lipids derived from PLD action [206]. The fatty acid composition was very different from that found in DAG generated by PLC action on PIP_2, in which polyunsaturated fatty acids predominate. The key issue is whether the DAG species derived from PC can activate PKC. Although studies indicate that this species is active on PKC *in vitro*, this has not been demonstrated *in vivo*. As described above [196], PC-derived DAG can activate Ca^{2+}-independent PKC isozymes but, because of the lack of a Ca^{2+} rise, it cannot activate Ca^{2+}-dependent PKC isozymes.

It is probable that additional physiological targets of PA will be found, and that the full range of PLD actions *in vivo* is yet to be revealed.

4.18 References

1 C. P. Ponting and I. D. Kerr, *Protein Sci.*, **1996**, *5*, 914–922.

2 E. V. Koonin, *Trends Biochem. Sci.*, **1996**, *21*, 242–243.

3 H. Interthal, J. J. Pouliot, and J. J. Champoux, *Proc. Natl. Acad. Sci. USA* **2001**, *98*, 12009–12014.

4 T.-C. Sung, R. I. Roper, Y. Zhang, S. A. Rudge, R. Temel, S. M. Hammond, A. J. Morris, B. Moss, J. A. Engebrecht, and M. A. Frohman, *EMBO J.*, **1997**, *16*, 4519–4539.

5 M. Frohman, T.-C. Sung, and A. J. Morris, *Biochim. Biophys. Acta*, **1999**, *1439*, 175–186.

6 C. Qin and X. Wang, *Plant Physiol.*, **2002**, *128*, 1057–1068.

7 H. A. Brown, S. Gutowski, C. R. Moomaw, C. Slaughter, and P. C. Sternweis, *Cell*, **1993**, *75*, 1137–1144.

8 V. A. Sciorra, S. A. Rudge, G. D. Prestwich, M. A. Frohman, J. A. Engebrecht, and A. J. Morris, *EMBO J.*, **1999**, *20*, 5911–5921.

9 M. N. Hodgkin, M. R. Masson, D. Powner, K. M. Saqib, C. P. Ponting, and M. J. O. Wakelam, *Curr. Biol.*, **1999**, *10*, 43–46.

10 S. Wang, Y. Banno, S. Nakashima, and Y. Nozawa, *K. Eukaryot. Microbiol.*, **2001**, *48*, 194–201.

11 S. M. Hammond, J. M. Jenco, S. Nakashima, K. Cadwallader, Q.-M. Gu, S. Cook, Y. Nozawa, G. D. Prestwich, M. A. Frohman, and A. J. Morris, *J. Biol. Chem.*, **1997**, *272*, 3860–3868.

12 W. C. Colley, T.-C. Sung, R. Roll, J. Jenco, S. M. Hammond, Y. Altshuller, D. Bar-Sagi, J. Morris, and M. A. Frohman, *Curr. Biol.*, **1997**, *7*, 191–201.

13 I. Leiros, F. Secundo, C. Zambonelli, S. Servi, and E. Hough, *Structure*, **2000**, *8*, 655–667.

14 J. A. Stuckey and J. E. Dixon, *Nat. Struct. Biol.*, **1999**, *6*, 278–284.

15 D. R. Davies, H. Interthal, J. J. Champoux, and W. G. J. Hol, *Structure*, **2002**, *10*, 237–248.

16 E. B. Gottlin, A. E. Rudolph, Y. Zhao, H. R. Matthews, and J. E. Dixon, *Proc. Natl. Acad. Sci. USA*, **1998**, *95*, 9202–9207.
17 Z. Xie, W.-T. Ho, and J. H. Exton, *J. Biol. Chem.*, **1998**, *273*, 34679–34682.
18 Z. Xie, W.-T. Ho, and J. H. Exton, *J. Biol. Chem.*, **2000**, *275*, 24962–24969.
19 Z. Xie, W.-T. Ho, and J. H. Exton, *Eur. J. Biochem.*, **2000**, *267*, 7138–7146.
20 M. Y. Liu, S. Gutowski, and P. C. Sternweis, *J. Biol. Chem.*, **2001**, *270*, 5556–5562.
21 S.-K. Park, D. S. Min, and J. H. Exton, *Biochem. Biophys. Res. Commun.*, **1998**, *244*, 364–367.
22 T.-C. Sung, Y. M. Altshuller, A. J. Morris, and M. A. Frohman, *J. Biol. Chem.*, **1999**, *274*, 494–502.
23 T.-C. Sung, Y. Zhang, A. J. Morris, and M. A. Frohman, *J. Biol. Chem.*, **1999**, *274*, 3659–3666.
24 Z. Xie, W.-T. Ho, and J. H. Exton, *J. Biol. Chem.*, **2001**, *275*, 24962–24969.
25 Z. Xie, W.-T. Ho, and J. H. Exton, *Biochem. Biophys. Acta*, **2002**, *1580*, 9–21.
26 J. M. Sugars, S. Cellek, M. Manifava, J. Coadwell, and N. T. Ktistakis, *J. Biol. Chem.*, **1999**, *274*, 30023–30027.
27 J. J. Provost, J. Fudge, S. Israelit, A. R. Siddiqi, and J. H. Exton, *Biochem. J.*, **1996**, *319*, 285–291.
28 M. Liscovitch, M. Czarny, G. Fiucci, Y. Lavie, and X. Tang, *Biochem. Biophys. Acta*, **1999**, *1439*, 245–263.
29 Z. Freyberg, D. Sweeney, A. Siddhanta, S. Bourgoin, M. Frohman, and D. Shields, *Mol. Biol. Cell*, **2001**, *12*, 943–955.
30 F. D. Brown, N. Thomas, K. M. Saqib, J. M. Clark, D. Powner, N. T. Thompson, R. Solari, and M. J. O. Wakelam, *Curr. Biol.*, **1998**, *8*, 835–838.
31 Y. Kim, J.-E. Kim, S. D. Lee, T. G. Lee, J. H. Kim, J. B. Park, J. M. Han, S. K. Jang, P.-G. Suh, and S. H. Ryu, *Biochim. Biophys. Acta*, **1999**, *1436*, 319–330.
32 Y. Kam and J. H. Exton, *Mol. Cell Biol.*, **2001**, *21*, 4055–4066.
33 M. Emoto, J. K. Klarlund, S. B. Waters, V. Hu, J. M. Buxton, A. Chawla, and M. P. Czech, *J. Biol. Chem.*, **2000**, *275*, 7144–7151.
34 N. Vitale, A.-S. Caumont, S. Golaz-Chasserot, G. Du, S. Wu, V. A. Sciorra, A. J. Morris, M. A. Frohman, and M.-F. Bader, *EMBO J.*, **2001**, *20*, 2424–2434.
35 J. B. Park, J. H. Kim, Y. Kim, S. H. Ha, J. H. Kim, J.-S. Yoo, G. Du, M. A. Frohman, P.-G. Suh, and S. H. Ryu, *J. Biol. Chem.*, **2000**, *275*, 21295–21301.
36 Z. Freyberg, S. Bourgoin, and D. Shields, *Mol. Biol. Cell*, **2002**, *13*, 3930–3942.
37 M. Czarny, Y. Lavie, G. Fiucci, and M. Liscovitch, *J. Biol. Chem.*, **1999**, *274*, 2717–2724.
38 M. Czarny, G. Fiucci, Y. Lavaie, Y. Banno, Y. Nozawa, and M. Liscovitch, *FEBS Lett.*, **2000**, *467*, 326–332.
39 S. S. Iyer and D. J. Kusner, *J. Biol. Chem.*, **1999**, *274*, 2350–2359.
40 V. A. Sciorra and A. J. Morris, *Mol. Biol. Cell*, **1999**, *10*, 3863–3876.
41 J. H. Kim, J. M. Han, S. Lee, Y. Kim, T. G. Lee, J. B. Park, S. D. Lee, P.-G. Suy, and S. H. Ryu, *Biochemistry*, **1999**, *38*, 3763–3769.
42 Y. Kim, J. M. Han, B. R. Han, K.-A. Lee, J. H. Kim, B. D. Lee, I.-H. Jang, P.-G. Suy, and S.H. Ryu, *J. Biol. Chem.*, **2000**, *275*, 13621–13627.
43 S. D. Lee, J. B. Park, J. H. Kim, Y. Kim, J. H. Kim, K.-J. Shin, J. S. Lee, S. H. Ha, P.-G. Suh, and S. H. Ryu, *J. Biol. Chem.*, **2001**, *276*, 28252–28260.
44 S. A. Rudge, A. J. Morris, and J. A. Engebrecht, *J. Cell Biol.*, **1998**, *140*, 81–90.
45 T. Hu and J. H. Exton, *J. Biol. Chem.*, **2003**, *278*, 2348–2355.
46 M. Manifava, J. Sugars, and N. T. Ktistakis, *J. Biol. Chem.*, **1999**, *274*, 1072–1077.
46 T. Kodaki and S. Yamashita, *J. Biol. Chem.*, **1997**, *272*, 11408–11413.
47 D. S. Min, S.-K. Park, and J. H. Exton, *J. Biol. Chem.*, **1998**, *273*, 7044–7051.
48 L. Zheng, J. Shan, R. Krishnamoorthi, and X. Wang, *Biochemistry*, **2002**, *41*, 4546–4553.

49 M. Liscovitch, V. Chalifa, P. Pertile, C.-S. Chen, and L.C. Cantley, *J. Biol. Chem.*, **1994**, *269*, 21403–21496.

50 P. Pertile, M. Liscovitch, V. Chalifa, and L.C. Cantley, *J. Biol. Chem.* **1995**, *270*, 5130–5135.

51 M. Schmidt, M. Rümenapp, C. Nehls, S. Ott, J. Keller, C. von Eichel-Streiber, and K.H. Jacobs, *Eur. J. Biochem.*, **1996**, *240*, 707–712.

52 N. Divecha, M. Roefs, J.R. Halstead, S. D'Andrea, M. Fernandez-Borga, I. Oomen, K.M. Saqib, M.J.P. Wakelam, and C. D'Santos, *EMBO J.*, **2000**, *19*, 5440–5449.

53 J.H. Exton, *Rev. Physiol. Biochem. Pharmacol.*, **2002**, *144*, 1–94.

54 K.M. Conricode, K.A. Brewer, and J.H. Exton, *J. Biol. Chem.*, **1992**, *267*, 7199–7202.

55 K.M. Conricode, J.L. Smith, D.J. Burns, and J.H. Exton, *FEBS Lett.*, **1994**, *342*, 149–153.

56 W.D. Singer, H.A. Brown, X. Jiang, and P.C. Sternweis, *J. Biol. Chem.*, **1996**, *271*, 4504–4510.

57 V.A. Sciorra, S.M. Hammond, and A.J. Morris, *Biochemistry*, **2001**, *40*, 2640–2646.

58 D.S. Min and J.H. Exton, *Biochem. Biophys. Res. Commun.*, **1998**, *248*, 533–537.

59 T.G. Lee, J.B. Park, S.D. Lee, S. Hong, J.H. Kim, Y. Kim, K.S. Yi, S. Bae, Y.A. Hannun, L.M. Obeid, P.-G. Suh, and S.H. Ryu, *Biochim. Biophys. Acta*, **1997**, *1347*, 199–204.

60 Y. Zhang, Y.M. Altshuller, S.M. Hammond, A.J. Morris, and M.A. Frohman, *EMBO J.*, **1999**, *18*, 6339–6348.

61 Y. Kim, J.M. Han, J.B. Park, S.D. Lee, Y.S. Oh, C. Chung, T.G. Lee, J.H. Kim, S.-K. Park, J.-S. Yoo, P.-G. Suh, and S.H. Ryu, *Biochemistry*, **1999**, *38*, 10344–10351.

62 A.R. Siddiqi, G.E. Srajer, and C.C. Leslie, *Biochim. Biophys. Acta*, **2000**, *1497*, 103–114.

63 Z. Xie, W.-T. Ho, R. Spellman, S. Cai, and J.H. Exton, *J. Biol. Chem.*, **2002**, *277*, 11979–11986.

64 J.H. Exton, *Biochim. Biophys. Acta*, **1999**, *1439*, 121–133.

65 M. Oka, T. Hitomi, T. Okada, S.-I. Nakamura, H. Nagai, M. Ohba, T. Kuroki, U. Kikkawa, and M. Ichihashi, *Biochem. Biophys. Res. Commun.*, **2002**, *294*, 1109–1113.

66 J.M. Han, J.H. Kim, B.D. Lee, S.D. Lee, Y. Kim, Y.W. Jung, S. Lee, W. Cho, M. Ohba, T. Kuroki, P.-G. Suh, and S.H. Ryu, *J. Biol. Chem.*, **2002**, *277*, 8290–8297.

67 J.H. Exton, *Physiol. Rev.*, **1999**, *77*, 303–320.

68 E.-J. Yeo, A. Kazlauskas, and J.H. Exton, *J. Biol. Chem.*, **1994**, *269*, 27823–27826.

69 J.A. Hess, Q.-S. Ji, G. Carpenter, and J.H. Exton, *J. Biol. Chem.*, **1998**, *273*, 20517–20524.

70 Y.H. Lee, H.S. Kim, J.-K. Pai, S.H. Ryu, and P.-G. Suh, *J. Biol. Chem.*, **1994**, *269*, 26842–26847.

71 J.-K. Pai, E.A. Dobek, and W.R. Bishop, *Cell Regul.*, **1991**, *2*, 897–903.

72 J.A. Pachter, J.-K. Pai, R. Mayer-Ezell, J.M. Petrin, E. Dobek, and W.R. Bishop, *J. Biol. Chem.*, **1988**, *267*, 9826–9830.

73 H. Eldar, P. Ben-Av, P.U.-S. Schmidt, E. Livneh, and M. Liscovitch, *J. Biol. Chem.*, **1993**, *268*, 12560–12564.

74 M.A. Balboa, B.L. Firestein, C. Godson, K.S. Bekk, and P.A. Insel, *J. Biol. Chem.*, **1994**, *269*, 10511–10516.

75 V.A.T. Thorsen, B. Bjorndal, G. Nolan, M.H. Fukami, O. Bruland, J.R. Lillehaug, and H. Holmsen, *Biochim. Biophys. Acta*, **2000**, *1487*, 163–176.

76 S. Jaken, *Molecular Biology Intelligence Unit*, R.G. Landes Company, Austin, **1997**.

77 S.-K. Park, J.J. Provost, C.D. Bae, W.-T. Hos, and J.H. Exton, *J. Biol. Chem.*, **1997**, *272*, 29263–29271.

78 C.D. Bae, D.S. Min, I.N. Fleming, and J.H. Exton, *J. Biol. Chem.*, **1998**, *273*, 11596–11604.

79 S.J. Walker and H.A. Brown, *J. Biol. Chem.*, **2002**, *277*, 26260–26267.

80 G. Du, Y.M. Altshuller, Y. Kim, J.M. Han, S.H. Ryu, A.J. Morris, and M.A. Frohman, *Mol. Biol. Cell*, **2000**, *11*, 4359–4368.

81 S. Cai and J.H. Exton, *Biochem. J.*, **2001**, *355*, 779–785.

82 J.A. Hess, A.H. Ross, R.-G. Qiu, M. Symons, and J.H. Exton, *J. Biol. Chem.*, **1997**, *272*, 1615–1620.

83 S.G. Plonk, S.-K. Park, and J.H. Exton, *J. Biol. Chem.*, **1998**, *273*, 4823–4826.

84 E. Meacci, V. Vasta, J.P. Moorman, D.A. Bobak, P. Bruni, J. Moss, and M. Vaughan, *J. Biol. Chem.*, **1999**, *274*, 18605–18612.

85 S.E. Senogles, *Mol. Pharmacol.*, **2000**, *58*, 455–462.

86 K.S. Murthy, H. Zhou, J.R. Grider, and G.M. Makhlouf, *Am. J. Physiol.* **2001**, *280*, G381–G388.

87 M. Schmidt, U. Rümenapp, C. Bienek, J. Keller, C. von Eichel-Streiber, and K.H. Jakobs, *J. Biol. Chem.*, **1996**, *271*, 2422–2426.

88 M. Schmidt, M. Voss, M. Thiel, B. Baur, A. Grannas, E. Tapp, R.H. Cool, J. de Gunzburg, C. von Eichel-Streiber, and K.H. Jakobs, *J. Biol. Chem.*, **1998**, *273*, 7413–7422.

89 K. Ojio, Y. Banno, S. Nakashima, N. Kato, K. Watanabe, D.M. Lyerly, H. Miyata, and Y. Nozawa, *Biochem. Biophys. Res. Commun.*, **1996**, *224*, 591–596.

90 H. Kuribara, K. Tago, T. Yokozeki, T. Sasaki, Y. Takai, N. Morii, S. Narumiya, T. Katadda, and Y. Kanaho, *J. Biol. Chem.*, **1995**, *270*, 25667–25671.

91 L.D. Chong, A. Traynor-Kaplan, G.M. Bokoch, and M.A. Schwartz, *Cell*, **1994**, *79*, 507–513.

92 X.-D. Ren, G.M. Bokoch, A. Traynor-Kaplan, G.H. Jenins, R.A. Anderson, and M.A. Schwartz, *Mol. Biol. Cell*, **1996**, *7*, 435–442.

93 P.A. Oude Weernink, P. Schulte, Y. Guo, J. Wetzel, M. Amano, K. Kaibuchi, S. Haverland, M. Vob, M. Schmidt, G.W. Mayr, and K.H. Jakobs, *J. Biol. Chem.*, **2000**, *275*, 10168–10174.

94 M. Yamazaki, Y. Zhang, H. Watanabe, T. Yokozeki, S. Ohno, K. Kaibuchi, H. Shibata, H. Mukai, Y. Ono, M.A. Frohman, and Y. Kanaho, *J. Biol. Chem.*, **1999**, *274*, 6035–6038.

95 M.N. Hodgkin, J.M. Clark, S. Rose, K. Saqib, and M.J.O. Wakelam, *Biochem. J.*, **1999**, *339*, 87–93.

96 S. Cockcroft, G.M.H. Thomas, A. Fensome, B. Geny, E. Cunningham, I. Gout, I. Hiles, N.F. Totty, O. Truong, and J.J. Hsuan, *Science*, **1994**, *263*, 523–526.

97 D. Massenburg, J.-S. Han, M. Liyanage, W.A. Patton, S.G. Rhee, J. Moss, and M. Vaughan, *Proc. Natl. Acad. Sci. USA*, **1994**, *91*, 11718–11722.

98 H.A. Brown, S. Gutowski, R.A. Kahn, and P.C. Sternweis, *J. Biol. Chem.*, **1995**, *270*, 14935–14943.

99 S.-C. Tsai, R. Adamik, J.-X. Hong, J. Moss, M. Vaughan, H. Kanoh, and J.H. Exton, *J. Biol. Chem.*, **1998**, *273*, 20697–20701.

100 A. Fensome, E. Cunningham, S. Prosser, S.K. Tan, P. Swigart, G. Thomas, J. Hsuan, and S. Cockcroft, *Curr. Biol.*, **1996**, *6*, 730–738.

101 I. Lopez, R.S. Arnold, and J.D. Lambeth, *J. Biol. Chem.*, **1998**, *273*, 12846–12852.

102 R. Slaaby, G. Du, Y.M. Altshuller, M.A. Frohman, and K. Seedorf, *Biochem. J.*, **2000**, *351*, 613–619.

103 G.-F. Zhang, W.A. Patton, F.-J.L. Lee, M. Lliyanage, J.-S. Han, S.G. Rhee, J. Moss, and M. Vaughan, *J. Biol. Chem.*, **1995**, *270*, 21–24.

104 D.H. Jones, B. Bax, A. Fensome, and S. Cockcroft, *Biochem. J.*, **1999**, *341*, 183–190.

105 M.M. Cavenagh, J.A. Whitney, K. Carroll, C.-J. Zhang, A.L. Bowman, A.G. Rosenwald, I. Mellman, and R.A. Kahn, *J. Biol. Chem.*, **1996**, *271*, 21767–21774.

106 O. Al-Awar, H. Radhakrishna, N.N. Powell, and J.G. Donaldson, *Mol. Cell Biol.*, **2000**, *20*, 5998–6007.

107 J. Gaschet and V.W. Hsu, *J. Biol. Chem.*, **1999**, *274*, 20040–20045.

108 H. Radhakrishna, J.G. Donaldson, *J. Cell Biol.* **1997**, *139*, 49–61.

109 A.-S. Caumont, M.-C. Galas, N. Vitale, D. Aunis, and M.-F. Bader, *J. Biol. Chem.*, **1998**, *273*, 1373–1379.

110 H.L. Le Stunff, L. Dokhac, S. Bourgoin, M.-F. Bader, and S. Harbon, *Biochem. J.*, **2000**, *352*, 491–499.

111 U. Rümenapp, M. Geiszt, F. Wahn, M. Schmidt, and K.H. Jakobs, *Eur. J. Biochem.*, **1995**, *234*, 240–244.

112 K. Shome, Y. Nie, and G. Romero, *J. Biol. Chem.*, **1998**, *273*, 30836–30841.

113 K. Shome, M.A. Rizzo, C. Vasudevan, B. Andresen, and G. Romero, *Endocrinol.*, **1999**, *141*, 2200–2208.

114 B.T. Andresen, E.K. Jackson, and G.G. Romero, *Hypertension*, **2001**, *37*, 635–639.

115 R. Mitchell, D. McCulloch, E. Lutz, M. Johnson, C. MacKenzie, M. Fennell, G. Fink, W. Zhou, and S.C. Sealfon, *Nature*, **1998**, *392*, 411–414.

116 I. Guillemain and J.H. Exton, *Eur. J. Biochem.*, **1997**, *249*, 812–819.

117 M.G. Houle, R.A. Kahn, P.H. Naccache, and S. Bourgoin, *J. Biol. Chem.*, **1995**, *270*, 22795–22800.

118 M.G. Houle, P.H. Naccache, and S. Bourgoin, *J. Leukoc. Biol.*, **1999**, *66*, 1021–1030.

119 J. Whatmore, C.P. Morgan, E. Cunningham, K.S. Collison, K.R. Willison, and S. Cockcroft, *Biochem. J.*, **1996**, *320*, 785–794.

120 A. Fensome, J. Whatmore, C. Morgan, D. Jones, and S. Cockcroft, *J. Biol. Chem.*, **1998**, *273*, 13157–13164.

121 A. Schürmann, M. Schmidt, M. Asmus, S. Bayer, F. Fliegert, S. Koling, S. Massmann, Schilf, M.C. Subauste, M. Voss, K.H. Jakobs, and H.-G. Joost, *J. Biol. Chem.*, **1999**, *274*, 9744–9751.

122 H. Kanoh, B.-T. Williger, and J.H. Exton, *J. Biol. Chem.*, **1997**, *272*, 5421–5429.

123 B.-T. Williger, J.J. Provost, W.-T. Ho, J. Milstine, and J.H. Exton, *FEBS Lett.*, **1999**, *454*, 85–89.

124 B.-T. Williger, J. Ostermann, and J.H. Exton, *FEBS Lett.*, **1999**, *443*, 197–200.

125 D.H. Jones, J.B. Morris, C.P. Morgan, H. Kondo, R.F. Irvine, and S. Cockcroft, *J. Biol. Chem.*, **2000**, *275*, 13962–13966.

126 A. Honda, M. Nogami, T. Yokozeki, M. Yamazaki, H. Nakamura, H. Watanabe, K. Kawamoto, K. Nakayama, A.J. Morris, M.A. Frohman, and Y. Kanaho, *Cell*, **1999**, *99*, 521–532.

127 V. Natarajan, W.M. Scribner, and S. Vepa, *Chem. Phys. Lipids*, **1996**, *80*, 103–116.

128 M.G. Houle and S. Bourgoin, *Biochim. Biophys. Acta*, **1999**, *1439*, 135–150.

129 D.S. Min, E.-G. Kim, and J.H. Exton, *J. Biol. Chem.*, **1998**, *273*, 29986–29994.

130 J. Gomez-Cambronero, *J. Int. Cyto. Res.*, **1995**, *15*, 877–885.

131 H. Jiang, J.-Q. Luo, T. Urano, P. Frankel, Z. Lu, D.A. Foster, and L.A. Feig, *Eur. J. Pharmacol.*, **1996**, *302*, 183–190.

132 J.-Q. Luo, X. Liu, P. Frankel, T. Rotunda, M. Ramos, J. Flom, H. Jiang, L.A. Feig, A.J. Morris, R.A. Kahn, and D.A. Foster, *Proc. Natl. Acad. Sci. USA*, **1998**, *95*, 3632–3637.

133 J.H. Kim, S.D. Lee, J.M. Ham, T.G. Lee, Y. Kim, J.B. Park, J.D. Lambeth, P.-G. Suh, and S.H. Ryu, *FEBS Lett.*, **1998**, *430*, 231–235.

134 M.R. Popoff, E. Chaves-Olarate, E. Lemichez, C. von Eichel-Streiber, M. Thelestam, P. Chardin, D. Cussac, B. Antonny, P. Chavrier, G. Flatau, M. Giry, J. de Gunzburg, and P. Boquet, *J. Biol. Chem.*, **1996**, *271*, 10217–10224.

135 N.B. El Hadj, M.R. Popoff, J.-C. Marvaud, B. Payrastre, P. Boquet, and B. Geny, *J. Biol. Chem.*, **1999**, *274*, 14021–14031.

136 Z. Lu, A. Hornia, T. Joseph, T. Sukezane, P. Frankel, M. Zhong, S. Bychenok, L. Xu, L.A. Feig, and D.A. Foster, *Mol. Cell Biol.*, **2000**, *20*, 462–467.

137 M. Voss, P.A. Weernink Oude, S. Haupenthal, U. Moller, R.H. Cool, B. Bauer, J.H. Camonis, K.H. Jakobs, and M. Schmidt, *J. Biol. Chem.*, **2000**, *274*, 34691–34698.

138 K. Rose, S.A. Rudge, M.A. Frohman, A.J. Morris, and J. A. Engebrecht, *Proc. Natl. Acad. Sci. USA*, **1995**, *92*, 12151–12155.

139 S.A. Rudge, A.J. Morris, and J.A. Engebrecht, *J. Cell Biol.*, **1998**, *140*, 81–90.

140 X. Wang, *Prog. Lipid Res.*, **2000**, *39*, 109–149.

141 X. Wang, *Annu. Rev. Plant Physiol. Plant Mol. Biol.*, **2001**, *52*, 211–231.

142 K. Fukami and T. Takenawa, *J. Biol. Chem.* **1992**, *267*, 10988–10993.

143 I. Kusaka, S.-E. Ishikawa, M. Higashiyama, T. Saito, S. Nagasaka, and T. Saito, *Endocrinology*, **1996**, *137*, 5421–5428.

144 N. Wilkie, C. Morton, L.I. Ng, and M.R. Boarder, *J. Biol. Chem.*, **1996**, *271*, 32447–32453.

145 J.-H. Hong, S.-O. Oh, M. Lee, Y.-R. Kim, D.-U. Kim, G.M. Hur, J.H. Lee, K. Lim, B.-D. Hwang, and S.-K. Park, *Biochem. Biophys. Res. Commun.*, **2001**, *281*, 1337–1342.

146 N.T. Ktistakis, H.A. Brown, P.C. Sternweis, and M.G. Roth, *Proc. Natl. Acad. Sci. USA*, **1995**, *92*, 4952–4956.

147 Y.-G. Chen, A. Siddhanta, C.D. Austin, S.M. Hammond, T.-C. Sung, M.A. Frohman, J. Morris, and D. Shields, *J. Cell Biol.*, **1997**, *138*, 495–504.

148 N.T. Ktistakis, H.A. Brown, M.G. Waters, P.C. Sternweis, and M.G. Roth, *J. Cell Biol.*, **1996**, *134*, 295–306.

149 K. Bi, M.G. Roth, and N.T. Ktistakis, *Curr. Biol.* **1997**, *7*, 301–307.

150 A. Siddhanta and D. Shield, *J. Biol. Chem.* **1998**, *273*, 17995–17998.

151 M. Stamnes, G. Schiavo, G. Stenbeck, T.H. Sollner, and J.E. Rothman, *Proc. Natl. Acad. Sci. USA*, **1998**, *95*, 13676–13680.

152 J. Kuai, A.L. Boman, R.S. Arnold, X. Zhu, and R.A. Kahn, *J. Biol. Chem.*, **2000**, *274*, 4022–4032.

153 B.-T. Williger, W.-T. Ho, and J.H. Exton, *J. Biol. Chem.*, **1999**, *274*, 735–738.

154 C.P. Morgan, H. Sengelov, J. Whatmore, N. Borregaard, and S. Cockcroft, *Biochem. J.*, **1997**, *325*, 581–585.

155 A.-S. Caumont, N. Vitale, M. Gensse, M.-C. Galas, J.E. Casanova, and M.-F. Bader, *J. Biol. Chem.*, **2000**, *275*, 15637–15644.

156 Y. Humeau, N. Vitale, S. Chasserot-Golaz, J.-L. Dupont, G. Du, M.A. Frohman, M.-F. Bader, and B. Poulain, *Proc. Natl. Acad. Sci. USA*, **2001**, *98*, 15300–15305.

157 Y. Shen, L. Xu, and D.A. Foster, *Mol. Cell. Biol.*, **2001**, *21*, 595–602.

158 A.V. Vieira, C. Lamaze, and S.L. Schmid, *Science*, **1996**, *274*, 2086–2089.

159 A.T. Jones and M.J. Clague, *Biochem. Biophys. Res. Commun.*, **1997**, *236*, 285–288.

160 L.S. Arneson, J. Kunz, R.A. Anderson, and L.M. Traub, *J. Biol. Chem.*, **1999**, *274*, 17794–17805.

161 M. Emoto, J.K. Klarlund, S.B. Waters, V. Hu, J.M. Buxton, A. Chawla, and M.P. Czech, *J. Biol. Chem.*, **2000**, *275*, 7144–7151.

162 M.M. Billah, J.-K. Pai, T.J. Mullmann, R.W. Egan, and M.I. Siegel, *J. Biol. Chem.*, **1989**, *264*, 9069–9076.

163 D.E. Agwu, L.C. McPhail, M.C. Chabot, L.W. Daniel, R.L. Wykle, and C.E. McCall, *J. Biol. Chem.*, **1989**, *264*, 1405–1413.

164 J.-K. Pai, M.I. Siegel, R.W. Egan, and M.M. Billah, *J. Biol. Chem.*, **1988**, *263*, 12472–12477.

165 P. Bellavite, F. Corso, S. Dusi, M. Grzeskowiak, V. Della-Bianca, and F. Rossi, *J. Biol. Chem.*, **1988**, *263*, 8210–8214.

166 T. Ohtsuka, M. Ozawa, N. Okamura, and S. Ishibashi, *J. Biochem.*, **1989**, *106*, 259–263.

167 S.A. Bauldry, D.A. Bass, S.L. Cousart, and C.E. McCall, *J. Biol. Chem.*, **1991**, *266*, 4173–4179.

168 D. Qualliotine-Mann, D.E. Agwu, M.D. Elenburg, C.E. McCall, and L.C. McPhail, *J. Biol. Chem.*, **1993**, *268*, 23843–23849.

169 L.C. McPhail, D. Qualliotine-Mann, and K.A. Waite, *Proc. Natl. Acad. Sci. USA*, **1995**, *92*, 7931–7935.

170 K.A. Waite, R. Wallin, D. Qualliotine-Mann, and L.C. McPhail, *J. Biol. Chem.*, **1997**, *272*, 15569–15578.

171 D.S. Regier, D.G. Green, S. Sergeant, A.J. Jesaitis, and L.C. McPhail, *J. Biol. Chem.* **1999**, *274*, 36601–36608.

172 K.-S. Ha and J.H. Exton, *J. Cell Biol.*, **1993**, *123*, 1789–1796.

173 K.-S. Ha, E.-J. Yeo, and J.H. Exton, *Biochem. J.*, **1994**, *303*, 55–59.

174 M.J. Cross, S. Roberts, A.J. Ridley, M.N. Hodgkin, A. Stewart, L. Claesson-Welsh, and M.J.O. Wakelam, *Curr. Biol.*, **1996**, *6*, 588–597.

175 G.H. Jenkins, P.L. Fisette, and R.A. Anderson, *J. Biol. Chem.*, **1994**, *269*, 11547–11554.

176 A. Moritz, P.N.E. De Graan, W.H. Gispen, and K.W.A. Wirtz, *J. Biol. Chem.*, **1992**, *267*, 7207–7210.

177 N. O'Luanaigh, R. Pardo, A. Fensome, V. Allen-Baume, D. Jones, M.R. Holt,

and S. Cockcroft, *Mol. Biol. Cell*, **2002**, *13*, 3730–3746.
178 D.J. Kusner, S.J. Schomisch, and G.R. Dubyak, *J. Biol. Chem.*, **1993**, *268*, 19973–19982.
179 W.H. Moolenaar, *Exp. Cell Res.* **1999**, *253*, 230–238.
180 A.H. Yang, I. Ishii, and J. Chun, *Biochim. Biophys. Acta*, **2002**, *1582*, 197–203.
181 H.N. Higgs, M.H. Han, G.E. Johnson, and J.A. Glomset, *J. Biol. Chem.*, **1998**, *273*, 5468–5477.
182 H. Sonoda, J. Aoki, T. Hiramatsu, M. Ishida, K. Bandoh, Y. Nagai, R. Taguchi, K. Inoue, and H. Arai, *J. Biol. Chem.*, **2002**, *277*, 34254–34263.
183 F.J. Thomson and M.A. Clark, *Biochem. J.*, **1995**, *306*, 305–309.
184 M.C.M. van Dijk, F. Postma, H. Hilkmann, K. Jalink, W.J. van Blitterswijk, and W.H. Moolenaar, *Curr. Biol.*, **1998**, *8*, 386–392.
185 M. Umezu-Goto, Y. Kishi, A. Taira, K. Hama, N. Dohmae, K. Takio, T. Yamori, G.B. Mills, K. Inoue, J. Aoki, and H. Arai, *J. Cell Biol.*, **2002**, *158*, 227–233.
186 S. Ghosh, J.C. Strum, V.A. Sciorra, L. Daniel, and R.M. Bell, *J. Biol. Chem.*, **1996**, *271*, 8472–8480.
187 Z. Zhao, C. Wallasch, H. Daub, and A. Ullrich, *Proc. Natl. Acad. Sci. USA*, **90**, 4251–4255.
188 S. El Bawab, O. Macovschi, C. Sette, M. Conti, M. Lagarde, G. Némoz, and A.-F. Prigent, *J. Biochem.*, **1997**, *247*, 1151–1157.
189 M. Grange, C. Sette, M. Cuomo, M. Conti, M. Lagarde, A.-F. Prigent, and G. Némoz, *J. Biol. Chem.*, **2000**, *275*, 33379–33387.
190 K. Kishikawa, C.E. Chalfant, D.K. Perry, A. Bielawska, and Y.A. Hannun, *J. Biol. Chem.*, **1999**, *274*, 21335–21341.
191 M.-H. Tsai, C.-L. Yu, F.-S. Wie, and D.W. Stacey, *Science*, **1989**, *243*, 522–526.
192 T.-H. Chuang, B.B. Bohl, and G.M. Bokoch, *J. Biol. Chem.*, **1993**, *268*, 26206–26211.
193 G.S. Baillie, E. Huston, G. Scotland, M. Hodgkin, I. Gall, A.H. Peden, C. MacKenzie, E.S. Houslay, R. Currie, T.R. Pettit, A.R. Walmsley, M.J.O. Wakelam, J. Warwicker, and M.D. Houslay, *J. Biol. Chem.*, **2002**, *277*, 28298–28309.
194 S. Ahmed, J. Lee, R. Kozma, A. Best, C. Monfries, and L. Lim, *J. Biol. Chem.*, **1993**, *268*, 10709–10712.
195 Y. Fang, M. Vilella-Bach, R. Bachmann, A. Flanigan, and J. Chen, *Science*, **2001**, *294*, 1942–1945.
196 K-S. Ha and J.H. Exton, *J. Biol. Chem.*, **1993**, *268*, 10534–10539.
197 S. Cockcroft and D. Allan, *Biochem. J.*, **1984**, 222, 557–559.
198 S.B. Bocckino, P.F. Blackmore, and J.H. Exton, *J. Biol. Chem.*, **1987**, *262*, 15309–15315.
199 S.B. Bocckino, P.B. Wilson, and J.H. Exton, *FEBS Lett.* **1987**, *225*, 201–204.
200 G. Augert, S.B. Bocckino, P.F. Blackmore, and J.H. Exton, *J. Biol. Chem.*, **1989**, *264*, 21689–21698.
201 T.M. Wright, L.A. Rangan, H.S. Shin, and D.M. Raben, *J. Biol. Chem.*, **1988**, *263*, 9374–9380.
202 M.S. Pessin, J.J. Baldassare, and D.M. Raben, *J. Biol. Chem.*, **1989**, *264*, 8729–8738.
203 M.S. Pessin, J.J. Baldassare, and D.M. Raben, *J. Biol. Chem.*, **1990**, *265*, 7959–7966.
204 C. Lee, S.K. Fisher, B.W. Agranoff, and A.K. Hajra, *J. Biol. Chem.*, **1991**, *266*, 22837–22846.
205 R. Plevin, S.J. Cook, S. Palmer, and M.J.O. Wakelam, *Biochem. J.*, **1991**, *279*, 559–565.
206 T.R. Pettit, A. Martin, T. Horton, C. Liossis, J.M. Lord, and M.J.O. Wakelam, *J. Biol. Chem.*, **1997**, *272*, 17354–17359.

5
Sphingomyelinases and Their Interaction with Membrane Lipids

Félix M. Goñi and Alicia Alonso

5.1
Introduction and Scope

This chapter is devoted to sphingomyelinases, enzymes that catalyze the hydrolysis of sphingomyelin (ceramide phosphorylcholine) into ceramide and phosphorylcholine. The reaction is formally similar to that of a phospholipase C. Sphingomyelinases have long been known [1, 2], but the last decade has seen renewed interest after the discovery of the sphingomyelin signal transduction pathway [3–6]. Ceramide appears to be a lipid second messenger in programmed cell death (apoptosis), cell differentiation and cell proliferation [reviews in 7–12; for the opposite point of view see 13], and sphingomyelinase is probably a major source of ceramide in the cells, although this is also a debated point. This contribution is based on a recent review [14] that we have modified according to several suggestions and criticisms received, and to which we have added what we consider as the most significant novel data to have appeared since.

Two main aspects of sphingomyelinases are covered here, one corresponds to "classical enzymology", with a description of the known sphingomyelinases, and of their properties. The second is more directly related to our experimental work, and describes the interaction of sphingomyelinases with phospholipid bilayers, and the mutual influence of enzyme and substrate properties. One important function of sphingomyelinases may be related to changing membrane properties (charge, fluidity, permeability) by transforming sphingomyelin into ceramide. The present work does not cover the cell physiological effects of sphingomyelinases and/or ceramides, which are described in detail in the above-mentioned reviews.

Lipases and Phospholipases in Drug Development
Edited by G. Müller and S. Petry

ISBN: 3-527-30677-3

5.2 Sphingomyelinases

5.2.1 Types of Sphingomyelinases

Samet and Barenholz [15] proposed a useful classification of sphingomyelinases into five categories. We shall follow rather closely their description, adding a sixth group for a specific kind of bacterial sphingomyelinases.

5.2.1.1 Acid Sphingomyelinase (aSMase)

This soluble glycoprotein was the first described sphingomyelinase [1]. Its name refers to its optimum activity at pH ≈ 5, as befits its localization in lysosomes. It is number EC 3.1.4.12 in the Enzyme Commission classification. The absence of this enzyme is responsible for the neurological disorder known as the Niemann-Pick syndrome. It is the best characterized of sphingomyelinases [15, 16], and has been cloned from human DNA and expressed in *E. coli* [17–19] and in insect Sf21 cells [20, 21]. In the latter case the recombinant protein is secreted as a 72 kDa protein, as opposed to the 75 kDa molecule purified from human placenta, both proteins having similar sphingomyelinase activity. The pattern of the six disulfide bonds in human aSMase has been described recently [22]. The recombinant and native (placental) human proteins possess an identical disulfide structure. Human aSMase has also been overexpressed in Chinese hamster ovary cells [23], and the authors have described the preparation and use of a fluorescent FITC-labelled aSMase. Acid sphingomyelinases have been characterized from leukemia HL-60 cells [14], and thrombin-activated human platelets [24, 25]. The latter enzyme (70–73 kDa) may not originate in the lysosomes [25]. A 58 kDa aSMase has been located in the lesional skin of atopic dermatitis patients [26].

Grassme et al. [27] described the overexpression of human aSMase in JY B cells and in aSMase-deficient lymphocytes from patients with Niemann-Pick disease type A. These authors made the important observation that, upon activation of the surface receptor Fas (CD95), a SMase was translocated from its intracellular location to the plasma membrane outer leaflet, which is rich in sphingomyelin. aSMase activity induced clustering of CD95 in sphingolipid-rich membrane rafts, and subsequent apoptosis. Clustering was probably the result of lateral segregation of ceramide (Section 5.3.2). In a related observation Cremesti et al. [28] showed that the anti-Fas antibody-induced lateral segregation ("capping") of Fas into one pole of the cell membrane was ceramide-dependent, and that SMase$^{-/-}$ hepatocytes were defective in antibody-induced ceramide generation, capping and apoptosis. aSMase in lysosomes appears to play a role in the defense against infection by intracellular pathogens. Ceramide generated by the enzyme activates proteases like cathepsin D, and aSMase may be a mediator of the antibacterial effects of the tumor necrosis factor (TNF) and of interferon-gamma. Mice deficient in aSMase were completely incapable of restricting the growth of *Listeria mono-*

cytogenes, with which they had been infected [29]. The role of aSMase in ceramide signaling and apoptosis has been reviewed by Gulbins and Kolesnick [30].

5.2.1.2 Secretory Sphingomyelinase (sSMase)

This enzyme arises from the acid sphingomyelinase gene via differential protein trafficking of a common protein precursor; this precursor can be targeted to either lysosomes (aSMase) or to the Golgi secretory pathway (sSMase) [31, 32]. The secretory enzyme is typically activated by physiological concentrations of Zn^{2+}, but Zn^{2+}-independent forms may exist [32]. sSMases may be involved in signal transduction via the sphingomyelin pathway. sSMases has been reviewed in Ref. [33].

Secretory sphingomyelinase has been proposed to have a role in atherogenesis. Serum low-density lipoprotein (LDL), when treated with bacterial sphingomyelinase, gives rise to LDL aggregates that became retained on the extracellular matrix and stimulate macrophage foam cell formation [34]. Schissel et al. [31] showed that sSMase from arterial wall cells could hydrolyze and aggregate LDL modified by oxidation, or extracted from human atherosclerotic lesions, at pH 7.4, while native LDL could only be cleaved by sSMase at pH 5.5. Thus sSMase is, probably, the arterial wall enzyme responsible for the hydrolysis of sphingomyelin in LDL and subendothelial LDL aggregation, a key event in atherogenesis. In addition, macrophages from aSMase knockout mice, thus deprived of both aSMase and sSMase, displayed defective cholesterol trafficking and efflux [35]. This may implicate sSMase deficiency as a new atherosclerosis risk factor.

5.2.1.3 Neutral, Mg^{2+}-dependent Sphingomyelinases (nSMase)

Neutral Mg^{2+}-dependent sphingomyelinases are integral membrane proteins in mammals, and soluble proteins in bacteria. The mammalian enzymes have pH optima near 7, and require millimolar concentrations of Mg^{2+} or Mn^{2+} for activity. nSMase appears to be ubiquitous in mammalian tissues, with the highest activities in the brain. The molecular mass is of about 92 kDa, although smaller isoenzymes are found, which may result either from proteolysis or from alternative processing of common transcripts. nSMase mediates several cellular processes including differentiation, cell cycle arrest, and programmed cell death (apoptosis) through the generation of ceramide [36, 37]. nSMase activation has been observed after ligation of CD95 and tumor necrosis factor (TNF) receptors [38, 39], after irradiation [40], and after triggering of mechanosensors [41], among many examples.

The first nSMases of mammalian origin were characterized by Gatt [42] and by Rao and Spence [43]. Liu et al. [44] purified by 3030-fold a Mg^{2+}-dependent, nSMase activity (ca. 60 kDa) from rat brain membranes. Enzyme activity was highly dependent on phosphatidylserine, and it was 95% inhibited by 3 mM glutathione [45]. Glutathione inhibition suggested that nSMase was activated by stress [46]. More recently, Bernardo et al. [47] have purified by 23 330-fold a Mg^{2+}-dependent nSMase from bovine brain that was inhibited by glutathione. The enzyme consists of two polypeptides, of 97 kDa and 46 kDa, that appeared to be structurally related, the latter probably being a degradation product. The bovine enzyme shares many properties with the one purified by Liu et al. [44] from rat brain (Mg^{2+}-dependence, substrate specificity, activation by phosphatidylserine, inhibition by glutathione). However, the respective molecular masses differ, 60 vs. 97 kDa, thus the relationship between both enzymes remains to be established.

Ghosh et al. [48] have isolated two isoforms of nSMase from rabbit skeletal muscle (92 and 53 kDa). Peptide mapping revealed important structural similarities, and the catalytic activities were also similar, except that the 53 kDa protein was Mg^{2+}-independent. These nSMases are located in the transverse tubules of the muscle cells, which may be related to the observation that sphingosine modulates calcium release from sarcoplasmic reticulum membranes [49]. Two Mn^{2+}- and Mg^{2+}-dependent nSMases located in the microsomal membranes of seminiferous tubes of immature Wistar rats have been characterized [50] whose properties do not appear to differ significantly from other mammalian nSMases. Finally, we mention two other nSMases purified from eukaryotic natural sources, namely the Mg^{2+}-dependent nSMase isolated from *Saccharomyces cerevisiae* [51], and that obtained from membrane fractions of intraerythrocytic *Plasmodium falciparum*, the malaria parasite [52]. The latter enzyme was activated by phosphatidylserine and other anionic phospholipids, and was sensitive to scyphostatin, an inhibitor of mammalian nSMase (see below).

The cloning of mammalian nSMase has aroused debate. Tomiuk et al. [53] announced the cloning, overexpression and functional characterization of murine and human Mg^{2+}-dependent nSMase. Cloning was achieved by searching for sequences similar to those of bacterial nSMase. Overexpressing cells were treated with TNF-α but no clearly elevated ceramide levels or apoptosis were observed, which argued against the proposed role of nSMase in apoptosis. However, several concerns were soon raised, particularly with respect to the specificity of the cloned protein, i.e. it was not clear whether sphingomyelin was the natural substrate. Shortly afterwards Sawai et al. [54] demonstrated that although the cloned nSMase had sphingomyelinase activity *in vitro*, cells overexpressing the protein did not show marked changes in sphingomyelin metabolism. Furthermore 1-0-alkyl-lyso-PC (lyso-Platelet Activating Factor, or lyso-PAF) was metabolized into 1-0-alkylglycerol in overexpressing cells, but not in vector cells. Sawai et al. [54] concluded that the protein acted as lyso-PAF phospholipase C rather than as nSMase in cells. Neuberger et al. [55] then demonstrated the localization of the enzyme cloned by Tomiuk et al. [53] in the endoplasmic reticulum of cells, which argues against a role of this enzyme in sphingomyelin signaling.

Subsequently, Tomiuk et al. [56] also concluded that the enzyme they had previously cloned (nSMase 1 in their nomenclature) was predominantly localized in the microsomal fraction (endoplasmic reticulum and Golgi). anti-nSMase 1 antibodies did not affect the nSMase activity in membrane fractions from murine brain [56], indicating that nSMase 1 was one of at least two mammalian nSMases. Indeed, shortly afterwards, the same group published the cloning and characterization of a 71 kDa mammalian brain-specific, Mg^{2+}-dependent nSMase [57], which they called nSMase 2. This enzyme colocalized with a Golgi marker in several cell lines, according to immunofluorescence methods.

The enzyme with lyso-PAF phospholipase C and nSMase activities (nSMase 1) has been examined by several groups. Katan and co-workers confirmed its localization in the endoplasmic reticulum, and showed that the enzyme has a strong requirement for reducing agents, and is reversibly inhibited by reactive oxygen species and by oxidized glutathione [58]. They also showed that nSMase 1 is irreversibly inactivated by peroxynitrite, a nitric oxide-derived oxidant, presumably through thiol group oxidation [59]. The same laboratory reported an essential role for two His residues (136 and 272 in nSMase 1) that were conserved in sequences from bacteria to human, and demonstrated that the transmembrane region contained determinants for the endoplasmic reticulum localization [60]. The same group have also identified nSMase 1 as the major neutral SMase activity in the DT40 B-cell line, and have pointed out that the enzyme does not appear to play a role in the apoptotic response to endoplasmic reticulum stress [61]. Independently, Mizutani et al. [62] cloned and expressed an essentially similar enzyme from rat, which they located afterwards in the nuclear matrix of ascites hepatoma AH7974 cells [63].

In turn Marchesini et al. [64] have characterised the mouse nSMase2 overexpressed in either yeast cells or MCF7 breast cancer cells. nSMase2 recognised endogenous sphingomyelin as substrate, leading to ceramide production. Stable overexpression of nSMase2 resulted in a 30–40% decrease in the mammalian cell growth rate at the late exponential phases, but not in cell death. Moreover, TNF induced a significant activation of nSMase2 in MCF7 cells overexpressing the enzyme. These results suggest that nSMase2 corresponds to a physiologically meaningful mammalian neutral SMase that is involved in the regulation of cell growth and in cell signaling.

Sphingomyelin signaling in response to external agents (TNF-α, oxidized LDL) results in a SMase at the plasma membrane level. It is thus disappointing that none of the cloned nSMases are localized there. In contrast, two recent reports characterize plasma membrane-associated nSMases. Veldman et al. [65] described a nSMase in caveolae of human skin fibroblasts. This enzyme was inhibited by a peptide that corresponded to the scaffolding domain of caveolin, suggesting a direct molecular interaction of both proteins. Such interaction would in turn imply a cytosolic orientation of the caveolar nSMase. Stimulation of overexpressing fibroblasts with TNF-α resulted in caveolin-sensitive nSMase activity in the non-caveolar fractions. These results suggest that caveolar nSMase may be involved in TNF-α-signaling. More recently, mechanical (haemodynamic) stress has been shown to stimulate the activity of nSMase, but not of aSMase, in caveolae, leading to the generation of ceramide [41]. Working on a different system, namely pig liver plasma membranes, Martin et al. [66, 67] isolated a Mg^{2+}-dependent nSMase that was not inhibited by glutathione but was inhibited by ubiquinol and other lipophilic antioxidants. Also relevant here is the somewhat forgotten report of a latent nSMase in chicken erythrocyte membranes [68].

The identification of a nSMase activity *in* LDL is significant [69, 70]. These authors have demonstrated that apolipoprotein B-100, the protein moiety of LDL, has a nSMase activity, which is also reflected by partial sequence homology between apolipoprotein B-100 and bacterial SMases. The physiological role of this activity could be related to nonreceptor-mediated endocytotic entry of LDL into cells. This observation may also open a new mechanistic link between elevated plasma levels of LDL, apoptosis and atherosclerosis.

Although not strictly a sphingomyelinase, since its natural substrate appears to be ceramide phosphoinositol, i.e. the inositol analogue of sphingomyelin, a lipid found in yeast, it is worth mentioning the cloning and characterisation of yeast *ISC1*, an inositol phosphosphingolipid phospholipase C [71]. The enzyme, which has homology to bacterial SMase, is activated by phosphatidylserine and cardiolipin. A study using site-directed and deletion mutants [72] has shown that one of the two transmembrane domains, and the C-terminus of the enzyme are required for phosphatidylserine binding. The authors suggest that, in the presence of anionic phospholipids, the catalytic domain (located at the N-terminus of the polypeptide chain) is "pulled" towards the membrane to interact with its substrate.

A whole group of sphingomyelinases found in bacteria are structurally related to nSMase 1 [53, 60]. In this sense they can be called neutral sphingomyelinases, although in practice they show a rather wide range of pH optima, and hydrolyze phospholipids other than sphingomyelin. These enzymes have been found in *Bacillus cereus*, *Staphylococcus aureus* (β-toxin), *Listeria ivanovi*, *Leptospira interrogans*, *Chromobacterium violaceum*, *Helicobacter pylori* and *Pseudomonas TK4* [73–77]. Of these, the nSMase from *B. cereus* is the best known. A distant evolutionary relationship has been noted between bacterial sphingomyelinase and mammalian DNase I [78]; residues of DNase I that are involved in the active center (including His 134 ad His 252) are conserved in bacterial sphingomyelinase, and are essential for activity. They correspond to the essential His 136 and His 272 in nSMase 1 [60]. Bacterial sphingomyelinases may be descendants from the eukaryotic enzymes, as proposed by Heinz et al. [79] for certain phospholipases C. In favour of this hypothesis is that these enzymes are not usually required for the bacterial life cycle, and are often found in pathogens functioning as bacterial toxins [60]. The similarity between bacterial and mammalian nSMases has permitted workers to subclone chimeras of bacterial SMase and green fluorescent protein into mammalian vectors containing sequences that target the enzyme to several subcellular organelles, with the result that only when bacterial SMase was targeted to mitochondria did cells undergo apoptosis [80].

5.2.1.4 Mg^{2+}-independent Neutral Sphingomyelinases

This small and barely known group of sphingomyelinases includes an enzyme found in the cytosol of HL-60 cells by Okazaki et al. [81], and the 53 kDa isoform of nSMase found by Ghosh et al. [48] in rabbit skeletal muscle.

5.2.1.5 Alkaline Sphingomyelinase from the Intestinal Tract

The best characterized enzyme in this group is an alkaline sphingomyelinase (bSMase), purified ca. 1600-fold from rat intestine [82]. The enzyme is not expressed in other organs, requires bile salts for activity, is Mg^{2+}-independent, and not inhibited by glutathione. The lack of this enzyme may be related to colon carcinogenesis. Its activity appears to be influenced by dietary fats other than sphingolipids [83]. For a review on alkaline SMase, see Nilsson and Duan [84].

5.2.1.6 Bacterial Sphingomyelinase-phospholipase C

This category was not included in the classification of Samet and Barenholz [15]. At last two representatives are known, namely the so-called "α-toxin" from *Clostridium perfringens* [85], and the product of *plcB* gene from *Listeria monocytogenes* [86]. Both can use phosphatidylcholine and sphingomyelin as substrates, and perhaps other glycerophospholipids as well [87]. Both enzymes are structurally related to the phosphatidylcholine-preferring phospholipase C from *B. cereus* but the latter has no SMase activity.

5.2.2 Sphingomyelinase Mechanism

5.2.2.1 Binding of Magnesium Ions

Activation by Mg^{2+} is observed in most, but not all, nSMases of mammalian and bacterial origin. The Mg^{2+} effect on these enzymes becomes saturated at ca. 10^{-3} M. The role of Mg^{2+} in the activation of *B. cereus* SMase was analyzed in detail by Fujii et al. [73]. The enzyme had two classes of binding sites for Mg^{2+}, with binding constants 2.1×10^2 and 5.0×10^6 M^{-1} at pH 6.0. The high-affinity site was saturated with a single Mg^{2+} ion. The functional meaning of the high-affinity binding is unclear. Low-affinity Mg^{2+} binding is essential for catalysis, but unrelated to substrate binding. Mg^{2+} binding also protects the enzyme against alkaline denaturation.

5.2.2.2 Binding of Substrate

As with all lipases, substrate binding is only one part of a more complex phenomenon, involving surface adsorption and perhaps other precatalytic stages. The interaction of SMase with its substrate has been carefully studied by Fanani and Maggio [88] using sphingomyelin monolayers at an air–water interface. The enzyme (*B. cereus*) was injected into the aqueous phase, from which it could interact with the substrate. Three stages could be distinguished in the enzyme activity, (i) a latency period, during which interfacial adsorption and other precatalytic steps took place; (ii) a period of steady-state enzyme activity, i.e. a constant rate of substrate degradation (pseudo zero-order kinetics); and (iii) a period of gradual halting of product formation.

During the latency period the amount of SMase adsorbed onto the lipid monolayer increased up to a plateau, the latter marking the start of the constant velocity region. The amount of adsorbed enzyme did not vary with the presence of product. An interface/subphase partition coefficient $K_P=7\times10^{-4}$ was calculated for the enzyme, indicating a high tendency of SMase to associate with the lipid interface. Enzyme partition to the interface and interfacial activation were essentially irreversible.

Fanani and Maggio [88] analyzed quantitatively the activation phase of the SMase reaction, and compared the experimental results with six different kinetic models. The best fit was found for the so-called "model V", (Scheme 5.1):

$$2E \underset{}{\overset{K_P}{\rightleftharpoons}} 2E_i + 2S \overset{K}{\rightleftharpoons} 2E_iS \overset{K_d}{\rightleftharpoons} (E_iS)_2 \xrightarrow{k_{act}} (E_i^*S)_2 + S \overset{K_S}{\rightleftharpoons} (E_i^*S)_2S$$

$$\xrightarrow{k_{act}} (E_i^*S)_2 + P$$

Scheme 5.1

After the partition step the model involves association of the enzyme with the substrate and subsequent dimerization before activation. The proportion of activated enzyme (E_i^*) was assumed to be very small in relation to the total amount of enzyme at the interface (E_i + E_i^*). Enzyme activation was the rate-limiting step for the reaction. The activation was slow ($k_{act}=3\times10^{-3}$ s^{-1}) compared with the rate of catalysis in the steady-state regime ($k_{cat}=4\times10^2$ s^{-1}), the being latter estimated on the basis of classic Michaelis-Menten analysis. The authors conclude that the lag time in SMase catalysis is due to an equilibrium between monomolecular and bimolecular activated states of the enzyme in the interface, and to a slow activation process.

5.2.2.3 **Mechanism of Catalysis**

A model for the interaction of SMase with sphingomyelin (Fig. 5.1) and a catalytic mechanism (Fig. 5.2) have been proposed for *B. cereus* SMase by Matsuo et al. [78] and by Fujii et al. [73], respectively, on the basis of four lines of evidence:

(i) Mutational analysis suggesting that Asp 126 and Asp 156 are involved in substrate recognition, and that Asp 295, His 151 and His 296 are essential for the hydrolytic activity [89]. Note that His 136 and His 272 are essential for the activity of nSMase 1 [60, 62]. His 134 and His 252 of DNase correspond to histidines 136 and 272 of nSMase1, and to histidines 151 and 296 of *B. cereus* SMase.
(ii) The prediction by Matsuo et al. [78] of the three-dimensional structure of the enzyme using a protein fold recognition method, which suggested that SMase adopts a structure similar to that of bovine pancreatic DNase I. This prediction included the proposal of several amino acid residues involved in catalysis and enzyme recognition, in agreement with mutational analysis.

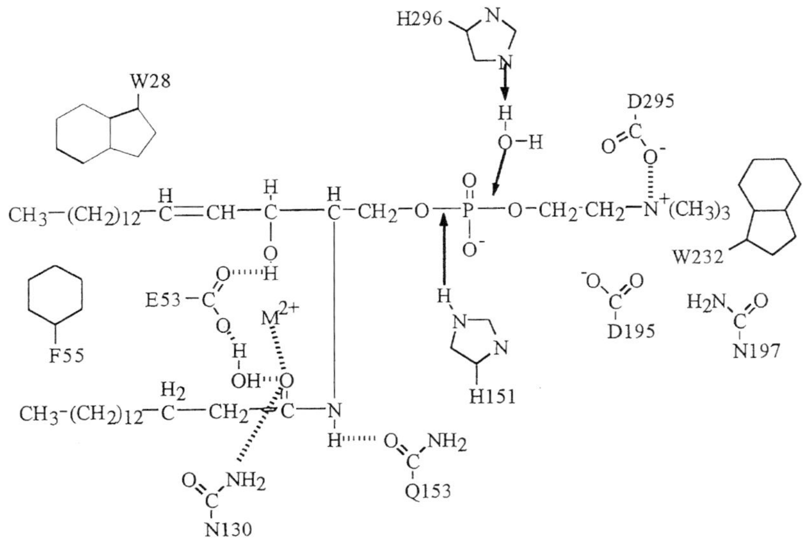

Fig. 5.1 A model for the interaction between SMase and SM. These residues are all conserved in homologous bacterial SMases. Residues W28, E53, F55, N130, Q153, W232 and D295 may be involved in substrate recognition. E53 and D295 would be involved in hydrolysis of the phosphodiester bond. M^{2+} denotes the metal ion that is essential for catalysis. Redrawn from Ref. [78].

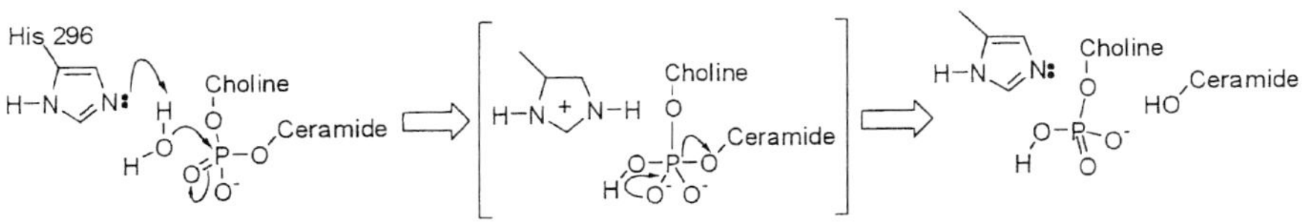

Fig. 5.2 General-base catalytic mechanism for bacterial sphingomyelinase. Redrawn from Ref. [73].

(iii) The suggestion by Weston et al. [90], on the basis of their X-ray studies of a DNase I-substrate complex, that His 252 of DNase I acts as a general base to accept a proton from a water molecule, which then, as a nucleophile, attacks the phosphoryl group, and His 134 functions as a general acid to protonate the leaving oxygen of the phosphoester. His 252 and His 134 from DNase I correspond to His 296 and His 151, respectively, in *B. cereus* SMase.

(iv) The studies by Fujii et al. [73] on the pH dependence of kinetic parameters of the enzyme, in which they conclude that the mechanism of SMase appears to be one of general-base catalysis, with His 296 acting as a general-base (Fig. 5.2).

5.2.3
Sphingomyelinase Assay

Two situations must be distinguished, (i) assays of pure enzyme and (ii) assays of cell extracts. When purified enzyme preparations are available, no labeled substrate is required. Natural or synthetic sphingomyelin is prepared, pure or mixed with other lipids, in the form of extruded large unilamellar vesicles (LUV) ca. 100 nm in diameter. When pure sphingomyelin vesicles are used extrusion must take place at a temperature close to or above the gel-fluid transition temperature of the lipid, i.e. often 45–50 °C. LUV and enzyme are mixed in the appropriate assay buffer and aliquots are removed at fixed time intervals. The aliquots are mixed with chloroform–methanol and, after phase separation, phosphorous (from phosphorylcholine) is assayed in the aqueous phase. The procedure has been described in detail by Ruiz-Argüello et al. [91].

When LUV are assayed with sphingomyelinase, ceramide production in the bilayers leads to vesicle aggregation, which in turn produces an increase in turbidity or light scattering in the suspension. Thus the reaction can be followed in real time simply by measuring the increase in turbidity (absorbance at 500 nm) or in light scattering (e.g. with a fluorometer with both the excitation and emission monochromators adjusted at 500 nm). The chemical assay of the water-soluble product phosphorylcholine and the turbidity assay give superimposable results [91, 92].

Some commercial preparations of purified sphingomyelinase from *B. cereus* may be contaminated with phospholipase C. This is irrelevant when the substrate consists of pure sphingomyelin, but when the LUV contain glycerophospholipids, e.g. phosphatidylcholine, contamination is serious because both enzymes will give off the water-soluble end-product phosphorylcholine. In such cases, the assays include *o*-phenanthroline, a specific inhibitor of phospholipase C that is innocuous for SMase [92–94].

Assaying SMase activity in cell extracts requires that sphingomyelin be labeled, radioactively, fluorescently or otherwise. Also, and because of the difficulty in bringing together the enzyme and substrate molecules in the presence of cell homogenates, assays are often carried out in Triton X-100 or other suitable detergents. However, as well as from emulsifying the substrate, the detergents bind and modify the enzyme activity and thus they are best avoided. If they are required, the detergent concentration and the initial detergent to substrate ratio must be kept constant for reproducibility of assays. After catalysis, the labeled enzyme products are separated, usually by thin-layer chromatography, and quantitated. A general description of SMase assays in cell extracts has been provided by Santana et al. [95], and by Gatt [96]. Samet and Barenholz [15] describe in detail assay conditions for aSMase and nSMase using a fluorescent sphingomyelin derivative. An alternative fluorescence assay is given by Loidl et al. [97]. He et al. [98] use a fluorescent (BODIPY) derivative of sphingomyelin as a substrate, and the resulting fluorescent ceramide is separated by HPLC within 4 min. Liu and Hannun [99] have detailed a SMase assay using a radioactively labelled substrate, namely *N*-[methyl-^{14}C]sphingomyelin. In a novel non-radioactive approach Zhou et al. [100] use biotinylated substrate conjugates and then purify the enzyme products by bioaffinity capture on streptavidin-agarose beads; after release, they then analyze them by electrospray ionization mass spectroscopy. High-throughput methods for sphingomyelinase screening, using *N*-[methyl-^{14}C]sphingomyelin, have also been published [101, 102].

A SMase assay kit is commercially available, based on the use of 10-acetyl-3,7-dihydroxyphenoxazine (Amplex Red), a sensitive fluorogenic probe for H_2O_2. First, SMase hydrolyzes the sphingomyelin to yield ceramide and phosphorylcholine. After the action of alkaline phosphatase, which hydrolyzes phosphorylcholine, choline is then oxidized by choline oxidase to betaine and H_2O_2. Finally, H_2O_2, in the presence of a peroxidase, reacts with Amplex Red to generate a highly fluorescent product. However, any enzyme generating phosphorylcholine will give off the fluorescent product, thus the assay does not discriminate between SMase and phospholipase C.

5.2.4 Sphingomyelinase Inhibitors

Some physiological SMase inhibitors, e.g. glutathione, are mentioned in Section 5.2.1. Among the pharmacological inhibitors scyphostatin, obtained from mycelial extracts from the discomycete *Trichopeziza mollissima*, is becoming widely accepted as a neutral sphingomyelinase inhibitor [47, 52, 103–105]. Other natural inhibitors of mammalian nSMase include macquarimicin A [106], alutenusin [107], chlorogentisylquinone [108], and manumycin A [109]. Some synthetic substrate analogues can also be used as SMase inhibitors [110–112]. Arenz and Giannis [113] have produced the first synthetic irreversible inhibitor of nSMase. Recently, GW4869 was developed as an inhibitor of nSMase, and mammalian nSMase 2 was found to be a molecular target for this inhibitor [64, 114]. Hexadecylphosphocholine has pronounced antiproliferative effects on neoplastic cells and acts through interference of cell signaling via phospholipases, but it does not modify SMase activity [115].

Inhibitors of aSMase are also known, such as the xanthone α-mangostin, isolated from the bark of *Garcinia speciosa* [116]; α-mangostin has recently been obtained by organic synthesis [117].

5.3 Sphingomyelinase–Membrane Interactions

At least two obvious aspects must be considered here, namely the influence of the membrane lipids on sphingomyelinase activity, and the effects of the enzyme on the membrane properties. For the latter, we shall deal separately with enzyme effects on membrane permeability and effects on membrane architecture.

5.3.1
Lipid Effects on Sphingomyelinase Activity

Studies on this subject are scarce, particularly for substrates in bilayer form. Linke et al. [118] have published data that show stimulation of recombinant human acid sphingomyelinase by lipids that occur frequently in lysosomes. The substrate was present as liposomes, and no detergents were used. The liposomes contained, basically, phosphatidylcholine, cholesterol and sphingomyelin. To this mixture other lipids were added as required. Lysosomally occurring lipids, e.g. phosphatidylinositol and bis(monoacylglycero)phosphate, were effective stimulators (by up to 3-fold) of aSMase activity. Dolichol, and in particular dolicholphosphate, which occur in both Golgi and lysosomal membranes, were also effective aSMase activators. Interestingly, Linke et al. [118] also studied the effect of membrane curvature on aSMase activity, by preparing unilamellar liposomes of varying diameters (sub-50 to 200 nm). Enzyme activity increased remarkably with increasing curvature (decreasing size) of the vesicles. However, in the smaller liposomes tested, the vesicles are subjected to a considerable lateral tension, which could by itself influence enzyme activity. Finally, the authors demonstrated [118], using surface plasmon resonance, that binding of a SMase to lipid bilayers was actually increased by anionic lipids such as bis(monoacylglycero)phosphate.

Ruiz-Argüello et al. [119] have noted that, unusually, most molecular species of sphingomyelin are in the gel state in the physiological range of temperatures, i.e. most naturally occurring sphingomyelins have gel-to-fluid transition temperatures within 35 to 45 °C. These authors used liposomes (large unilamellar vesicles) composed of pure egg sphingomyelin (gel-to-fluid transition temperature ca. 39 °C) or of binary mixtures of sphingomyelin and glycerophospholipids (phosphatidylcholine or phosphatidylethanolamine). The enzyme was of bacterial (*B. cereus*) origin. SMase was active on pure sphingomyelin bilayers but activity was exceedingly low when the lipid was in the gel state. It increased by over one order of magnitude when the substrate went from the gel to the fluid phase. SMase activity showed lag times, followed by bursts of activity. Lag times decreased markedly when the substrate went across the phase transition to the fluid state. When egg phosphatidylcholine or egg phosphatidylethanolamine were included in the bilayer composition together with sphingomyelin, sphingomyelinase activity at 37 °C, which was negligible for the pure sphingolipid bilayers, increased with the proportion of glycerophospholipid, while the latency times became progressively shorter. A differential scanning calorimetry study of the mixed-lipid vesicles revealed that both phosphatidylcholine and phosphatidylethanolamine decreased in a dose-dependent way the transition temperature of sphingomyelin. Thus, as those glycerophospholipids were added to the membrane composition, the proportion of sphingomyelin in the fluid state at 37 °C increased accordingly, in this way becoming amenable to rapid hydrolysis by the enzyme. Thus sphingomyelinase requires the substrate in bilayer form to be in the fluid state, irrespective of whether this is achieved through a thermotropic transition or by modulating bilayer composition.

Also relevant in this context is the observation by Jaffrezou et al. [120] that ceramide enhances neutral sphingomyelinase activity in myeloid leukemia cells and in normal skin fibroblasts. However, these results were obtained with short-chain, cell-permeant ceramides, whose physical effects on membranes may differ from the more frequent long-chain ones [9]. The subject is interesting enough to deserve a more detailed investigation.

Fanani et al. [121] have described a beautiful example of the effect of the lipid environment on SMase activity. In particular, they have examined the effect of lipid lateral organization on SMase activity in lipid monolayers by epifluorescence microscopy. The enzyme alters surface topography by inducing phase separation into condensed (ceramide-enriched) and expanded (sphingomyelin-enriched) domains. The ceramide-enriched phase grows steadily as the reaction proceeds at a constant rate, and when it becomes continuous in the monolayer plane, i.e. beyond the percolation point, the reaction rate in the now discontinuous expanded, sphingomyelin-rich domains decreases rapidly. Thus the long-range topographic changes induced by the enzyme feed back on its kinetics.

5.3.2 Effects of Sphingomyelinase Activity on Membrane Properties

5.3.2.1 Effects on Membrane Lateral Organization

One important effect of the action of SMase on membranes is the generation of ceramide-rich domains due to the lateral segregation of the enzyme lipidic end-product. This phenomenon is interesting in terms of lipid biophysics, and may have important physiological implications. Separation into ceramide-rich domains was first observed by Huang et al. [122], who examined the structure of bilayers composed of fully deuterated dipalmitoylphosphatidylcholine and bovine brain ceramide using ^{2}H NMR spectroscopy. These authors observed that addition of ceramide induced lateral phase separation of fluid phospholipid bilayers into regions of gel and liquid crystalline (fluid) phases, ceramide partitioning largely into the gel phase of d_{62}-DPPC. These studies were later extended to different chain length ceramides [123], with the result that C16-ceramide, but not C2-, C6- or C8-ceramides, gave rise to in-plane phase separations. A similar behavior of long-chain ceramides has been detected in phosphatidylcholine/phosphatidylserine mixtures [124]. The use of a pyrene-labeled phospholipid, a fluorescent probe that is sensitive to lateral mobility and to the local concentration of fluorophore in the membrane, allowed Holopainen et al. [125, 126] to detect ceramide-enriched microdomains in fluid phosphatidylcholine membranes. A parallel kinetic study of ceramide-rich domains [126] revealed that, under conditions allowing fast ceramide synthesis, changes in the lateral organization of the membrane took significantly longer time (one order of magnitude, under the experimental conditions). The same group found that putting a small area of the membrane of giant liposomes containing sphingomyelin in contact with immobilized SMase [127] resulted in ceramide-rich domains in the area of interaction.

Domain formation by ceramides was also described by a combination of differential scanning calorimetry and IR spectroscopy, using natural ceramides (brain, egg) and several synthetic phospholipids [128]. Calorimetry was used to detect gel-fluid transitions. Different domains, when formed, "melt" at different temperatures, so that they could be easily detected. Veiga et al. [128] found lateral separation of ceramide-rich domains with as little as 5 mol% ceramide. Although disparity of chain lengths between lipid species ("chain mismatch") has been mentioned as the origin of certain cases of lateral phase separation [129], in the present example ceramides of different lengths (brain, egg) gave almost equivalent results in phospholipid mixtures examined by calorimetry. IR spectroscopy, together with the use of deuterated lipids, can be applied to the resolution of the gel-fluid transition in ceramide-rich and -poor domains. The above authors prepared mixtures of deuterated phospholipid (d_{54}-DPPC) with natural ceramide. IR spectroscopy clearly reveals a gradual melting of DPPC along a wide temperature range (covering the various domains), while ceramide exhibits a much sharper transition, corresponding to the high-temperature melting, ceramide-enriched domains [128]. Also, Wang and Silvius [130], using fluorescence-quenching measurements, observed that ceramide exhibits a much higher affinity for ordered lipid domains in both cholesterol-free and cholesterol-containing bilayers than do other sphingolipids.

Carrer and Maggio [131] have studied mixtures of bovine brain ceramide, which contain mostly C18:0 and C24:1 fatty acids, with DPPC, both by differential scanning calorimetry and in lipid monolayers extended at the air–water interface. The calorimetric results are essentially coincident with those by Veiga et al. [128], with in-plane phase separation being hinted at with only 1% ceramide. Measurements of surface potentials in monolayers reveal that molecular dipole potentials may play an important role in lateral phase separation, with favorable ceramide-PC dipolar matching in the liquid state being one local determinant for close molecular interactions, and unfavorable matching explaining domain segregation of ceramide-enriched phases (see also Moore et al. [132]). Studies in Langmuir troughs, in which sphingomyelin is extended at an air–water interface and SMase is added to the water subphase, provide a different and valuable tool to observe enzyme effects. In particular, epifluorescence microscopy may reveal ceramide-rich condensed domains coexisting with SM-rich expanded domains. Slotte [133] and more recently Fanani et al. [121] have quantified the area occupied by the different domains, and have correlated domain formation with enzyme activity.

In the latter cases chain mismatch appears to be an important factor regulating in-plane phase separation. However, such a phenomenon has also been observed in the absence of chain-length disparity in ceramide–phospholipid mixtures [126, 128]. The causes of domain formation by ceramides may be manifold, but the high hydrogen-bonding capacity of ceramides (in fact, of sphingolipids) should not be overlooked. While glycerophospholipids can act only as hydrogen-bond acceptors, sphingolipids such as sphingomyelin and ceramides can act as both acceptors and donors through their hydroxyl and amide groups [126, 134]. A study [135] has provided evidence for hydrogen bonding between sphingomyelin and cholesterol as the basis for the detergent insolubility of certain membrane fractions.

The capacity of ceramide to segregate laterally into ordered domains may be crucial in certain physiological events. Cremesti et al. [136] have suggested that this is the basis of "capping", in which cell stimulation results in the clustering of certain membrane proteins, typically receptor molecules, within membrane domains. According to these authors, when an extracellular signal, e.g. Fas engagement by ligand or agonistic antibody, elicits SMase activity at the plasma membrane level, ceramide released as an enzyme end-product leads to the reorganization of rafts into larger platforms or macrodomains within which Fas clusters. Receptor clustering is supposed to activate downstream effectors inside the cell.

5.3.2.2 **Effects on Membrane Permeability**

Ruiz-Argüello et al. [91] were the first to point out that generation of ceramides in the bilayer induced efflux of vesicle or cell contents. They treated either large unilamellar vesicles consisting of sphingomyelin, phosphatidylethanolamine and cholesterol (at a 2:1:1 mole ratio), or resealed erythrocyte ghosts, both loaded with low-molecular mass fluorescent dyes, with bacterial SMase. In both cases, rapid efflux ensued in parallel with enzyme activity. This is in contrast with phosphatidylcholine:phosphatidylethanolamine:cholesterol (2:1:1) vesicles treated with phospholipase C, for which aqueous contents were not released.

In a more detailed study Montes et al. [137] entrapped in liposomes fluorescent solutes of varying sizes, up to 20 kDa, and complemented the experiments in which ceramide was generated by SMase with others in which ceramide was added in organic solvent to the pre-formed liposome suspension. Small proportions of ceramide (10 mol% of total lipid) that may exist under physiological conditions of ceramide-dependent signaling were used. When long-chain (egg-derived) ceramides were used, both externally added and enzymatically produced ceramides induced release of vesicle contents. However, the same proportion of ceramides generated by SMase induced faster and more extensive efflux than when added in organic solution to the vesicles. Remarkably, SMase treatment of bilayers containing 50 mol% sphingomyelin gave rise to release of fluorescein-derivatised dextrans of $\approx$ 20 kDa, i.e. larger than cytochrome c. Furthermore, some experiments in the same paper were designed to explore the mechanism of ceramide-induced efflux. Two properties of ceramide appear to be important in the process of membrane restructuring that leads to solute efflux. One is the capacity of ceramide to induce a "negative" monolayer curvature, i.e. a curvature of the opposite sign to that found in cell membranes [9]. The other is its tendency to segregate into ceramide-rich domains (Section 5.3.2.1). The interface between ceramide-rich and -poor domains is probably the high-efflux region (Fig. 5.3). Recently, Siskind et al. [138] have shown that ceramide increases the permeability of the mitochondrial outer membrane to proteins of molecular mass of up to 60 kDa.

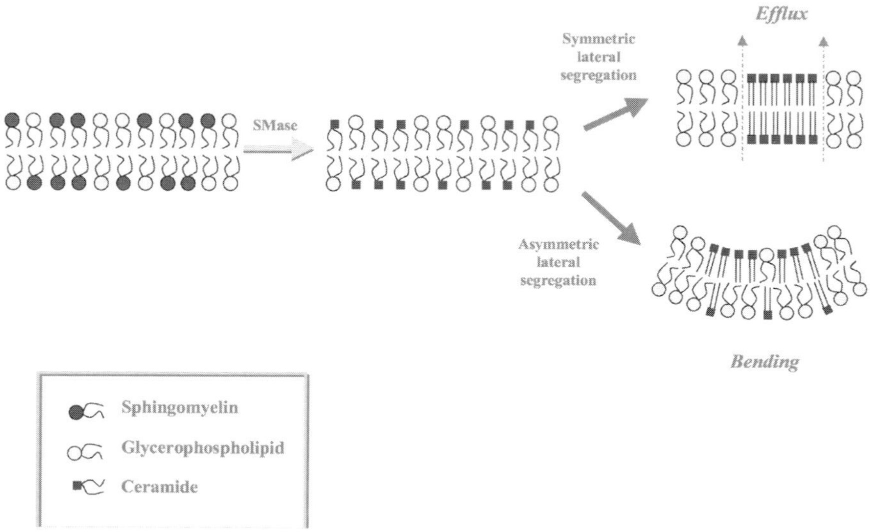

Fig. 5.3 Schematic drawing of the effects of SMase on lipid bilayers. In a bilayer containing both sphingomyelin and glycerophospholipids, sphingomyelin will be in the fluid state [119]. After SMase action, ceramides are still mixed with the glycerophospholipids, but the bilayer is in a metastable state. Even if SMase is acting only on one side of the bilayer, ceramide flip-flop tends to distribute ceramide molecules more or less evenly across the bilayer [94, 126]. Simultaneously, and perhaps competing with flip-flop, lateral segregation forms patches of solid ordered ceramide-rich phases. The outcome of these events is increased membrane *efflux* through the solid–fluid interfaces [91, 137] and/or the *bending* of the bilayer, perhaps contributing to the formation of "stalk"-type fusion/fission intermediates [94, 126].

5.3.2.3 **Effects on Membrane Aggregation and Fusion**

Previous studies [139, review in 140] had shown that phospholipase C induced aggregation and fusion of vesicles composed of phosphatidylcholine, phosphatidylethanolamine and cholesterol (2 : 1 : 1 mole ratio). Interestingly, bacterial SMase acting on vesicles containing sphingomyelin, phosphatidylethanolamine and cholesterol (2 : 1 : 1 mole ratio) produced (apart from solute efflux, as discussed above) vesicle aggregation, but not fusion, the latter being defined as intervesicular mixing of lipids and of aqueous contents. However, when the sphingomyelin-containing vesicles were treated with low amounts of SMase, so that the rate of sphingomyelin hydrolysis was reduced to considerably [141] lower than in the previously discussed experiments, about one-third of the vesicles were clearly larger (by ca. six-fold) than the original ones. Fusion had taken place together with extensive leakage; for this and other reasons SMase-induced fusion might occur via a different mechanism than phospholipase C-induced fusion [140, 141].

Other lines of evidence also point towards a fusogenic capacity of SMase. Holopainen et al. [142] showed microscopic images of "endocytic" budding of vesicles composed of phosphatidylcholine and sphingomyelin, upon addition of sphingomyelinase. Budding and fission are probably mirror-image processes of fusion, and they may share similar intermediates. The authors attributed the observed budding to the tendency of ceramide to separate into domains, and to its negative spontaneous curvature that would lead to membrane invagination (Fig. 5.3). An interesting development of the previous experiments was described by Nurminen et al. [127], who coupled SMase to amino-derivatised acrylate microspheres in such a way that the enzyme retained its catalytic activity. The spheres were held inside a micropipette, and could be brought into contact with a giant liposome membrane composed of phosphatidylcholine and sphingomyelin. Thus the enzyme could be used topically, i.e. on a defined region of the membrane, and the consequences of the enzyme reaction examined by light microscopy. The enzyme induced microdomain formation, followed by domain clustering (capping) in the membrane, and subsequent shedding of small vesicles from the membrane into the interior of the giant liposome.

Also relevant is the observation by Zha et al. [143] that endocytotic vesicles are formed in the absence of ATP when fibroblasts and macrophages are treated with exogenous SMase (or ceramides). According to those authors, addition of *B. cereus* SMase to ATP-depleted cells induces the rapid formation of vesicles, ca. 400 nm in diameter, not enriched in clathrin or caveolin, which pinch off from the plasma membrane and go into the cytosol. They speculate that hydrolysis of sphingomyelin in the plasma membrane causes inward curvature and subsequent formation of sealed vesicles, in agreement with the biophysical data [140]. Moreover, entry of *Neisseria gonorrhoeae* into non-phagocytic cells is mediated by the activation of an acidic sphingomyelinase [144], while SMases in the genus *Listeria* mediate bacterial escape from phagocytic vacuoles and cell-to-cell propagation [75, 86]. These examples of bacterial movements across membranes require either true membrane fusion or events mechanistically related to fusion, all of them induced by a SMase. In a non-membranous system, SMase was seen to induce aggregation and fusion of LDL particles [145]. In turn, the modified LDL particles bound the human aortic proteoglycans with increased strength.

Returning to model membranes, Ruiz-Argüello et al. [94] have observed the cooperative action of ceramide and diacylglycerol, generated in situ by SMase and phospholipase C, respectively, to induce aggregation and fusion of vesicles containing equimolar mixtures of sphingomyelin, phosphatidylcholine, phosphatidylethanolamine, and cholesterol. Neither enzyme could produce either aggregation or fusion on its own under those conditions, but when both were added their activities appeared to be mutually potentiated. In a conceptually related study on lipid monolayers at the air–water interface, Fanani and Maggio [146] demonstrated that the activities of SMase and phospholipase A_2 could be mutually modulated. The activity of one enzyme was affected by its own reaction products and by substrates and products of the other enzyme. These authors concluded that the mutual lipid-mediated interfacial modulation between both phosphohydrolytic pathways indicated that phospholipid degradation might be self-amplified or dampened depending on subtle changes of surface pressure and composition.

5.4 Acknowledgments

The authors are grateful to the Ministerio de Ciencia y Tecnología (grant no. BMC2001-0791) and the Universidad del País Vasco (grant no. 13552/2001) for continuing support. The assistance of Dr M.I. Collado with the reaction mechanisms is greatly appreciated.

5.5 References

1 Barenholz, Y., Roitman, A., and Gatt, S., *J. Biol. Chem.*, **1966b**, *241*, 3731–3737.
2 Schneider, P.B. and Kennedy, E.P., *J. Lipid Res.*, **1967**, *8*, 202–209.
3 Okazaki, T., Bell, R.M., and Hannun, Y.A., *J. Biol. Chem.*, **1989**, *264*, 19076–19080.
4 Okazaki, T., Bielawska, A., Bell, R.M., and Hannun, Y.A., *J. Biol. Chem.*, **1990**, *265*, 15823–15831.
5 Kolesnick, R.N., *Prog. Lipid Res.*, **1991**, *30*, 1–38.
6 Mathias, S., Dressler, K.A., and Kolesnick, R.N., *Proc. Natl. Acad. Sci. USA*, **1991**, *88*, 10009–10013.
7 Gómez-Muñoz, A., *Biochim. Biophys. Acta*, **1998**, *1391*, 92–109.
8 Kronke, M., *Chem. Phys. Lipids*, **1999**, *102*, 157–166.
9 Kolesnick, R.N., Goñi, F.M., and Alonso, A., *J. Cell Physiol.*, **2000**, *184*, 285–300.
10 Venkataraman, K. and Futerman, A.H., *Trends Cell Biol.*, **2000**, *10*, 408–412.
11 Ohanian, J. and Ohanian, V., *Cell Mol. Life Sci.*, **2001**, *58*, 2053–2068.
12 Pettus, B.J., Chalfant, C.E., and Hannun, Y.A., *Biochim. Biophys. Acta*, **2002**, *1585*, 114–125.
13 Hofmann, K. and Dixit, V.M., *Trends Biochem. Sci.*, **1998**, *23*, 374–377.
14 Goñi, F.M. and Alonso, A., *FEBS Lett.*, **2002**, *531*, 38–46.
15 Samet, D. and Barenholz, Y., *Chem. Phys. Lipids*, **1999**, *102*, 65–77.
16 Stoffel, W., *Chem. Phys. Lipids*, **1999**, *102*, 107–121.
17 Quintern, L.E., Schuchman, E.H., Levran, O., Suchi, M., Ferlinz, K., Reinke, H., Sandhoff, K., and Desnick, R.J., *EMBO J.*, **1989**, *8*, 2469–2473.
18 Schuchman, E.H., Levran, O., Suchi, M., and Desnick, R.J., *Nucleic Acids Res.*, **1991**, *19*, 3160.
19 Newrzella, D. and Stoffel, W., *Biol. Chem. Hoppe Seyler*, **1992**, *373*, 1233–1238.
20 Bartelsen, O., Lansmann, S., Nettersheim, M. Lemm, T., Ferlinz, K., and Sandhoff, K., *J. Biotechnol.*, **1998**, *63*, 29–40.
21 Lansmann, S., Bartelsen, O., and Sandhoff, K., *Methods Enzymol.*, **2000**, *311*, 149–156.
22 Lansmann, S., Schuette, C.G., Bartelsen, O., Hoernschemeyer, J., Linke, T., Weisgerber, J., and Sandhoff, K., *Eur. J. Biochem.*, **2003**, *270*, 1076–1088.
23 He, X., Miranda, S.R., Xiong, X., Dagan, A., Gatt, S., and Schuchman, E.H., *Biochim. Biophys. Acta*, **1999**, *1432*, 251–264.
24 Simon, C.G. Jr., Chatterjee, S., and Gear, A.R., *Thromb. Res.*, **1998**, *90*, 155–161.
25 Romiti, E., Vasta, V., Meacci, E., Farnararo, M., Linke, T., Ferlinz, K., Sandhoff, K., and Bruni, P., *Mol. Cell Biochem.*, **2000**, *205*, 75–81.
26 Kusuda, S., Cui, C.Y., Takahashi, M., and Tezuka, T., *J. Invest. Dermatol.* **1998**, *111*, 733–738.
27 Grassme, H., Jekle, A., Riehle, A., Schwartz, H., Berger, J., Sandhoff, K., Kolesnick, R., and Gulbins, E., *J. Biol. Chem.*, **2001**, *276*, 20589–20596.
28 Cremesti, A., Paris, F., Grassme, H., Holler, N., Tschopp, J., Fucks, Z., Gul-

bins, E., and Kolesnick, R., *J. Biol. Chem.*, **2001**, *276*, 23954–23961.

29 Utermohlen, O., Karow, U., Lohler, J., and Kronke, M., *J. Immunol.*, **2003**, *170*, 2621–2628.

30 Gulbins, E. and Kolesnick R., *Subcell. Biochem.*, **2002**, *36*, 229–244.

31 Schissel, S.L., Jiang, X., Tweedie-Hardman, J., Jeong, T., Camejo, E.H., Najib, J., Rapp, J.H., Williams, K.J., and Tabas, I., *J. Biol. Chem.*, **1998**, *273*, 2738–2746.

32 Schissel, S.L., Kesler, G.A., Schuchman, E.H., Williams, K.J., and Tabas, I., *J. Biol. Chem.*, **1998**, *273*, 18250–18259.

33 Tabas, I., *Chem. Phys. Lipids*, **1999**, *102*, 123–130.

34 Xu, X. and Tabas, I., *J. Biol. Chem.*, **1991**, *266*, 24849–24858.

35 Leventhal, A.R., Chen, W., Tall, A.R., and Tabas, I., *J. Biol. Chem.*, **2001**, *276*, 44976–44983.

36 Chatterjee, S., *Chem. Phys. Lipids*, **1999**, *102*, 79–96.

37 Hannun, Y.A. and Luberto, C., *Trends Cell Biol.*, **2000**, *10*, 73–80.

38 Liu, B., Andrieu-Abadie, N., Levade, T., Zhang, P., Obeid, L.M., and Hannun, Y.A., *J. Biol. Chem.*, **1998**, *273*, 11313–11320.

39 Seguí, B., Cuvillier, O., Adam-Klages, S., García, V., Malagarie-Cazenave, S., Leveque, S., Caspar-Bauguil, S., Coudert, J., Salvayre, R., Kronke, M., and Levade, T., *J. Clin. Invest.*, **2001**, *108*, 143–151.

40 Haimowitz-Friedman, A., Kan, C.C., Ehleiter, D., Persand, R.S., McLoughlin, M., Fuks, Z., and Kolesnick, R.N., *J. Exp. Med.*, **1994**, *180*, 525–535.

41 Czarny, M., Liu, J., Oh, P., and Schnitzer, J.E., *J. Biol. Chem.*, **2003**, *278*, 4424–4430.

42 Gatt, S., *Biochem. Biophys. Res. Commun.*, **1976**, *68*, 235–241.

43 Rao, B.G. and Spence, M.W., *J. Lipid Res.*, **1976**, *17*, 506–515.

44 Liu, B., Hassler, D.F., Smith, G.K., Weaver, K., and Hannun, Y.A., *J. Biol. Chem.*, **1998**, *273*, 34472–34479.

45 Liu, B. and Hannun, Y.A., *Methods Enzymol.*, **2000**, *311*, 156–164.

46 Sawai, H. and Hannun, Y.A., *Chem. Phys. Lipids*, **1999**, *102*, 141–147.

47 Bernardo, K., Krut, O., Wiegmann, K., Kreder, D., Micheli, M., Schafer, R., Sickman, A., Schmidt, W.E., Schroder, J.M., Meyer, H.E., Sandhoff, K., and Kronke, M., *J. Biol. Chem.*, **2000**, *275*, 7641–7647.

48 Ghosh, N., Sabbadini, R., and Chatterjee, S., *Mol. Cell Biochem.*, **1998**, *189*, 161–168.

49 Sabbadini, R.A., Betto, R., Teresi, A., Fachechi-Cassano, G., and Salviati, G., *J. Biol. Chem.*, **1992**, *267*, 15475–15484.

50 Raimann, P.E., Custodio de Souza, I.C., Bernard, E.A., and Guma, F.C., *Mol. Cell Biochem.*, **1999**, *201*, 125–129.

51 Ella, K.M., Qi, C., Dolan, J.W., Thompson, R.P., and Meier, K.E., *Arch. Biochem. Biophys.*, **1997**, 101–110.

52 Hanada, K., Mitamura, T., Fukasawa, M., Magistrado, P.A., Horii, T., and Nishijima, M., *Biochem. J.*, **2000**, *346*, 671–677.

53 Tomiuk, S., Hofmann, K., Nix, M., Zumbansen, M., and Stoffel W., *Proc. Natl. Acad. Sci. USA*, **1998**, *95*, 3638–3643.

54 Sawai, H., Domae, N., Nagan, N., and Hannun, Y.A., *J. Biol. Chem.*, **1999**, *274*, 38131–38139.

55 Neuberger, Y., Shogomori, H., Levy, Z., Fainzilber, M., and Futerman, A.H., *FEBS Lett.*, **2000**, *469*, 44–46.

56 Tomiuk, S., Zumbansen, M., and Stoffel, W., *J. Biol. Chem.*, **2000**, *275*, 5710–5717.

57 Hofmann, K., Tomiuk, S., Wolff, G., and Stoffel, W., *Proc. Natl. Acad. Sci. USA*, **2000**, *97*, 5895–5900.

58 Fensome, A.C., Rodrigues-Lima, F., Josephs, M., Paterson, H.F., and Katan, M., *J. Biol. Chem.*, **2000**, *275*, 1128–1136.

59 Josephs, M., Katan, M., and Rodrigues-Lima, F., *FEBS Lett.*, **2002**, *531*, 329–334.

60 Rodrigues-Lima, F., Fensome, A.C., Josephs, M., Evans, J., Veldman, R.J., and Katan, M., *J. Biol. Chem.*, **2000**, *275*, 28316–28325.

61 Fensome, A.C., Josephs, M., Katan, M., and Rodrigues-Lima, F., *Biochem. J.*, **2002**, *365*, 69–77.

62 Mizutani, Y., Tamiya-Koizumi, K., Irie, F., Hirabayashi, Y., Miwa, M., and Yoshida, S., *Biochim. Biophys. Acta*, **2000**, *1485*, 236–246.

63 Mizutani, Y., Tamiya-Koizumi, K., Nakamura, N., Kobayashi, M., Hirabayashi, Y., and Yoshida, S., *J. Cell Sci.*, **2001**, *114*, 3727–3736.

64 Marchesini, N., Luberto, C., and Hannun, Y.A., *J. Biol. Chem.*, **2003**, *278*, 13775–13783.

65 Veldman, R.J., Maestre, N., Aduib, O.M., Medin, J.A., Salvayre, R., and Levade, T., *Biochem. J.*, **2001**, *355*, 859–868.

66 Martin, S.F., Navarro, F., Forthoffer, N., Navas, P., and Villalba, J.M., *J. Bioenerg. Biomembr.*, **2001**, *33*, 143–153.

67 Martin, S.F., Gómez-Díaz, C., Navas, P., and Villalba, J.M., *Biochem. Biophys. Res. Commun.*, **2002**, *297*, 581–586.

68 Rousseau, A., Livni, N., and Gatt, S., *Arch. Biochem. Biophys.*, **1986**, *244*, 838–845.

69 Holopainen, J.M., Medina, O.P., Metso, A.J., and Kinnunen, P.K., *J. Biol. Chem.*, **2000**, *275*, 16484–16489.

70 Kinnunen, P.K. and Holopainen, J.M., *Trends Cardiovasc. Med.*, **2002**, *12*, 37–42.

71 Sawai, H., Okamoto, Y., Lubert, C., Mao, C., Bielawska, A., Domae, M., and Hannun, Y.A., *J. Biol. Chem.*, **2000**, *275*, 39793–39798.

72 Okamoto, Y., Vaena de Avalos, S., and Hannun, Y.A., *J. Biol. Chem.*, **2002**, *277*, 46470–46477.

73 Fujii, S., Inoue, B., Yamamoto, H., Ogata, K., Shinki, T., Inoue, S., Tomita, M., Tamura, H., Tsukamoto, K., Ikezawa, H., and Ikeda, K., *J. Biochem.*, **1998**, *124*, 1178–1187.

74 Fujii, S., Ogata, K., Inoue, S., Murakami, M., Iwama, S., Katsumura, S., Tomita, M., Tamura, H., Tsukamoto, K., Ikezawa, H., and Ikeda, K., *J. Biochem.*, **1999**, *126*, 90–97.

75 Gonzalez-Zorn, B., Dominguez-Bernal, G., Suarez, M., Ripio, M.T., Vega, Y., Novella, S., and Vazquez-Boland, J.A., *Mol. Microbiol.*, **1999**, *33*, 510–523.

76 Chan, E.C., Chang, C.C., Li, Y.S., Chang, C.A., Chiou, C.C., and Wu, T.Z., *Biochemistry*, **2000**, *39*, 4838–4845.

77 Sueyoshi, N., Kita, K., Okino, N., Sakaguchi, K., Nakamura, T., and Ito, M., *J. Bacteriol.*, **2002**, *184*, 540–546.

78 Matsuo, Y., Yamada, A., Tsukamoto, K., Tamura, H., Ikezawa, H., Nakamura, H., and Nishikawa, K., *Protein Sci.*, **1996**, *5*, 2459–2467.

79 Heinz, D.W., Essen, L.O., and Williams, R.L., *J. Mol. Biol.*, **1998**, *275*, 635–650.

80 Birbes, H., El Bawab, S., Hannun, Y.A., and Obeid, L.M., *FASEB J.*, **2001**, *14*, 2669–2679.

81 Okazaki, T., Bielawska, A., Domae, N., Bell, R.M., and Hannun, Y.A., *J. Biol. Chem.*, **1994**, *269*, 4070–4077.

82 Cheng, Y., Nilsson, A., Tomquist, E., and Duan, R.D., *J. Lipid Res.*, **2002**, *43*, 316–324.

83 Liu, J.J., Nilsson, A., and Duan, R.D., *Lipids*, **2002**, *37*, 469–474.

84 Nilsson, A. and Duan, R.D., *Chem. Phys. Lipids*, **1999**, *102*, 97–105.

85 Titball, R.W., *Microbiol Rev.*, **1993**, *57*, 347–366.

86 Zückert, W.R., Marquis, H., and Goldfine, H., *Infect Immun.*, **1998**, *66*, 4823–4831.

87 Vazquez-Boland, J.A., Kuhn, M., Berche, P., Chakraborty, T., Dominguez-Bernal, G., Goebel, W., Gonzalez-Zorn, B., Wehland, J., and Kreft, J., *Clin. Microbiol Rev.*, **2001**, *14*, 584–640.

88 Fanani, M.L. and Maggio, B., *J. Lipid Res.*, **2000**, *41*, 1832–1840.

89 Tamura, H., Tameishi, K., Yamada, A., Tomita, M., Matsuo, Y., Nishikawa, K., and Ikezawa, H., *Biochem. J.*, **1995**, *309*, 757–764.

90 Weston, S.A., Lahm, A., and Suck, D., *J. Mol. Biol.*, **1992**, *226*, 1237–1256.

91 Ruiz-Argüello, M.B., Basañez, G., Goñi, F.M., and Alonso, A., *J. Biol. Chem.*, **1996**, *271*, 26611–26621.

92 Goñi, F.M., Villar, A.V., Nieva, J.L., and Alonso, A., *Methods Enzymol.*, **2003**, *372*, 3–19.

93 Little, C. and Otnass, A.B., *Biochim. Biophys. Acta*, **1975**, *391*, 326–333.

94 Ruiz-Argüello, M.B., Goñi, F.M., and Alonso, A., *J. Biol. Chem.*, **1998**, *273*, 22977–22982.

95 Santana, P., Fanjul, L.F., and Ruiz de Galarreta, C.M., *Methods Mol. Biol.*, **1998**, *105*, 217–221.

96 Gatt, S., *Chem. Phys. Lipids*, **1999**, *102*, 45–53.

97 Loidl, A., Claus, R., Deigner, H.P., and Hermetter, A., *J. Lipid Res.*, **2002**, *43*, 815–823.

98 He, X., Chen, F., Dagan, A., Gatt, S., and Schuchman, E.H., *Anal. Biochem.*, **2003**, *314*, 116–120.

99 Liu, B. and Hannun, Y.A., *Methods Enzymol.*, **2000**, *311*, 164–167.

100 Zhou, X., Turecek, F., Scott, C.R., and Gelb, M.H., *Clin. Chem.*, **2001**, *47*, 874–881.

101 Barbone, A.G., Jackson, A.C., Ritchie, D.M., and Argentieri, D.C., *Methods Enzymol.*, **2000**, *311*, 168–176.

102 Hassler, D.F., Laethem, R.M., and Smith, G.K., *Methods Enzymol.*, **2000**, *311*, 176–184.

103 Nara, F., Tanaka, M., Hosoya, T., Suzuki-Konagai, K., and Ogita, T., *J. Antibiot. (Tokyo)*, **1999**, *52*, 525–530.

104 Nara, F., Tanaka, M., Masuda-Inoue, S., Yamasato, Y., Doi-Yoshioka, H., Suzuki-Honagai, K., Kumakura, S., and Ogita, T., *J. Antibiot. (Tokyo)*, **1999**, *52*, 531–535.

105 Arenz, C., Gartner, M., Wascholowshi, V., and Giannis, A., *Bioorg. Med. Chem.*, **2001**, *9*, 2901–2904.

106 Tanaka, M., Nara, F., Yamasato, Y., Ono, Y., and Ogita, T., *J. Antibiot. (Tokyo)*, **1999**, *52*, 827–830.

107 Uchida, R., Tomoda, H., Dong, Y.S., and Omura, S., *J. Antibiot.*, **1999**, *52*, C4.

108 Uchida, R., Tomoda, H., Arai, M., and Omura, S., *J. Antibiot.*, **2001**, *54*, 882–889.

109 Arenz, C., Thutewohl, M., Block, O., Waldmann, H., Altenbach, H.J., and Giannis, A., *Chembiochem*, **2001**, *2*, 141–143.

110 Hakogi, T., Monden, Y., Iwama, S., and Katsumura, S., *Org. Lett.*, **2000**, *2*, 2627–2629.

111 Yokomatsu, T., Takechi, H., Akiyama, T., Shibuya, S., Kominato, T., Soeda, S., and Shimeno, H., *Bioorg. Med. Chem. Lett.*, **2001**, *11*, 1277–1280.

112 Hakogi, T., Monden, Y., Taichi, M., Iwama, S., Fujii, S., Ikeda, K., and Katsumura, S., *J. Org. Chem.*, **2002**, *67*, 4839–4846.

113 Arenz, C. and Giannis, A., *Angew. Chem. Int. Ed.*, **2000**, *39*, 1440–1442.

114 Luberto, C., Hassler, D.F., Signorelli, P., Okamoto, Y., Sawai, H., Boros, E., Hazen-Martin, D.J., Obeid, L.M., Hannun, Y.A., and Smith, G.K., *J. Biol. Chem.*, **2002**, *277*, 41128–41139.

115 Berkovic, D., Berkovic, K., Binder, C., Haase, D., and Fleer, E.A., *J. Exp. Ther. Oncol.*, **2002**, *2*, 213–218.

116 Okudaira, C., Ikeda, Y., Kondo, S., Furuya, S., Hirabayashi, Y., Koyano, T., Saito, Y., and Umezawa, K., *J. Enzyme Inhib.*, **2000**, *15*, 129–138.

117 Iikubo, K., Ishikawa, Y., Ando, N., Umezawa, K., and Nishiyama, S., *Tetrahedron Lett.*, **2002**, *43*, 291–293.

118 Linke, T., Wilkening, G., Lansmann, S., Moczall, H., Bartelsen, O., Weisgerber, J., and Sandhoff, K., *Biol. Chem.*, **2001**, *382*, 283–290.

119 Ruiz-Argüello, M.B., Veiga, M.P., Arrondo, J.L., Goñi, F.M., and Alonso, A., *Chem. Phys. Lipids*, **2002**, *114*, 11–20.

120 Jaffrezou, J.P., Maestre, N., de Mas-Mansat, V., Bezombes, C., Levade, T., and Laurent, G., *FASEB J.*, **1998**, *12*, 999–1006.

121 Fanani, M.L., Hartel, S., Oliveira, R.G., and Maggio, B., *Biophys. J.*, **2002**, *83*, 3416–3424.

122 Huang, H., Goldberg, E.M., and Zidovetzki, R., *Biochem. Biophys. Res. Commun.*, **1996**, *220*, 834–838.

123 Huang, H., Goldberg, E.M., and Zidovetzki, R., *Eur. Biophys. J.*, **1998**, *27*, 361–366.

124 Huang, H., Goldberg, E.M., and Zidovetzki, R., *Biophys. J.*, **1999**, *77*, 1489–1497.

125 Holopainen, J.M., Lehtonen, J.Y.A., and Kinnunen, P.K.J., *Chem. Phys. Lipids*, **1997**, *88*, 1–13.

126 Holopainen, J.M., Subramanian, M., Kinnunen, P.K.J., *Biochemistry*, **1998**, *37*, 17562–17570.

127 Nurminen, T.A., Holopainen, J.M., Zhao, H., and Kinnunen, P.K., *J. Am. Chem. Soc.*, **2002**, *124*, 12129–12134.

128 Veiga, M.P., Arrondo, J.L.R., Goñi, F.M., and Alonso, A., *Biophys. J.*, **1999**, *76*, 342–350.

129 Lehtonen, J.Y.A., Rytömaa, M., and Kinnunen, P.J.K., *Biophys. J.*, **1996**, *70*, 2185–2194.

130 Wang, T.Y. and Silvius, J.R., *Biophys. J.*, **2003**, *84*, 367–378.

131 Carrer, D.C. and Maggio, B., *J. Lipid Res.*, **1999**, *40*, 1978–1989.

132 Moore, D.J., Rerek, M.E., and Mendelsohn, R., *J. Phys. Chem. B*, **1997**, *101*, 8933–8940.

133 Slotte, J.P., *Biochim. Biophys. Acta*, **1995**, *1235*, 419–427.

134 Shah, J., Atienza, J.M. Jr., Duclos, R.I., Rawlings, A.V., Dong, Z., and Shipley, G.G., *J. Lipid Res.*, **1995**, *36*, 1936–1944.

135 Patra, S.K., Alonso, A., Arrondo, J.L.R., and Goñi, F.M., *J. Liposome Res.*, **1999**, *9*, 247–260.

136 Cremesti, A.E., Goñi, F.M., and Kolesnick, R., *FEBS Lett.*, **2002**, *531*, 47–53.

137 Montes, L.R., Ruiz-Argüello, M.B., Goñi, F.M., and Alonso, A., *J. Biol. Chem.*, **2002**, *277*, 11788–11794.

138 Siskind, L.J., Kolesnick, R.N., and Colombini, M., *J. Biol. Chem.*, **2002**, *277*, 26796–26803.

139 Nieva, J.L., Goñi, F.M., and Alonso, A., *Biochemistry*, **1989**, *28*, 7364–7367.

140 Goñi, F.M. and Alonso, A., *Biosci. Rep.*, **2000**, *20*, 443–463.

141 Basañez, G., Ruiz-Argüello, M.B., Alonso, A., Goñi, F.M., Karlsson, G., and Edwards, K., *Biophys. J.*, **1997**, *72*, 2630–2637.

142 Holopainen, J.M., Angelova, M.I., and Kinnunen, P.K., *Biophys. J.* **2000**, *78*, 830–838.

143 Zha, X., Pierini, L.M., Leopold, P.L., Skiba, P.J., Tabas, I., and Maxfield, F.R., *J. Cell Biol.*, **1998**, *140*, 39–47.

144 Grassme, H., Gulbins, E., Brenner, B., Ferlinz, K., Sandhoff, K., Harzer, K., Lang, F., and Meyer, T.F., *Cell*, **1997**, *91*, 605–615.

145 Oorni, K., Hakala, J.K., Annila, A., Ala-Korpela, M., and Kovanen, P.T., *J. Biol. Chem.*, **1998**, *273*, 29127–29134.

146 Fanani, M.L. and Maggio, B., *Lipids*, **1998**, *33*, 1079–1087.

6 Glycosyl-phosphatidylinositol Cleavage Products in Signal Transduction

Yolanda León and Isabel Varela-Nieto

6.1 Introduction

Glycosyl-phosphatidylinositol (GPI) molecules have two well-defined functions within living cells: GPIs anchor proteins to the plasma membrane and free GPI participates in cellular activation. The hydrolysis of GPI generates a diffusible mediator, the inositol phosphoglycan (IPG) that has biological activity. The first agonist reported to induce GPI hydrolysis, insulin, raised much interest as IPG molecules can mimic insulin actions, opening an unexpected field for insulin and diabetes researchers. However, the results of the past decade have not fulfilled the initial expectations. IPG research has been impaired by the difficulties found in the structural characterization of these molecules and also by the deficient or absent characterization of the enzymes involved in free-GPI and IPG metabolism [1]. In contrast, the characterization of the biological actions of both natural and synthetic IPGs has advanced considerably. Based on compositional data obtained from metabolically labelled cells and on the structural information available on GPI anchors, Müller's group carried out the synthesis of IPG analogues and demonstrated that synthetic IPG molecules present a complete panel of insulin-like activities [2–6]. This complete and elegant work backed many of the results reported on natural IPG insulin-like activities *in vivo* and on cultured cells. Parallel work by several research groups has uncovered other aspects of the GPI/IPG signaling pathway that will be discussed in this chapter. However, further research on the localization, metabolism and signaling cross talk with other pathways is urgently required to fully understand the *in vivo* relevance of these molecules in insulin regulation.

In addition to GPI/IPG regulation by insulin, there is much evidence that the hydrolysis of GPI and the concomitant release of IPG are early steps in the cellular response to other growth factors, hormones and cytokines in various cell types and species [7, 8]. These results raise the question of the specificity in the response to extracellular factors. In this regard, the data available on IPG composition suggest that there is a group of closely related molecular species, termed generically IPG, that are released in a tissue-specific manner and have cell-specific targets. Indeed, chemically synthesised IPG possess distinct biological activities, depending on the oligosaccharide structure [4, 9–15]. Treatment with natural or

Lipases and Phospholipases in Drug Development
Edited by G. Müller and S. Petry

ISBN: 3-527-30677-3

synthetic IPG of cellular extracts, intact cells or organotype cultures elicits both short-term metabolic actions and long-term effects on cell growth and differentiation [1–5, 7, 9, 15–17]. The intracellular targets for GPI/IPG signaling include other second messenger molecules, such as cAMP, calcium ions and PI3K, therefore the GPI/IPG pathway is closely connected with the intracellular circuit of molecules that participate in cell activation.

6.2
GPI Structure and Hydrolysis by Specific Phospholipases

GPIs are an abundant and ubiquitous class of eukaryotic glycolipids. Although these structures were originally discovered in the form of GPI-anchored cell surface glycoproteins, a significant proportion of the GPI synthetic output of the cell is not directed to protein anchoring but to free GPIs [18]. GPI anchors have been the subject of several reviews [1, 8, 18–21].

Preliminary structural analysis showed that the rat liver free glycolipid contained a phosphatidylinositol molecule glycosidically linked to a glucosamine residue, which was in turn bound to a glycan tail, whose composition and size have not yet been fully elucidated. Free-GPI phospholipids participating in signal transduction and GPI protein anchors share some structural motifs but have a different glycan composition. Free-GPI lacks the trimannose motif and the ethanolamine phosphate. A simplified model of the similarities and differences between the GPI anchors of membrane proteins and free-GPI phospholipids is shown in Fig. 6.1.

Several phospholipases are able to cleave free-GPI *in vitro*: bacterial phosphatidylinositol phospholipase C (PI-PLC), *Trypanosoma brucei* GPI-PLC and mammalian GPI-PLD. It was initially proposed that a PLC was the enzyme involved in GPI hydrolysis in mammalian cells as there was a rapid and transient intracellular rise in the level of diacylglycerol and on the ability of bacterial PI-PLC to partially mimic the effects of insulin, among other extracellular ligands, through the generation of IPG. However, a mammalian GPI-specific phospholipase C (PLC) has not yet been purified nor cloned. In contrast, a mammalian GPI-PLD has been cloned and it is an abundant, although inactive, serum protein [22]. Furthermore, mammalian GPI-phospholipase D generates *in vitro* biologically active IPG from GPI [23, 24]. The relevance of GPI-PLD in insulin signaling *in vivo* has not been clearly demonstrated; however, the presence of GPI-PLD in serum and in mammalian cells together with its release from pancreatic islet cells in response to insulin secretagogues suggest that GPI-PLD might be one of the enzymes responsible for free-GPI hydrolysis *in vivo* [25–29]. An appealing hypothesis is that differential GPI hydrolysis by phospholipases may give rise to specific IPG species in different cellular contexts, thus providing a further level of regulation in the cellular response. The cloning and characterization of mammalian GPI-lipases will certainly be a step forward in understanding this system. The recent release of the genome data of several species should help in the identification of candidate enzymes.

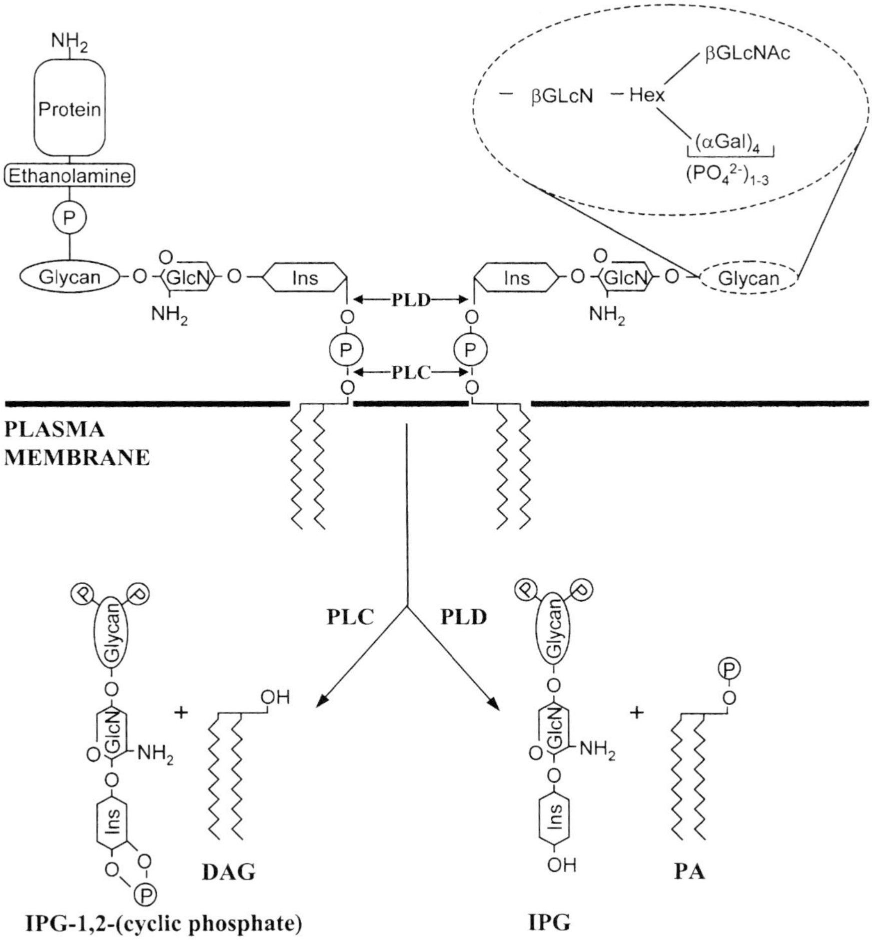

Fig. 6.1 Scheme representing the structure and hydrolysis of glycosyl-phosphatidylinositol (GPI) (adapted from Refs. [1] and [7]). Free GPI and GPI-anchored proteins are represented, showing their similarity in structure. The points for phospholipase C (PLC) and phospholipase D (PLD) hydrolysis are indicated. The glycan component of free GPI is drawn and consists of galactose (Gal), phosphate (PO_4^{2-}), an unidentified residue of hexose (Hex), *N*-acetylglucosamine (GlcNAc) and glucosamine (GlcN). Below, free-GPI products of PLC and PLD hydrolysis: Ins, *myo*-inositol; IPG, inositol phosphoglycan; DAG, diacylglycerol; PA, phosphatidic acid.

Another open issue in GPI hydrolysis by phospholipases is how intracellular receptor activation couples to the hydrolysis of an extracellular plasma membrane lipid. There are many hypotheses that have not been fully explored, some of which have an experimental basis – for example, there are evidences of the presence of both free-GPI and GPI-PLD in intracellular vesicles [27, 30]. Although intracellular GPI-PLD has not been reported to hydrolyse GPI, an intracellular release of IPG would accord with different experimental data, for example with those reporting IPG actions on cell extracts. However, the hypothesis with the best experimental support is the presence of GPI-rich plasma membrane compartments, caveolae, where the activation of receptors could be coupled to PL-mediated GPI hydrolysis and to IPG downstream signaling. Data supporting caveolae as the key compartment for the activation of GPI/IPG signaling include: The presence of GPI in caveolae, the modulation of GPI-PLs activity by the lipidic environment, the activation of different insulin-regulated signaling molecules such as PI-3K and other kinases in caveolae and the existence of IPG carrier proteins in the plasma membrane [4, 16, 24, 32–34].

In summary, there has been a considerable advance in the knowledge of the nature of the specific phospholipases involved in GPI hydrolysis and the mechanism by which these enzymes are regulated and may couple to different receptors. But still many essential aspects of GPI-PLs structure, localization and function have to be elucidated.

6.3 Diffusible Factors and the Regulation of GPI Levels

Hormones, growth factors, neurotrophic factors and cytokines have been reported to induce GPI hydrolysis and/or IPG release in different cell types [8]. Tab. 6.1 [7, 35–41] summarises the diffusible factors reported to stimulate GPI hydrolysis. Among the best-studied GPI agonists are insulin and insulin-like growth factor-I (IGF-I). Insulin, IGFs I and II, their cellular receptors and the specific IGF binding proteins form the insulin-related growth-factor system [42–45], which is involved in the modulation of many physiological processes, including development, growth, reproduction, aging and metabolism. Distantly related factors and receptors have been described in non-mammalian vertebrates and invertebrates [46, 47]. IGF-I, initially known as somatomedin-C, was sequenced in the 1970s and renamed on the basis of its amino acid sequence homology with human proinsulin. A second, structurally distinct insulin-like peptide was found in serum and termed IGF-II. Insulin is a crucial anabolic hormone, produced and secreted to the plasma by the adult pancreas. The IGFs are important growth mediators, produced and secreted to the plasma by the adult liver, among other organs [48, 49]. The cellular actions of these factors are mediated by binding to membrane-bound tyrosine kinase receptors, which were initially characterized during the 1970s. The classic receptors include the insulin receptor (IR) and the IGF type-1 receptor, both consisting of disulfide-linked heterotetramers conformed by two extracellular alpha subunits (binding activity) and two transmembrane beta subunits (tyrosine kinase and signaling activities); these receptors seem to be coupled to overlapping signaling pathways [50, 51].

Insulin-stimulated GPI hydrolysis is strictly dependent on the tyrosine kinase activity of the insulin receptor. The lack of tyrosine kinase activity in the insulin recep-

Tab. 6.1 Stimulation of GPI hydrolysis by growth factors, hormones and cytokines. Adapted and updated from Ref. [7].

Factor	*Cell type*
Insulin	BC_3H-1 myocytes
	H35 hepatoma
	T-lymphocytes
	Rat hepatocytes and adipocytes [35]
	CHO cells
	Human fibroblasts
	Chick embryo fibroblasts [36]
IGF-I	BC_3H-1 myocytes
	Otic vesicle
	Fibroblasts
EGF	BC_3H-1 myocytes
	EGFRT17 fibroblasts
NGF	Chicken cochleovestibular ganglion cells
	PC-12 cells
BDNF and NT-3	Chicken cochleovestibular ganglion cells
TGF-β1	Rabbit articular chondrocytes [37]
	Keratinocytes [38]
Prolactin, FSH and chorionic gonadotropin	Rat granulosa cells
Epo	Rat erythroid progenitor cells [39]
ACTH	Adrenal glomerulosa cells
	Calf and adrenocortical cells [40]
TRH	Pituitary lactotrophes
TSH	Pig thyroid cells [41]
IL-2	T- and B-lymphocytes
IL-4[a]	B-lymphocytes

Abbreviations: ACTH, adrenocorticotropic hormone; BDNF, brain-derived neurotrophic factor; EGF, epidermal growth factor; Epo, erythropoietin; FSH, follicle-stimulating hormone; IGF-I, insulin-like growth factor-I; IL, interleukin; NGF, nerve growth factor; NT-3, neurotrophin-3; TGF-β1, Transforming growth factor-β1; TRH, thyrotropin releasing hormone; TSH, thyrotropin. [a]Blockade of interleukin-2-dependent GPI hydrolysis.

tor is associated with alterations in GPI metabolism and, in turn, a normal GPI metabolism is required for insulin actions [1, 8, 52]. GPI levels and/or hydrolysis are impaired in metabolically altered or aged cells when compared with normal cells, as for example in the streptozotocin-induced rat model of diabetes (reviewed in Ref. [1]). Furthermore, cells overexpressing insulin receptors with mutated tyrosine residues or with a deficient tyrosine kinase domain do not present insulin-stimulated GPI hydrolysis [53, 54]. The activation of GPI-PLs is coupled not only to the insulin receptor but also to other receptors. For example, tyrosine kinase inhibitors can inhibit GPI hydrolysis stimulated by epidermal growth factor [55] and loss-of-activity mutants of the receptor for transforming growth factor-beta cannot generate IPG [56]. Finally, the culture of chicken otic vesicle explants has provided a physiologically relevant model to test the ability of neurotrophins and IGFs to induce the hydrolysis of GPI leading to cellular proliferation [11, 57, 58].

In summary, during cellular activation growth factors, neurotrophins, cytokines and classical hormones induce the rapid and reversible hydrolysis of GPI and the generation of bioactive IPG molecules. The specificity of the response could be explained by the multiplicity of free-GPI and IPG types (see next section), which suggests that different agonists may stimulate the hydrolysis of tissue-specific or structurally distinct GPIs, the generation of signal-specific IPG molecules would induce differential cellular responses. Conversely, IPG alone cannot achieve the complete scope of actions of this panel of diffusible factors, or even of just one. Therefore, the GPI/IPG system should be considered as a novel pathway to be incorporated into the cellular signaling network, whose *in vivo* relevance in the intracellular response to these factors remains an issue of study.

6.4 IPG Structure and Biological Activities

The data available suggest that IPG-like molecules are widely distributed throughout the plant and animal kingdoms, and that IPG can act on cell systems from phylogenetically distinct species [4, 47]. For example, pea IPG and rat, bovine or human liver IPG are active on mollusc neurones, human fibroblasts or chicken embryo epithelium, suggesting that IPG main structural motifs and functions have been conserved throughout evolution [7, 11, 59, 60]. Mammalian IPG has been purified and subjected to chemical analysis from different cell types, organs and biological fluids. These analyses suggest that there are two related oligosaccharidic structures, which have distinct chemical compositions, biological activities and tissue distribution. These two general IPG subtypes have been termed IPG-A and IPG-P [1, 8, 15, 17]. IPG type A was defined as an inhibitor of protein kinase A and adenylate cyclase but an activator of lipogenesis, whereas IPG type P stimulates pyruvate dehydrogenase and glycogen synthase activities [15, 17, 29, 61–63]. Several groups have reported partial analyses of the sequence and structure of natural IPG; however, there are no available data on the structure of any mammalian IPG. This frustrating lack of results could be in part due to the limited amounts and insufficient purification of the IPG preparations from the mammalian tissues analyzed. Taken into account these important limitations, there is a general agreement on that IPG type A contains *myo*-inositol and glucosamine, whereas IPG-P contains α-pinitol (3-*O*-methyl-D-*chiro*-inositol) and galactosamine as core components, both types of IPG contain neutral sugars and phosphate residues. Bioactive IPG-A can be generated by either GPI-PLC or PLD-mediated hydrolysis of *myo*-inositol-containing GPI (Fig. 6.1). In contrast, the metabolic precursor of IPG-P is undefined and a direct generation from a phospholipidic precursor has been disregarded because no reported phospholipase can cleave a phospholipid containing a D-*chiro*-inositol. Despite the metabolic effects described for

IPG-P [63–65], less compositional information is available and its metabolic origin remains undefined.

Because the structures of natural IPG have been elusive, efforts have focussed on the search for the minimal structural motif with insulin-like biological activity. Oligosaccharides containing *myo-* or *chiro*-inositol with a variable degree of complexity have been synthesised using organic chemistry methods [4, 9, 10, 12–14, 66–71] and tested for biological activity [2, 4, 6, 9–14, 16]. In this context, the organotype culture of chick otic vesicle explants has provided a useful model system to rapidly screen the insulin-like activity of minute amounts of synthetic compounds [9, 11, 12, 57, 59]. The availability of synthetic IPGs has allowed the characterization of the structure–function relationship within IPG molecules. For example, the synthetic disaccharide 6-*O*-(2-amino-2-deoxy-*α*-D-glucopyranosyl)-D-*myo*-inositol-1,2-cyclic phosphate has insulin-like biological activity, but the case for the related acyclic-inositol-P disaccharides, such as 6-*O*-(2-amino-2-deoxy-*α*-D-glucopyranosyl)-D-*myo*-inositol-1-phosphate, do not [4, 9, 10, 12, 68]. Inositol, inositol-P, glucosamine, other monosaccharides or disaccharides with a hexosamine-(1→4)-inositol linkage did not present activity either. These results indicate that the glucosaminyl-(1→6)-inositol-1,2-cyclic phosphate is the minimal structural motif required to achieve biological responses. However, some of the actions of natural IPG are not mimicked by this minimal structure and others are mimicked but only up to a level of 30–40% of that induced by insulin at maximal concentrations. The group of Müller has published an extensive characterization of the insulin-like activity of synthetic IPG molecules in adipose, cardiac and muscle cells [2–6, 10, 16, 32–34].

The design and use of insulin-like molecules acting downstream insulin receptor might be useful in treating insulin-resistance states; in this context, IPG is a good candidate molecule as an alternative to insulin treatment. The ability of IPG to mimic insulin actions is certainly striking. Tab. 6.2 summarises insulin-like effects of IPG on both intact cells and cellular extracts [2–5, 8, 10, 32, 34, 72–74]. The biological effects of IPG include tissue-specific regulation of lipolysis, lipogenesis, glycolytic flux, protein synthesis and/or phosphorylation, DNA and RNA synthesis and also long-term actions such as cellular proliferation. IPG-P activates glycerol-3-phosphate acyltransferase [17], but almost all the other insulin-mimetic activities reported were tested for IPG-A prepared *in vitro* by hydrolysis of highly purified GPI [1, 7, 8, 75, 76].

It is worth noting that GPI/IPG alterations have been associated with diabetes and that IPG has been proposed as candidate cofactor to treat diabetes [1]. Besides data obtained in experimental models of diabetes, subjects with non-insulin-dependent diabetes mellitus (type II) appear to have an abnormal regulation of IPG production *in vivo* [17, 65, 77–82]. Dietary treatment with D-*chiro*-inositols may be effective in lowering symptoms of non-insulin-dependent diabetes mellitus (type II) [83]. Therefore, the urinary *chiro*-inositol (IPG-P) to *myo*-inositol (IPG-A) ratio has been proposed as a clinical marker to predict the onset of diabetes at an early stage [1, 4, 15, 84, 85]. IPG has been also proposed to have a regulatory role in placental function during pregnancy, and more specifically in the onset of preeclampsia as well as in polycystic ovarian disease [86–89].

Tab. 6.2 Insulin-mimetic effects of purified IPG, PIG-P and various chemically synthesized analogues in the control of glucose homeostasis, lipid, protein and nucleic acid metabolism (adapted and updated from Ref. [8]).

Intact cells		***Cellular extracts***	
Biological activity	***Effect***	***Enzymatic activity***	***Effect***
Short-term effects			
Glucose transport	Stimulation/ No effect [34]	Glucose-6-phosphatase	Inhibition
Glucose oxidation	Stimulation	Fructose-1,6-biphosphatase	Inhibition
Glucose production	Inhibition	Pyruvate dehydrogenase (PDH)	Stimulation
Fructose-2,6-biphosphate levels	Stimulation	PDH phosphatase	Stimulation
Glycogen synthesis	Stimulation	cAMP kinase	Inhibition
Glycogen synthase kinase-3	Inhibition	cAMP phosphodiesterase	Stimulation
Lactate accumulation	Stimulation	Adenylate cyclase	Inhibition
Glycogen phosphorylase *a*	Inhibition	Glycogen synthase phosphatase 2C	Stimulation [74]
Pyruvate kinase activity	Stimulation	Casein kinase II	Biphasic
cAMP levels	Inhibition	Glycerol-3P acyltransferase	Stimulation
Insulin secretion	Inhibition	ATP citrate lyase	Stimulation
Acetyl CoA carboxylase activity	Stimulation	Galactolipid sulfotransferase	Inhibition
Lipolysis	Inhibition		
Lipogenesis	Stimulation		
Phospholipid methyltransferase	Inhibition		
Steroidogenesis	Stimulation		
Amino acid transport	Stimulation		
Tyrosine aminotransferase activity	No effect		
Protein phosphorylation	Stimulation/ Inhibition [4, 5]		
PI-3 kinase	Stimulation [10, 34, 72]		
Mitogen activated kinase	Stimulation [2, 3, 4, 73]		
Protein kinase B phosphorylation	Stimulation [2, 3, 4, 73]		
Myelin basic protein kinase	Stimulation [2, 3, 4, 73]		
Ion channels	Modulation		
$Ca^{2+}Mg^{2+}$-ATPase	Stimulation		
Calcium entry	Inhibition		
IRS1/2 phosphorylation	Stimulation [4, 32, 34]		

Tab. 6.2 (continued)

Intact cells		*Cellular extracts*	
Biological activity	***Effect***	***Enzymatic activity***	***Effect***
Long-term effects			
Protein synthesis	Stimulation		
Specific mRNA levels	Stimulation/ Inhibition		
DNA and RNA synthesis	Stimulation		
Cellular proliferation	Stimulation		
Cellular differentiation	Stimulation [a)]		

a) Authors unpublished results.

In summary, the participation of the GPI/IPG system in cell activation by insulin is supported by the following evidences: First, insulin induces the hydrolysis of GPI and the release of IPG; second, impaired GPI metabolism or treatment with anti-IPG antibodies neutralize insulin actions; and, third, natural and synthetic IPG have insulin-like activity. Several other diffusible factors activate GPI hydrolysis with the concomitant release of IPG. The next question is at which points IPG second messengers cross talk with the established signal transduction pathways.

6.5 GPI/IPG Pathway and the Intracellular Signaling Circuit

Most of our knowledge on the GPI/IPG signaling pathway has come up from studies on insulin signal transduction [1, 90]. Insulin interacts with the α-subunits of the insulin receptor and thereby activates the tyrosine kinase activity of the β-subunits, which in turn phosphorylate a number of intracellular docking proteins, leading to the activation of intracellular signaling pathways (reviewed in Refs. [45, 90]). Classical pathways activated by insulin include activation of a distinct subset of G proteins including Ras, activation of kinases such as PKC and the Raf/MAP kinase cascade, the activation of phospholipases including PC-PLD and activation of PI3Kinase. Insulin stimulation of glucose utilisation and other metabolic effects seem to be independent of the activation of mitogen-activated-protein kinase or mitogen-activated-protein kinase, two key kinases of the phosphorylation cascade pathway. In contrast, in response to insulin there are rapid changes in phospholipid metabolism and thereby generation of biologically active lipids that serve as intracellular signaling factors that regulate downstream enzymes, including Akt, leading to glucose transport and glycogen synthesis [45, 91–94]. As discussed above, upstream of insulin-sensitive GPI hydrolysing enzymes lies the insulin receptor, which implies that in some way insulin-stimulated tyrosine phosphorylation events control GPI hydrolysis. This tyrosine-dependent GPI hydrolysis should be related to IRS protein expression and function. IPG regulates the translocation of GLUT4 glucose transporters to the plasma membrane and glucose metabolism, whereas IRS-1 and -2 deficient mice present attenuated insulin signal transduction and/or hyperglycaemia [2, 95–97]. The analysis of the GPI/IPG system in these mouse models of insulin resistance has not been reported; however, there are evidences directly linking IRS activation with IPG [4, 16, 32, 34]. For example, PIG-P, an IPG type-A-like molecule isolated from *Saccharomyces cerevisiae*, and synthetic IPG can increase the tyrosine phosphorylation state of IRS-1 without increasing insulin receptor phosphorylation, indicating a clear bypass of the insulin receptor [4, 73]. IPG-mediated tyrosine phosphorylation of IRS could constitute a mechanism of cross talk with the insulin receptor–tyrosine kinase cascade. GPI hydrolysis has also been associated with the activation of G proteins. By using specific inhibitors and other cell signaling tools it has been proposed that insulin binding to its receptor produces the activation of small G-proteins, which in turn activate membrane phospholipases C and/or D to cleave GPIs to water-soluble IPGs [15, 98]. Although the mechanisms linking insulin receptor activation with GPI/IPG downstream signaling are not still elucidated, there is mounting evidence that both natural and synthetic IPGs mimic insulin actions in the modulation of the activity of intracellular protein kinases and phosphatases; therefore, as insulin, IPG regulates the general phosphorylation state of the cell (Tab. 6.2) [15–17, 74, 99–104].

Finally, insulin induces a rapid activation of PI3K through its interaction with the IRS proteins and, interestingly, insulin-dependent activation of PI3K is compartmentalised within cells [35, 105–107]. Several authors have proposed caveolaee as the membrane compartment where a number of insulin-dependent events occur, including IPG generation and action. Evidences favouring this hypothesis include: First, the localization in caveolae of the elements required for signaling, including GPI and IPG, GLUT4 transporters, an IPG transporter and several downstream effectors, including PI3K subtypes and other kinases; and, second, the cross-activation of these elements within caveolae, for example IPG activation of PI3K and other kinases leading to GLUT4 translocation [2, 4, 16, 31, 98, 108]. In addition, the activity and substrate affinity of GPI-PLs are modulated *in vitro* by lipid composition and by the action of PI3K on GPI substrates [24, 109 and unpublished data]. Taken together, these results represent a significant advance in that they suggest that PI3K activation may represent the point at which insulin-stimulated signal transduction via its receptor and GPI/IPG signaling converge. IPG targets include the regulation of the intracellular levels of other signaling molecules (Tab. 6.2) and the induction of cyclins and Fos and Jun transcription factors [11, 57]. AP-1 activation is part of the mechanism that activates cell division in response to insulin and IGF-I during early organogenesis of the avian inner ear. The hydrolysis of membrane glycolipids with the generation of soluble IPG has been characterised as part of the intracellular response to IGF-I and neurotrophins in cultures of developing chicken otic vesicle and cochleo-vestibular

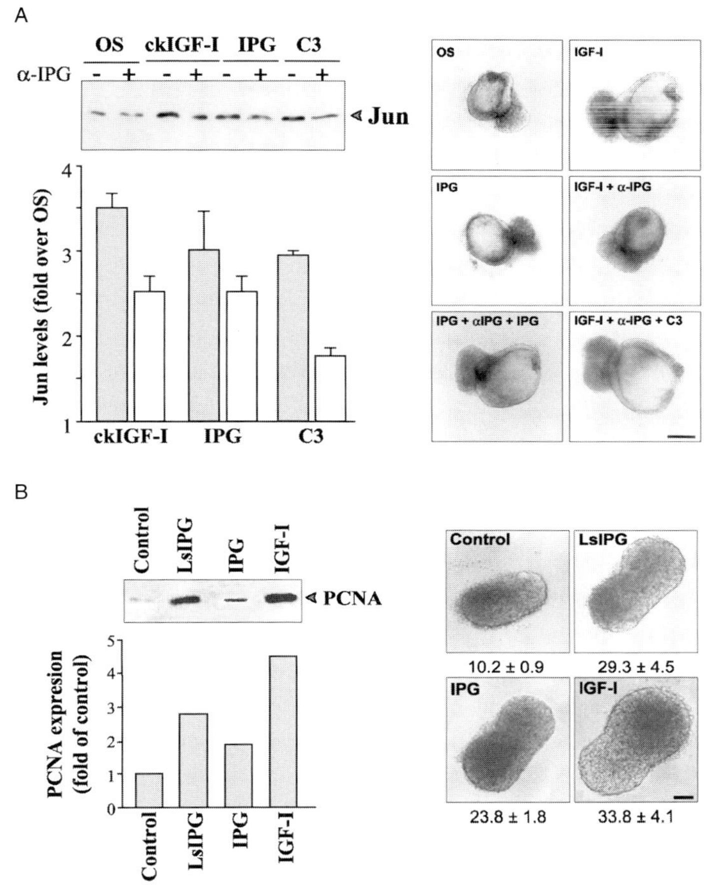

Fig. 6.2 Biological effects of IPG from different sources on otic vesicle and cochleovestibular ganglion. (A) Purified IPG mimicked the effects of IGF-I on Jun expression and could be involved in the mechanism that triggers proliferation in response to insulin and IGF-I during early organogenesis of the avian inner ear. C3, synthetic analogue of IPG; α-IPG, anti-IPG antibody. Scale bar, 125 µm. Reprinted with permission from Ref. [11]: *J. Comparative Neurol.*, © 2003. (B) Purified IPG from rat (IPG) or from *Lathyrus sativus* (LsIPG) showed similar proliferative effects on organotype cultures of cochleovestibular ganglion (CVG). Proliferation levels were estimated by the measurement of PCNA, a nuclear protein essential for DNA synthesis. The numbers below each photomicrograph represent CVG volume size (arbitrary units). Scale bar, 100 µm. Reprinted with permission from Ref. [59]: *Mol. Med.*, © 2003.

ganglia (Fig. 6.2) [11, 57, 110, 111]. IGF-I-induced cell proliferation and Fos/Jun expression are blocked by anti-IPG antibodies, and blockage is rescued by the addition of rat liver purified IPG-A or a *myo*-inositol-containing analogue [11, 57].

In summary, the activation of tyrosine kinase receptors induces the hydrolysis of GPI and the concomitant release of IPG, the mechanisms by which these two

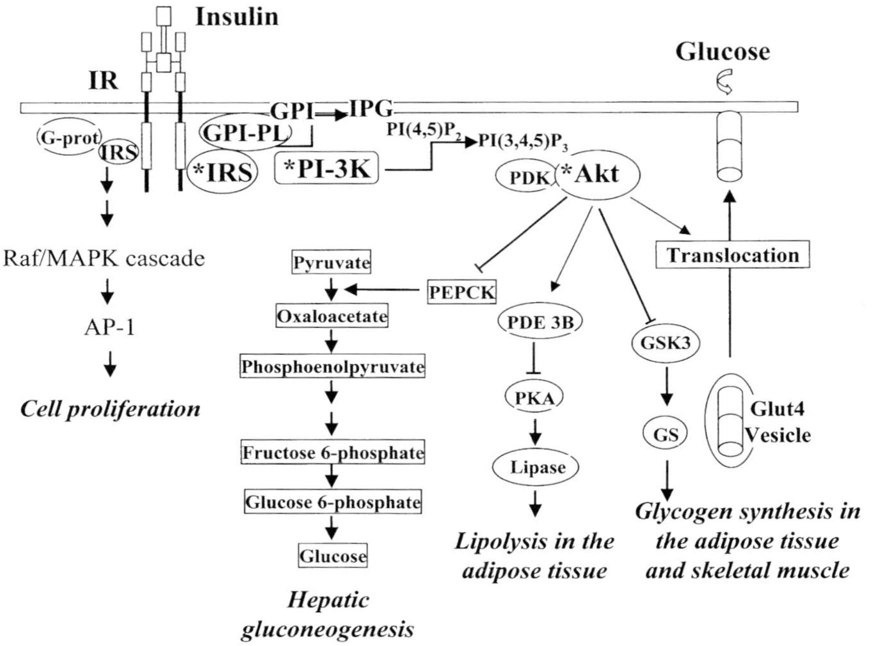

Fig. 6.3 Insulin signaling pathways. Possible routes through which insulin can promote its diverse biological actions: Cell proliferation, gluconeogenesis, lipolysis and glycogen synthesis. Arrows imply facilitating actions whereas crossbars indicate inhibitory ones. The asterisk represents the target molecules for IPG.

early events couple within membrane compartments are currently under study. IPG modulates the intracellular levels and/or activation of nodal signaling molecules, the general protein phosphorylation state of the cell and the levels of cyclins and transcription factors, leading in a cell type-specific manner to the regulation of metabolism homeostasis and to cellular proliferation (Fig. 6.3).

6.6 Acknowledgments

We thank Dr D.R. Jones for critical reading of this manuscript and for helpful suggestions. This work was supported by a grant from Dirección General de Investigación, Ciencia y Tecnología (PM99-0094).

6.7 References

1 JONES, D.R. and VARELA-NIETO, I. Diabetes and the role of inositol-containing lipids in insulin signaling. *Mol. Med.*, **1999**, *5*, 505–514.

2 MÜLLER, G., WIED, S., CRECELIUS, A., KESSLER, A., and ECKEL, J. Phosphoinositolglycan-peptides from yeast potently induce metabolic insulin actions in isolated rat adipocytes, cardiomyocytes, and diaphragms. *Endocrinology*, **1997**, *138*, 3459–3475.

3 MÜLLER, G., WIED, S., PIOSSEK, C., BAUER, A., BAUER, J., and FRICK, W. Convergence and divergence of the signaling pathways for insulin and phosphoinositolglycans. *Mol. Med.*, **1998**, *4*, 299–323.

4 MÜLLER, G. Insulin-like and insulin-mimetic molecules and signalling in lower eucaryontes. *Recent Res. Devel. Endocrinol.*, **2000**, *1*, 1–36.

5 MÜLLER, G., GREY, S., JUNG, C., and BANDLOW, W. Insulin-like signaling in yeast: modulation of protein phosphatase 2A, protein kinase A, cAMP-specific phosphodiesterase, and glycosyl-phosphatidylinositol-specific phospholipase C activities. *Biochemistry*, **2000**, *39*, 1475–1488.

6 MÜLLER, G., HANEKOP, N., KRAMER, W., BANDLOW, W., and FRICK, W. Interaction of phosphoinositolglycan(-peptides) with plasma membrane lipid rafts of rat adipocytes. *Arch. Biochem. Biophys.*, **2002**, *408*, 17–32.

7 VARELA-NIETO, I., LEON, Y., and CARO, H.N. Cell signalling by inositol phosphoglycans from different species. *Comp. Biochem. Physiol. B Biochem. Mol. Biol.*, **1996**, *115*, 223–241.

8 JONES, D.R. and VARELA-NIETO, I. The role of glycosyl-phosphatidylinositol in signal transduction. *Int. J. Biochem. Cell Biol.*, **1998**, *30*, 313–326.

9 ZAPATA, A., LEON, Y., MATO, J. M., VARELA-NIETO, I., PENADES, S., and MARTIN-LOMAS, M. Synthesis and investigation of the possible insulin-like activity of 1D-4-*O*- and 1D-6-*O*-(2-amino-2-deoxy-*α*-D-glucopyranosyl)-*myo*-inositol 1-phosphate and 1D-6-*O*-(2-amino-2-deoxy-*α*-D-glucopyranosyl)-*myo*-inositol 1,2-(cyclic phosphate). *Carbohydr. Res.*, **1994**, *264*, 21–31.

10 FRICK, W., BAUER, A., BAUER, J., WIED, S., and MÜLLER, G. Structure–activity relationship of synthetic phosphoinositolglycans mimicking metabolic insulin action. *Biochemistry*, **1998**, *37*, 13421–13436.

11 LEÓN, Y., SANZ, C., GIRALDEZ, F., and VARELA-NIETO, I. Induction of cell growth by insulin and insulin-like growth factor-I is associated with Jun expression in the otic vesicle. *J. Comp. Neurol.*, **1998**, *398*, 323–332.

12 DIETRICH, H., ESPINOSA, J.F., CHIARA, J.L., JIMENEZ-BARBERO, J., LEÓN, Y., VARELA-NIETO, I., MATO, J.M., CANO, F.H., FOCES-FOCES, C., and MARTÍN-LOMAS, M. Glycosyl inositol derivatives related to inositolphosphoglycan mediators: synthesis, structure and biological activity. *Chem. Eur. J.*, **1999**, *5*, 320–336.

13 MARTÍN-LOMAS, M., KHIAR, N., GARCÍA, S., KOESSLER, J.L., NIETO, P.M., and RADEMACHER, T.W. Inositolphosphoglycan mediators structurally related to glycosyl phosphatidylinositol anchors: synthesis, structure and biological activity. *Chem. Eur. J.*, **2000**, *6*, 3608–3621.

14 JAWOREK, C.H., IACOBUCCI, S., CALIAS, P., and D'ALARCAO, M. Synthesis of inositol glycan cyclic phosphates. *Carbohydr. Res.*, **2001**, *331*, 375–391.

15 LARNER, J. and HUANG, L.C. Identification of a novel inositol glycan signalling pathway with significant therapeutic relevance to insulin resistance: an insulin signalling model using both tyrosine kinase and G-proteins. *Diabetes Rev.*, **1999**, *7*, 217–231.

16 MÜLLER, G. Dynamics of plasma membrane microdomains and cross-talk to the insulin signalling cascade. *FEBS Lett.*, **2002**, *531*, 81–87.

17 LARNER, J. D-chiro-inositol – Its functional role in insulin action and its deficit in insulin resistance. *Int. J. Exp. Diab. Res.*, **2002**, *3*, 47–60.

18 MCCONVILLE, M.J. and MENON, A.K. Recent developments in the cell biology and biochemistry of glycosylphosphatidylinositol lipids (review). *Mol. Membr. Biol.*, **2000**, *17*, 1–16.

19 Ikezawa, H. Glycosylphosphatidylinositol (GPI)-anchored proteins. *Biol. Pharm. Bull.*, **2002**, *25*, 409–417.

20 Chatterjee, S. and Mayor, S. The GPI-anchor and protein sorting. *Cell Mol. Life Sci.*, **2001**, *58*, 1969–1987.

21 Lederkremer, R.M. and Bertello, L.E. Glycoinositolphospholipids, free and as anchors of proteins, in *Trypanosoma cruzi. Curr. Pharm. Des.*, **2001**, *7*, 1165–1179.

22 Scallon, B.J., Fung, W.J., Tsang, T.C., Li, S., Kado-Fong, H., Huang, K.S., and Kochan, J.P. Primary structure and functional activity of a phosphatidylinositol-glycan-specific phospholipase D. *Science*, **1991**, *252*, 446–448.

23 Jones, D.R., Avila, M.A., Sanz, C., and Varela-Nieto, I. Glycosyl-phosphatidylinositol-phospholipase type D: a possible candidate for the generation of second messengers. *Biochem. Biophys. Res. Commun.*, **1997**, *233*, 432–437.

24 Villar, A.V., Goni, F.M., Alonso, A., Jones, D.R., Leon, Y., and Varela-Nieto, I. Phospholipase cleavage of glycosylphosphatidylinositol reconstituted in liposomal membranes. *FEBS Lett.*, **1998**, *432*, 150–154.

25 Davitz, M.A., Hereld, D., Shak, S., Krakow, J., Englund, P.T., and Nussenzweig, V. A glycan-phosphatidylinositol-specific phospholipase D in human serum. *Science*, **1987**, *238*, 81–84.

26 Huang, L.C., Fonteles, M.C., Houston, D.B., Zhang, C., and Larner, J. Chiroinositol deficiency and insulin resistance. III. Acute glycogenic and hypoglycemic effects of two inositol phosphoglycan insulin mediators in normal and streptozotocin-diabetic rats *in vivo. Endocrinology*, **1993**, *132*, 652–657.

27 Hari, T., Kunze, H., Bohn, E., Brodbeck, U., and Butikofer, P. Subcellular distribution of glycosylphosphatidylinositol-specific phospholipase D in rat liver. *Biochem. J.*, **1996**, *320*, 315–319.

28 Deeg, M.A. and Verchere, C.B. Regulation of glycosylphosphatidylinositol-specific phospholipase D secretion from beta TC3 cells. *Endocrinology*, **1997**, *138*, 819–826.

29 Kunjara, K., Caro, H.N., McLean, P., and Rademacher, T.W. Tissue specific release of inositol phosphoglycans. *Biopolymers and Bioproducts: Structure, Function and Applications.* J. Svasti et al. (ed.) DOKYA publishing Co., Ltd., Bangkok, Thailand, **1995**, pp. 301–305.

30 Kristiansen, S. and Richter, E.A. GLUT4-containing vesicles are released from membranes by phospholipase D cleavage of a GPI anchor. *Am. J. Physiol. Endocrinol. Metab.*, **2002**, *283*, E374–382.

31 Parpal, S., Gustavsson, J., and Stralfors, P. Isolation of phosphooligosaccharide/phosphoinositol glycan from caveolae and cytosol of insulin-stimulated cells. *J. Cell Biol.*, **1995**, *131*, 125–135.

32 Müller, G. and Frick, W. Signalling via caveolin: involvement in the cross-talk between phosphoinositolglycans and insulin. *Cell Mol. Life Sci.*, **1999**, *56*, 945–970.

33 Müller, G., Hanekop, N., Wied, S., and Frick, W. Cholesterol depletion blocks redistribution of lipid raft components and insulin-mimetic signaling by glimepiride and phosphoinositolglycans in rat adipocytes. *Mol. Med.*, **2002**, *8*, 120–136.

34 Müller, G., Jung, C., Wied, S., Welte, S., Jordan, H., and Frick, W. Redistribution of glycolipid raft domain components induces insulin-mimetic signaling in rat adipocytes. *Mol. Cell Biol.*, **2001**, *21*, 4553–4567.

35 Johnson, S.A. and Denton, R.M. Insulin stimulation of pyruvate dehydrogenase in adipocytes involves two distinct signalling pathways. *Biochem. J.*, **2003**, *369*, 351–356.

36 Kajiyoshi, M. Incorporation of ethanolamine into insulin-sensitive glycosylated phosphatidylinositol of chick embryo fibroblasts. *Biochim. Biophys. Acta*, **2000**, *15*, 255–267.

37 Bogdanowicz, P. and Pujol, J.P. Glycoylphosphatidylinositol (GPI) hydrolysis by transforming growth factor-beta 1 (TGF-beta1) as a potential early step in the inhibition of epithelial cell proliferation. *Mol. Cell Biochem.*, **2000**, *208*, 143–150.

38 Tam, B.Y.Y., Larouche, D., Germain, L., Hooper, N.M., and Philip, A. Characterization of a 150 kDa accessory receptor for TGF-β1 on keratinocytes: Direct evidence for a GPI anchor and ligand binding of the released form. *J. Cell Biochem.*, **2001**, *83*, 494–507.

39 Boudot, C., Kadri, Z., Petitfrere, E., Lambert, E., Chretien, S., Mayeux, P., Haye, B., and Billat, C. Phosphatidylinositol 3-kinase regulates glycosylphosphatidylinositol hydrolysis through PLC-gamma (2) activation in erythropoietin-stimulated cells. *Cell Signal*, **2002**, *14*, 869–878.

40 de Avalos, S.V., Lima, C., Martín, C., Lederkremer, R.M., and Vila, M.C. ACTH-mediated glucocorticoid and mineralocorticoid production is inhibited by an inositolphosphoglycan and a glycosylphosphatidylinositol-phospholipase C is activated by the hormone in mammalian adrenocortical cells. *Steroids*, **1998**, *63*, 70–75.

41 Petitfrere, E., Sartelet, H., Vivien, D., Varela-Nieto, I., Elbtaouri, H., Martiny, L., and Haye, B. Glycosylphosphatidylinositol (GPI)/inositolphosphate glycan (IPG): an intracellular signalling system involved in the control of thyroid cell proliferation. *Biochimie*, **1998**, *80*, 1063–1067.

42 Varela-Nieto, I., de la Rosa, E.J., Valenciano, A.I., and León, Y. Cell death in the nervous system: Lessons from insulin and insulin-like growth factors. *Mol. Neurobiol.*, **2003**, *28*, 23–49.

43 Anlar, B., Sullivan, K.A., and Feldman, E.L. Insulin-like growth factor-I and central nervous system development. *Horm. Metab. Res.*, **1999**, *31*, 120–125.

44 Allan, G.J., Flint, D.J., and Patel, K. Insulin-like growth factor axis during embryonic development. *Reproduction*, **2001**, *122*, 31–39.

45 Bevan, P. Insulin signalling. *J. Cell Sci.*, **2001**, *114*, 31–39.

46 Holzenberger, M. and Lapointe, F. Expression of insulin-like growth factor-I (IGF-I) and IGF-II in the avian brain: relationship of in situ hybridisation patterns with IGF type 1 receptor expression. *Int. J. Dev. Neurosci.*, **2000**, *18*, 69–82.

47 Tatar, M., Kopelman, A., Epstein, D., Tu, M.P., and Garofalo, R.S. A mutant Drosophila insulin receptor homolog that extends life-span and impairs neuroendocrine function. *Science*, **2001**, *292*, 107–110.

48 Jones, J.I. and Clemmons, D.R. Insulin-like growth factors and their binding proteins: biological actions. *Endocrinol. Rev.*, **1995**, *16*, 3–34.

49 LeRoith, D. and Bondy, C. In: *Growth Factors and Cytokines in Health and Disease* (LeRoith, D., Bondy, C., eds.), JAI Press, Greenwich, Connecticut, **1996**, pp. 1–26.

50 De Meyts, P. Insulin and insulin-like growth factors: the paradox of signaling specificity. *Growth Horm. IGF Res.*, **2002**, *12*, 81–83.

51 Saltiel, A.R. and Pessin, J.E. Insulin signalling pathways in time and space. *Trends Cell Biol.*, **2002**, *12*, 65–71.

52 Lazar, D.F., Knez, J.J., Medof, M.E., Cuatrecasas, P., and Saltiel, A.R. Stimulation of glycogen synthesis by insulin in human erythroleukemia cells requires the synthesis of glycosyl-phosphatidylinositol. *Proc. Natl. Acad. Sci. USA.*, **1994**, *91*, 9665–9669.

53 Villalba, M., Alvarez, J.F., Russell, D.S., Mato, J.M., and Rosen, O.M. Hydrolysis of glycosyl-phosphatidylinositol in response to insulin is reduced in cells bearing kinase-deficient insulin receptors. *Growth Factors*, **1990**, *2*, 91–97.

54 Suzuki, S., Taneda, Y., Hirai, S., Yamamoto-Honda, R., and Toyota, T. Mutated insulin receptor Val996 reduces insulin-dependent generation of inositol glycan and diacylglycerol. *Diabetes*, **1992**, *41*, 1373–1379.

55 Clemente, R., Jones, D.R., Ochoa, P., Romero, G., Mato, J.M., and Varela-Nieto, I. Role of glycosyl-phosphatidylinositol hydrolysis as a mitogenic signal for epidermal growth factor. *Cell Signal*, **1995**, *7*, 411–421.

56 Bogdanowicz, P., Vivien, D., Felisaz, N., Leon, V., and Pujol, J.P. An inositolphosphate glycan released by TGF-beta mimics the proliferative but not the transcriptional effects of the factor and requires functional receptors. *Cell Signal*, **1996**, *8*, 503–509.

57 León, Y., Vazquez, E., Sanz, C., Vega, J.A., Mato, J.M., Giraldez, F., Represa, J., and Varela-Nieto, I. Insulin-like growth factor-I regulates cell proliferation in the developing inner ear, activating glycosyl-phosphatidylinositol hydrolysis and Fos expression. *Endocrinology*, **1995**, *136*, 3494–3503.

58 Represa, J., Avila, M.A., Miner, C., Giraldez, F., Romero, G., Clemente, R., Mato, J.M., and Varela-Nieto, I. Glycosyl-phosphatidylinositol/inositol phosphoglycan: a signaling system for the low-affinity nerve growth factor receptor. *Proc. Natl. Acad. Sci. USA*, **1991**, *88*, 8016–8019.

59 Paneda, C., Villar, A.V., Alonso, A., Goni, F.M., Varela, F., Brodbeck, U., León, Y., Varela-Nieto, I. and Jones, D.R. Purification and characterization of insulin-mimetic inositol phosphoglycan-like molecules from grass pea (Lathyrus sativus) seeds. *Mol. Med.*, **2001**, *7*, 454–460.

60 Caro, H.N., Kunjara, S., Rademacher, T.W., Leon, Y., Jones, D.R., Avila, M.A., and Varela-Nieto, I. Isolation and partial characterisation of insulin-mimetic inositol phosphoglycans from human liver. *Biochem. Mol. Med.*, **1997**, *61*, 214–228.

61 Mato, J.M., Kelly, K.L., Abler, A., and Jarett, L. Identification of a novel insulin-sensitive glycophospholipid from H35 hepatoma cells. *J. Biol. Chem.*, **1987**, *262*, 2131–2137.

62 Mato, J.M., Kelly, K.L., Abler, A., Jarett, L., Corkey, B.E., Cashel, J.A., and Zopf, D. Partial structure of an insulin-sensitive glycophospholipid. *Biochem. Biophys. Res. Commun.*, **1987**, *146*, 764–770.

63 Larner, J., Huang, L.C., Schwartz, C.F., Oswald, A.S., Shen, T.Y., Kinter, M., Tang, G.Z., and Zeller, K. Rat liver insulin mediator which stimulates pyruvate dehydrogenase phosphate contains galactosamine and D-chiroinositol. *Biochem. Biophys. Res. Commun.*, **1988**, *151*, 1416–1426.

64 Fonteles, M.C., Huang, L.C., and Larner, J. Infusion of pH 2.0 D-*chiro*-inositol glycan insulin putative mediator normalizes plasma glucose in streptozotocin diabetic rats at a dose equivalent to insulin without inducing hypoglycaemia. *Diabetologia*, **1996**, *39*, 731–734.

65 Fonteles, M.C., Almeida, M.Q., and Larner, J. Antihyperglycemic effects of 3-*O*-methyl-D-*chiro*-inositol and D-*chiro*-inositol associated with manganese in streptozotocin diabetic rats. *Horm. Metab. Res.*, **2000**, *32*, 129–132.

66 Tagliaferri, F., Wang, S.N., Berlin, W.K., Outten, R.A., and Shen, T.Y. Glycosyl-inositol derivatives I. Synthesis of 1-substituted *chiro*-inositol derivatives. *Tetrahedron Lett.*, **1990**, *31*, 1105–1108.

67 Berlin, W.K., Zhang, W.S., and Shen, T.Y. Glycosyl-inositol derivatives III. Synthesis of hexosamine-inositol-phosphates related to putative insulin mediators. *Tetrahedron*, **1991**, *47*, 1–20.

68 Plourde, R., D'Alarcao, M., and Saltiel, A.R. Synthesis and characterization of an insulin-mimetic disaccharide. *J. Org. Chem.*, **1992**, *57*, 2606–2610.

69 Cobb, J.E. and Johnson, M.R. Synthesis of 6-*O*-(2-aminoethyl)-D,L-*myo*-inositol-1,2-cyclic phosphate: A model of a putative insulin second messenger. *Tetrahedron*, **1991**, *47*, 21–30.

70 Reddy, K.K., Falck, J.R., and Capdevila, J. Insulin second messengers:synthesis of 6-*O*-(2-amino-2-deoxy-*α*-D-glucopyranosyl)-D-*chiro*-inositol-1-phosphate. *Tetrahedron, Lett.*, **1993**, *34*, 7869–7872.

71 Jaramillo, C., Chiara, J.L., and Martín-Lomas, M. An effective strategy for the synthesis of 6-*O*-(2-amino-2-deoxy-*α*-D-glucopyranosil)-D-*chiro*- and -D-*myo*-inositol 1-phosphate related to putative insulin mimetics. *J. Org. Chem.*, **1994**, *59*, 3135–3141.

72 Frick, W., Bauer, A., Bauer, J., Wied, S., and Müller, G. Insulin-mimetic signalling of synthetic phosphoinositolglycans in isolated rat adipocytes. *Biochem. J.*, **1998**, *336*, 163–181.

73 Kessler, A., Müller, G., Wied, S., Crecelius, A., and Eckel, J. Signalling pathways of an insulin-mimetic phosphoinositolglycan-peptide in muscle and adipose tissue. *Biochem. J.*, **1998**, *330*, 277–286.

74 Huang, L.C., Heimark, D., Linko, J., Nolan, R., and Larner, J. A model phosphatase 2C→phosphatase 1 activation cascade via dual control of inhibitor-1 (INH-1) and DARPP-32 dephosphorylation by two inositol glycan putative insulin mediators from beef liver. *Biochem. Biophys. Res. Commun.*, **1999**, *255*, 150–156.

75 Alvarez, J.F., Sanchez-Arias, J.A., Guadano, A., Estevez, F., Varela, I., Feliu, J.E., and Mato, J.M. Transport in isolated rat hepatocytes of the phospho-

oligosaccharide that mimics insulin action. Effects of adrenalectomy and glucocorticoid treatment. *Biochem. J.*, **1991**, *274*, 369–374.

76 Alvarez, L., Avila, M.A., Mato, J.M., Castano, J.G., and Varela-Nieto, I. Insulin-like effects of inositol phosphate-glycan on messenger RNA expression in rat hepatocytes. *Mol. Endocrinol.*, **1991**, *5*, 1062–1068.

77 Ostlund, R.E., Jr., McGill, J.B., Herskowitz, I., Kipnis, D.M., Santiago, J.V., and Sherman, W.R. D-chiro-inositol metabolism in diabetes mellitus. *Proc. Natl. Acad. Sci. USA*, **1993**, *90*, 9988–9992.

78 Kennington, A.S., Hill, C.R., Craig, J., Bogardus, C., Raz, I., Ortmeyer, H.K., Hansen, B.C., Romero, G. and Larner, J. Low urinary chiro-inositol excretion in non-insulin-dependent diabetes mellitus. *New Engl. J. Med.*, **1990**, *323*, 373–378.

79 Pak, Y., Hong, Y., Kim, S., Piccariello, T., Farese, R.V., and Larner, J. In vivo chiro-inositol metabolism in the rat: a defect in *chiro*-inositol synthesis from *myo*-inositol and an increased incorporation of chiro-[^{3}H]inositol into phospholipid in the Goto-Kakizaki (GK) rat. *Mol. Cells*, **1998**, *8*, 301–309.

80 Asplin, I., Galasko, G., and Larner, J. *chiro*-Inositol deficiency and insulin resistance: a comparison of the *chiro*-inositol- and the *myo*-inositol-containing insulin mediators isolated from urine, hemodialysate, and muscle of control and type II diabetic subjects. *Proc. Natl. Acad. Sci. USA*, **1993**, *90*, 5924–5928.

81 Shashkin, P.N., Shashkina, E.F., Fernqvist-Forbes, E., Zhou, Y.P., Grill, V., and Katz, A. Insulin mediators in man: effects of glucose ingestion and insulin resistance. *Diabetologia*, **1997**, *40*, 557–563.

82 Sun, T.H., Heimark, D.B., Nguyen, T., Nadler, J.L., and Larner, J. Both *myo*-inositol to *chiro*-inositol epimerase activities and *chiro*-inositol to *myo*-inositol ratios are decreased in tissues of GK type 2 diabetic rats compared to Wistar controls. *Biochem. Biophys. Res. Commun.*, **2002**, *293*, 1092–1098.

83 Ortmeyer, H.K., Larner, J., and Hansen, B.C. Effects of D-chiroinositol added to a meal on plasma glucose and insulin in hyperinsulinemic rhesus monkeys. *Obes. Res.*, **1995**, *3*(Suppl 4), 605S–608S.

84 Larner, J. and Craig, J.W. Urinary *myo*-inositol-to-*chiro*-inositol ratios and insulin resistance. *Diabetes Care*, **1996**, *19*, 76–78.

85 Kunjara, S., Wang, D.Y., Greenbaum, A.L., McLean, P., Kurtz, A., and Rademacher, T.W. Inositol phosphoglycans in diabetes and obesity: urinary levels of IPG A-type and IPG P-type, and relationship to pathophysiological changes. *Mol. Genet. Metab.*, **1999**, *68*, 488–502.

86 Arkwright, P.D., Rademacher, T.W., Dwek, R.A., and Redman, C.W. Pre-eclampsia is associated with an increase in trophoblast glycogen content and glycogen synthase activity, similar to that found in hydatidiform moles. *J. Clin. Invest.*, **1993**, *91*, 2744–2753.

87 Kunjara, S., Greenbaum, A.L., Wang, D.Y., Caro, H.N., McLean, P., Redman, C.W., and Rademacher, T.W. Inositol phosphoglycans and signal transduction systems in pregnancy in preeclampsia and diabetes: evidence for a significant regulatory role in preeclampsia at placental and systemic levels. *Mol. Genet. Metab.*, **2000**, *69*, 144–158.

88 Kunjara, S., Wang, D.Y., McLean, P., Greenbaum, A.L., and Rademacher, T.W. Inositol phosphoglycans and the regulation of the secretion of leptin: in vitro effects on leptin release from adipocytes and the relationship to obesity. *Mol. Genet. Metab.*, **2000**, *70*, 61–68.

89 Nestler, J.E., Jakubowicz, D.J., Reamer, P., Gunn, R.D., and Allan, G. Ovulatory and metabolic effects of D-*chiro*-inositol in the polycystic ovary syndrome. *New Engl. J. Med.*, **1999**, *340*, 1314–1320.

90 Farese, R.V. Insulin-sensitive phospholipid signaling systems and glucose transport. Update II. *Exp. Biol. Med. (Maywood)*, **2001**, *226*, 283–295.

91 Ohmichi, M., Pang, L., Ribon, V., and Saltiel, A.R. Divergence of signaling pathways for insulin in PC-12 pheochromocytoma cells. *Endocrinology*, **1993**, *133*, 46–56.

92 Saltiel, A. R. The paradoxical regulation of protein phosphorylation in insulin action. *FASEB J.*, **1994**, *8*, 1034–1040.

93 Lazar, D. F., Wiese, R. J., Brady, M. J., Mastick, C. C., Waters, S. B., Yamauchi, K., Pessin, J. E., Cuatrecasas, P. and Saltiel, A. R. Mitogen-activated protein kinase inhibition does not block the stimulation of glucose utilization by insulin. *J. Biol. Chem.*, **1995**, *270*, 20801–20807.

94 Denton, R. M. and Tavare, J. M. Does mitogen-activated-protein kinase have a role in insulin action? The cases for and against. *Eur. J. Biochem.*, **1995**, *227*, 597–611.

95 Kerouz, N. J., Horsch, D., Pons, S., and Kahn, C. R. Differential regulation of insulin receptor substrates-1 and -2 (IRS-1 and IRS-2) and phosphatidylinositol 3-kinase isoforms in liver and muscle of the obese diabetic (ob/ob) mouse. *J. Clin. Invest.*, **1997**, *100*, 3164–3172.

96 Paz, K., Hemi, R., LeRoith, D., Karasik, A., Elhanany, E., Kanety, H., and Zick, Y. A molecular basis for insulin resistance. Elevated serine/threonine phosphorylation of IRS-1 and IRS-2 inhibits their binding to the juxtamembrane region of the insulin receptor and impairs their ability to undergo insulin-induced tyrosine phosphorylation. *J. Biol. Chem.*, **1997**, *272*, 29911–29918.

97 Withers, D. J., Gutierrez, J. S., Towery, H., Burks, D. J., Ren, J. M., Previs, S., Zhang, Y., Bernal, D., Pons, S., Shulman, G. I. et al. Disruption of IRS-2 causes type 2 diabetes in mice. *Nature*, **1998**, *391*, 900–904.

98 Sleight, S., Wilson, B. A., Heimark, D. B., and Larner, J. $G_{q/11}$ is involved in insulin-stimulated inositol phosphoglycan putative mediator generation in rat liver membranes: co-localizatiom of $G_{q/11}$ with the insulin receptor in membrane vesicles. *Biochem. Biophys. Res. Commun.*, **2002**, *295*, 561–569.

99 Alemany, S., Mato, J. M., and Strålfors, P. Phospho-dephospho-control by insulin is mimicked by a phospho-oligosaccharide in adipocytes. *Nature*, **1987**, *330*, 77–79.

100 Alemany, S., Puerta, J., Guadao, A., and Mato, J. M. Modulation of casein kinase II activity by the polar head group of an insulin sensitive glycosyl-phosphatidylinositol. *J. Biol. Chem.*, **1990**, *265*, 4849–4852.

101 Strålfors, P. Insulin second messengers. *Bioessays*, **1997**, *19*, 327–335.

102 Ortmeyer, H. K. Insulin increases liver protein phosphatase-1 and protein phosphatase-2C activities in lean, young adult rhesus monkeys. *Horm. Metab. Res.*, **1998**, *30*, 705–710.

103 Misek, D. E. and Saltiel, A. R. An inositol phosphate glycan derived from a Trypanosoma brucei glycosyl-phosphatidylinositol mimics some of the metabolic actions of insulin. *J. Biol. Chem.*, **1992**, *267*, 16266–16273.

104 Misek, D. E. and Saltiel, A. R. An inositol phosphate glycan derived from a Trypanosoma brucei glycosyl phosphatidylinositol promotes protein dephosphorylation in rat epididymal adipocytes. *Endocrinology*, **1994**, *135*, 1869–1876.

105 Shepherd, P. R., Withers, D. J., and Siddle, K. Phosphoinositide 3-kinase: the key switch mechanism in insulin signalling. *Biochem. J.*, **1998**, *333*, 471–490.

106 Alessi, D. R. and Downes, C. P. The role of PI 3-kinase in insulin action. *Biochim. Biophys. Acta*, **1998**, *1436*, 151–164.

107 Anai, M., Ono, H., Funaki, M., Fukushima, Y., Inukai, K., Ogihara, T., Sakoda, H., Onishi, Y., Yazaki, Y., Kikuchi, M. et al. Different subcellular distribution and regulation of expression of insulin receptor substrate (IRS)-3 from those of IRS-1 and IRS-2. *J. Biol. Chem.*, **1998**, *273*, 29686–29692.

108 Gustavsson, J., Parpal, S., and Strålfors, P. Insulin-stimulated glucose uptake involves the transition of glucose transporters to a caveolae-rich fraction within the plasma membrane: implications for type II diabetes. *Mol. Med.*, **1996**, *2*, 367–372.

109 Villar, A. V., Alonso, A., Pañeda, C., Varela-Nieto, I., Brodbeck, U., and Goñi, F. M. Towards the in vitro reconstitution of caveolae. Asymmetric incorporation of glycosylphosphatidylinositol (GPI) and gangliosides into liposomal membranes. *FEBS Lett.*, **1999**, *457*, 71–74.

110 Avila, M.A., Varela-Nieto, I., Romero, G., Mato, J.M., Giraldez, F., Van de Water, T.R., and Represa, J. Brain-derived neurotrophic factor and neurotrophin-3 support the survival and neuritogenesis response of developing cochleovestibular ganglion neurons. *Dev. Biol.*, **1993**, *159*, 266–275.

111 Represa, J., Avila, M.A., Romero, G., Mato, J.M., Giraldez, F., and Varela-Nieto, I. Brain-derived neurotrophic factor and neurotrophin-3 induce cell proliferation in the cochleovestibular ganglion through a glycosyl-phosphatidylinositol signaling system. *Dev. Biol.*, **1993**, *159*, 257–265.

7
High-throughput Screening of Hormone-sensitive Lipase and Subsequent Computer-assisted Compound Optimization

Stefan Petry, Karl-Heinz Baringhaus, Karl Schoenafinger, Christian Jung, Horst Kleine, and Günter Müller

7.1
Introduction

7.1.1
Lipases in Metabolism

Lipases (EC 3.1.1.3) play a key role in the metabolism of lipids. They not only break down lipids, but also initiate their mobilization (Fig. 7.1), and they degrade the lipids in the duodenum. They act on stored or wrapped lipids in the blood stream, the liver, the adipose and muscle tissue and β-cells and they are, presum-

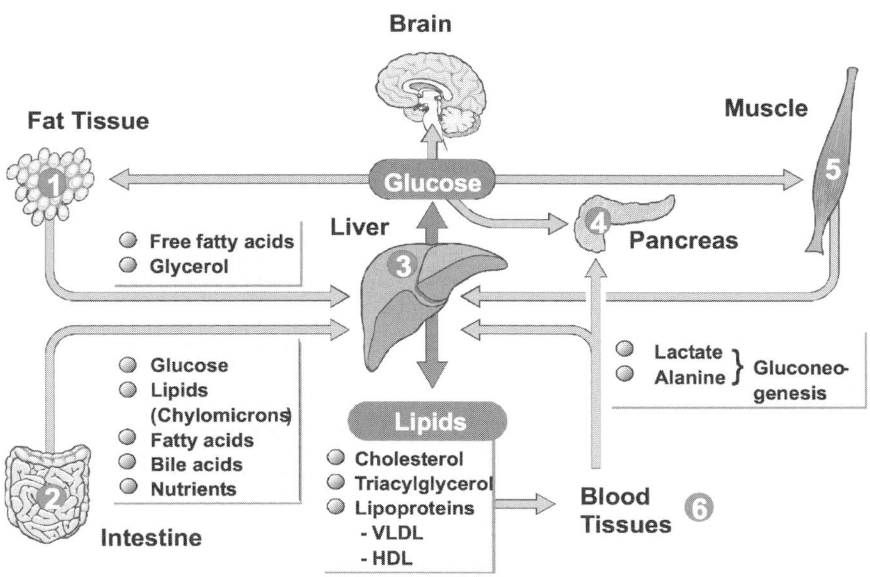

Fig. 7.1 Lipases control the degradation, mobilization and trafficking of lipids in humans. Hormone-sensitive lipase (HSL, 1, 4, 5), monoacylglyceride lipase (MGL, 1); pankreatic lipase (PL, 2); hepatic lipase (HL, 3); lipoprotein lipase (LPL, 6); endothelial lipase (EL, 6).

Lipases and Phospholipases in Drug Development
Edited by G. Müller and S. Petry

ISBN: 3-527-30677-3

ably, also present in other peripheral cells. Consequently, they are potential targets for the development of drugs to treat obesity, diabetes and atherosclerosis.

Pancreatic lipase degrades dietary fat to non-esterified fatty acids (NEFA) and monoacylglycerol, which are then transported into the blood stream. Conversely, hormone-sensitive lipase (HSL) controls the hydrolysis of stored fat in adipocytes, concomitantly regulating the mobilization of NEFA and glycerol into the blood stream. In addition to these two big players, which control the blood level of NEFAs, other lipases degrade lipoproteins such as HDL, LDL and VLDL. Hepatic lipase (HL) is responsible for the degradation of lipoproteins in the liver [1]. Lipoprotein lipase (LPL) [2] and endothelial lipase (EL) [3–5] are involved in the cleavage of phospholipids as well as neutral lipids with different preferences in the endothelial luminal surface. Remarkably, the activities of LPL and HSL are under tight hormonal control by a complex interplay both of transcriptional and post-transcriptional mechanisms [2, 6].

7.2
Lipases Show Unique Differences in Comparison to Other Drug Targets

Lipases are water-soluble enzymes that hydrolyze water-insoluble substrates such as long-chain mono-, di-, triacylglycerols (TAGs), phospholipids and cholesterol esters. Thus, the cleavage reaction has to occur at the lipid–water interface [7–9]. This distinguishes lipases from other hydrolytic enzymes and makes them unique target proteins with regard to selectivity and inhibition. The water-insolubility of TAGs has several striking consequences. Neutral lipids are neither in the serum compartment nor in the cytoplasm of cells in direct contact with the aqueous milieu. Instead, they are organized in two ways (1) as high-molecular-mass aggregates in emulsified droplets consisting of an interior TAG core and surrounded by a monolayer shell of phospholipids, cholesterol, and embedded proteins (cytoplasmic lipid droplets and serum lipoprotein particles); (2) in smaller complexes bound to serum or intracellular fatty acid/lipid-binding proteins such as albumin and caveolin [10, 11]. Lipases interact with these special lipid complexes or "super-substrates" via hydrophobic domains. These hydrophobic domains become exposed to the substrate upon contact with the lipid complexes due to a conformational change of the protein ("interfacial activation") [12, 13]. Thus, the apparently "two-dimensional" reaction of lipases does not simply follow Michaelis-Menten kinetics and critically depends on the interface [9, 12, 13]. Substrate-specific measurements of lipase activity and the development of reliable lipase assay systems have to account for these unique features.

7.3
Lipase Assays

The inherent difficulty in studying enzymatic reactions on lipophilic and insoluble substrates has given rise to many assays designed to measure lipase activity, using various substrates and experimental designs. Hendrickson [14], and more recently Beisson et al. [15] and Gupta et al. [16] have summarized these assays. The most sensitive and reliable assays are discontinuous "time stop" assays using radioactively labeled TAG [17, 18] or phosphoglycerides [19] as substrates. The substrate has to be separated from the product by chromatographic methods, partitioning, or filtration. Non-radioactive enzymatic or chromatographic methods to determine the products of lipolysis are highly sensitive, but are discontinuous like the radioactive methods. Titrimetric assays are continuous, but are relatively insensitive, and tedious in daily practise. They may also interfere with acidic or basic compounds.

Verger and coworkers [20] have developed a continuous lipase assay using naturally occurring fluorescent TAG isolated from *Parinari glaberrimum* as substrates. We have synthesized octadeca-9,11,13,15-tetronic-3-hydroxy-octadecyloxypropylester, a 1 acyl-2 alkyl glycerol from parinaric acid. This compound is a substrate of HSL, but it is too sensitive to oxidation to be used under high-throughput assay conditions. Recently, the same group has reported an ultraviolet spectrophotometric assay using less oxidation-sensitive TAGs extracted from *Aleurites fordii* seeds [21].

Continuous assays are rarely described for lipases, but are frequently described for esterases. Similar to lipases, esterases hydrolyze ester bonds. However, in contrast to lipases, their substrates are water-soluble and thus water-soluble fluorescent substrates can be used to measure their enzymatic activity. Some of these water-soluble substrates have been proposed for the measurement of lipolytic ac-

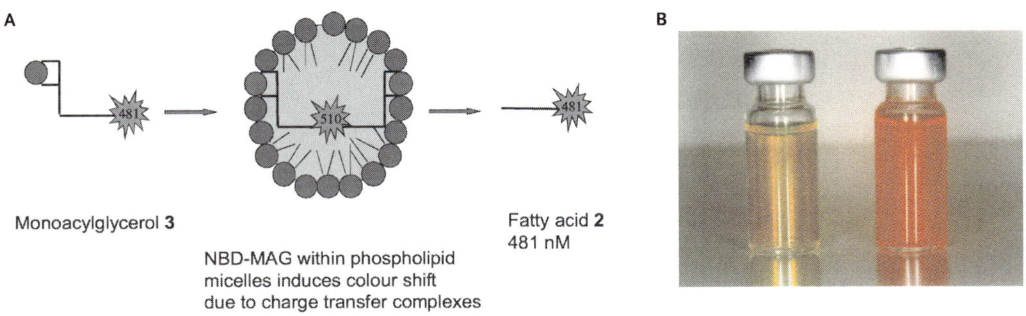

Fig. 7.2 (A) Incorporation of the artificial fluorescently labeled substrate (NBD-MAG) into mixed micelles is associated by a change from yellow (481 nm) to red (510 nm), presumably due to interaction of the aromatic ring systems. (B) Left-hand side: $CHCl_3$ solution of **3**; right-hand side: mixed micelles after ultrasonification in the presence of phospholipids and **3**.

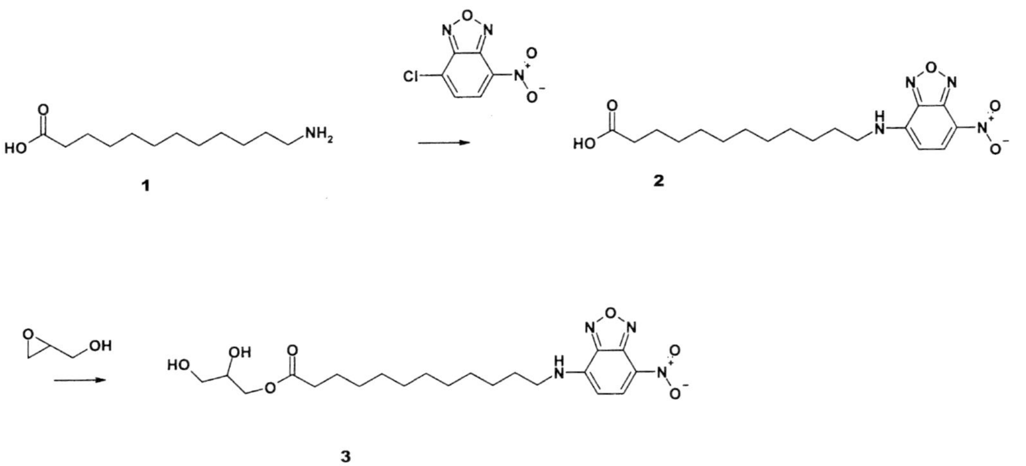

Scheme 7.1 Synthesis of a non-soluble fluorescent lipase substrate.

tivity [14, 22]. The assay systems for esterases are simple, easy to handle and are sufficiently sensitive. However, they do not account for interfacial activation and therefore do not discriminate lipase from esterase activity. This may lead to overestimation of the esterase activity and therefore the results have to be interpreted with caution, particularly if the assays are used to determine lipase activity in protein mixtures or body fluids.

Consequently, we decided to use a lipidic substrate equipped with an appropriate fluorogenic moiety to measure lipase activity. Importantly, even huge moieties, such as BODIPY [23], rhodamine [24] or pyrene [25] coupled to typical lipase substrates usually do not interfere with their cleavage by lipolytic enzymes. The size of the fluorophore does not seem to be very critical. In general, it should be as small as possible, lipophilic, stable and insensitive to oxidation. For these reasons we preferred *p*-nitrobenzofurazan (NBD) as a fluorescent label.

Starting from amino acid **1** we have synthesized the fluorescently labeled, non-soluble lipase substrate **3** (Scheme 7.1).

The monoacylglycerol **3** is insoluble in water and has to be presented in micelles or lipid complexes (e.g. droplets) to be acceptable as lipase substrate. To generate the substrate, **3** was incorporated into mixed phospholipid micelles. To prepare these micelles, a solution of **3** in $CHCl_3$ was added to a solution of phospholipids (phosphocholine, phosphoinositol) in a buffer system. The organic solvent was purged with nitrogen and the organic mixture was sonicated. To our surprise, the development of mixed micelles was accompanied by a color shift from yellow to red (Fig. 7.2, see p. 123).

When a lipase (e.g. hormone-sensitive lipase, HSL) is added, the substrate is hydrolyzed to yield NBD-labeled free fatty acid **2** and glycerol. Concomitantly, the solution regains a yellow fluorescent color. The hydrolytic reaction is followed by measurement of the absorption at 481 nm (Fig. 7.3).

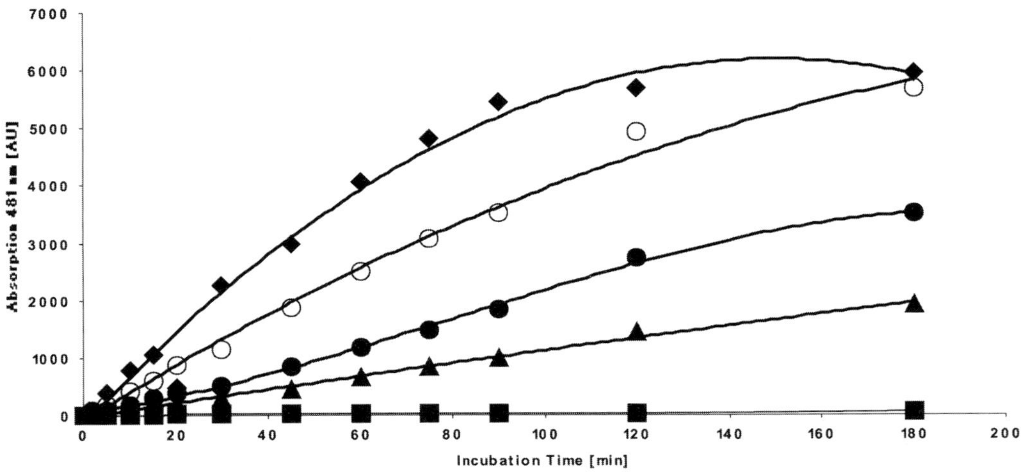

Fig. 7.3 Time course of cleavage of NBD-MAG. NBD-MAG incorporated in PI/PC micelles was incubated (30 °C) in the presence of 40 (◆), 20 (○), 10 (●) and 5 µg (▲) or absence (■) of extract prepared from basal or isoproterenol-treated (1 µM) adipocytes. At the times indicated absorption was measured at 481 nm using a microtiter plate reader ($n=8$, mean). Alternatively, portions of the incubation mixtures were extracted with acidic methanol–chloroform–heptane and then subjected to phase separation. The organic phase was dried, suspended in THF and analyzed by TLC (diethyl ether/light petroleum/acetic acid; 78/22/1, by vol.) and fluorescence imaging ($n=8$, data not shown). The two analytical methods yielded very similar results.

The assay is robust, easily handled, has a high signal-to-noise ratio and can be performed under high-throughput screening conditions using a robotic pipetting machine. We applied this assay in a 384-format to screen the Aventis compound collection for inhibitors of several lipases. One example, HSL is described below.

7.4 Hormone-sensitive Lipase (HSL) as a Drug Target in Diabetes

7.4.1 Biological Role of HSL

TAGs from stored lipid droplets of adipocytes are hydrolyzed into NEFAs and glycerol through the concerted action of at least two lipases, HSL and monoacylglycerol lipase [26], and additional proteins are involved in the recruitment and trafficking of HSL to lipid droplets (book chapter 11). The NEFAs are released to the blood stream, where they bind to albumin and are transported to the peripheral tissues and the liver. In muscle tissue, NEFAs are used as an energy source, whereas in the liver they are transformed into VLDL. The lipolytic activity in adipocytes is under strict hormonal control and HSL is one of the major target proteins for this regulation. HSL activity is controlled by the status of phosphorylation. HSL is phosphorylated and activated by PKA, a cyclic AMP-dependent kinase [27, 28]. The degree of phosphorylation of HSL depends directly on the cAMP level since cAMP is an allosteric activator of PKA. Hormones, triggered by the sympathic nervous system, such as adrenalin, noradrenalin and glucagon bind to adrenergic receptors, which are coupled to adenylate cyclase and thus enhance the intracellular cAMP level and PKA activity. Activated HSL leads to an increase of lipolysis in adipose tissue and to elevated levels of NEFAs in the blood stream. Since the β-oxidation of fatty acids in liver and muscle cells depends on the level of NEFAs, β-oxidation is activated under these conditions. In addition, c-AMP-dependent kinase deactivates acetyl-CoA carboxylase (a key enzyme in fatty acid biosynthesis). In summary, the hormones associated with the sympathic nervous system activate lipolysis and release of NEFAs from adipocytes to the blood stream. NEFAs are transported to liver and muscle cells where they are absorbed, oxidized and degradated to yield ATP. Concurrently, lipogenesis is blocked by inhibition of acetyl-CoA carboxylase.

Insulin, a hormone associated with the parasympathic nervous system, has the opposite effect. It decreases the intracellular level of c-AMP. Under these conditions PKA is deactivated and HSL is dephosphorylated by the corresponding phosphatase. Lipolysis in adipocytes is blocked and the amount of NEFAs available for β-oxidation is reduced. At the same time lipogenesis is activated by the stimulation of cAMP-independent kinases, which phosphorylate and activate acetyl-Co carboxylase.

7.4.2
Characteristics of HSL

HSL activity was first identified in 1964 as an epinephrine-sensitive lipolytic activity in adipose tissue. It was named according to its property of being activated by hormones such as catecholamines, ACTH, and glucagon [29]. In 1981 the isolation of a relatively pure and biologically active HSL from rat adipocytes was reported [30]. The first cDNA clones were isolated for both rat and human HSL in 1988 [31] and the gene sequence for human HSL was elucidated in 1993 [32].

The rat and human cDNAs predict 768 and 775 amino acids, respectively, which is in good agreement with the molecular mass of 84 000 [30]. Despite 82% identity between the amino acid sequence of rat and human HSL and their similar molecular masses, human HSL exhibits slightly lower electrophoretic mobility during SDS-PAGE compared to rat HSL, i.e. 88 vs. 84 kDa [30]. HSL has a uniquely broad substrate specificity, hydrolyzing with varying efficacy tri-, di- and monoacylglycerols, as well as cholesteryl, steroid fatty acid, retinyl and *p*-nitrophenyl esters [33]. However, HSL lacks detectable phospholipase activity. HSL prefers cleavage of diacylglycerol (10- and 5-fold vs. tri- and monoacylglycerol) and the 1- or 3-ester bond of its acylglycerol substrate (4-fold vs. 2-ester bond). Polyunsaturated fatty acids are preferred over saturated fatty acids and shorter vs. longer fatty acids. Phosphorylation under standard *in vitro* assay conditions leads to moderate increases in activity against TAG and cholesterol esters, but not against di- and monoacylglycerols [30]. It is unclear whether phosphorylation of HSL also provokes differential effects on its activity *in vivo*. Rat adipocyte HSL is a dimer in solution, according to gel chromatography and sucrose gradient centrifugation [34] and, subsequently, human recombinant HSL and sedimentation equilibrium analysis. Kinetic studies finally revealed that the HSL functional dimer is characterized by greater maximal activity with no difference in substrate affinity compared with its monomeric subunits [35].

Reflecting its unique catalytic and regulatory (see Chapter 11) properties. HSL exhibits only minimal sequence similarity with enzymes of the lipase gene family. Based on inhibition of HSL activity by diisopropylphosphofluoridate and the covalent linkage of this inhibitor to the HSL protein, HSL has been categorized as a serine esterase [36]. The Gly-X-Ser-X-Gly motif contains the catalytic serine in most lipases, and HSL contains this consensus pentapeptide (Gly-Asp-Ser-Asp-Gly). Site-directed mutagenesis has confirmed that the serine of this pentapeptide is within the catalytic site of HSL [37]. Human HSL also exhibits a region of similarity with lipase 2 of *Moraxella* TA144, an Antarctic bacterium that hydrolyzes lipids at low temperature. Strikingly, HSL also manages to cleave TAGs with higher efficacy at lower temperatures compared to other mammalian lipases [32].

The higher order domain structure of HSL could not be derived from the amino acid sequence. Two different functional domains – the catalytic domain and regulatory domain (see below) – are encoded by different exons of the human gene, suggesting the mosaic nature of the HSL protein [32]. However, the overall lack of homology with other lipases precludes assignment of specific functions to different regions of the protein. Other lipases with little primary sequence homology display highly similar structural features based on crystal structure, but the crystal structure of HSL is urgently needed to gain further information. Detailed biochemical studies point to a distinct lipid-binding domain that is separate from the catalytic region. Mild proteolysis of HSL with trypsin almost completely destroyed its activity against trioleoylglycerol with only marginal reductions against water-soluble *p*-nitrophenylbutyrate, indicating that HSL contains a lipid-binding domain [33]. As trypsin treatment generates a stable 17.6 kDa fragment, HSL can reasonably be assumed to consist of separately folded protease-resistant domains joined by short protease-sensitive linker regions [38]. Studies on interfacial activation, i.e. enhancement of lipolytic activity by the presence of a lipid–water interface, provided additional evidence for a separate lipid-binding domain. Intact HSL is activated by phospholipid vesicles or emulsified trioleoylglycerol, but these lipids have no effect on activity of the 17.6 kDa fragment on the water-soluble substrate, strongly suggesting that proteolysis has separated the lipid-binding domain from the catalytic domain. A more stable peptide composed of amino acids 1–323 can be generated by prolonged incubation of HSL with endoproteinase Lys-C, suggesting that this fragment represents a separate structural domain [39]. Circular dichroism and fluorescence spectroscopy analyses during denaturation with guanidine hydrochloride and heat revealed two major steps of unfolding [40], which is in agreement with the existence of two distinct structural domains.

The amino-terminal 323 amino acid domain of possibly globular structure is encoded by exons 1–4 and has no primary or secondary structural similarity with known proteins, but interacts with the adipocyte lipid-binding protein (ALBP) [41]. The carboxy-terminal portion of HSL is similar to acetylcholinesterase, bile salt-stimulated lipase and several fungal lipases, and is composed of α/β-hydrolase folds that constitute the catalytic site. The secondary structure elements that constitute this scaffold are expected to be within regions formed by amino acids 332–535 and 688–766 (rat). Despite the lack of significant primary sequence homology, the organization of the secondary structure predicted for the carboxy-terminal 450 amino acids of HSL is similar to the secondary structure of two fungal lipases from *Geotrichum candidum* and *Candida rugosa* [41]. Secondary structure prediction in this region allowed the identification of the central elements of these α/β-hydrolase folds, according to which a three-dimensional model for this domain was proposed [41]. Elucidation of the crystal structure for the Brefeldin A esterase from *Bacillus subtilis*, a protein with significant homology to HSL, confirmed the usefulness of this approach [42]. On basis of molecular modeling it was proposed, and later confirmed by site-directed mutagenesis, that Ser423, Asp703, and His733 (rat) constitute the catalytic triad for HSL and are found within this carboxy-terminal region [43]. This portion is interrupted by the 150 amino acid regulatory module, which according to secondary structure prediction does not contain α helices and β sheets but harbours the known phosphorylation sites for PKA, AMPK, and ERKs (see Chapter 11) and therefore has been assigned a regulatory function. Recently the crystal structure of another member of the subfamily of HSL, a novel thermophilic esterase (EST2) from *Alicyclobacillus acidocaldarius*, has been solved with 2-[4-(2-hydroxyethyl)-1-piperazinol]ethanesulfonic acid (HEPES) covalently bound to the catalytic active serine 155 [44]. Whereas several crystal structures of “HSL related” esterases are available, the crystallization of HSL is still a challenge.

7.4.3 Inhibitors of HSL

Several inhibitors of lipases have been described (for a review see Patkar et al. [45]). HSL was first described in 1964, but it was not until 2000 that the first inhibitors were reported. Several companies have recently investigated HSL inhibitors to treat diabetes and lipid disorders. In 2000 Alteon [46] published a patent on eudesmannolides, a natural product isolated from Ivoa microcephala nut. One year later Aventis [47] reported a series of natural products from *Streptomyces* called cyclipostines. In the same year Aventis [48] and Novo Nordisk [49] published the first synthetic HSL inhibitors (3,4-dihydro-1*H*-isoquinolin-2-yl)carbamates, Novo Nordisk, and 3-phenyl-5-alkoxy-1,3,4-oxadiazol-2-ones, Aventis. In 2003, Ontogene [50] described pyrrolopyrazinediones and Novo Nordisk [51] reported another class

Tab. 7.1 Compounds described as inhibitors of HSL.

Company	***Structure***	***Company***	***Structure***
Alteon	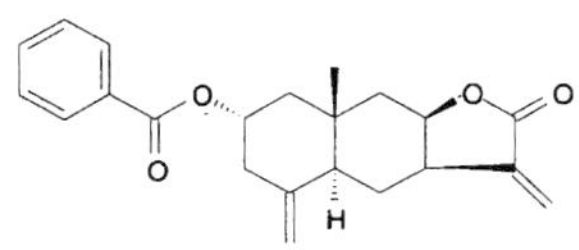 Eudesmanolide	Aventis	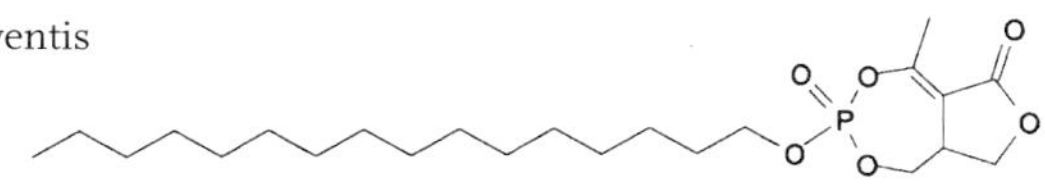 Cyclipostine
Aventis	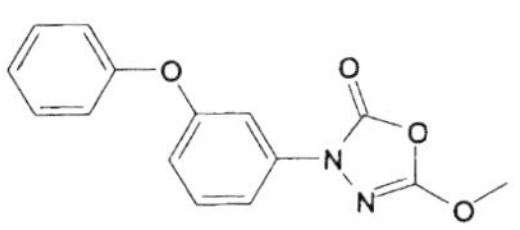 Oxadiazolone	Novo Nordisk	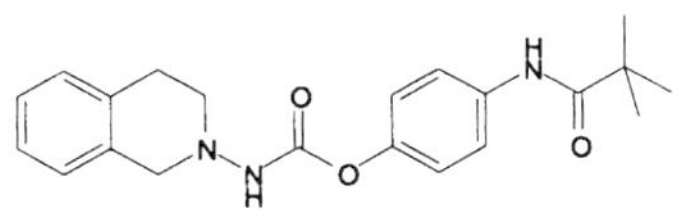 (3,4-dihydro-1H-isoquinolin-2-yl)carbamates
Ontogene	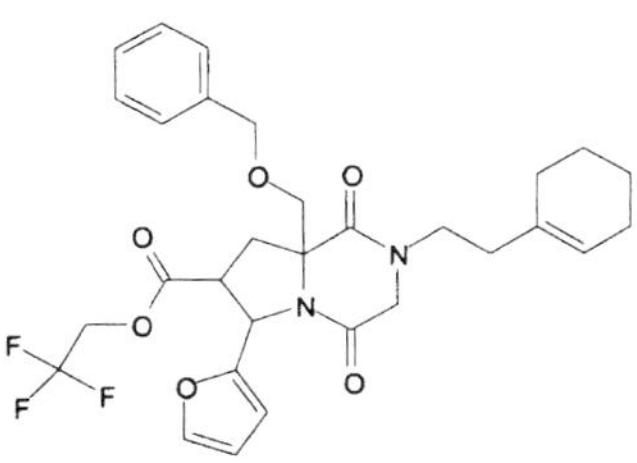 Pyrrolopyrazinediones	Novo Nordisk	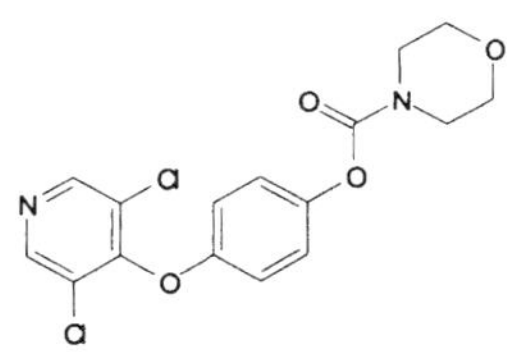 Carbamates

of carbamates. ISIS [52] reported antisense–RNA-like compounds that modulate HSL expression. Taken together, within only three years, several companies have reported progress towards the inhibition of HSL as a strategy to treat diabetes and lipid disorders (Tab. 7.1).

High-throughput screening with fluorescent monoacylglycerol [53] at Aventis revealed several classes of HSL inhibitors, with oxadiazolones as one of the most interesting. To improve the biological activity, we applied a computer-assisted experimental design strategy (CAED). A weak structure–activity relationship (SAR) from screening was used to design new compounds to broaden and strengthen SAR. Chemical synthesis, model refinement and the design of new compounds were performed in iterative cycles (Fig. 7.4).

For the first cycles we left the oxadiazolone moiety unchanged and focused on optimizing the R2-part of the molecule in Fig. 7.4. We were interested which type of hydrophobic interactions favors inhibition of HSL. Therefore, we selected R2 substituents with different aliphatic, aromatic, and alicyclic moieties. Only aromatic moieties provoked a reasonable activity. For example, **4** is inactive, whereas

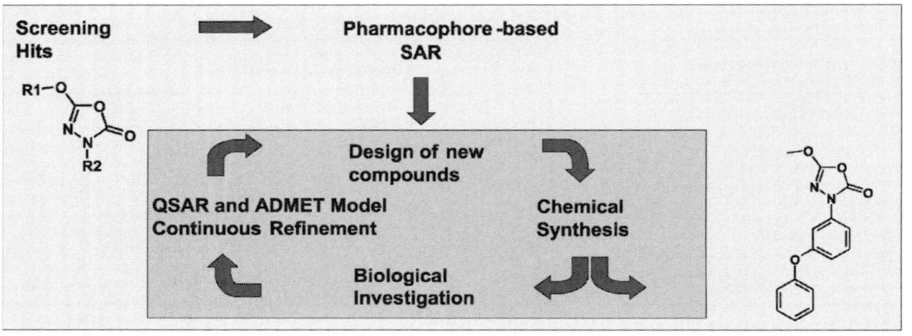

Fig. 7.4 Improvement of inhibitory potency using computer-assisted experimental design (CAED). Starting from initial "hits" (IC_{50}s in the range of 70 μM) the activity has been improved by a factor of 2800 (compound 7600, 25 nM) (QSAR: quantitative structure–activity relationship; ADMET, adsorption, distribution, metabolism, extrusion, toxicity).

the corresponding aromatic compound (**5**) inhibits HSL with considerable potency (IC_{50} 10 μM).

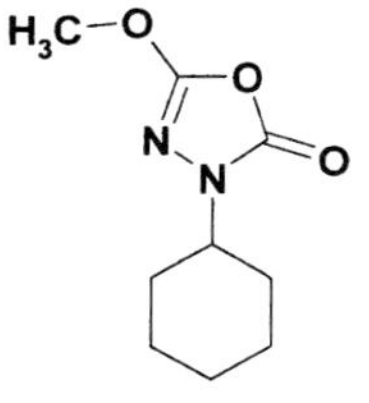

4

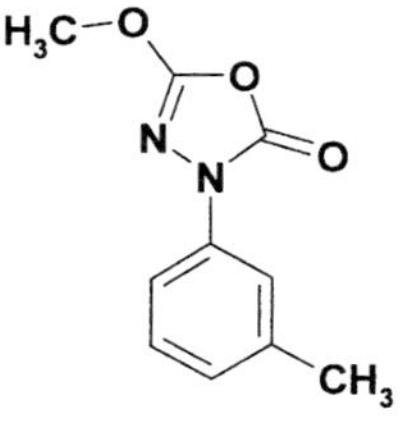

5

Both meta- and para-substitutions on the phenyl ring are well tolerated (**6**, **7**), whereas ortho-substitution seems to hamper biological activity (**8**).

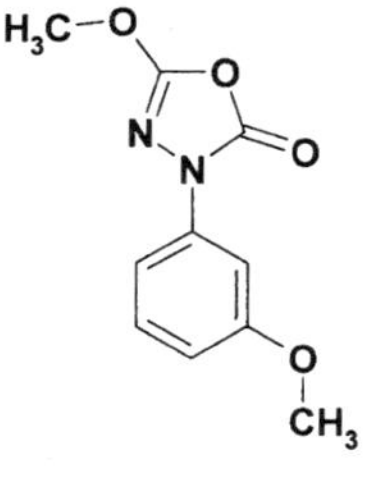

6 (10 μM)

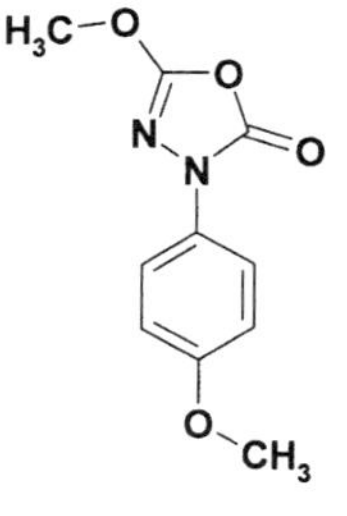

7 (30 μM)

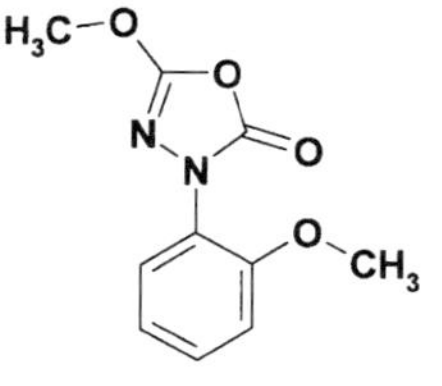

8 (inactive)

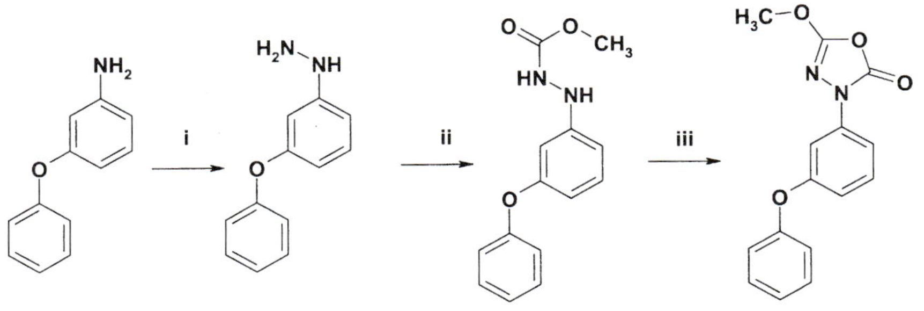

Scheme 7.2 Synthesis of HSL inhibitor 7600: i, HCl, $NaNO_2$, $SnCl_2$; ii, methoxycarbonylchloride, Py; Iii, $COCl_2$, Py

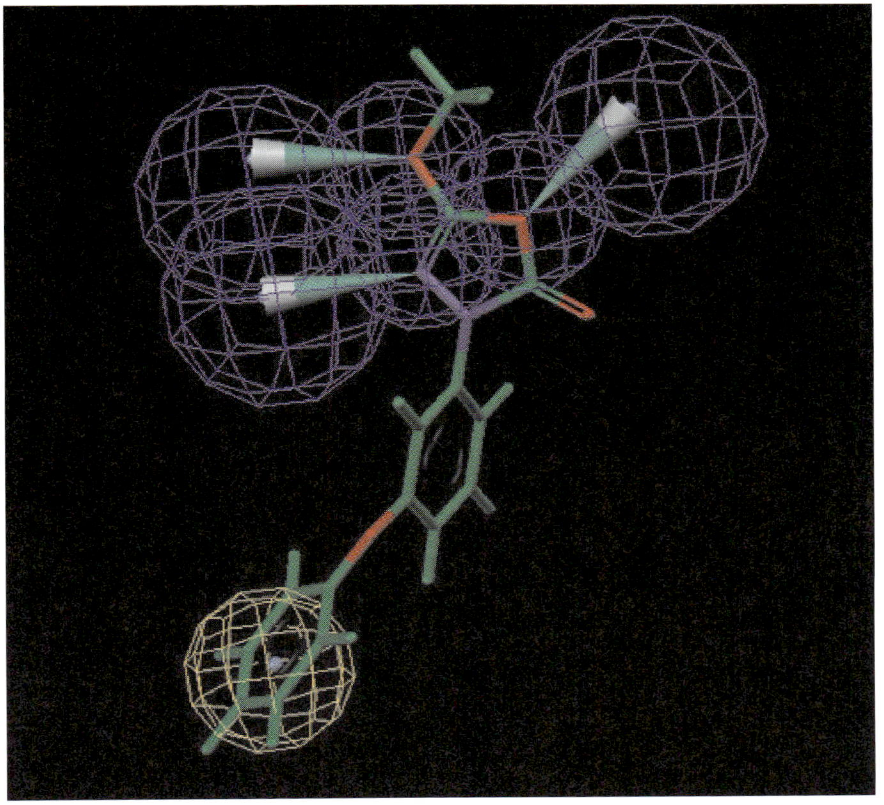

Fig. 7.5 HSL-Pharmacophore model with Aventis inhibitor 7600: Three electron-pair donors are recognized by HSL. In addition, the second phenyl ring system interacts with a hydrophobic protein domain.

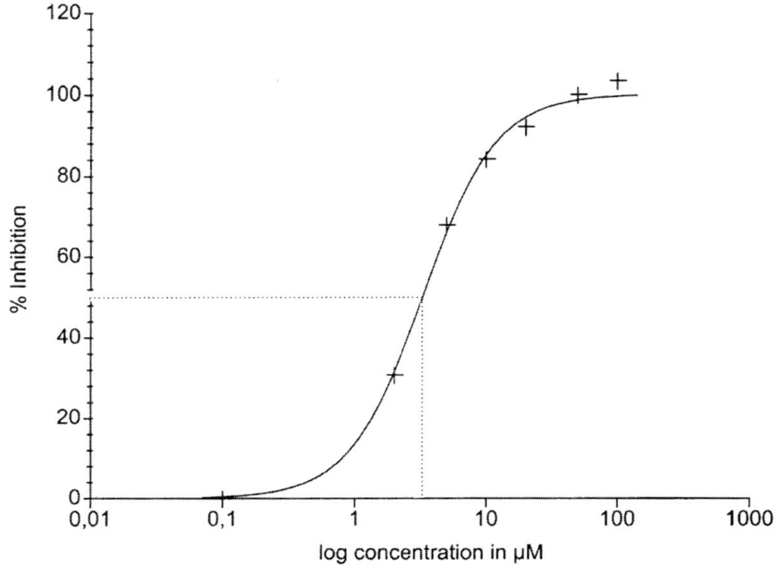

Fig. 7.6 Analysis of Aventis inhibitor 7600 using the radioactivity-based assay: Partially purified HSL from rat epididymal adipose tissue was incubated with lipid droplets prepared by ultrasonic treatment of a mixture of radiolabeled trioleoylglycerol, tripalmitin, phosphatidylcholine and phosphatidylinositol in the presence of increasing concentrations of Aventis inhibitor 7600. After incubation (2 h at 37 °C), the radiolabeled oleate released was separated from neutral lipids by alkaline partitioning and determined by liquid scintillation counting of the aqueous phase. To fit the concentration–response curves and calculate IC_{50}s, the fatty acids recovered from a control reaction performed in the absence of HSL were subtracted from each value. Details of this assay have been described previously [47].

Further optimization of **6**, with various types of substituents, revealed the 3-phenoxy derivative 7600 as the most promising molecule, the synthesis of which is shown in Scheme 7.2.

Within these two optimization cycles the initial activity has been improved (initial hit: 70 µM) by a factor of 2800 (compound 7600: 25 nM). R1 in Fig. 7.4 can also be varied, and has a minor influence on the biological activity.

We used the generated data to create a first model of the active site of HSL. After two additional cycles including further variation of the hydrophobic moiety and the electron pair-donating system of the oxadiazolone, a clear pharmacophore model was obtained (Fig. 7.5). The inhibition curve of 7600 with HSL isolated from rat adipocytes is shown in Fig. 7.6.

The higher IC_{50} (i.e. lower potency) determined for Aventis Inhibitor 7600 using the radioactive compared to the NBD-MAG assay design may be explained by different substrate presentations and/or accessibility of the lipid droplets and mixed micelles respectively for the inhibitor.

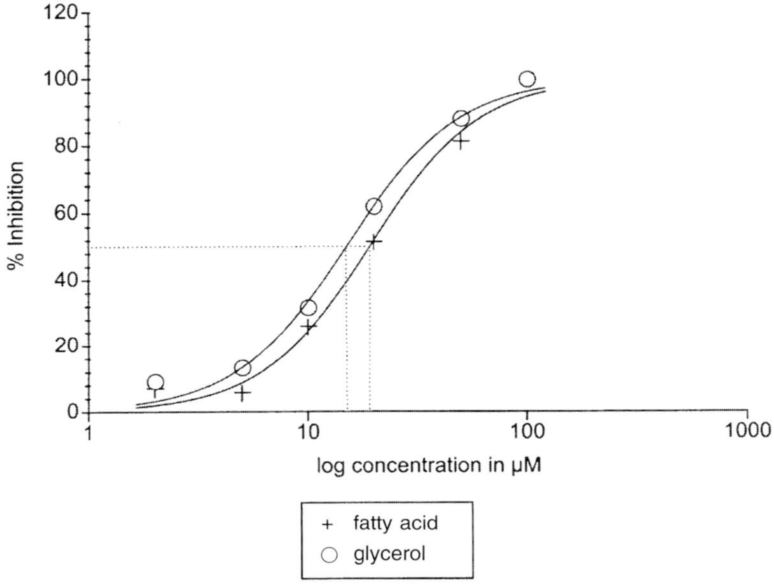

Fig. 7.7 Analysis of Aventis inhibitor 7600 using the cellular lipolysis assay: Isolated rat adipocytes were incubated with NBD-labeled dodecanoic acid (for incorporation into neutral lipid droplets) and then challenged with 1 μM isoproterenol (to stimulate lipolysis) prior to addition of increasing concentrations of inhibitor 7600. After incubation (1 h at 37 °C), the adipocytes were separated from the medium by flotation through an oil layer. Portions of the medium were enzymatically analyzed for glycerol. Portions of the perchloric acid extracts from the adipocytes were analyzed in duplicate by thin-layer chromatography and fluorescence imaging using a phosphor imager. The amount of NBD-dodecanoic acid released (comigrating with a standard) was calculated using ImageQuant software. To fit concentration–response curves and calculate IC_{50}s for glycerol and fatty acid release, the fatty acids/glycerol recovered from a control reaction performed in the presence of 100 μM inhibitor 7600 (corresponding to maximal inhibition) were substrated from each value. IC_{50}s for glycerol and fatty acid release do not differ significantly. Details of this assay have been described previously [55].

Fig. 7.7 illustrates the reduction of TAG lipolysis in isolated rat adipocytes. In both cases the isoproterenol-stimulated lipolysis is inhibited completely.

IC_{50}s for the enzyme- (radioactive design) and cellular lipolysis inhibition by Aventis inhibitor 7600 were within one order of magnitude, strongly arguing for efficient penetration of the adipocyte plasma membrane and access to cytoplasmic lipid droplets.

7.5 Perspective

A new lipase assay has been developed and used to screen the Aventis compound library. Several classes of "hits" have been elucidated – one of which, the oxadiazolones, has been selected for a chemical optimization program. Computer-assisted experimental design has enabled optimization with respect to activity and selectivity. Small potent inhibitors were obtained that are "drug-like" and follow the Lipinsky rules of five, which are often useful for a first theoretical prediction of oral bioavailability [54]. Ongoing pharmacological trials will show if lipase inhibitors can successfully treat diabetes and atherosclerosis.

7.6 References

1 S. Santamarina-Fojo, C. Haudenschild, and M. Amar, The role of hepatic lipase in lipoprotein metabolism and atherosclerosis, *Curr. Opin. Lipidol.*, **1998**, *9*, 211–219.

2 I. J. Goldberg, Lipoprotein lipase and lipolysis: central roles in lipoprotein metabolism and atherogenesis, *J. Lipid Res.*, **1996**, *37*, 693–707.

3 M. Jaye, K. J. Lynch, J. Krawiec, D. Marchadier, C. Maugeais, K. Doan, V. South, D. Amin, M. Perrone, and D. J. Rader, A novel endothelial-derived lipase that modulates HDL metabolism, *Nat. Genet.*, **1999**, *21*, 424–428.

4 K.-I. Hirata, H. L. Dichek, J. A. Cioffi, S. Y. Choi, N. J. Leeper, I. Quintana, G. S. Kronmal, A. D. Cooper, and T. Quertermous, Cloning of a unique lipase from endothelial cells extends the lipase gene family, *J. Biol. Chem.*, **1999**, *274*, 14170–14175.

5 W. Jin, S. John, U. Broedl, J. M. Glick, and D. J. Rader, Inhibition of endothelial lipase causes increased HDL cholesterol levels in vivo, *J. Clin. Invest.*, **2003**, *111*, 357–362.

6 C. Holm, T. Osterlund, H. Laurell, and J. A. Contreras, Molecular mechanisms regulating hormone-sensitive lipase and lipolysis, *Annu. Rev. Nutr.*, **2000**, *20*, 365–393.

7 G. Bengtsson-Olivecrona and T. Olivrecrona. In: *Lipases* (P. Wooley and S. B. Petersen, eds.), Cambridge University Press, Cambridge, **1994**, pp. 315–336.

8 W.A. Hide, L. Chain, and W. H. Li, Structure and evolution of the lipase superfamily, *J. Lipid Res.*, **1992**, *33*, 167–178.

9 R. Verger, Interfacial activation of lipases: facts and artifacts, *TIBS TECH*, **1997**, *15*, 32–38.

10 T. Fujimoto, H. Kogo, K. Ishiguro, K. Tauchi, and R. Nomura, Caveolin-2 is targeted to lipid droplets, a new "membrane domain" in the cell, *J. Cell. Biol.*, **2001**, *152*, 1079–1085.

11 A. G. Ostermeyer, J. M. Paci, Y. Zeng, D. M. Lublin, S. Munro, and D. A. Brown, Accumulation of caveolin in the endoplasmic reticulum redirects the protein to lipid storage droplets, *J. Cell. Biol.*, **2001**, *152*, 1071–1078.

12 I. Panaiotov and R. Verger, in: *Physical Chemistry of Biological Interfaces* (A. Baszkin and W. Norde, eds.), Dekker, New York, **2000**, pp. 359–400.

13 R. D. Schmid and R. Verger, Lipases: interfacial enzymes with attractive applications, *Angew. Chem. Int. Ed.*, **1998**, *37*, 1609–1633.

14 H. S. Hendrickson, Fluorescence-based assays of lipases, phospholipases, and

other lipolytic enzymes, *Anal. Biochem.*, **1944**, *219*, 1–8.

15 F. Beisson, A. Tiss, C. Riviere, and R. Verger, Methods for lipase detection and assay: a critical review, *Eur. J. Lipid Sci. Technol.*, **2000**, pp. 133–153.

16 R. Gupta, P. Rathi, N. Gupta, and S. Bradoo, Lipase assays for conventional and molecular screening: an overview, *Biotechnol. Appl. Biochem.*, **2003**, *37*, 63–71.

17 M. J. Hills and K. D. Mukherjee, Assay for triacylglycerol lipase by a rapid thin-layer chromatographic technique, *J. Lipid Res.*, **1988**, *29*, 1397–1399.

18 A. Kaplan, Kaplan, Arnold. Simple radioactive assay for triglyceride lipase, *Anal. Biochem.*, **1970**, *33*, 218–225.

19 A. K. Das and A. K. Hajra, A novel chemical synthesis of 1-O-hexadecyl-rac-[2-3H] glycero-3-phosphorylethanolamine and a simple assay for plasmanyl desaturase, *J. Lipid Res.*, **1996**, *37*, 2706–2714.

20 F. Beisson, N. Ferte, J. Nari, G. Noat, V. Arondel, and R. Verger, Use of naturally fluorescent triacylglycerols from Parinari glaberrimum to detect low lipase activities from Arabidopsis thaliana seedlings, *J. Lipid Res.*, **1999**, *40*, 2313–2331.

21 G. Pencreac'h, J. Graille, M. Pina, and R. Verger, An ultraviolet spectrophotometric assay for measuring lipase activity using long-chain triacyglycerols from Aleurites fordii seeds. *Anal. Biochem.*, **2002**, *303*, 17–24.

22 T. de Laborde Monpezat, B. de Jeso, J.-L. Butour, L. Chavant, and M. Sancholle, A fluorimetric method for measuring lipase activity based on umbelliferyl ester, *Lipids*, **1990**, *25*, 661–664.

23 T. Meshulam, H. Herscovitz, D. Casavant, J. Bernardo, R. Roman, R. P. Haugland, G. S. Strohmeier, R. D. Diamond, and E. R. Simons, Flow cytometric kinetic measurements of neutrophil phospholipase A activation, *J. Biol. Chem.*, **1992**, *267*, 21465–21470.

24 V. Agmon, S. Cherbu, A. Dagan, M. Grace, G. A. Grabowski, and S. Gatt, Synthesis and use of novel fluorescent glycosphingolipids for estimating β-glucosidase activity in vitro in the absence of detergents and subtyping Gaucher disease variants following administration into intact cells, *Biochim. Biophys. Acta*, **1993**, *1170*, 72–79.

25 H. Scholze, H. Stutz, F. Paltauf, and A. Hermetter, Fluorescent inhibitors for the qualitative and quantitative analysis of lipolytic enzymes, *Anal. Biochem.*, **1999**, *276*, 72–80.

26 P. Belfrage, G. Fredrikson, P. Strålfors, and H. Tornquist, *Lipases* (B. Borgström and H. L. Brockman, eds.), Elsevier Science, Amsterdam **1984**, pp. 365–416.

27 P. Strålfors and P. Belfrage, Phosphorylation of hormone-sensitive lipase by cyclic AMP-dependent protein kinase, *J. Biol. Chem.*, **1983**, *258*, 15146–15152.

28 P. Strålfors, P. Björgell, and P. Belfrage, Hormonal regulation of hormone-sensitive lipase in intact adipocytes: identification of phosphorylated sites and effects on the phosphorylation by lipolytic hormones and insulin, *Proc. Natl. Acad. Sci. USA*, **1984**, *81*, 3317–3321.

29 M. Vaughan, J. E. Berger, and D. Steinberg, Hormone-sensitive lipase and monoglycerol lipase activities in adipose tissue, *J. Biol. Chem.*, **1964**, *239*, 401–409.

30 G. Fredrikson, P. Strålfors, N. O. Nilsson, and P. Belfrage, Hormone-sensitive lipase of rat adipose tissue: Purification and some properties, *J. Biol. Chem.*, **1981**, *256*, 6311–6319.

31 C. Holm, T. G. Kirchgessner, K. L. Svenson, G. Fredrikson, S. Nilsson, C. G. Miller, J. E. Shively, C. Heinzmann, R. S. Sparkes, and T. Mohandas, Hormone-sensitive lipase: sequence, expression, and chromosomal localization to 19 cent-q13.3, *Science*, **1988**, *241*, 1503–1506.

32 D. Langin, H. Laurell, L. S. Holst, P. Belfrage, and C. Holm, Gene organization and primary structure of human hormone-sensitive lipase: possible significance of a sequence homology with a lipase of *Moraxella* TA144, an Antarctic bacterium, *Proc. Natl. Acad. Sci. USA*, **1993**, *90*, 4897–4901.

33 S. J. Yeaman, Hormone-sensitive lipase – a multipurpose enzyme in lipid metabolism, *Biochem. Biophys. Acta*, **1990**, *1052*, 128–132.

34 P. Belfrage, B. Jergil, P. Strålfors, and H. Tornqvist, Hormone-sensitive lipase of rat adipose tissue: identification and some properties of the enzyme protein, *FEBS Lett.*, **1977**, *75*, 259–264.

35 W.-J. Shen, S. Patel, and F. B. Kraemer, Hormone-sensitive lipase functions as an oligomer, *Biochem. J.*, **2000**, *328*, 137–143.

36 P. Strålfors, H. Olsson, and P. Belfrage, Hormone-sensitive lipase (P. D. Boyer, E. G. Krebs eds.), *The Enzymes,* 3rd edn, Vol. 18, Part B, Academic Press, London, **1987**, pp. 147–177.

37 C. Holm, R.C. Davis, and T. Osterlund, Identification of the active site serine of hormone-sensitive lipase by site-directed mutagenesis, *FEBS Lett.*, **1994**, *344*, 234–238.

38 T. Osterlund, B. Danielsson, E. Degerman, J. A. Contreras, G. Edgren, R. C. Davis, M. C. Schotz, and C. Holm, Domain-structure analysis of recombinant rat hormone-sensitive lipase, *Biochem. J.*, **1996**, *319*, 411–420.

39 T. Osterlund, D. J. Beussman, K. Julenius, P. H. Poon, and S. Linse, Domain identification of hormone-sensitive lipase by circular dichroism and fluorescence spectroscopy, limited proteolysis, and mass spectroscopy, *J. Biol. Chem.*, **1999**, *274*, 15382–15388.

40 W.-J. Shen, K. Sridhar, D.A. Bernlohr, and F.B. Kraemer, Interaction of rat hormone-sensitive lipase with adipocyte lipid-binding protein, *Proc. Natl. Acad. Sci. USA*, **1999**, *96*, 5528–5532.

41 J. A. Contreras, M. Karlsson, T. Osterlund, H. Laurell, A. Svensson, and C. Holm, Hormone-sensitive lipase is structurally related to acetylcholinesterase, bile salt-stimulated lipase, and several fungal lipases: building of a three-dimensional model for the catalytic domain of hormone-sensitive lipase, *J. Biol. Chem.*, **1996**, *271*, 31426–31430.

42 Y. Wei, J. A. Contreras, P. Sheffield, T. Osterlund, and U. Derewenda, Crystal structure of brefeldin A esterase, a bacterial homolog of the mammalian hormone-sensitive lipase, *Nat. Struct. Biol.*, **1999**, *6*, 340–345.

43 T. Osterlund, J. A. Contreras, and C. Holm, Identification of essential aspartic and histidine residues of hormone-sensitive lipase: apparent residues of the catalytic triad, *FEBS Lett.*, **1997**, *403*, 259–262.

44 G. Simone De, S. Galdiero, G. Maco, D. Lang, M. Rossi, and C. Pedone, A snapshot of a transition state analogue of a novel thermophilic esterase belonging to the subfamily of mammalian hormone-sensitive lipase, *J. Mol. Biol.*, **2000**, *303*, 761–771.

45 S. Patkar and F. Bjoerkling, *Lipases* (P. Wooley and S. B. Petersen, eds.), Cambridge University Press, Cambridge, **1994**, pp. 207–224.

46 D. Wagle, S. D. Fang, I. Terlecky, J. Resek, and J. Egan, Preparation and formulation of eudesmanolides for use as glucose and lipid lowering compounds, *PCT Int. Appl.*, **2000**, WO 0027388 a1, 41 pp.

47 (a) L. Vertesy, K. Ehrlich, M. Kurz, and J. Wink, Method for the production of cyclipostins obtained by the cultivation of Streptomyces species HAG 004107 (DSM 13381) and their use as inhibitors of lipases, *PCT Int. Appl.*, **2001**, WO 0183497 A1, 56 pp. (b) L. Vertesy, B. Beck, M. Bronstrup, K. Ehrlich, M. Kurz, G. Müller, D. Schummer, G. Seibert, Cyclipostins, novel hormone-sensitive lipase inhibitors from Streptomyces sp. DSM 13381: II. Isolation, structure elucidation and biological properties, *J. Antibiot.*, **2002**, *55*, 480–494.

48 K. Schoenafinger, S. Petry, G. Müller, and K.-H. Baringhaus, Preparation of 3-phenyl-5-alkoxy-1,3,4-oxadiazol-2-ones as hormone-sensitive lipase inhibitors, *PCT Int. Appl.*, WO 0166531 A1, **2001**, 44 pp.

49 H. Beltrandelrio, P. Jacobsen and J. Cornelis De Jong (Novo Nordisk A/S, Den.), Preparation of (3,4-dihydro-1H-isoquinolin-2-yl)carbamates for treating disorders where a decreased level of plasma FFA is desired, *PCT Int. Appl.*, WO 2001087843 A1, **2001**, 80 pp.

50 D. H. Slee, A. S. Bhat, T. N. Nguyen, M. Kish, K. Lundeen, M. J. Newman, and S. J. McConnel, Pyrrolopyrazinedione-based inhibitors of human hormone-sen-

sitive lipase, *J. Med. Chem.*, **2003**, *46*, 1120–1122.

51 S. Ebdrup, H.C. Hansen, P. Vedso, J. Cornelis De Jong, and P. Jacobsen, Preparation of carbamates as hormone-sensitive lipase inhibitors for the treatment of diabetes and related disorders, *PCT Int. Appl.*, WO 0351842 A2, **2003**, 390 pp.

52 M.M. Butler, A.T. Watt, S.M. Freier, and J.R. Wyatt, Antisense modulation of hormone-sensitive lipase expression, *PCT Int. Appl.*, WO 0310139 A2, **2003**, 167 pp.

53 G. Müller, S. Petry, H. Jordan, H. Kleine, H. Wenzel, Fluorescence-labeled acylglycerides forming charge-transfer complexes in association with phospholipids and their usage as substrates in lipase and lipase inhibition assays, *PCT Int. Appl.*, WO 2000067025 A1, **2000**, 44 pp.

54 C.A. Lipinski, F. Lombardo, B.W. Dominy, and P.J. Feeney, Experimental and computational approaches to estimate solubility and permeability in drug discovery and development settings, *Adv. Drug. Delivery Rev.*, **1997**, *23*, 3–25.

55 G. Müller, H. Jordan, C. Jung, H. Kleine, and S. Petry, Analysis of lipolysis in adipocytes using a fluorescent fatty acid derivative, *Biochimie*, in press.

8
Endothelial Lipase: A Novel Drug Target for HDL and Atherosclerosis?

Karen Badellino, Weijun Jin, and Daniel J. Rader

8.1
Introduction

The production, remodeling and clearance of cholesterol, triglyceride and phospholipid in the form of plasma lipoproteins are part of the complex, regulated process of lipid metabolism. Two major categories of enzymes are involved in the process of lipoprotein lipid hydrolysis and clearance: triglyceride lipases and phospholipases. Until recently, two triglyceride lipases were known that acted upon lipoproteins in the plasma compartment: lipoprotein lipase (LPL) and hepatic lipase (HL). They have been well characterized for their lipid substrate preference, lipoprotein association and ability to mediate binding and uptake of lipoproteins by vascular cells. Both LPL and HL have well-established roles in lipid metabolism and in modulating atherosclerosis [1, 2]. LPL prefers triglycerides as substrates and hydrolyzes phospholipids very poorly. HL has a greater ability to hydrolyze phospholipids but also prefers triglycerides. Interventions that influence the activity of these two enzymes markedly affect plasma triglyceride levels, have more moderate effects on LDL-C (low density lipoprotein) and HDL-C (high density lipoprotein) levels, and can influence atherosclerosis [1, 2].

In 1999, a gene encoding endothelial lipase was reported as a new member of the triglyceride lipase gene family [3, 4]. This gene was cloned simultaneously by two different groups using two different strategies. Jaye et al. [3] exposed the monocyte cell line, THP-1, to oxidized LDL-C after differentiation with phorbol 12-myristate 13-acetate (PMA). An RNA amplification product was found in the oxidized LDL-treated cells but not in untreated cells. This product was used to probe a human placental complementary DNA (cDNA) library. A clone was obtained with an open reading frame of 500 amino acids with significant sequence homology with hepatic lipase, pancreatic lipase and lipoprotein lipase. Subsequent Northern and Western blotting showed that endothelial lipase (EL) is unique in that it is constitutively expressed by human coronary artery endothelial cells (HCAEC) and human umbilical vein endothelial cells (HUVEC). Hirata et al. [4] independently cloned EL through subtractive hybridization from HUVECs undergoing tube formation compared to growth-arrested cells in monolayers. The cDNA that was upregulated in the early stages of tube formation was found to be a lipase with significant sequence identity with hepatic lipase and lipoprotein lipase.

Lipases and Phospholipases in Drug Development
Edited by G. Müller and S. Petry

ISBN: 3-527-30677-3

8.2 Structure of Endothelial Lipase

The human gene encoding for endothelial lipase (LIPG) is located on chromosome 18q21.1 (*http://www.ncbi.nlm.nih.gov*) and spans 10 exons and 9 introns. The open reading frame of the human EL sequence is 1500 nucleotides, encoding a protein of 500 amino acids and a predicted 18 amino acid signal peptide. By comparison of the primary amino acid sequence, endothelial lipase has 45% homology with lipoprotein lipase and 40% homology with hepatic lipase. The location of 10 cysteine residues is conserved (Fig. 8.1), suggesting that it shares a similar tertiary structure with both HL and LPL, including a 19 amino acid lid region. The EL sequence contains the same potential catalytic residues, ^{149}serine, ^{173}aspartic acid, and ^{254}histidine, found in the other lipase family members. The GXSXG lipase motif surrounding the active site serine is also retained. Based on the presence of positively charged residues that resemble heparin-binding consensus sequences, there are four possible heparin-binding regions in endothelial lipase, ^{14}KLHKPK^{19}, ^{282}RFKK^{285} and ^{292}RKNR^{295}, ^{304}KKMRNKR^{310}, and ^{427}RRIRVK^{432}. By comparison to both the crystal structure of pancreatic lipase [5, 6] and a molecular model of lipoprotein lipase [7], there are two potential lipid-binding regions:

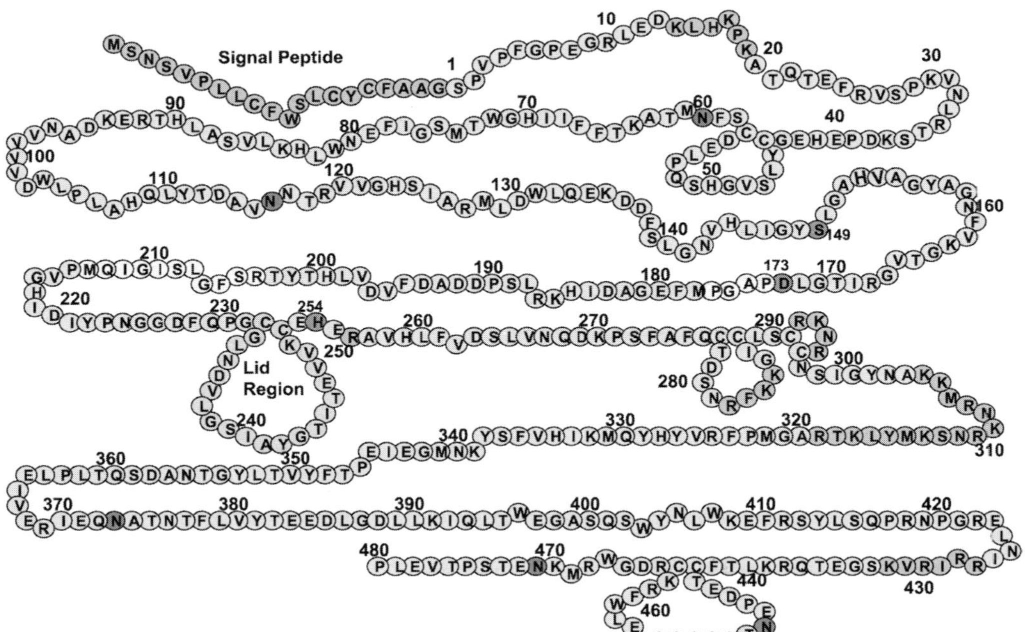

Fig. 8.1 A model of endothelial lipase was created using the reported amino acid sequence [3]. The structure was based on the crystal structure of pancreatic lipase [5] and a published molecular model of lipoprotein lipase [7].

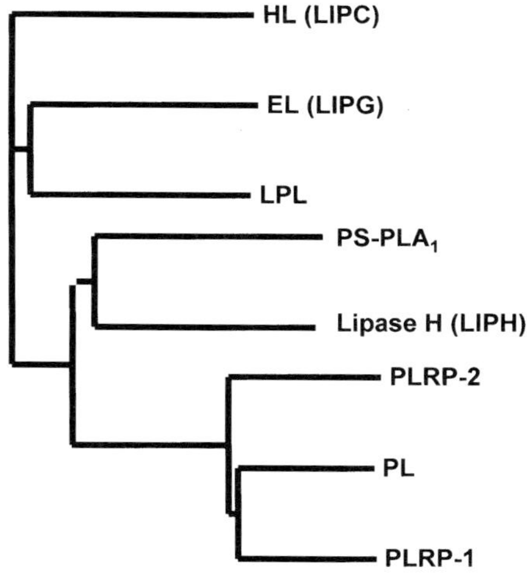

Fig. 8.2 Endothelial lipase (EL) is a member of the triglyceride lipase gene family. It more closely resembles lipoprotein lipase (LPL) and hepatic lipase (HL) and is predicted to have arisen from a common precursor.

^{170}GLDPAGP^{177} and ^{204}RSFGLSIGIQM^{214}. There are five potential N-linked glycosylation sites. Within the triglyceride lipase gene family, EL most closely resembles LPL and HL and is predicted to have arisen from a common precursor (Fig. 8.2).

The size of expressed endothelial lipase was determined by immunoblot analysis of both HUVEC and HCAEC. Bands of 68 and 40 kDa were detected using an antibody directed against the amino-terminus of EL [3]. The expressed protein, based on amino acid translation of the EL cDNA, is predicted to be approximately 55 kDa. The 68 kDa form may represent the glycosylated full-length monomer, whereas the 40 kDa form may be the product of alternative splicing or a proteolytic cleavage product or both.

8.3 Tissue Expression of Endothelial Lipase and Its Implications

The tissue expression of endothelial lipase is unique in comparison to both LPL and HL. LPL is predominantly expressed by cardiac muscle, skeletal muscle and adipocytes with lesser amounts produced by macrophages, kidney, adrenal gland, liver, pancreas, lung and brain [8, 9]. HL is primarily expressed in the liver [10]. Analysis of various human tissues has detected EL RNA in placenta, lung, liver, kidney, testis and abundantly in the thyroid in both human and mouse tissues [3, 4]. In a transgenic mouse model of human endothelial lipase under control of its own promoter, Ishida et al. [11] detected EL protein by immunohistochemistry on the surface of endothelium of the aorta and large vessels in the kidney, lung, and spleen as well as the microcirculation in the spleen and lung. EL is expressed by endothelial cells in culture as detected at both the mRNA level [3] and the protein level [12].

The functional significance of the high EL expression in lung, placenta and thyroid is undetermined. Hirata et al. [4] noted that the distribution of EL in lung, as detected by in situ hybridization, suggested that EL expression may be localized to type II alveolar epithelial cells, which synthesize surfactant, a phosphatidylcholine-rich substance. EL might be involved in surfactant metabolism.

The character of HDL in cord blood is distinct from adult blood. Fetal HDL is apolipoprotein E-rich, with much lower esterified cholesterol. HDL represents the major lipoprotein class in cord blood, a marked change from the adult plasma lipoprotein profile. The phospholipid content was the same as the HDL of adults [13]. A plausible hypothesis is that EL is directly involved in lipid metabolism and cholesterol transport in the developing fetus.

Thyroid hormone decreases hepatic lipase activity [14]; deficiency of thyroid hormone does not result in significant effects on HDL levels, however. Individuals with hypothyroidism have either unchanged or slightly elevated HDL levels [15]. It is possible that EL expression in the thyroid is regulated by thyroid hormone or that EL is involved in thyroid lipid metabolism.

8.4 Enzymatic Activity and Effects on Cellular Lipid Metabolism of Endothelial Lipase

EL can hydrolyze both triglyceride and phospholipids. In the initial reports on this enzyme, two groups reported the ability of EL in conditioned medium to hydrolyze phosphatidylcholine. Hirata et al. originally reported an inability of EL to hydrolyze triglyceride [4], whereas Jaye et al. reported that EL hydrolyzed a radiolabeled triolein substrate, albeit at a low level [3].

McCoy et al. further characterized EL lipolytic activity and directly compared it to both hepatic lipase and lipoprotein lipase using various substrates [16]. In the absence of serum as a source of apolipoprotein CII, the ability of endothelial lipase to hydrolyze tributyrin, containing 3-carbon acyl chains, and triolein, an unsaturated octadecenoic acid, was similar. The effects of the addition of heat-inactivated serum on the triglyceride lipase activity of endothelial lipase and lipoprotein lipase were investigated using a glycerol-stabilized emulsion of ^{3}H-triolein and egg phosphatidylcholine. While addition of serum (as a source of apoC-II) to lipoprotein lipase potentiated its activity nearly 1000%, as expected, addition of the same amount of serum to endothelial lipase completely abolished its ability to hydrolyze triolein. The effect of serum on the phospholipase activity of EL was found to have a similar inhibitory effect. The source of this inhibition is not a result of plasma lipoproteins acting as competitive substrates, since lipoprotein-deficient serum also inhibited EL triglyceride lipase activity by 84%.

McCoy et al. also compared EL to LPL and HL in their substrate preferences [16]. Using lipases expressed by adenoviral gene transfer into Cos-1 cells, the triglyceride lipase to phospholipase activity ratios were determined to be 140 for LPL (plus serum), 24 for HL and 0.65 for EL (Fig. 8.3). These ratios demonstrate the preference of endothelial lipase for phospholipid substrates versus triolein substrates as compared to LPL and HL. The effect of purified apolipoprotein CII, at 10 μg ml^{-1}, on both triglyceride lipase and phospholipase activities of LPL was compared to the effect on EL activity. As expected, apolipoprotein CII potentiated both the triglyceride lipase and phospholipase activities of LPL. The addition of apolipoprotein CII to either activity assay had no effect on EL. The addition of 1 M NaCl potentiates the activity of hepatic lipase but inhibits lipoprotein lipase. In a manner similar to LPL inhibition, EL triglyceride lipase and phospholipase activities were completely abolished in the presence of 1 M NaCl.

The significance of the ability of HL and LPL to hydrolyze synthetic triglyceride and phospholipids lies in its modeling of the hydrolysis of naturally occurring lipoprotein substrates. McCoy et al. compared EL with HL and LPL in their preference for lipoproteins isolated from non-fasted hyperlipemic subjects (Fig. 8.4). Lipoprotein fractions containing 1.25 mM phospholipid were incubated with conditioned media containing EL, HL or LPL. A comparison of the free fatty acid released per reaction tube showed that, while EL hydrolyzes triglyceride-containing lipoproteins, it is more efficient in hydrolyzing HDL than either HL or LPL. Therefore, EL lies at the opposite end of the lipolytic spectrum from LPL: whereas LPL

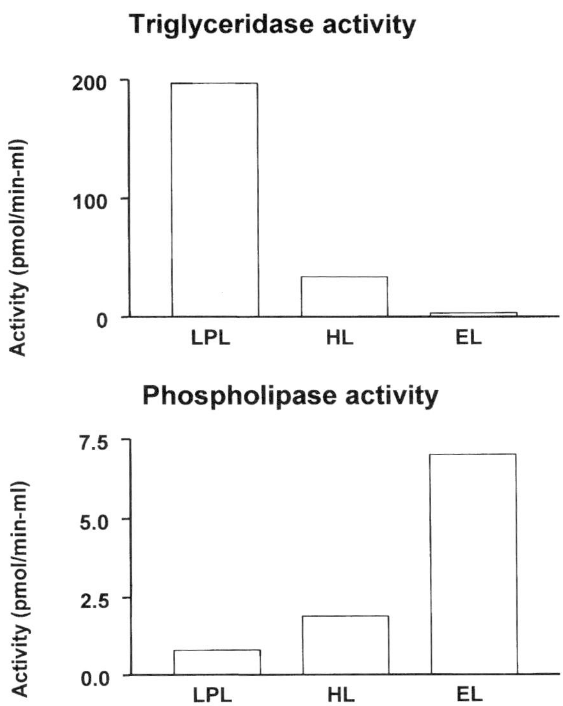

Fig. 8.3 Aliquots of EL, HL and LPL medium were assayed for triglyceride lipase and phospholipase activities [16]. The ratio of triglyceride lipase to phospholipase activity was then determined.

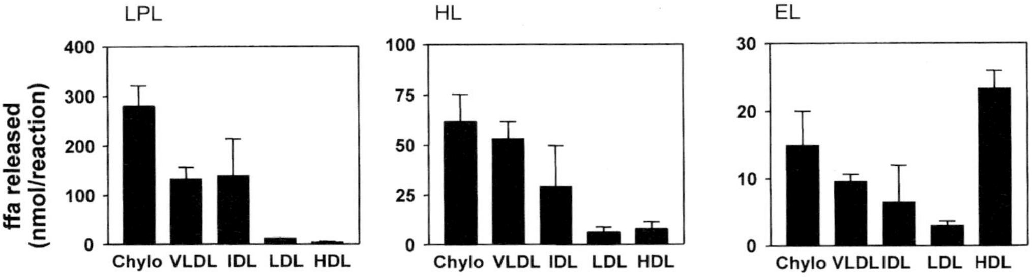

Fig. 8.4 Comparison of the hydrolysis of isolated lipoprotein fractions by conditioned media containing LPL, HL or EL as indicated [16].

preferentially hydrolyzes triglycerides in triglyceride-rich lipoproteins, EL preferentially hydrolyzes phospholipid-rich HDL, with hepatic lipase in between (Fig. 8.5).

The effect of endothelial lipase hydrolysis of HDL phospholipids on cellular lipid metabolism was examined by Strauss et al. [17]. Using either HepG2 cells infected with adenovirus encoding EL, or 293-HEK cells transiently transfected with EL, they studied the fate of ^{14}C-dipalmitoylphosphatidylcholine that had been incorporated into HDL isolated from normolipemic human plasma. In contrast to control LacZ adinfected cells, there was a 1.6-fold increase in ^{14}C-phospholipids and a 2-fold increase in ^{14}C-triglycerides in the lysate of cells overexpressing endothelial lipase. Infection of HepG2 cells with a catalytically inactive mutant EL also resulted in an increase in amount of ^{14}C-phospholipid in those cells as compared with LacZ infected control cells. This suggests that endothelial lipase can mediate binding and uptake of HDL to cells *in vitro* in a process independent of its enzymatic activity. The ability of endothelial lipase to mediate HDL binding and selective uptake of cholesterol esters to HepG2 cells was also studied [17]. HepG2 cells infected with adenovirus encoding EL had a 1.5-fold increase in binding of HDL and a 1.8-fold increase in selective uptake of HDL-cholesterol ester as compared with adLacZ control cells. Catalytically inactive EL-expressing cells had 1.7-fold higher HDL binding and 1.8-fold higher holoparticle HDL uptake, as compared to wild-type EL expressing cells, but an equivalent level of selective uptake of cholesteryl ester, independent of SR-BI. In the presence of tetrahydrolipstatin, a lipase inhibitor, there was a 5.2-fold increase in HDL binding, a 2.6-fold increase in holoparticle uptake and a 1.1-fold increase in selective uptake.

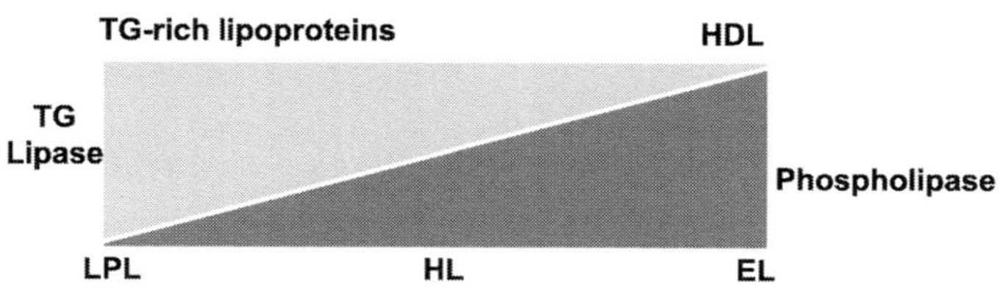

Fig. 8.5 Continuum of lipid and lipoprotein preferences for LPL, HL and EL.

8.5 Regulation of Endothelial Lipase Expression

Hirata et al. [4] have noted that activated endothelial cells had increased endothelial lipase mRNA. Jaye et al. found that, upon stimulation with phorbol ester, the intensity of both 68 and 40 kDa bands increased in HCAEC and new bands of approximately 55 kDa, 38 and 36 kDa appeared in both HCAEC and HUVEC upon stimulation [3]. These new bands were not characterized, but may represent alternative splice products or an enzymatic cleavage product. Hirata et al. demonstrated that the expression of endothelial lipase is regulated by cytokines, by fluid shear stress and by cyclic stretch [18]. Using both HUVEC and HCAEC, they examined the effects of IL-1β and TNF-α on endothelial lipase mRNA levels. Exposure to IL-1β for 24 h resulted in a 6.5-fold increase in EL mRNA in both HUVEC and HCAEC. TNF-α produced a ca. 4-fold increase in EL mRNA. The same group examined the response of both HUVEC and HCAEC to shear stress, and found a 2–3-fold increase in EL mRNA after 6 h of exposure. Additionally, cyclic stretch resulted in a 2.7-fold increase in EL mRNA. They concluded that EL expression is highly regulated by factors implicated in the development of vascular disease.

Jin et al. have examined the effects of cytokines on EL mRNA, protein, and activity in HUVEC and two microvascular cell lines [12]. Stimulation of HUVEC with either IL-1β or TNF-α resulted in a marked increase in both triglyceride lipase and phospholipase activities of culture medium. Immunoprecipitation of this culture medium with rabbit polyclonal anti-endothelial lipase antibody resulted in loss of most phospholipase and triglyceride lipase activity from cytokine-stimulated cells, indicating that EL is the source of most of the secreted lipase activity. There is also a dose-dependent increase in both protein expression (Fig. 8.6) and mRNA levels in response to increasing doses of either IL-1β or TNF-α, an effect that was also seen in both dermal microvascular cells and uterine microvascular cells. Using various agents that antagonize the transcriptional effects of cytokines on gene expression, they found that only SN50, an inhibitor of the NF-κB pathway, partially abolished EL mRNA expression.

The cytokine-induced increase in endothelial lipase expression and secretion by endothelial cells is in marked contrast to the effect of cytokines on hepatic lipase and lipoprotein lipase expression. Feingold et al. found that cytokines down-regulate lipoprotein lipase [19] and hepatic lipase [20] expression. Both acute and chronic inflammatory conditions are associated with decreased HDL levels [21]. A decrease in HL levels in inflammatory states would be expected to increase HDL levels. These results support the possibility that endothelial lipase has a significant role in regulating HDL levels, especially in inflammatory states.

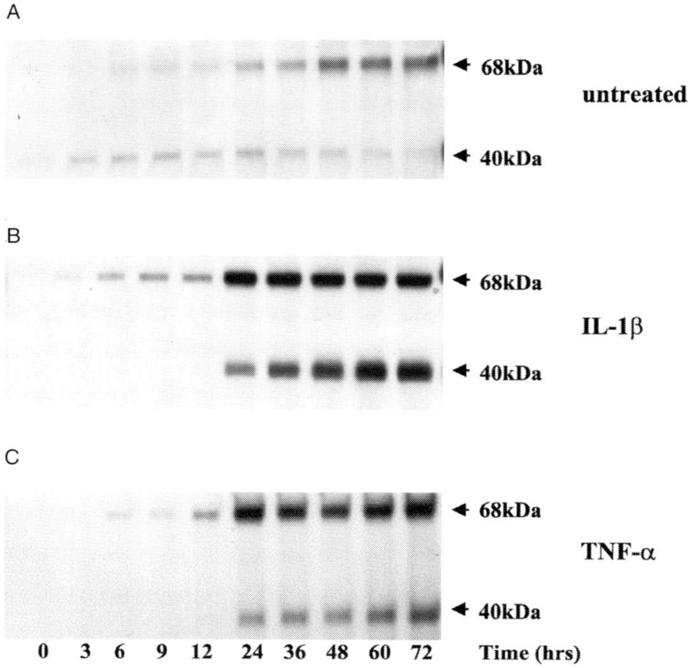

Fig. 8.6 EL protein secretion in HUVEC increases in response to increasing amounts of either TNFα or IL-1β [12]. Unstimulated HUVEC represents baseline expression and PMA-stimulated cells serve as a positive control.

8.6 Physiology of Endothelial Lipase

The physiological role of endothelial lipase (EL) in regulating lipoprotein metabolism has been demonstrated by overexpression and loss of function studies *in vivo*. Jaye et al. [3], using somatic gene transfer with adenovirus encoding human EL cDNA in wild-type mice, showed that hepatic overexpression of EL resulted in a dramatic reduction of HDL-C and apoA-I levels (Fig. 8.7). Three days after vector injection, HDL-C levels were barely detectable, and remained extremely low for up to 4 weeks. These results were the first indication that EL expression has major effects on HDL metabolism *in vivo*. To a lesser extent, overexpression of EL decreased the amount of VLDL/LDL cholesterol to 50% in LDL$^{-/-}$ mice [3]. Ishida et al. found that transgenic mice using a BAC clone containing the human EL gene had a 19% reduction in HDL-C, a 26% decrease in apoA1, but otherwise similar composition of their HDL particles [11]. These transgenic mice tended to have lower levels of apoB-containing lipoproteins which did not reach statistical significance.

Loss-of-function studies have been critical to establishing a physiological role for EL in lipoprotein metabolism. Jin et al. used a neutralizing antibody specific for

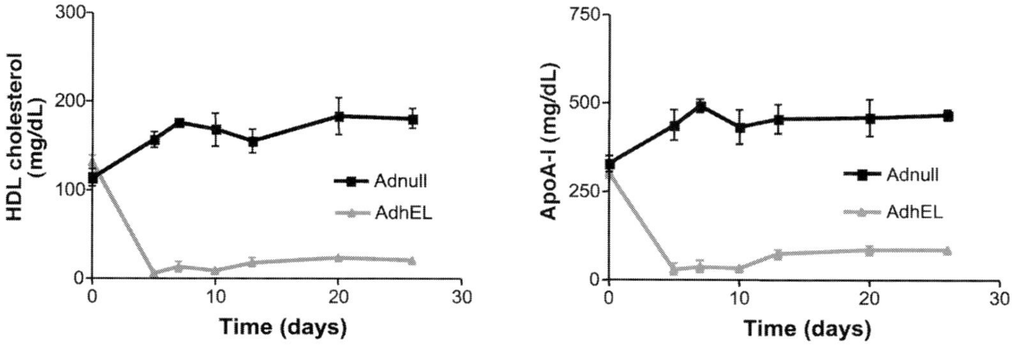

Fig. 8.7 Adenoviral-mediated EL gene transfer into wild-type mice resulted in a dramatic reduction in HDL-C and Apo-AI levels [3]. Three days after vector injection, HDL-C levels were barely detectable, and remained extremely low for up to 4 weeks.

murine EL and found that the acute inhibition of EL in wild-type mice caused a ca. 25% increase in HDL cholesterol and a 47% increase in phospholipids [22]. In human apoA-I transgenic mice, acute EL inhibition resulted in a 47% increase in HDL-C levels and a 45% increase in apoA-I levels (Fig. 8.8). The increase in HDL levels was a result of a 21% slower fractional catabolic rate of HDL particles (0.42 ± 0.03 vs. 0.53 ± 0.02 pools per hour) in the mice injected with murine EL antibody compared with mice injected with control antibody. Mice lacking hepatic lipase were also studied and found to have a significant increase in HDL-C levels and size after injection of the anti-EL antibody, proving that the results were not due to cross-inhibition of HL. The composition of HDL particles was not significantly changed by EL inhibition, indicating that the overall number of HDL particles was increased by inhibition of EL.

Two lines of EL knockout mice, generated by two independent groups [11, 23], were found to have a ca. 50% increase of HDL-C levels in homozygous mice and an intermediate 25% increase in heterozygous mice. Ishida et al. found no change in the lipoprotein particle composition and suggested that the increase in HDL levels was also due to an increased number of HDL particles [11]. Ma et al. reported that NMR analysis of the lipoprotein particles in these mice revealed an abundance of larger particles, H4 and H5, with mean HDL diameters that were larger in the knockouts compared to wild-type mice (10.25 ± 0.34 nm vs. 9.46 ± 0.37 nm) [23]. Using a dipalmitoyl phosphocholine emulsion assay, Ishida et al. observed that post-heparin phospholipase activity was reduced in $EL^{-/-}$ mice [11]. The hepatic expression of several genes involved in regulating HDL structure and metabolism was altered in $EL^{-/-}$ mice [23]. There was down-regulation of PLTP, up-regulation of HL and LPL, and no change in SR-BI and ABCA1 in the liver. Similar to the findings using antibody inhibition, the absence of EL delayed the clearance of HDL [23]. This suggests that HDL structure and other alterations in protein expression in the $EL^{-/-}$ mice contribute to the delay in HDL clearance in these mice.

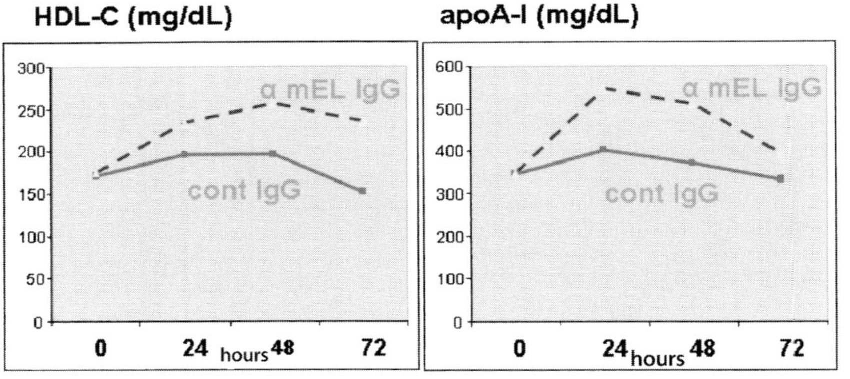

Fig. 8.8 A neutralizing antibody specific for murine EL injected into human apoA-I transgenic mice resulted in a 47% increase in HDL-C levels and a 45% increase in apoA-I levels.

There are currently no published data regarding EL mass or activity levels in human plasma. Indeed, there has been relatively little study of phospholipase activity in human plasma. Phospholipase activity increases after administration of heparin [24]. Some of the phospholipase activity in human plasma [25] has been attributed to lecithin-cholesterol acyltransferase (LCAT) [26] and hepatic lipase [27]. In the presence of inflammation, the secretory phospholipase A2 (sPLA2) may account for some of the plasma phospholipase activity and is also increased after heparin administration [28]. The contribution of endothelial lipase to plasma phospholipase activity is unknown, but the decrease in post-heparin phospholipase activity in EL knockout mice suggests that EL may contribute substantially to plasma phospholipase activity in humans.

In our current working model of endothelial lipase action (Fig. 8.9) we suggest that EL is synthesized by endothelial cells in various tissues, and is secreted and binds to HSPGs on the cell surface, where it binds to circulating lipoproteins (particularly HDL), hydrolyzing phospholipids and resulting in lipoprotein particles with reduced phospholipid content and altered systemic metabolism. The products of the lipolytic reaction are fatty acids and lyso-PC. Fatty acids may be utilized by the endothelial cells or underlying tissues as an energy source or for storage, or may serve other cellular functions. The phospholipid-depleted HDL particle may have increased uptake of cholesteryl ester by tissues. The loss of phospholipids results in dissociation of apoA-I, generating lipid-poor apoA-I that is rapidly catabolized. Many aspects of this working model have yet to be tested experimentally.

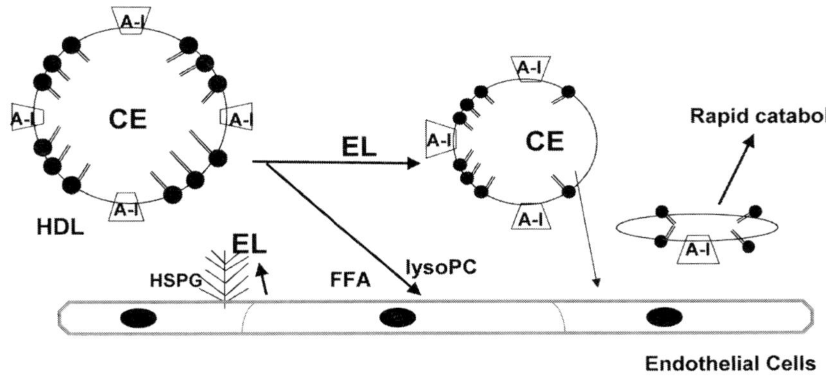

Fig. 8.9 Current working model of endothelial lipase function. EL is synthesized by endothelial cells in variety of various tissues, is secreted and binds to HSPGs on the cell surface. There it binds to circulating lipoproteins, hydrolyzing phospholipids and resulting in lipoprotein particles that are reduced in phospholipid content and have altered systemic metabolism. The products of the lipolytic reaction are fatty acids and lyso-PC. Fatty acids may be utilized by the endothelial cells or underlying tissues as an energy source or for storage, or may serve other cellular functions.

8.7
Variation in the Human Endothelial Lipase Gene

Approximately 50% of the influence on HDL levels is considered to be genetic [29, 30]. Genetic variation in endothelial lipase may influence the size and amount of HDL in human plasma. The potential that elevated HDL-C levels may be a result of functional polymorphisms and mutations in the EL gene has been examined by deLemos et al. [31]. 1200 base pairs of the promoter and all exons of the EL gene were sequenced in 20 individuals with HDL-C levels greater than the 90th percentile for age, sex and race. Seventeen polymorphic sites were identified in these individuals and confirmed (Tab. 8.1), six of which were determined "potentially functional": four resulting in amino acid changes (Gly26Ser, Thr111Ile, Thr298Ser and Asn396Ser) and two in the promoter (-303A/C and -410C/G). The genotypic frequencies of each variant was then determined in 176 black controls, 165 white controls and 123 whites with high HDL-C (Tab. 8.2). The Thr111Ile polymorphism was most common, occurring in 32.6% of high white HDL-C subjects, 31.2% of control white subjects but in 10.3% of control black subjects. There was no association between HDL-C levels and the presence of this polymorphism. Three variants, Gly26Ser, Thr298Ser, and -303A/C, were found in the cohort of high HDL-C whites and in blacks but not in the control white population. The Gly26Ser, Thr111Ile and promoter polymorphisms are the subject of further study.

Ma et al. [23] have determined the frequency of the single nucleotide polymorphism (SNP) 584C/T, which produces the Thr111Ile variant, in 372 members

Tab. 8.1 EL variants identified in 20 individuals with high HDL-C

Name	*Alleles*	*Location*	*Base position*	*a.a. position*	*a.a. change*	*No. alleles*
-410C/C	C/G	Promoter	–410	n/a	n/a	1
-303A/C	A/C	Promoter	–303	n/a	n/a	1
Gly26Ser	G/A	Exon 1	328	26	Gly-Ser	1
Thr298Ser	C/G	Exon 6	1145	298	Thr-Ser	1
Asn396Ser	A/G	Exon 8	1439	396	Asn-Ser	1
Thr111Ile	C/T	Exon 3	584	111	Thr-Ile	14
229.5′UTR	T/G	5′UTR	229	n/a	n/a	17
51.Intron1	C/T	Intron 1	51	n/a	n/a	14
55.Intron1	C/G	Intron 1	55	n/a	n/a	14
98.Intron4	G/A	Intron 4	98	n/a	n/a	13
Ser4Ser	C/A	Exon 1	264	4	Ser-Ser	3
Ser21Ser	C/T	Exon 1	315	21	Ser-Ser	3
1922.Intron2	T/C	Intron 2	1922	n/a	n/a	5
42.Intron5	C/T	Intron 5	42	n/a	n/a	9
840.Intron7	A/G	Intron 7	840	n/a	n/a	15
2725.Intron8	T/C	Intron 8	2725	n/a	n/a	16
2237.3′UTR	A/G	3′UTR	2237	n/a	n/a	11

Tab. 8.2 Frequency of selected EL variants

Name	*Control African Americans (n=176)*	*Control Caucasians (n=165)*	*High HDL-C Caucasians (n=123)*
-410C/G	0.0	0.3	0.0
-303A/C	1.8	0.0	0.4
Gly26Ser	5.7	0.0	0.8
Thr111ILe	10.3	31.2	32.6
Thr298Ser	2.3	0.0	0.4
Asn396Ser	0.6	1.2	2.4

of the Lipoprotein and Coronary Atherosclerosis Study. They found a significant univariate association between the 584C/T allele and mean plasma levels of HDL-C but did not report a multivariate analysis. There was an allele-dependent variation with the rank order TT > CT > CC for both HDL-C and apolipoprotein A-I/apolipoprotein B ratios. Over a 2.5 year period there was no difference between genotypes in progression or regression of disease. They reported no significant difference in the HDL-C response of individuals with the 584C/T SNP to fluvastatin therapy.

8.8 Endothelial Lipase as a Potential Pharmacologic Target

Epidemiologic studies have indicated a very strong inverse association between HDL-cholesterol levels and the incidence of arteriosclerotic cardiovascular disease [32]. HDL is thought to promote efflux of cholesterol from peripheral cells in the vascular wall and return of that cholesterol to the liver for excretion in bile, a process known as reverse cholesterol transport [21]. HDL has also been shown to have anti-inflammatory and anti-thrombotic activities [21]. Therefore, HDL levels may influence both the hyperlipidemia and inflammatory processes proposed as factors in the development of atherosclerosis. Although clinical data in support of HDL-raising are limited, in the VA-HIT trial gemfibrozil therapy in men with coronary artery disease and low HDL-cholesterol modestly increased HDL levels and reduced coronary events by 22% [33]. Presently, other methods of effectively and safely increasing HDL-cholesterol levels have not been discovered.

The gain-of-function and loss-of-function studies with endothelial lipase in mice reviewed above suggest that levels of EL are an important determinant of HDL-C levels and that inhibition of EL would raise HDL-C levels. By hydrolyzing HDL phospholipids, EL generates increased amounts of lipid-poor apoA-I, which are at greater risk of being catabolized by the kidneys. Thus, inhibition of EL might reduce HDL phospholipid hydrolysis and reduce apoA-I and HDL-C turnover, thus increasing apoA-I and HDL-C levels and HDL size. Importantly, EL has other physiologic effects that could promote atherogenesis. The products of the EL-catalyzed reaction are unesterified fatty acids and lysophosphatidylcholine. Free fatty acids induce endothelial dysfunction, and lysoPC has various pro-inflammatory effects that may promote atherosclerosis. Therefore, in addition to raising HDL-C levels, inhibition of EL could reduce the generation of FFA and lysoPC at the level of the vessel wall and therefore have direct anti-atherogenic effects.

More data are needed before EL can be regarded as a *bona fide* candidate for the development of new pharmacologic inhibitors. Data regarding the effects of modulating EL activity on atherosclerosis in mice and other animals are needed. Data in humans regarding the relationship between EL mass and activity and HDL-C levels and function, as well as with atherosclerosis, are also needed. Nevertheless, pharmacologic inhibition of endothelial lipase appears to be a potentially promising approach to increasing HDL levels and reducing atherosclerosis.

8.9 References

1 Goldberg, I. J. Lipoprotein lipase and lipolysis: central roles in lipoprotein metabolism and atherogenesis. *J. Lipid Res.*, **1996**, *37*, 693–707.

2 Santamarina-Fojo, S., Haudenschild, C., and Amar, M. The role of hepatic lipase in lipoprotein metabolism and

atherosclerosis. *Curr. Opin. Lipidol*, **1998**, *9*, 211–219.

3 Jaye, M., Lynch, K.J., Krawiec, J., Marchadier, D., Maugeais, C., Doan, K., South, V., Amin, D., Perrone, M., and Rader, D.J. A novel endothelial-derived lipase that modulates HDL metabolism. *Nat. Genet.*, **1999**, *21*, 424–428.

4 Hirata, K., Diechek, H.L., Cioffi, J.A., Choi, S.Y., Leeper, N.J., Quintana, L., Kronmal, G.S., Cooper, A.D., and Quertermous, T. Cloning of a unique lipase from endothelial cells extends the lipase gene family. *J. Biol. Chem.*, **1999**, *274*, 14170–14175.

5 Winkler, F.K., D'Arcy, A., and Hunziker, W. Structure of human pancreatic lipase. *Nature*, **1990**, *343*, 771–774.

6 van Tilbeurgh, H., Sarda, L., Verger, R., and Cambillau, C. Structure of the pancreatic lipase-procolipase complex. *Nature*, **1992**, *359*, 159–162.

7 van Tilbeurgh, H., Roussel, A., Lalouel, J.M., and Cambillau, C. Lipoprotein lipase. Molecular model based on the pancreatic lipase x-ray structure: consequences for heparin binding and catalysis. *J. Biol. Chem.*, **1994**, *269*, 4626–4633.

8 Zechner, R. The tissue-specific expression of lipoprotein lipase: implications for energy and lipoprotein metabolism. *Curr. Opin. Lipidol*, **1997**, *8*, 77–88.

9 Merkel, M., Eckel, R.H., and Goldberg, I.J. Lipoprotein lipase: genetics, lipid uptake, and regulation. *J. Lipid Res.*, **2002**, *43*, 1997–2006.

10 Wong, H. and Schotz, M.C. The lipase gene family. *J. Lipid Res.*, **2002**, *43*, 993–999.

11 Ishida, T., Choi, S., Kundu, R.K., Hirata, K., Rubin, E.M., Cooper, A.D., and Quertermous, T. Endothelial lipase is a major determinant of HDL level. *J. Clin. Invest.*, **2003**, *111*, 347–355.

12 Jin, W., Sun, G.S., Marchadier, D., Octtaviani, E., Glick, J.M., and Rader, D.J. Endothelial cells secrete triglyceride lipase and phospholipase activities in response to cytokines as a result of endothelial lipase. *Circ. Res.*, **2003**, *92*, 644–650.

13 Nagasaka, H., Chiba, H., Kikuta, H., Akita, H., Takahashi, Y., Yanai, H., Hui, S.P., Fuda, H., Fujiwara, H., and Kobayashi, K. Unique character and metabolism of high density lipoprotein (HDL) in fetus. *Atherosclerosis*, **2002**, *161*, 215–223.

14 Tan, K.C., Shiu, S.W., and Kung, A.W. Effect of thyroid dysfunction on high-density lipoprotein subfraction metabolism: roles of hepatic lipase and cholesteryl ester transfer protein. *J. Clin. Endocrinol. Metabol.*, **1998**, *83*, 2921–2924.

15 Scottolini, A.G., Bhagavan, N.V., Oshiro, T.H., and Abe, S.Y. Serum high-density lipoprotein cholesterol concentrations. *Clin. Chem.*, **1980**, *26*, 584–587.

16 McCoy, M.G., Sun, G.S., Marchadier, D., Maugeais, C., Glick, J.M., and Rader, D.J. Characterization of the lipolytic activity of endothelial lipase. *J. Lipid Res.*, **2002**, *43*, 921–929.

17 Strauss, J.G., Zimmermann, R., Hrzenjak, A., Zhou, Y., Kratky, D., Levak-Frank, S., Kostner, G.M., Zechner, R., and Frank, S. Endothelial cell-derived lipase mediates uptake and binding of high-density lipoprotein (HDL) particles and the selective uptake of HDL-associated cholesterol esters independent of its enzymic activity. *Biochem. J.*, **2002**, *368*, 69–79.

18 Hirata, K., Ishida, T., Matsushita, H., Tsao, P.S., and Quertermous, T. Regulated expression of endothelial cell-derived lipase. *Biochem. Biophys. Res. Commun.*, **2000**, *272*, 90–93.

19 Feingold, K.R., Marshall, M., Gulli, R., Moser, A.H., and Grunfeld, C. Effect of endotoxin and cytokines on lipoprotein lipase activity in mice. *Arterioscler. Thromb.*, **1994**, *14*, 1866–1872.

20 Feingold, K.R., Memon, R.A., Moser, A.H., Shigenaga, J.K., and Grunfeld, C. Endotoxin and interleukin-1 decrease hepatic lipase mRNA levels. *Atherosclerosis*, **1999**, *142*, 379–387.

21 Rader, D.J. High-density lipoproteins and atherosclerosis. *Am. J. Cardiol.*, **2002**, *90*, 62i–70i.

22 Jin, W., Millar, J.S., Broedl, U., Glick, J.M., and Rader, D.J. Inhibition of en-

dothelial lipase causes increased HDL cholesterol levels in vivo. *J. Clin. Invest.*, **2003**, *111*, 357–362.

23 Ma, K., Cilingiroglu, M., Otvos, J. D., Ballantyne, C. M., Marian, A. J., and Chan, L. Endothelial lipase is a major genetic determinant for high-density lipoprotein concentration, structure, and metabolism. *Proc. Natl. Acad. Sci. USA*, **2003**, *100*, 2748–2753.

24 Vogel, W. C. and Bierman, E. L. Evidence for in vivo activity of postheparin plasma lecithinase in man. *Proc. Soc. Exp. Biol. Med.*, **1968**, *127*, 77–80.

25 Vogel, W. C. and Bierman, E. L. Correlation between post-heparin lipase and phospholipase activities in human plasma. *Lipids*, **1970**, *5*, 385–391.

26 Shamburek, R. D., Zech, L. A., Cooper, P. S., Vandenbroek, J. M., and Schwartz, C. C. Disappearance of two major phosphatidylcholines from plasma is predominantly via LCAT and hepatic lipase. *Am. J. Physiol.*, **1996**, *271*, E1073–E1082.

27 Scagnelli, G. P., Cooper, P. S., Vandenbroek, J. M., Berman, W. F., and Schwartz, C. C. Plasma 1-palmitoyl-2-linoleoyl phosphatidylcholine. Evidence for extensive phospholipase A1 hydrolysis and hepatic metabolism of the products. *J. Biol. Chem.*, **1991**, *266*, 18002–18011.

28 Nakamura, H., Kim, D. K., Philbin, D. M., Peterson, M. B., Debros, F., Koski, G., and Bonventre, J. V. Heparin-enhanced plasma phospholipase A2 activity and prostacyclin synthesis in patients undergoing cardiac surgery. *J. Clin. Invest.*, **1995**, *95*, 1062–1070.

29 Namboodiri, K. K., Kaplan, E. B., Heuch, I., Elston, R. C., Green, P. P., Rao, D. C., Laskarzewski, P., Glueck, C. J., and Rifkind, B. M. The Collaborative Lipid Research Clinics Family Study: biological and cultural determinants of familial resemblance for plasma lipids, and lipoproteins. *Genet. Epidemiol.*, **1985**, *2*, 227–254.

30 Ellison, R. C., Myers, R. H., Zhang, Y., Djousse, L., Knox, S., Williams, R. R., and Province, M. A. Effects of similarities in lifestyle habits on familial aggregation of high density lipoprotein and low density lipoprotein cholesterol: the NHLBI Family Heart Study. *Am. J. Epidemiol.*, **1999**, *150*, 910–918.

31 deLemos, A. S., Wolfe, M. L., Long, C. J., Sivapackianathan, R., and Rader, D. J. Identification of genetic variants in endothelial lipase in persons with elevated high-density lipoprotein cholesterol. *Circulation*, **2002**, *106*, 1321–1326.

32 Gordon, D. J. and Rifkind, B. M. High-density lipoproteins – the clinical implications of recent studies. *New Engl. J. Med.*, **1989**, *321*, 1311–1316.

33 Rubins, H. B., Robins, S. J., Collins, D., Fye, C. L., Anderson, J. W., Elam, M. B., Faas, F. H., Linares, E., Schaefer, E. J., Schectman, G. et al. Gemfibrozil for the secondary prevention of coronary heart disease in men with low levels of high-density lipoprotein cholesterol. *New Engl. J. Med.*, **1999**, *341*, 410–418.

9
Digestive Lipases Inhibition: an *In vitro* Study

Ali Tiss, Nabil Miled, Robert Verger, Youssef Gargouri, and Abdelkarim Abousalham

9.1
Introduction

Under normal physiological conditions the digestion and absorption of dietary lipids are highly efficient processes. In industrialized countries, the daily human diet intake generally contains an average of 90–120 g of lipids, mostly triacylglycerides (TAG), more than 95% of which are absorbed, due to the action of the stomach, small intestine, liver and pancreas [1, 2]. Several steps can be distinguished in the processing of dietary lipids, including their emulsification, hydrolysis and micellization, and lastly, their uptake by the enterocytes. The emulsification of lipids starts in the stomach and is mediated by physical forces and facilitated by the partial lipolysis of the dietary lipids [2]. For a long time, the hydrolysis of dietary TAG was thought to begin in the intestinal lumen and to be catalyzed entirely by pancreatic lipase. The stomach was thought to be a transient storage organ, the role of which was limited to mixing lipids with the other nutrients and dispersing them. In human, it has been now clearly established that the hydrolysis of alimentary TAG begins in the stomach and is catalyzed by human gastric lipase (HGL) [3–5] which is able to hydrolyze short and long chain TAG at comparable rates. Under acidic pH conditions, HGL has been shown to be remarkably stable and active, whereas pancreatic lipase irreversibly loses its lipolytic capacity.

Human pancreatic lipase (HPL) alone is inactive *in vitro* on an emulsified TAG substrate in the presence of supramicellar concentrations of bile salts such as those found in the small intestine. Bile salts are amphiphilic molecules that bind to the oil–water interface and prevent pancreatic lipase adsorption, and thus lipolysis, from occurring [3, 4]. The inhibition by bile salts can, however, be reversed by the specific pancreatic lipase cofactor colipase [3, 5–7], *via* the formation of a specific 1 : 1 lipase–colipase complex.

HPL has been purified from the pancreatic juice by De Caro et al. [8]. In contrast to most other pancreatic enzymes, which are secreted as proenzymes and further activated by proteolytic cleavage in the small intestine, pancreatic lipase is directly secreted as a 50 kDa active enzyme consisting of 449 amino acid residues including a high mannose or complex-type glycan chain N-linked to Asn 166. *In vitro*, the max-

Lipases and Phospholipases in Drug Development
Edited by G. Müller and S. Petry

ISBN: 3-527-30677-3

imum specific activities, obtained using 4 mM NaTDC and colipase in excess, were 12000 units/mg at pH 8, 6000 units/mg at pH 7.8 and 4600 units/mg at pH 8 using tributyroylglycerol (tributyrin), trioctanoylglycerol and Intralipid™ as substrates, respectively. The cDNA has been cloned [9] and expressed in a baculovirus/insect cells system [10]. The recombinant HPL (rHPL) which was secreted in the culture medium at high level (40 mg/L) was obtained after a one-step purification procedure. The kinetic characterizations have shown that the kinetic and surface behavior of rHPL was similar to that of the native enzyme.

Bodmer et al. [11] have cloned, sequenced, and expressed active HGL in yeast. The amino acid sequence obtained from the cDNA consists of 379 residues as compared to 449 residues constituting HPL. No amino acid sequence homology exists between HGL and HPL, except for short regions of six residues containing serine in a position analogous to that of the essential serine 152. The cDNA sequence of HGL shows the presence of three cysteine residues in comparison to PPL which contains 14 cysteine residues (6 disulfide bridges and 2 free sulfhydryl groups).

Conventional treatments for obesity have focused largely on strategies to control energy intake, however, the long-term efficacy of such approaches is limited [12]. A reduction of dietary fat adsorption by an inhibitor of digestive lipases hold great promise as an anti-obesity agent.

9.1.1
3-D Structure of Human Pancreatic Lipase

The determination of the 3-D structure of HPL by Winkler et al. [13] confirmed the existence of two distinct domains in pancreatic lipase: a larger N-terminal domain comprising residues 1–336, and a smaller C-terminal domain made up of residues 337–449. The high degree of amino acid sequence homology observed within the lipase gene family supports the view that this particular architecture is also common to lipoprotein lipase and to hepatic lipase [14–19].

As illustrated in Fig. 9.1, and discussed in a review by Winkler and Gubernator [20], the large N-terminal domain is a typical α/β hydrolase fold dominated by a central parallel β-sheet [21]. It contains the active site with a catalytic triad formed by serine 152, aspartate 176, histidine 263, all of which are conserved in lipoprotein lipase and hepatic lipase. This catalytic triad is chemically analogous to that originally described in serine proteases such as chymotrypsin [22]. The nucleophilic elbow, β-strand/Ser/α-helix, structural motif including the Gly-X-Ser-X-Gly consensus sequence has been detected mostly in lipases and esterases [23–26]. *Bacillus* and *Candida antartica* B lipases have a variation to Ala-X-Ser-X-Gly and Thr-X-Ser-X-Gly, respectively. This consensus sequence is, however, homologous to that found in other enzymes such as haloalkane dehalogenase, bromoperoxidase and acyltransferase. The 3-D structure of HPL clearly demonstrated that serine 152 is the nucleophilic residue essential for catalysis, in agreement with the chemical modification of serine 152 in porcine pancreatic lipase (PPL) [27], and in contra-

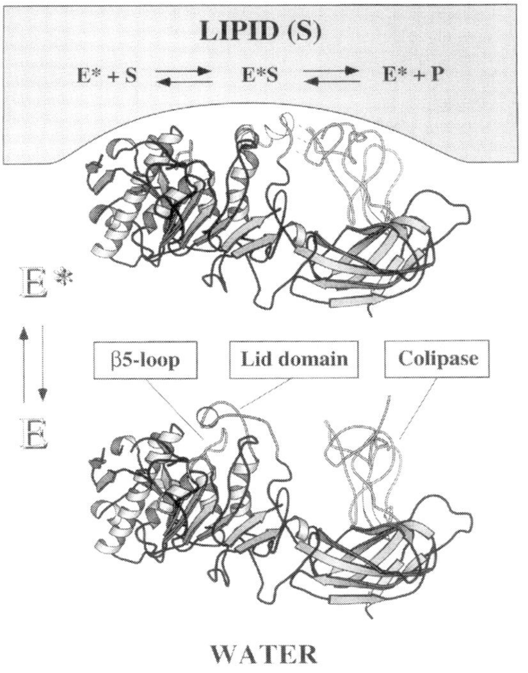

Fig. 9.1 3-D structure of HPL and conformational changes induced by the adsorption to a lipid interface. From [19], adapted from [32, 118].

diction with results suggesting a function of serine 152 in interfacial recognition [28]. The hydrolytic mechanism is directed by a nucleophilic attack by the serine 152, and the tetrahedral intermediate formed is stabilized by an oxyanion hole. An acyl-enzyme is then formed, and the alcohol moiety is leaving the active site. The action of a water molecule permits to regenerate the enzyme and the liberation of the acyl moiety.

In the 3-D structure resolved by Winkler et al. [13], the active site is covered by a surface loop between the disulfide bridge cysteine 237 and 261. This surface loop includes a short one-turn α-helix with a tryptophan residue (Trp 252) completely buried and sitting directly on top of the active site serine 152. Under this closed conformation, found mainly in aqueous solution, the "lid" prevents the dispersed substrate from gaining access to the active site. Spectroscopic studies of tryptophan fluorescence have shown that large spectral changes are induced by acylation of pancreatic lipase with the inhibitor tetrahydrolipstatin (THL, orlistat) in the presence of bile salt micelles [29]. By crystallizing the pancreatic lipase–procolipase complex in the presence of mixed lipid micelles, it was shown that the "lid" was shifted to one side, exposing both the active site and a larger hydrophobic surface [30]. This motion is induced when the binding to the lipid occurs and is probably the structural basis for "interfacial activation" of pancreatic lipase [31].

The β-sandwich C-terminal domain of pancreatic lipase is necessary for colipase binding to occur, as shown in the 3-D structure of the HPL–porcine procolipase complex [32]. Procolipase is a "three finger" protein which is topologically comparable to snake toxins, known to bind phospholipids, even though these proteins do not share any sequence homology. In the 3-D structure of the pancreatic lipase/procolipase complex, the N-terminal pentapeptide (activation peptide or "enterostatin") was not visible in the electron density map [32]. This is why we consider that pro- and colipase have basically identical 3-D structures. Colipase lacks any well-defined secondary structural elements. This small protein seems to be stabilized mainly by an extended network of five disulfide bridges that runs throughout the flat-shaped molecule, reticulating its finger-like loops. The colipase surface can be divided into a rather hydrophilic part, interacting with lipase, and a more hydrophobic part, formed by the tips of the fingers which are very mobile and constitute the lipid interaction surface. The interaction between colipase and the C-terminal domain of lipase is stabilized by eight hydrogen bonds and about 80 van der Waals contacts. Upon the opening of the lid, three more hydrogen bonds and about 28 van der Waals contacts are added, explaining in part the higher apparent affinity in the presence of a lipid/water interface. The fact that colipase and lipase are co-localized at the interface may contribute to a higher affinity constant. In the absence of an interface, no conformational change in the lipase molecule is induced by the binding of procolipase. The 3-D structure of the open form of the HPL–procolipase complex in the presence of mixed lipid micelles revealed, however, that the "lid" binds to the procolipase N-terminal domain when the complex is activated at an interface [33]. The open 3-D structure of the lipase–procolipase complex illustrates how colipase might anchor the lipase at the interface in the presence of bile salts: colipase binds to the non-catalytic β-sheet of the C-terminal domain of HPL and exposes the hydrophobic tips of its fingers at the opposite side of its lipase-binding domain. This hydrophobic surface, in addition to the hydrophobic backside of the lid, helps to bring the catalytic N-terminal domain of HPL into close contact with the lipid/water interface, as shown in Fig. 9.1. Horse pancreatic lipase has been solved and is almost identical to HPL [34].

9.1.2
3-D Structure of Human Gastric Lipase

The crystal structure of the HGL, expressed in baculovirus/insect cell system was obtained at 3 Å resolution [35]. This structure was the first to be resolved within the acidic lipases family. HGL is a globular protein that exhibits the α/β hydrolase fold (Fig. 9.2). The final model contains residues 9 to 53 and 57 to 379, six sugar residues located on the four potential N-glycosylation sites and a disulfide bridge Cys 227–Cys 236, whereas Cys 244 is free (Fig. 9.2). In lipases, as well as in serine proteases [36], the catalytic machinery consists of a triad and an oxyanion hole. In HGL, the nucleophilic serine (Ser 153) belonging to the usual consensus sequence G-X-S-X-G is located at the connection between an α helix and a β strand has an ε conformation, which are characteristic features of all enzymes within the α/β hydrolase fold family [21]. His 353 and Asp 324 are the other two residues forming the catalytic triad. The core domain, which is located between residues 9–183 and 309–379, contains a central *β*-sheet composed of eight strands, seven of which are parallel and one is anti-parallel (strand 2) with 1(-2)435678 connectivity and six helices, three on each side of the *β*-sheet (Fig. 9.2). Strikingly, an extrusion domain (residues 184–308), called cap domain, occurs at the same location in wheat serine carboxypeptidase II (residues 181 to 311) [37] and in human protective protein (residues 181 to 347) [38]. This domain is an intricate mixture of eight helices, turns and random coils (Fig. 9.2) covering the core domain and the catalytic triad. Protrusions have been observed in other lipases, generally constituting the device covering the active site and called the lid [13, 30, 39, 40].

Recently, Roussel et al. [41] reported the open 3-D structure of the recombinant dog gastric lipase (DGL) at 2.7 Å resolution in complex with the undecyl-butyl (C11Y4) phosphonate inhibitor (see chapter 3.3.3). HGL and DGL show 85.7% amino acid sequence identity, which makes it relevant to compare the forms from two different species. The open DGL structure confirms the previous description of the HGL catalytic triad, Ser(153), His(353), and Asp(324), with the catalytic serine buried and an oxyanion hole (NH groups of Gln(154) and Leu(67)). In DGL, the binding of the C11Y4 phosphonate inhibitor induces part of the cap domain, the lid, to roll over the enzyme surface and to expose a catalytic crevice measuring approximately $20 \times 20 \times 7$ Å^3. The C11Y4 phosphonate fits into this crevice, and a molecule of *β*-octyl glucoside fills up the crevice. The C11Y4 phosphonate inhibitor and the detergent molecule suggest a possible binding mode for the natural substrates, the TAG molecules.

9.2 Methods for Lipase Inhibition

In order to describe the kinetics of a lipolytic enzyme acting at an interface, a simple and versatile model has been proposed by Verger et al. [42]. This model consists basically of two successive equilibria. The first describes the reversible penetration of a water-soluble enzyme into an interface ($E \neq E^*$). This is followed by a second equilibrium in which one molecule of penetrated enzyme binds a single substrate molecule, forming the enzyme–substrate complex (E^*S). This is the two-dimensional equivalent of the classical Michaelis-Menten equilibrium. Once the complex (E^*S) has been formed, the catalytic steps take place, regenerating the enzyme in the form E^* and liberating the lipolysis products. An extension of the previous kinetic model was proposed by Ransac et al. for depicting the competitive inhibition [43], as well as the covalent inhibition of lipolytic enzymes at a lipid/water interface (Fig. 9.3) [44].

Achieving specific and covalent inhibition of lipolytic enzymes is a difficult task, because of mutually non-exclusive processes such as interfacial denaturation, changes in "interfacial quality" [45], and surface dilution phenomena [46]. Further-

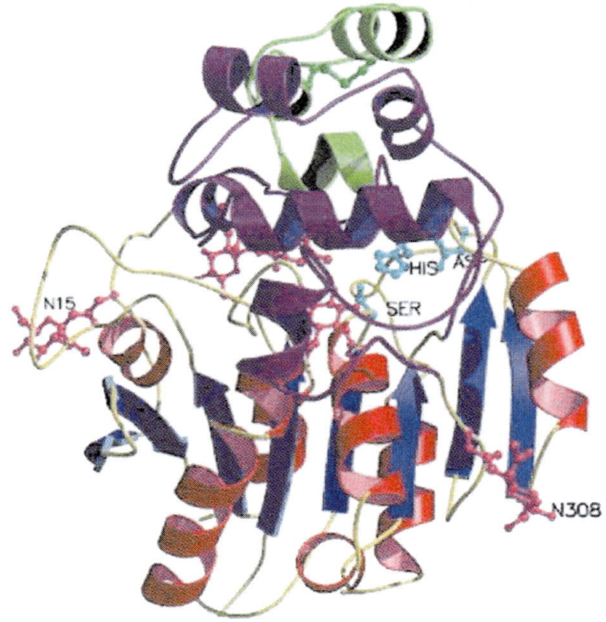

Fig. 9.2 Stereo ribbon representation of HGL. View with the central β-sheet parallel to the paper plane. The cap domain on top of the α/β hydrolase core domain is colored *magenta* and the putative lid, *green*. For the α/β hydrolase fold, the colour code is helices, *red*, strands, *blue*, turns and random coil, *yellow*. The catalytic triad, Ser-153, His-353, and Asp-324, the disulfide bridge, and the Asn-attached sugars are shown in ball and stick representation [35].

more, the interfacial enzyme binding and/or the catalytic turnover can be diversely affected by the presence of potential amphipathic inhibitors [47, 48].

In this chapter, we will present and discuss results concerning mostly the covalent inhibition of digestive lipases from different origins. Rather than presenting an exhaustive list of compounds tested so far with lipases of animal and microbial origin, we have selected experimental data illustrating the specific problems encountered during the covalent inhibition of digestive lipases. We have not reviewed here the non-covalent "inhibition" observed with proteins and tensio-active agents, which usually do not interact specifically with lipases but generally adsorb at the lipid/water interface, affecting the "interfacial quality" [49] and hence the lipase activity.

In Fig. 9.4 are depicted the chemical structures of all inhibitors we will discuss: diethyl *p*-nitrophenyl phosphate (E_{600}), orlistat and O-methyl-O-(*p*-nitrophenyl) *n*-undecylphosphonate (C11-P). In order to study the effects of these reagents, which can in principle modify covalently several essential amino acid residues of the lipases, we have proposed four different methods [50–52], depending upon the order of addition of lipase, substrate and inhibitor (Fig. 9.5).

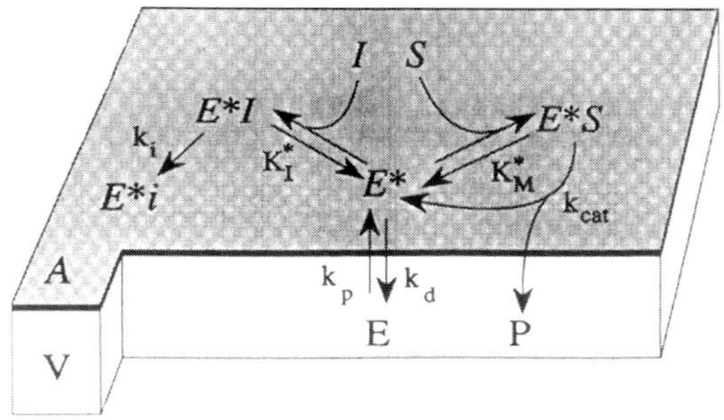

Fig. 9.3 Kinetic model illustrating the covalent inhibition of a lipolytic enzyme at a lipid/water interface. Symbols and abbreviations are as follows: *A*, total interfacial area (surface); *V*, total volume (volume); *E*, free enzyme concentration (molecule/volume); E^*, interfacial enzyme concentration (molecule/surface); *S*, interfacial concentration of substrate (molecule/surface); I, interfacial concentration of inhibitor (molecule/surface); *P*, product concentration (molecule/volume); E^*S, interfacial enzyme–substrate complex concentration (molecule/surface); E^*I, interfacial enzyme–inhibitor complex concentration (molecule/surface); E^*_i, covalently inactivated enzyme concentration (molecule/surface); k_d, desorption rate constant (time^{-1}); k_p, penetration rate constant (volume surface^{-1} time^{-1}); k_{cat}, catalytic rate constant (time^{-1}); k_i, inhibition rate constant (time^{-1}); K^*_M, interfacial Michaelis-Menten constant (molecule/surface); K^*_I, interfacial dissociation constant for the enzyme-inhibitor complex (molecule/surface). From [44].

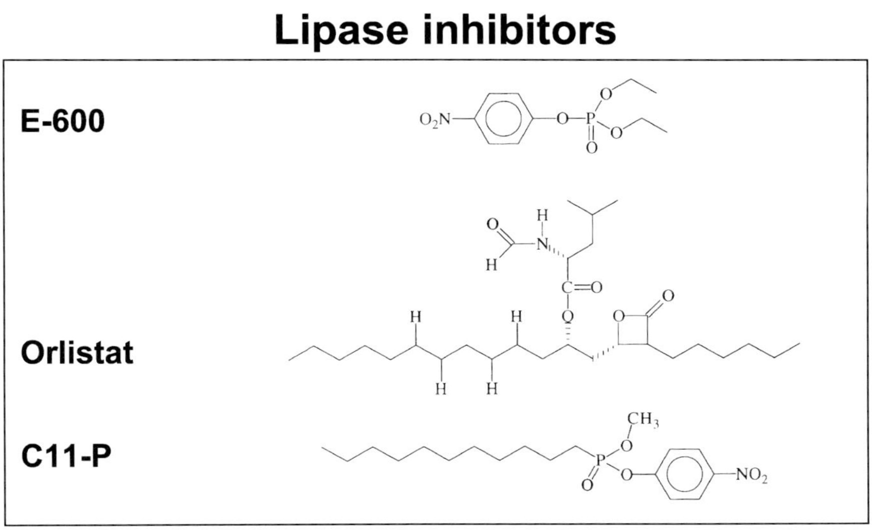

Fig. 9.4 Chemical structures of various lipase inhibitors: E_{600}; orlistat and C11-P.

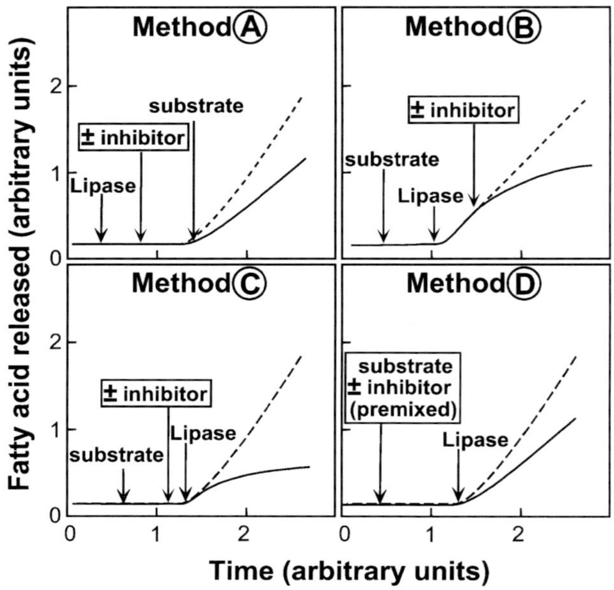

Fig. 9.5 Methods used to study the effects of inhibitors on lipase activity. Four methods A, B, C and D were used. The arrows indicate the order of the successive injections of inhibitor, lipases and substrate. Adapted from [57, 119].

9.2.1
Method A: Lipase/Inhibitor Pre-incubation

This method (Fig. 9.5 A) was set up to test, in aqueous medium and in the absence of substrate, the possible reactions between lipases and all the reagents to be tested [53, 54]. Residual lipase activity was measured separately using an emulsified substrate. Sometimes the inhibitor is incorporated into a solution of a tensio-active agent in order to create a mixed micellar interface, containing the incorporated inhibitor, and described as method A′ by Cudrey et al. [52].

9.2.2
Method B: Inhibition During Lipolysis

This method (Fig. 9.5 B) was designed to test whether any inhibition reaction occurred in the presence of a water-insoluble substrate during lipase hydrolysis. Lipase was injected into the reaction vessel of a pH-stat containing the emulsified substrate maintained under vigorous stirring. The inhibitor was injected a few minutes after lipase addition. Lipase activity was then continuously recorded (Fig. 9.5 B).

9.2.3
"Pre-poisoned" Interfaces

9.2.3.1 Method C

The inhibitor was injected before the enzyme addition at various concentrations into a lipolysis assay containing the emulsified substrate under vigorous stirring. Lipase activity was then continuously recorded (Fig. 9.5 C).

9.2.3.2 Method D

With method D, the inhibitor was previously dissolved into the substrate before carrying out the lipase assay (Fig. 9.5 D). This inhibition method was used with the pH-stat, monomolecular films and oil drop techniques. Lipase hydrolytic activity was then continuously recorded as described below.

Oil Emulsions

The inhibitor was previously dissolved at various concentrations into the lipid substrate before its emulsification under vigorous stirring in a lipolysis vessel (Fig. 9.5D). Lipase was then added and its activity was continuously recorded.

Monomolecular Films

With conventional emulsified systems, it is not possible to control the "interfacial quality" and to easily assess the distribution of soluble *versus* adsorbed amphiphilic molecules. This prompted us to use the monolayer technique, based upon surface pressure decrease due to lipid-film hydrolysis [55, 56]. This technique is applicable to those cases where the lipid forms a stable monomolecular film at the air/water interface and where reaction products are freely soluble and diffuse away rapidly into the aqueous phase.

The monolayer technique was also used to study the covalent inhibition of lipases using mixed monomolecular films of substrate containing amphiphilic inhibitor molecules and using a zero order trough with a reaction compartment containing the mixed substrate/inhibitor film whereas the reservoir compartment was covered with a film of pure substrate [44, 57].

Oil Drop

The kinetics of hydrolysis of purified soybean oil (PSO) by the lipase were monitored by recording the interfacial tension decrease with time, using the oil drop tensiometer (ITConcept, France) as described by Nury et al. [58]. The interfacial tension was measured by automatically analysing the drop profile on-line, using the Laplace equation [59]. The accumulation with time of tensio-active components, i.e., lipolytic products, at the surface of an oil drop results in a decrease in the interfacial tension, which in turn is correlated with changes in the oil drop profile [60, 61]. The main advantage of the drop tensiometer over other interfacial techniques is that it can be used to monitor the lipase activity on natural long-chain TAG, while controlling the interfacial parameters.

Inhibitor was previously dissolved at various concentrations in the PSO. Lipase was then injected into the aqueous phase a few minutes after the formation of the oil drop.

9.3
Inhibition of Lipases by E_{600} and Various Phosphonates

Organophosphorus compounds, such as diisopropylfluorophosphate and E_{600}, are phosphorylating agents known to be powerful inhibitors of serine proteases. They have been shown to specifically and irreversibly react with the active-site serine in these enzymes. Thus, these phosphorylating agents have been invaluable tools in the investigation of mechanistic and structural features of proteases. However, pancreatic lipase is not inhibited by classical serine esterases inhibitors such as diisopropylfluorophosphate, benzamidine and phenylmethylsulfonylfluoride; in fact, these serine protease inhibitors are currently used to protect pancreatic lipase during the purification procedures [62]. However, some aromatic sulfonylhalides such as pipsyl- and possibly also dansylchloride inhibit PPL by reacting with several residues probably including a serine [63, 64].

In previous studies, it has been shown that organophosphorus compounds such as diethyl-E_{600} are able to stoichiometrically inactivate PPL [63, 65, 66]. This chemical inhibition of pancreatic lipase occurred exclusively with mixed E_{600}/bile salt micelles and in the presence of colipase [66]. Guidoni et al. [27] have shown that serine 152 in PPL can be specifically labeled with E_{600}. Furthermore, it was reported by Chapus and Sémériva [28] that E_{600}-modified PPL or diethylphospho-lipase (DP-lipase) had a decreased interfacial binding capacity onto siliconized glass beads but was still able to hydrolyze a water-soluble substrate such as *p*-nitrophenyl acetate (*p*-NPA). Without any restriction, these authors extrapolated their results obtained with siliconized glass beads to lipid/water interfaces and speculated that the essential serine 152 of PPL was involved in the lipid binding domain, and not in the catalytic site. Along the same line, Sikk et al. [67] reported the irreversible inhibition of pancreatic lipase by bis-*p*-nitrophenyl methyl phosphonate. The covalently modified enzyme did not bind on siliconized glass beads. Tyrosine 49 was presumed to be the modification site. The conclusion has been made that this residue is involved in the interfacial recognition site of pancreatic lipase [67]. From these initial speculations, it came to be generally assumed that the homologous pentapeptide Gly-X-Ser-X-Gly which is to be found in all the known lipase sequences might be part of the lipid binding domain of these lipases [9, 47, 67–72]. It did seem rather puzzling however that the only preserved stretch of sequence detected in all lipases was apparently the lipid binding domain and not the catalytic site. This misinterpretation was probably due to the implicit assumption that siliconized glass beads were a good substitute and a valid model for the "interfacial quality" of TAG/water interfaces. Finally, the first three-dimensional crystallographic structures of human pancreatic [13] as well as several fungal and bacterial [73–80] lipases have been solved. From these lipase 3-D structures, it is now clear that essential serine 152 of the pancreatic lipase is part of the classical Asp/His/Ser triad and constitutes the nucleophilic residue essential for catalysis. The spatial arrangement of the proposed catalytic triad in these lipases is very similar to the catalytic triad of classical serine esterases, such as trypsin. The widespread distribution of Gly-X-Ser-X-Gly pentapeptide in the primary structures of all the known lipases and the apparent contradiction between previous interpretations of the interfacial binding data and the 3-D structure of pancreatic lipase have underlined the importance to investigate the potential role of serine 152. We therefore reinvestigated the inhibition of PPL using radiolabeled E_{600} and extended this study to gastric lipases such as HGL and rabbit gastric lipase (RGL) purified according to published methods [81, 82].

9.3.1 Inhibition of PPL, HGL and RGL by Radiolabeled E_{600}

Fig. 9.6 gives the time course of inhibition of PPL, HGL and RGL incubated at pH 6.0 with radiolabeled E_{600}/sodium taurodeoxycholate (NaTDC) mixed micelles at an E_{600} to lipase molar ratio of 84. PPL and HGL showed the same inhibition rates, reaching a plateau at a level of 20% remaining activity after 5 h of incubation. This partial inhibition of PPL by E_{600} was previously reported by Rouard et al. [66]. By contrast, RGL inhibition proceeded at a faster rate, and the lipolytic activity was completely abolished after 60 min of incubation. From previous data on gastric lipase inhibition by sulfhydryl reagents, RGL can be said to have shown a faster reaction rate than HGL [83, 84]. This difference in reaction rate might reflect a difference in active site accessibility.

After removing by gel filtration the excess micellar radiolabeled E_{600}, the labeling stoichiometry of the inactivated lipases was determined and correlated with the percentage of lipase inhibition [85]. One can note the existence of a direct correlation between the inhibition levels of the three digestive lipases tested and the stoichiometric labeling obtained with [^{14}C] E_{600}. In the case of PPL, it has been clearly shown that the classical serine reagent (E_{600}) inactivates the enzyme after reacting with the essential serine 152 [27]. Site-directed mutagenesis have confirmed the involvement of serine 152 in the catalytic activity [86]. Furthermore, we showed for the first time that gastric lipases were also stoichiometrically inactivated by mixed E_{600}/bile salt micelles under experimental conditions identical to those used with PPL.

Chemical modification of the single free sulfhydryl group of gastric lipases, with either 4,4′-dithiopyridine or 5,5′-dithiobis(2-nitrobenzoic acid) [83, 84], induced a complete loss of activity, when water-soluble *p*-NPA or emulsified tributyroylglycerol (tributyrin) were used as substrates [85]. Both catalytic activities were

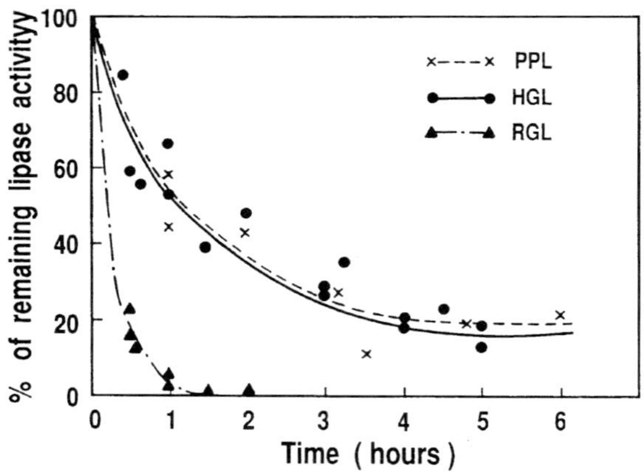

Fig. 9.6 Time course of inhibition of PPL, HGL, and RGL during incubation with radiolabeled E_{600} (E_{600} to lipase molar ratio: 84). Lipases (25 nmol) were incubated with 2.1 μmol of E_{600} (200 μCi) at 25 °C in 50 mM acetate buffer (pH 6.0), 50 mM NaCl, 25 mM $CaCl_2$, and 3 mM NaTDC. Residual lipase activity was measured on tributyrin as substrate. From [85].

also abolished by treatment with a serine reagent such as E_{600}, suggesting that *p*-NPA and tributyrin hydrolysis might take place at the same catalytic site of gastric lipases, in sharp contrast with PPL. Previous investigations [87, 88] showed that after a limited chymotrypsin treatment of PPL, two domains were cleaved with a concomitant loss of tributyrin activity. The isolated C-terminal domain of PPL (12 kDa) was still able to hydrolyze *p*-NPA and not tributyrin. The rate of hydrolysis of *p*-NPA was no longer affected by the presence of an interface. The published 3-D structure of HPL [13] shows that the essential serine 152 is located in the larger N-terminal domain, at the C-terminal edge of a doubly wound, parallel *β*-sheet, and that it is part of an Asp-His-Ser triad. This putative hydrolytic site is covered by a surface loop and therefore rendered inaccessible to solvents. Thus interfacial activation is likely to involve a substantial conformational change during adsorption at the lipid/water interface. A second potential "catalytic site", located in the C-terminal domain and containing histidine and aspartate residues, can be hypothesized to be responsible for the hydrolysis of *p*-NPA by pancreatic lipases. It should be stressed, however, that the specific activity of digestive lipases on *p*-NPA can be taken to be negligible since its amounts to less than one-thousandth of the catalytic activity measured on tributyrin as substrate [85]. The significance of *p*-NPA hydrolysis by pancreatic lipases has been overestimated in previous investigations, and there is no justification for extrapolating TAG hydrolysis mechanisms from the *p*-NPA ones [28, 89].

9.3.2
Interfacial Binding to Tributyrin Emulsion of Native and Chemically Modified Digestive Lipases

Several proteins (bearing lipolytic activity or otherwise) were incubated with a tributyrin emulsion. After centrifugation of the emulsified system, the non-lipolytic proteins were entirely located in the aqueous supernatants, whereas each protein bearing lipolytic activity was entirely located in the tributyrin phase. These data show that the experimental binding protocol in which a tributyrin emulsion is used and the oil and water phases are separated by centrifugation is an appropriate means of distinguishing between proteins able to specifically bind lipid/water interfaces such as lipolytic enzymes. With this experimental protocol, the binding of E_{600}-modified lipases was found to be comparable to that of native lipases [85]. Furthermore, when DP-PPL was used in the presence of a tributyrin emulsion, the interface-mediated labeling reaction with [^{3}H] sulfobenzoic cyclic anhydride was still possible [50], which also indicates that DP-PPL was present at the lipid/water interface.

All in all, the results suggest that in both gastric as in pancreatic lipases, an essential serine residue, which was stoichiometrically labeled with the organophosphorus reagent (E_{600}), is involved in catalysis and not in lipid binding.

9.3.3
Inhibition of Lipases by Phosphonates and the 3-D Structures of Lipase-Inhibitor Complexes

9.3.3.1 Synthesis of New Chiral Organophosphorus Compounds Analogous to TAG

New synthetic methods to obtain lipase inhibitors were described [90–94], replacing the carbonyl of the hydrolyzable ester bonds by a phosphonate group including a good leaving group, as shown in Fig. 9.7. These compounds, mimicking in both their charge distribution and configuration the transition state which occurs during carboxyester hydrolysis, were synthesized and investigated as potential inhibitors of HPL and HGL. Their efficiency as inhibitors was studied on the basis of the alkyl chain length, the nature of the leaving group and the influence of the ester substituent [91]. The released *p*-nitrophenol to enzyme ratio indicates that a 1:1 complex was formed. In the absence of substrate, the most powerful inhibitor was O-methyl O-(*p*-nitrophenyl) *n*-pentylphosphonate, which has a short alkyl chain, a small methoxy substituent and a good leaving group.

Stadler et al. [95] studied the influence of substrate hydrophobicity and steric hindrance by variation of the alkyl and acyl chain length at the *sn*-2 position of the glycerol backbone. Hydrolysis of these synthetic substrates demonstrated that minor structural variations at this *sn*-2 position of TAG strongly affect the stereoselectivity of the lipases tested. These authors concluded that the ester carbonyl in the (non-hydrolyzed) *sn*-2 position of a TAG was responsible for correct positioning of the substrate in the binding site of lipases. Taking into consideration the

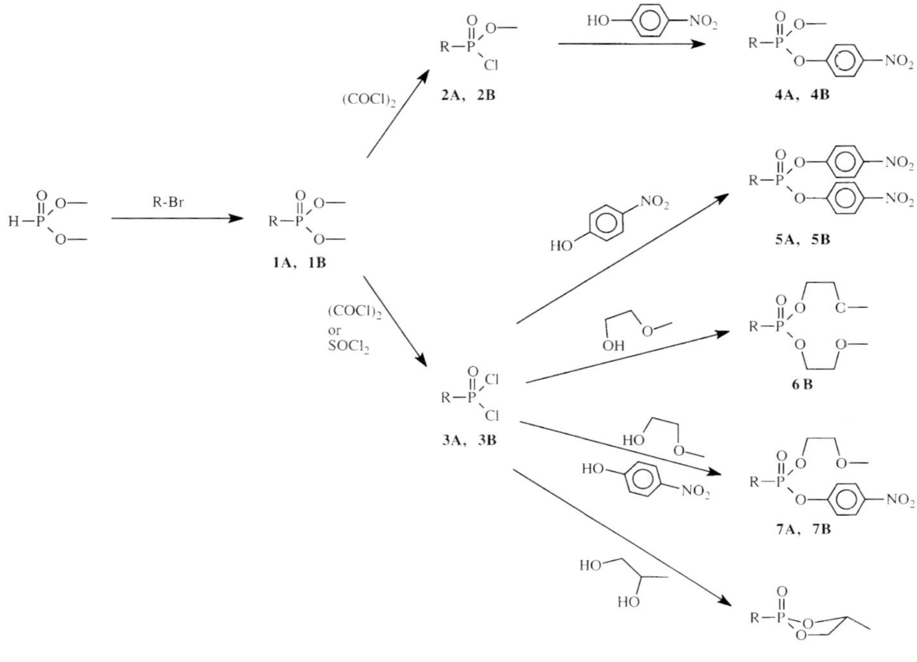

Fig. 9.7 Structure of various phosphonates and synthesis pathways. R = n-C_5H_{11} for A products, R=n-$C_{11}H_{23}$ for B products. From [91].

above mentioned results, 1,2-diacyl-3-phosphonoglycerides, which are true acylglycerol analogues, were therefore synthesized by Marguet [92]. The author intentionally conserved the two carboxyl ester linkages in order to keep as closely as possible to the structure of acylglycerols (natural substrates). This choice turned out *a posteriori* to be rational, since no or only negligible hydrolysis of these compounds occurred during the experiments with digestive lipases.

These organophosphorus compounds have been prepared in an optically pure form. The final step in the synthesis was the introduction of the phosphonate group leading to a pair of diastereoisomers which were separated by performing liquid chromatography. In this way, four pure stereoisomers, S_CR_P, S_CS_P, R_CS_P and R_CR_P, were obtained as shown in Fig. 9.8. Marguet et al. [92] have been studied the inhibition of HGL and HPL by the monomolecular film technique using mixed films of these chiral organophosphorus compounds, which are true TAG analogues, and 1,2-didecanoylglycerol (dicaprin). Interfacial lipase binding has been evaluated by means of ELISA tests with biotinylated lipases, with which it was possible to measure the surface density of enzymes in the nanogram range. With both enzymes, kinetic experiments were performed at various molar ratios of dicaprin premixed with each of the four chiral inhibitors. All the four stereoisomers investigated reduced the hydrolysis of dicaprin by HGL and HPL. With

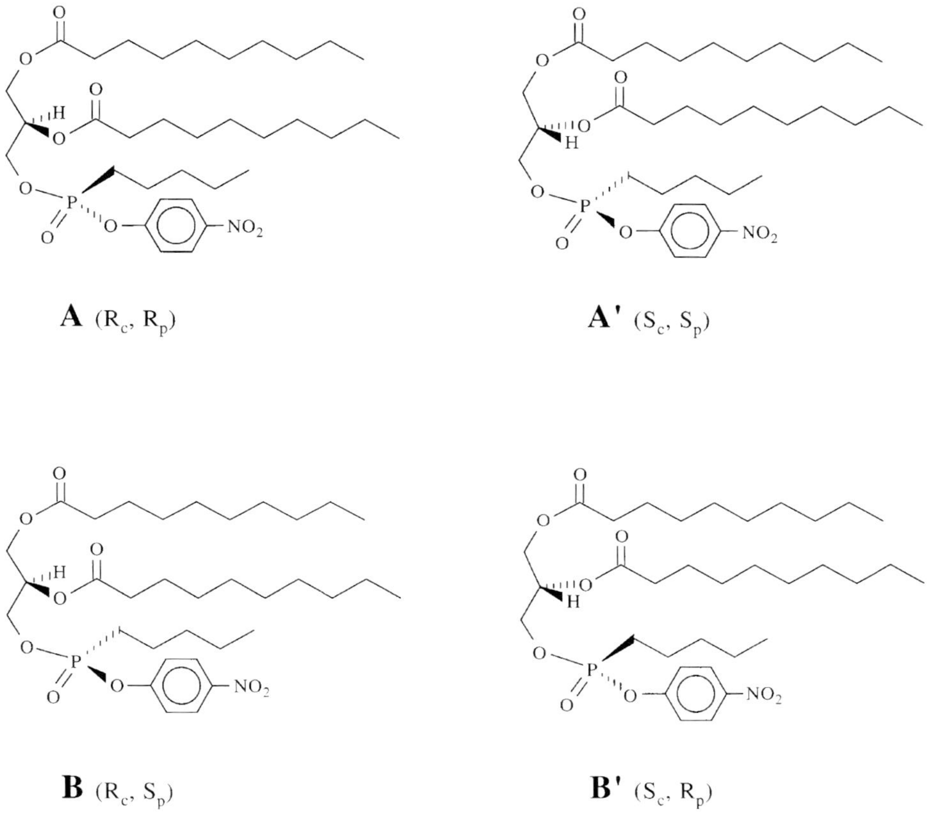

Fig. 9.8 Structure of organophosphorus inhibitors which are TAG analogues. From [92, 99].

HPL, the four stereoisomers exhibited a rather weak inhibition capacity, and no significant differences were observed among them. With each inhibitor tested, interfacial binding experiments using ELISA tests showed no significant difference in the surface density of HPL, which confirmed the low stereoselectivity of pancreatic lipases using either TAG [96] or TAG analogues [97]. With respect to gastric lipase, however, the enzyme adsorbed less on each stereoisomeric inhibitor that on the dicaprin substrate. Furthermore, the various organophosphorus enantiomers displayed differential inhibitory effects. The inhibition was strongly dependent upon the chirality on the *sn*-2 carbon of the glycerol backbone, while the chirality on the phosphorus atom had no influence. The $R_C S_P$ and $R_C R_P$, which both contain the phosphorus moiety at the *sn*-3 position, were found to be the best inhibitors. This latter finding correlates well with the *sn*-3 preference, during the hydrolysis of TAG catalyzed by gastric lipases [96–98]. Moreover, the levels of surface density of gastric lipase differed significantly with each enantiomeric inhibitor used. A clear correlation was observed between the molar ratio ($\alpha 50$) of inhibitor leading to half inhibition and the surface concentration of gastric lipase: the highest enzymatic inhibition was observed with films containing the enantiomeric inhibitor to which the HGL was best adsorbed [92, 99].

9.3.3.2 The 2.46 Å Resolution Structure of the Pancreatic/Procolipase Complex Inhibited by a C_{11} Alkylphosphonate

In an attempt to further characterize the active site and catalytic mechanism, a C_{11} alkyl phosphonate compound has been synthesized. This compound is an effective inhibitor of pancreatic lipase. The crystal structure of the pancreatic lipase-colipase complex inhibited by this compound was determined at a resolution of 2.46 Å and refined to a final *R*-factor of 18.3% [39]. As was observed in the case of the 3-D structure of the ternary pancreatic lipase–colipase–phospholipid complex [33], the binding of the ligand induces rearrangements of two surface loops in comparison with the closed structure of the enzyme [32]. The inhibitor, which could be clearly observed in the active site, was covalently bound to the active serine 152. A racemic mixture of the inhibitor was used in the crystallization, and there exists evidence that both enantiomers are bound at the active site. The C_{11} alkyl chain of the first enantiomer fits into a hydrophobic groove and is thought to thus mimic the interaction between the leaving fatty acid of a TAG substrate and the protein (Fig. 9.9 A). The alkyl chain of the second enantiomer also has an extended conformation and interacts with hydrophobic patches on the surface of the open amphipathic lid (Fig. 9.9 B). This may indicate the location of a second alkyl chain of a TAG substrate. The alkyl portions of the two C_{11} alkylphosphonate enantiomers superimpose well with the two fatty acyl chains of the phospholipid (Fig. 9.9 C) observed in the ternary phospholipid/lipase/colipase complex [33]. Some of the detergent molecules, needed for the crystallization, were also observed in the crystal. Some of them were located at the entrance of the active site, bound to the hydrophobic part of the lid. On the basis of this crystallographic study, a hypothesis about the binding mode of real substrates and the organization of the active site was proposed [39]. After partly leaving the lipid particle, the scissile acyl chain (*sn*-1 or *sn*-3) of the substrate may bind to the hydrophobic groove, implying that the substrate adopts a "fork" conformation at the active site.

Fig. 9.9 Polar interactions between the active site residues with the first enantiomer of C11-P (CONF1) (Panel A) and with the second enantiomer of C11-P (CONF2) (Panel B). In panel C, phosphatidylcholine molecule (thin lines) as refined in the ternary lipase/colipase/mixed micelle complex [30] superimposed on the two enantiomers of the C11-P inhibitor (thick lines) [39]. Carbon atoms of the glycerol moiety of the phosphatidylcholine are marked as *sn*-1, *sn*-2 and *sn*-3. The phosphorus atom and the choline moiety are indicated by arrows. Both C11-P inhibitors (CONF 1 and CONF 2) are shown covalently linked to the $O\gamma$ of the active site serine 152 (Panel C).

A

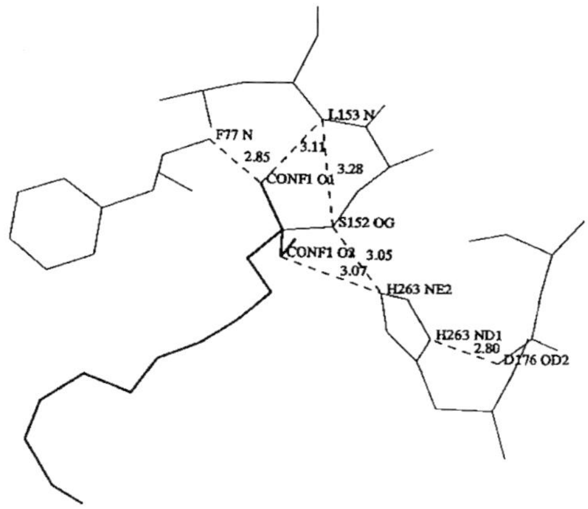
L153 N
F77 N
2.85
3.11
3.28
CONF1 O1
S152 OG
CONF1 O2
3.05
3.07
H263 NE2
H263 ND1
2.80
D176 OD2

B

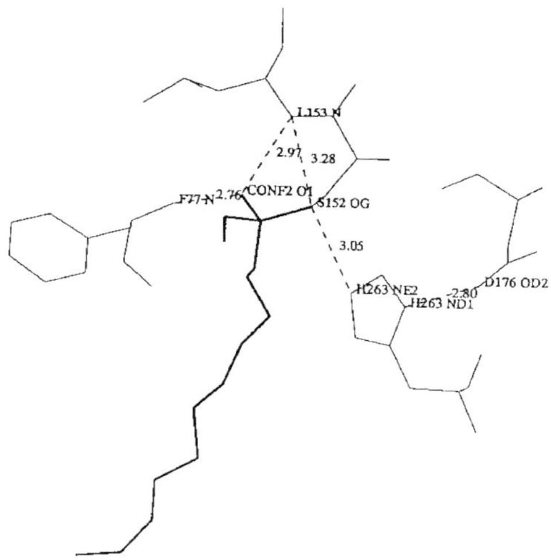
L153 N
2.97
3.28
F77 N
2.76
CONF2 O1
S152 OG
3.05
H263 NE2
H263 ND1
2.80
D176 OD2

C

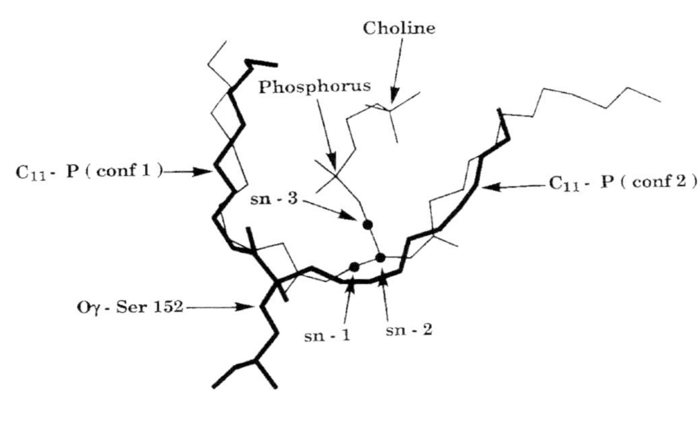
Choline
Phosphorus
C11 - P (conf 1)
C11 - P (conf 2)
sn - 3
Oγ - Ser 152
sn - 1
sn - 2

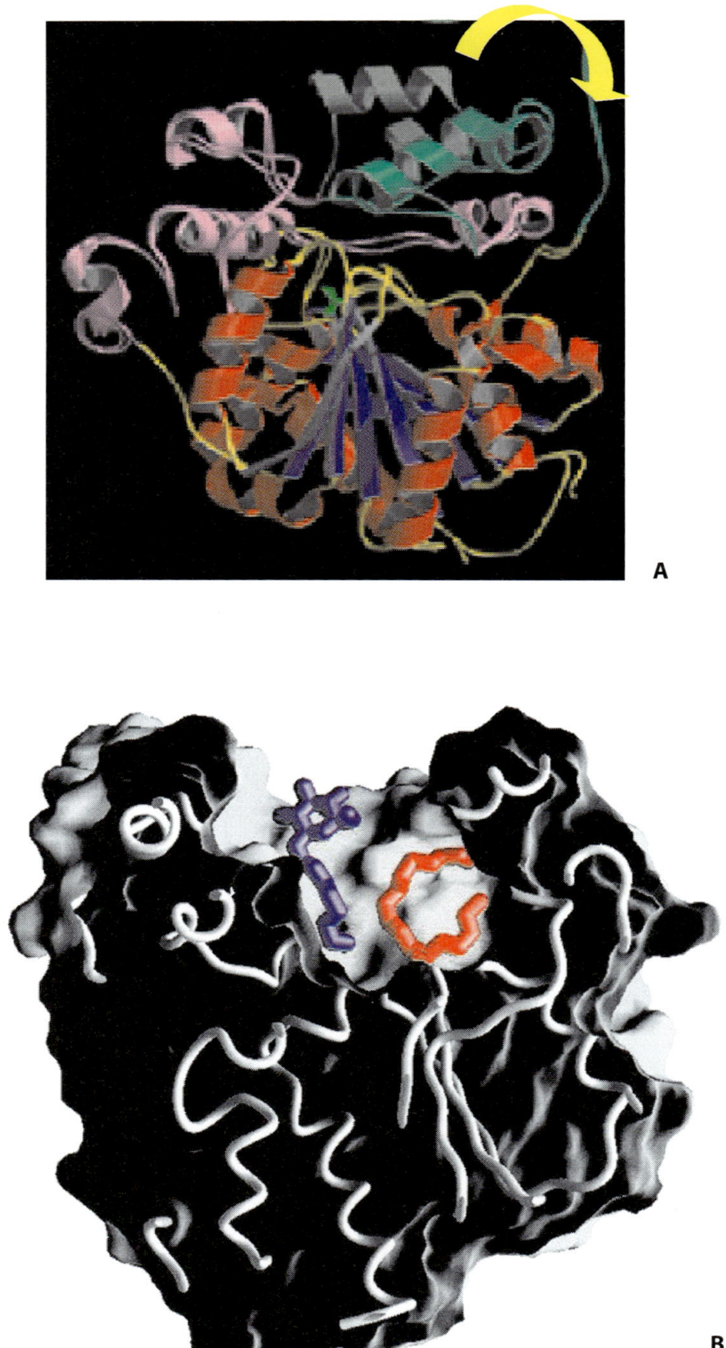

A

B

9.3.3.3 Crystal Structure of the Open Form of DGL in Complex with a Phosphonate Inhibitor

The crystal structure of the DGL, produced in transgenic maize, in complex with an alkyl phosphonate (C11Y4), was recently resolved at 2.7 Å [41]. The 3-D structure was resolved by molecular replacement using as a model the closed HGL form (the two lipases share 87% amino-acid identity). Like HGL, DGL consists of one globular domain and belongs to the α/β hydrolase fold family [21]. The polypeptide chain of DGL is visible in the electron density map from the N terminus to residue 377, which is located two residues before the C terminus. The core domain is located between residues 1–183 and 309–377 and contains a central β-sheet composed of eight strands, as observed for HGL (Fig. 9.10 A). Interestingly, a drastic conformation change affects the residues 212–251 (Fig. 9.10 A). This segment rolls over the enzyme surface and exposes a catalytic crevice measuring $20\times20\times7$ Å^3 (Fig. 9.10 B). This lid is part of the cap domain and the DGL structure reveals the first open, active form of an acidic lipase. When open, the lid domain interacts with the N-terminal segment which was not visible in the electron density map of the closed HGL (Fig. 9.10 A). The closed lid observed in the closed HGL is made out of three short helices. Upon opening, the lid of the DGL undergoes a conformational reorganization and a large movement. The open lid is formed by two long helices linked together by a disulfide bridge (Cys 227–Cys 236). The free cysteine (Cys 244) at the N-terminal end of one of these two helices remains at the vicinity of the active site residues, which explains the inhibitory effect of mercurial compounds. The reorganization of the secondary structural elements leads to a 3-D structure that looks more compact and rigid. The opening of the lid is achieved by a 180° rotation of the two helices around the axis of one of them. The open form of DGL was obtained after inhibition with the C11Y4, in the presence of β-octylglucoside at 30 mM, final concentration. Strikingly, the catalytic crevice was completely filled by the two chains of the alkyl phosphonates and one β-octylglucoside molecule. Most of the canyon wall residues are hydrophobic, the majority being leucines, isoleucines, and valines. The phosphonate inhibitor molecule, covalently bound to the catalytic serine (Ser 153) covers 287 Å^2 of the canyon water-accessible surface, and 86% of it consists of hydrophobic residues. An other canyon surface area of 247 Å^2 is covered by the detergent molecule with the same proportion of hydrophobic residues. It is worth noticing that the β-octyl-

Fig. 9.10 (A) Stereo ribbon view of DGL superimposed on HGL. The cores of both molecules are colored according the secondary structures (helix, *red*; strand, *blue*; remainder, *yellow*). The cap domain of both enzymes is depicted in magenta. The lid of the closed HGL is brown, and that of the open DGL is green. The yellow arrow indicates the direction of the lid motion from a closed to an open conformation. The missing parts in HGL (residues 1–9 and 53–56) are identified in the structure. The nucleophilic active site residue, Ser 153, is represented by green sticks. (B) GRASP representation of recombinant DGL ribbon and molecular surface slabbed at the active crevice level. The active site crevice is visible laterally (~20 × 20 Å) with the bound C11Y4 phosphonate (*red*) and β-octyl glycoside (*blue*).

glucoside is stacked against the open lid, with which most of its interactions occur. Consequently, the lid area covered by the *β*-octylglucoside molecule is 156 Å^2, whereas the latter detergent molecule covers only 73 and 45 Å^2 of the cap and the core domains, respectively. The inhibitor is located opposite the lid and covers only 32 Å^2 of the lid in a single interaction with Leu 243, whereas the inhibitor covers 103 and 152 Å^2 of the cap and the core domains, respectively. The orientation of the two chains of the alkyl phosphonate along with that of the detergent molecule suggested a binding mode of a TAG in the active site of the DGL (Fig. 9.10 B).

More recently, Miled et al. [100] have performed a kinetic study with DGL and HGL using several phosphonate inhibitors by varying the absolute configuration of the phosphorus atom and the chain length of the alkyl/alkoxy substituents. Using the 3-D structure of gastric lipases in the apo-form or in complex with the R-undecyl butyl phosphonate ($C_{11}Y_4(+)$) and that of a new crystal structure obtained with the other (S_P)-phosphonate enantiomer, C(11)Y(4)(–), the authors constructed models of phosphonate inhibitors fitting into the active site crevices of DGL and HGL [100]. All inhibitors with a chain length of fewer than 12 carbon atoms were found to be completely buried in the catalytic crevice, whereas longer alkyl/alkoxy chains were found to point out of the cavity. The main stereospecific determinant explaining the stronger inhibition of the S_P enantiomers is the presence of a hydrogen bond involving the catalytic histidine as found in the DGL-C(11)Y(4)(–) complex. On the basis of these results, a model of the first tetrahedral intermediate corresponding to the tristearoyl-lipase complex was proposed. The TAG molecule completely fills the active site crevice of DGL, in contrast with what is observed with other lipases such as pancreatic lipases which have a shallower and narrower active site. For substrate hydrolysis, the supply of water molecules to the active site might be achieved through a lateral channel identified in the protein core.

9.4 Inhibition of Digestive Lipases by Orlistat

9.4.1 Introduction

Orlistat (tetrahydrolipstatin, THL), a hydrogenated lipstatin analogue isolated from *Streptomyces toxytricini*, is a potent inhibitor of gastro-intestinal lipases [44, 51, 101–103]. It has been suggested that a stoichiometric covalent enzyme–inhibitor complex of a long-lived acyl-enzyme type is formed by a nucleophilic attack of the catalytic serine of pancreatic lipase on the *β*-lactone acyl group of orlistat (Fig. 9.11) [104, 105]. Several reports on the clinical application of orlistat to the treatment of human obesity have been published during the last decade (e.g., [106–109]). Orlistat has been approved since 1998 in the USA and in Europe as a

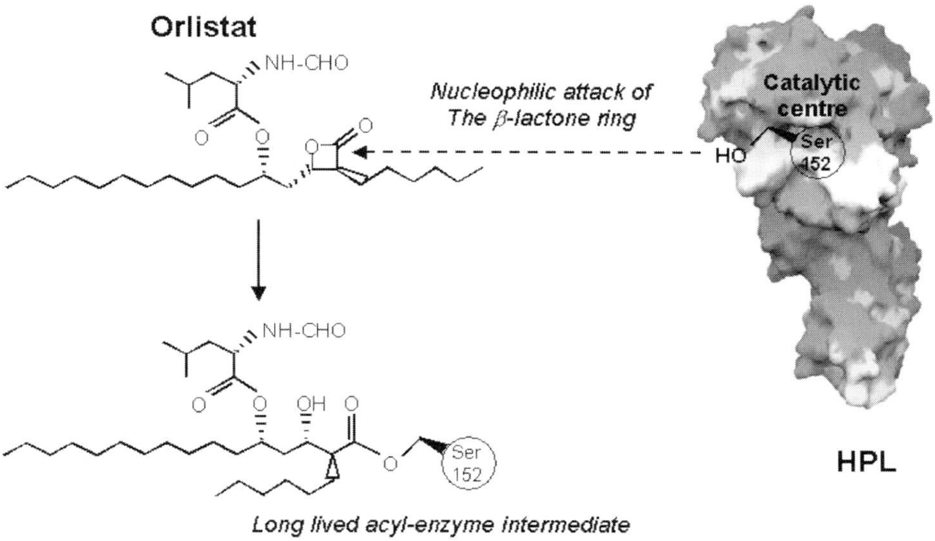

Fig. 9.11 Schematic illustration of the nucleophilic attack of the β-lactone by the active site serine (152) of HPL (upper panel) and the resulting covalent adduct between the serine 152 and the β-lactone ring of THL in its opened conformation (lower panel).

weight loss medication for use in the treatment of obesity, and it is marketed under the trade name Xenical®.

Studies of the partitioning of orlistat between the micellar and oil phases were performed at our laboratory [52, 110]. Even in the presence of bile salts, orlistat was found to be mostly associated with the oil phase. This interesting physicochemical property is probably a requirement for prototypic lipase inhibitors to be effective under physiological conditions, i.e. in the presence of bile and dietary lipids.

Based on *in vitro* studies, it was reported that bile salts above their critical micellar concentration (CMC) are required for orlistat to be able to effectively inhibit HPL [29] and lipoprotein lipase [111]. The fact that the inhibited HPL could be reactivated by reducing the bile salt concentration below its CMC suggested that bile salts (above their CMC) may stabilize the acyl-lipase complex [29].

9.4.2 Inhibition of Digestive Lipases by Pre-incubation with Orlistat (Method A)

9.4.2.1 Inhibition of Gastric Lipases

The incubation in a buffer solution at pH 6.0 of HGL and RGL (Fig. 9.12) in the presence of orlistat at a molar excess of 100 was performed and the residual lipase activities were measured [51]. In the absence of substrate (Method A), it can be seen from Fig. 9.12 that orlistat partially inhibits HGL and RGL. After 240 min of incubation, 50% and 25% of residual HGL and RGL activity were measured, re-

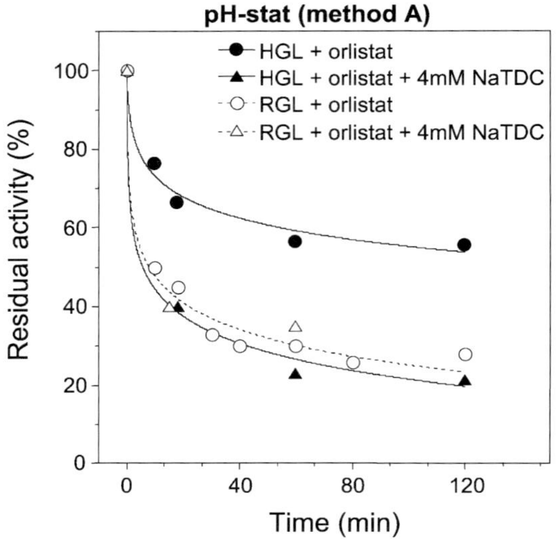

Fig. 9.12 Variations in HGL (full symbols) and RGL (open symbols) activity during incubation with orlistat in the absence (circles) or presence (triangles) of 4 mM NaTDC. Lipases (26 μM) were incubated at 37 °C, in a final volume of 150 μL of 50 mM acetate buffer (pH 6.0), 150 mM NaCl with orlistat (2.6 mM). Residual lipase activity was measured at pH 6.0, 37 °C by tributyrin (0.2 mL) hydrolysis using 10 mL of 150 mM NaCl, 2 mM NaTDC, 1.5 μM bovine serum albumin. Each kinetics represents one experiment. Adapted from [51].

spectively. The presence of NaTDC (final concentration 4 mM) in the incubation medium did not affect RGL inhibition, whereas the remaining HGL activity decreased from 50% to 20% when orlistat was used in the absence or presence of NaTDC, respectively.

The increase of HGL inhibition by orlistat in the presence of bile salt is due to the fact that orlistat probably formed mixed bile salt/orlistat micelles. As in the case of lipase substrates, the presence of mixed orlistat bile salt micelles may give rise to a better "interfacial quality", thus improving the lipase adsorption and/or increasing in orlistat accessibility. This finding is reminiscent of the effect of E_{600} on PPL. In the latter case, Desnuelle et al. [65] showed that emulsified E_{600} irreversibly inactivates PPL, whereas aqueous solutions of this organo-phosphorous compound do not. Maylié et al. [63] and Rouard et al. [66] described the covalent modification of a serine residue of PPL (in the presence of colipase) induced by mixed E_{600}/bile salt micelles.

9.4.2.2 **Inhibition of Pancreatic Lipases**

The incubation of HPL with orlistat (at a molar excess of 100) was performed in the presence and absence of bile salts at 25 °C at pH 8.0. The residual lipase activity was measured with the pH-stat technique on either tributyrin or PSO emulsions at 37 °C [112].

After a few minutes of incubation with orlistat, the HPL initial activity measured on both tributyrin (Fig. 9.13) and PSO (data not shown) emulsions was drastically reduced. The lipolytic activity gradually increased with time, however, reaching a steady state regime after a lag period of a few minutes. This reactivation phenomenon is probably due to a partial deacylation, during the lipase assay, of the covalent HPL–orlistat complex.

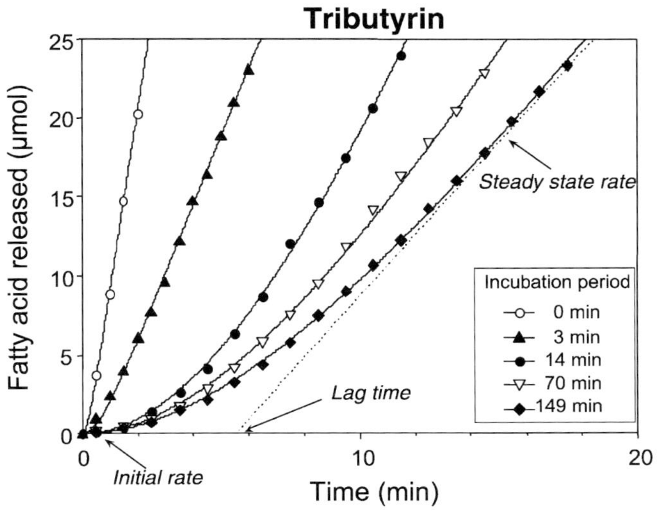

Fig. 9.13 Typical kinetic recordings of the hydrolysis of tributyrin emulsions by HPL, using the pH-stat technique. HPL was incubated with orlistat using Method A′ (in the presence of 4 mM NaTDC). At various time intervals, 10 µL of the incubation medium (~2 µg of HPL) were injected into the pH-stat vessel after recording the background hydrolysis for 2 min. Kinetic assays were performed in a thermostated (37 °C) vessel containing 0.5 mL tributyrin mechanically emulsified in 14.5 mL of 1 mM Tris-HCl buffer, 150 mM NaCl and 5 mM $CaCl_2$. Colipase was added to the assay vessel at a five-fold molar excess prior to HPL injection. The kinetic recordings shown are representative of three independent experiments. The dashed line corresponds to the asymptote reached under steady state conditions (curve fitting).

The residual HPL activity measured under the steady state conditions decreased rapidly as a function of the incubation time and reached a plateau at around 40 to 60% of its initial value both when tributyrin and PSO were used as substrate. These plateau values did not depend very greatly on the presence of 4 mM NaTDC in either the incubation medium or the assay medium [112].

In the presence of 4 mM NaTDC and colipase in the incubation medium, a rapid and near-complete decrease in the HPL residual activity was observed when either a tributyrin emulsion (Fig. 9.14 A) or a PSO emulsion (Fig. 9.14 B) were used as the substrates. In other words, after a 1-min period of incubation with orlistat, only 50% of the initial activity of HPL was detected using either tributyrin or PSO emulsions, and this value decreased further to less than 20% after 10 min of incubation.

In the absence of NaTDC in the incubation medium, orlistat partially inhibits HPL and a plateau (around 60%) in the residual activity was rapidly reached when using tributyrin as substrate, both in the absence and the presence of 4 mM NaTDC in the assay medium (Fig. 9.14 A). However, a different pattern was observed when PSO emulsion was used as the substrate (Fig. 9.14 B). In the absence of NaTDC in both the incubation medium and the assay medium, HPL was

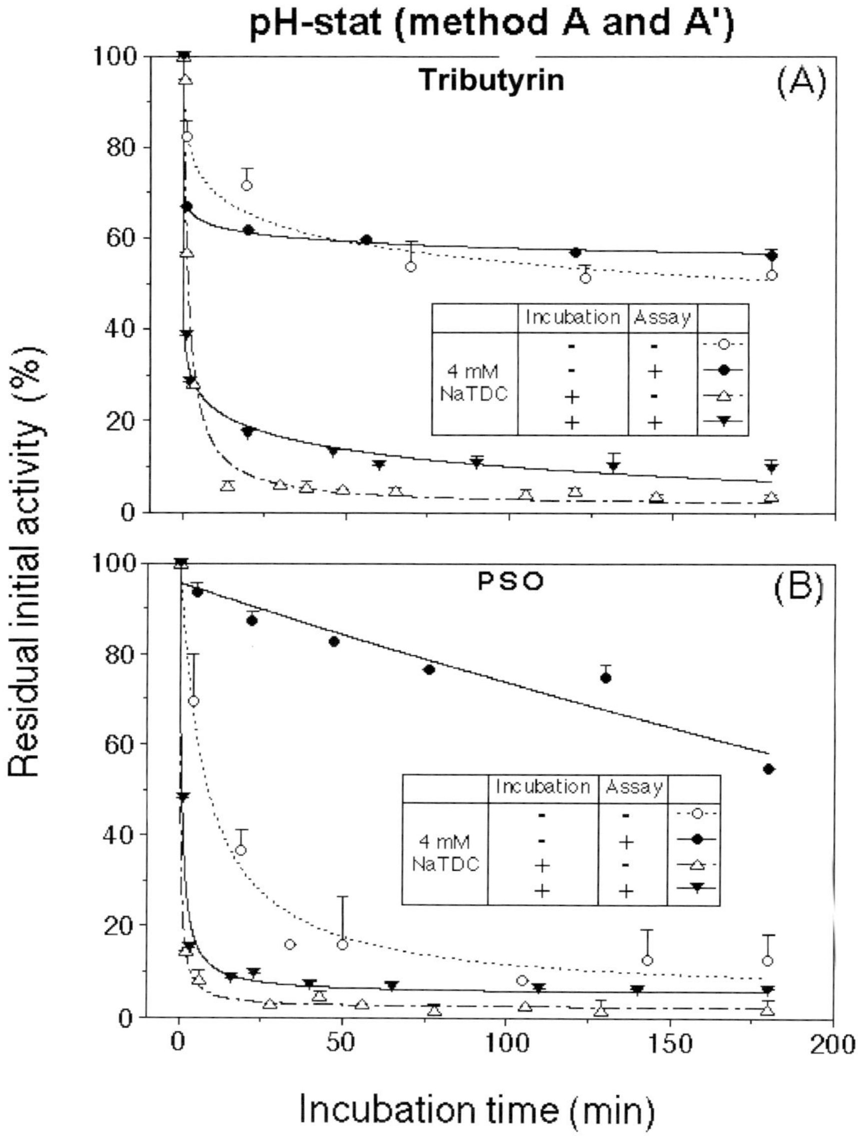

Fig. 9.14 Initial residual activity of HPL acting on a tributyrin emulsion (A) or a PSO emulsion (B), as a function of the incubation time with orlistat, measured using the pH-stat technique. The experiments were carried out in the absence (circles) or presence (triangles) of 4 mM NaTDC in the incubation medium. The full symbols (λ, τ) illustrate the experiments performed in the presence of 4 mM NaTDC in the assay and the open symbols (μ, Δ), in its absence. The other experimental conditions used are the same as those described in Fig. 9.13. The residual activity values are means ± SD (n≥3 for each experiment). From [112].

found to be strongly inhibited by orlistat although the inhibition rate was slower than that observed in the presence of bile salts in the incubation medium. The absence of NaTDC in the incubation medium and its presence in the lipase assay medium resulted in a moderate inhibition of HPL by orlistat.

This dependence of HPL inhibition by orlistat in the presence of bile salts in the incubation medium might be attributable to a better 'interfacial quality' of orlistat by the formation of mixed micelles with bile salts, which may improve the lipase adsorption onto these mixed micelles and/or interaction with orlistat.

The increase in the inhibitory efficacy of orlistat induced by bile salts over their CMC was previously reported in the case of pancreatic lipases (in humans [29], dog and guinea pig [52]) as well with HGL [51] and lipoprotein lipase [111]. Furthermore, upon studying the transfer of radiolabeled orlistat from an oil drop to the aqueous phase, we noted that the transfer rate increased in the presence of bile salts above their CMC [110], reflecting the solubilization of orlistat in the aqueous phase *via* the formation of mixed micelles with bile salts.

The mass spectra, under denaturing conditions, of a sample of the orlistat–HPL incubation mixtures were recorded at various incubation times ranging from 5 min to 24 h. As can be seen from Fig. 9.15, an increase of around 500 Da was observed in the molecular mass of pure HPL (Fig. 9.14, spectra I) when incubated with orlistat (Fig. 9.14, spectra II). This mass increase reflects the covalent binding

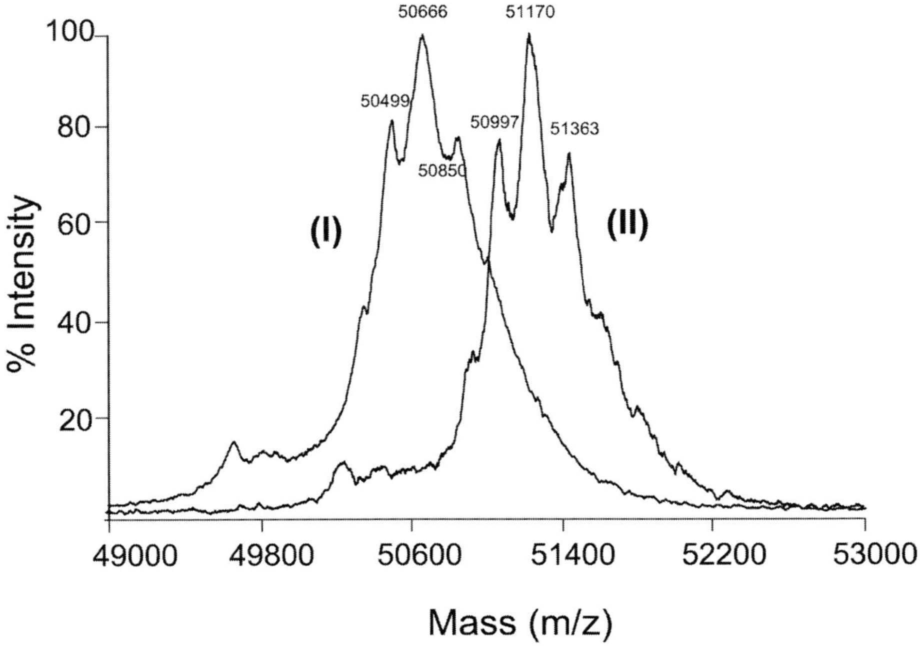

Fig. 9.15 MALDI-TOF mass spectra of HPL (I) or HPL incubated 1 h with orlistat (II) at a HPL:orlistat molar ratio of 1:100. Adapted from [112].

of orlistat (molecular mass, 496 Da) to HPL, as previously reported by Lüthi-Peng et al. [105].

9.4.2.3 Kinetic Model Illustrating the Covalent Inhibition of HPL in the Aqueous Phase

On the basis of the above mentioned kinetic and mass spectrometry analysis, a model was proposed for the covalent inhibition of HPL by orlistat in the aqueous phase as well as its partial reactivation at the lipid–water interface (Fig. 9.16). This model takes into account the putative existence of two different forms of the covalent orlistat/lipase complex.

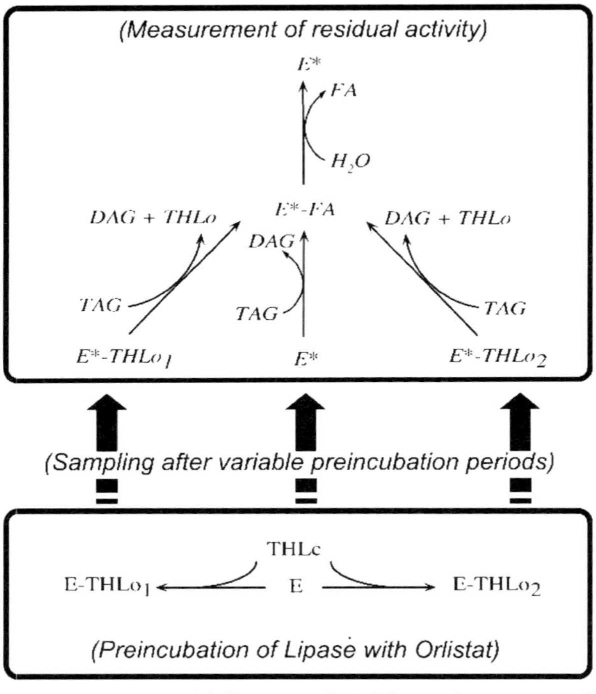

Fig. 9.16 Kinetic model illustrating the inhibition of HPL by orlistat in the aqueous phase and its reactivation at a lipid–water interface. The following symbols and abbreviations are used here: *E*, free enzyme (molecule/volume); *E**, interfacial enzyme (molecule/surface); *FA*, fatty acid at the interface (molecule/surface); *E**-FA, interfacial enzyme–fatty acid complex (molecule/surface); THLc, closed reactive orlistat in the bulk (molecule/volume); THLo, open non-reactive orlistat at the interface (molecule/surface); *E*-$THLo_1$, form 1 of covalent enzyme-orlistat complex in the bulk (molecule/volume); *E*-$THLo_2$, form 2 of covalent enzyme–orlistat complex in the bulk (molecule/volume); *E*-$THLo_1$*, interfacial form 1 of covalent enzyme–orlistat complex (molecule/surface); *E*-$THLo_2$*, interfacial form 2 of covalent enzyme–orlistat complex (molecule/surface). TAG, triacylglycerol at the interface (molecule/surface); DAG, diacylglycerol at the interface (molecule/surface). From [112].

One can hypothesize that the covalent binding of orlistat to HPL occurs in two possible ways. One fraction of the lipase molecules may form a covalent complex in the aqueous phase with the open β-lactone ring of orlistat in a given form (E-THLo_1), whereas the other fraction of the lipase molecules may also form a covalent complex with orlistat, but in another form (E-THLo_2). The hypothesis that a fraction of the inhibited lipase molecules could form a non-covalent complex with orlistat in its closed form (E-THLc) can be ruled out, since the mass spectra analysis (Fig. 9.15) showed the presence of only the HPL–orlistat covalent complex.

In the lipase assay system, when the (E, THLc, E-THLo_1 and E-THLo_2) mixture was added to a TAG emulsion, the 'weaker' complex (E-THLo_1) was hydrolyzed, explaining the reactivation process as revealed by the existence of lag times. The hydrolysis of this 'weak' complex was enhanced when bile salts were present in the assay medium, probably due to the formation of a mixed orlistat/lipase/bile salt complex. The fact that the reactivation of HPL did not reach 100% of its initial value (before incubation with THL) at the end of the kinetic experiments (Fig. 9.13) might be attributable to the existence of a second fraction of lipase molecules, which is covalently and irreversibly inhibited by orlistat (E-THLo_2). The coexistence of two different forms of orlistat covalently bound to HPL may be due to the inhibitor molecule being differently oriented in the catalytic cavity of the lipase molecule. The existence in the catalytic cleft of lipases of two different orientations in the case of molecules of substrate analogues was previously reported [113, 114].

9.4.3
Inhibition of Digestive Lipases During Lipolysis (Method B)

The hydrolysis of tributyrin by HGL and RGL was recorded at pH 6.0 and at pH 8.0 with PPL. Two minutes after enzyme addition, orlistat was injected at variable final concentrations. As can be seen from the kinetic curves in Fig. 9.17, the rate of tributyrin hydrolysis by PPL (Fig. 9.17 A), RGL (Fig. 9.17 B) and HGL (Fig. 9.17 C) decreased to various extents. This decrease was more rapid at higher orlistat concentrations. With all lipases tested, more than 90% of the enzyme inhibition occurred at a final orlistat concentration of 200 μM [51].

9.4.4
Inhibition of Digestive Lipases on Oil Emulsions "Poisoned" with Orlistat (Method C)

The rate of tributyrin hydrolysis by HGL, RGL or PPL was determined as a function of the orlistat concentration (Fig. 9.18 A) [51]. Orlistat was injected into a tributyrin emulsion 1 min prior to the enzyme addition and the initial lipase hydrolysis rate was measured. As shown from the curve in Fig. 9.18 A, the rate of tributyrin hydrolysis by HGL, RGL or PPL decreased similarly with increasing the final orlistat concentration. The inhibitor concentration which reduced the lipase activity to 50% (I_{50}) of its initial value was around 10 μM.

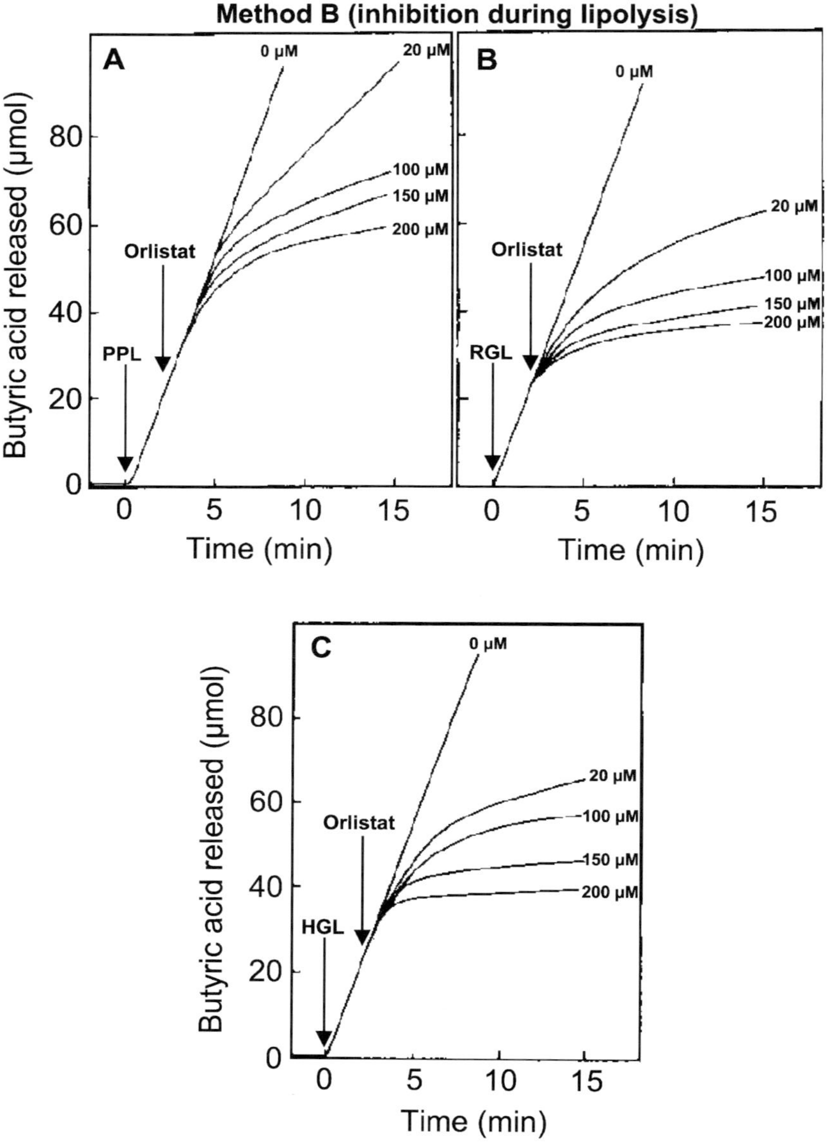

Fig. 9.17 Effect of variable orlistat concentrations on the rate of hydrolysis of tributyrin by: (A) PPL (5 nM) + colipase (100 nM); (B) RGL (24 nM); (C) HGL (24 nM). Assays were carried out at 37 °C, pH 8.0 with PPL and pH 6.0 with HGL and RGL using 0.2 mL tributyrin emulsified in 10 mL of 150 mM NaCl, 2 mM NaTDC, 1.5 μM bovine serum albumin. Each kinetics represents one typical experiment. From [51].

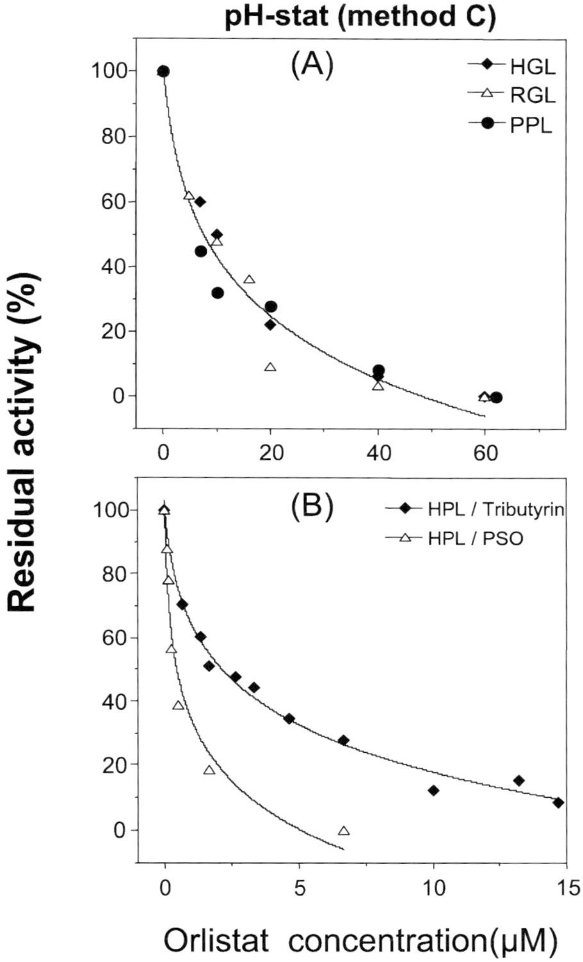

Fig. 9.18 (A) Effect of increasing concentrations of orlistat on the rate of hydrolysis of tributyrin by PPL (5 nM) in the presence of colipase (100 nM) (●), or RGL (24 nM) (△) or HGL (24 nM) (◆). Experiments were performed at 37 °C (pH 8.0), with PPL and pH 6.0 with HGL and RGL using 0.2 mL tributyrin emulsified in 10 mL of 150 mM NaCl, 2 mM NaTDC, 1.5 μM bovine serum albumin. Orlistat, from a 20 mM ethanolic solution, was injected 1 min prior to the enzyme addition. Initial velocity measurements are given. Each point represents an independent assay. From [51]. (B) Effects of increasing orlistat concentration on HPL residual activity as measured, using the pH-stat technique, on tributyrin (◆) or PSO (△) according to method C. Kinetic assays were performed in a thermostated (37 °C) vessel containing 0.5 mL TAG substrate mechanically emulsified in 14.5 mL of 1 mM Tris-HCl buffer (pH 8.0), 150 mM NaCl and 5 mM $CaCl_2$. The final lipase concentration was 2 nM. Colipase was added to the assay vessel at a five-fold molar excess prior to HPL injection.

The inhibition effects of orlistat on HPL activity were studied in the absence or in the presence of NaTDC, using tributyrin or PSO as substrate (Tiss et al., manuscript in preparation). In the absence of bile salts, as shown in Fig. 9.18 B, the rate of tributyrin hydrolysis by HPL decreased rapidly according to a similar pattern, with increasing orlistat concentration and the I_{50} value was found to be around 2.2 μM. When using PSO as substrate, orlistat was found to be more efficient in inhibiting HPL activity (I_{50}=0.4 μM).

For both substrates used (tributyrin and PSO), the presence of bile salts (4 mM NaTDC, final concentration) improved the orlistat inhibition of HPL (data not shown) in agreement with the results found when using method A′ and showing an enhancement of the orlistat effect in the presence of NaTDC [112].

9.4.5 Inhibition of Digestive Lipases on Oil Substrate "Poisoned" with Orlistat (Method D)

9.4.5.1 Inhibition of Pancreatic Lipase on Emulsion "Pre-poisoned" with Orlistat

Variable amounts of orlistat were dissolved directly in pure tributyrin or PSO. As shown in Fig. 9.19, the rate of hydrolysis by HPL, in the absence of bile salts, was determined as a function of orlistat concentration.

The hydrolysis rate of tributyrin or PSO by HPL decreased sharply with increasing orlistat concentration. The I_{50} values were found to be around 2.8 μM when using tributyrin and 0.2 μM when using PSO as substrate, respectively (Tiss et al. in preparation). It is worth noticing that using either method C (Fig. 9.18 B) or method D (Fig. 9.19) comparable IC_{50} values were obtained when using both tributyrin or PSO as substrates.

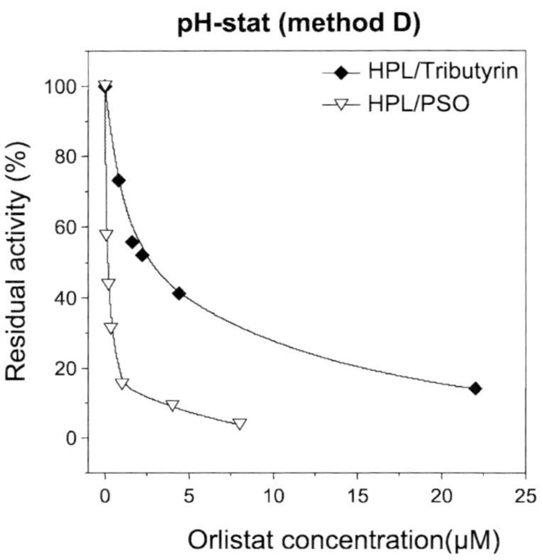

Fig. 9.19 Effects of increasing orlistat concentration on HPL residual activity as measured, using the pH-stat technique, on tributyrin (◆) or PSO (▽) according to method D. Kinetic assays were performed in a thermostated (37 °C) vessel containing 0.5 mL TAG substrate mechanically emulsified in 14.5 mL of 1 mM Tris-HCl buffer (pH 8.0), 150 mM NaCl and 5 mM $CaCl_2$. The final lipase concentration was 2 nM. Colipase was added to the assay vessel at a five-fold molar excess prior to HPL injection.

9.4.5.2 Inhibition of Gastric and Pancreatic Lipases on Mixed Films Containing Orlistat

As indicated in Fig. 9.20, lipase activities were measured as a function of the orlistat surface molar fraction. With all lipases tested, the hydrolysis of dicaprin decreased sharply as the surface molar fraction of orlistat increased. The surface molar fractions of orlistat, which reduces lipase activity to 50% (α_{50}), are 0.013%, 0.025% and 0.25% with PPL, RGL and HGL, respectively (Fig. 9.20 A) and 0.00025% with HPL (Fig. 9.20 B). On the one hand, it should be remembered that the turnover rates of gastric and pancreatic lipases are quite different [115]. Consequently, the amounts of each lipase injected into the reaction compartment of the zero-order trough were different (see legend of Fig. 9.20), leading to various orlistat/lipase molar ratios. On the other hand, we know that monolayer systems are characterized by a low specific surface (around 1 cm^2/cm^3). As a consequence, only a small fraction of the injected enzyme is bound to the monomolecular film. In the particular case of PPL and HGL, using pure dicaprin films, it was found that around 1.5% of the total amount of injected enzyme was recovered after film aspiration [115, 116]. As judged by this percentage of lipid-bound enzyme, the effective surface molar excesses of orlistat to film-bound lipase, inducing a 50% reduction in catalytic activity, were estimated as 10- to 20-fold with PPL, RGL and HGL. The stoichiometry at the interface can be described as follows: one lipase molecule embedded within 105 substrate molecules will be inactivated to half its initial rate by the presence of 10 orlistat molecules.

With HPL, the kinetic studies were carried out at 12 mN m^{-1}, using mixed films of orlistat/dicaprin, orlistat/diolein and orlistat/PSO. At this surface pressure, which was below the collapse pressure of all the components (35, 29, 25 and 13 mN m^{-1} for diC10, diolein, orlistat and PSO, respectively), all the above films were stable with time, i.e., no significant decrease in the surface pressure was observed within 20 min. Under these conditions, the α_{50} values, leading to half inhibition of HPL activity on dicaprin, diolein and PSO were found to be 0.0002% (see Fig. 9.20 A), 0.45% and 0.2% (data not shown).

With all the substrates used, HPL was thus found to be highly inhibited by orlistat as compared to other lipases. Indeed, one molecule of orlistat, embedded within 500000 dicaprin molecules at the interface, was found to be sufficient to inhibit HPL to half its initial activity. When one molecule of orlistat within 15000 molecules of dicaprin is present at the interface, HPL activity was totally abolished. When using PSO as substrate, an interfacial stoichiometry of one orlistat molecule to 500 PSO molecules was needed to achieve 50% of HPL inhibition.

9.4.5.3 Inhibition of Pancreatic Lipase on Oil Drop "Pre-poisoned" with Orlistat

Tiss et al. [117] studied the partitioning of orlistat between the core and the surface of an oil drop, as a function of the orlistat concentration in the oil. The molar fraction of orlistat at the oil drop surface was estimated using the Gibbs surface excess equation. With these data it becomes possible to express orlistat con-

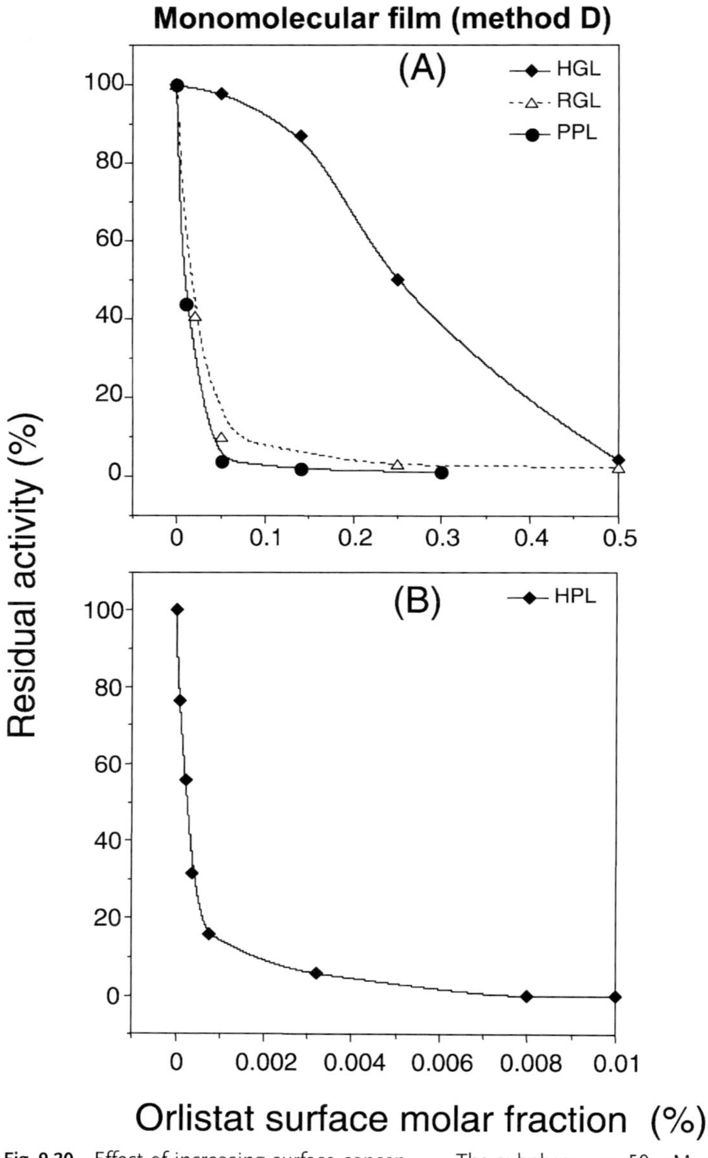

Fig. 9.20 Effect of increasing surface concentration of orlistat on the rate of hydrolysis of dicaprin monolayers at a constant surface pressure (A) by HGL (final concentration 2 nM) (◆), RGL (final concentration 0.4 nM) (△), PPL (final concentration 0.2 nM) (●), and (B) by HPL (final concentration 0.16 nM). The subphase was 50 mM glycine-HCl (pH 4.0), 100 mM NaCl and 10 mM $CaCl_2$ in the HGL and RGL assays, and 10 mM Tris-HCl (pH 8.0), 100 mM NaCl, 21 mM $CaCl_2$ and 1 mM EDTA in the PPL and HPL assays. From [44].

centrations as surface concentrations (or molar fraction) instead of volumic concentrations.

HPL residual activity on "poisoned" drops of PSO was measured as a function of the orlistat molar fraction at the surface of the oil drop. The α_{50} value was found to be around 3% (Tiss et al., manuscript in preparation).

The efficacy of orlistat in inhibiting HPL acting on PSO was found to be higher when using the monomolecular film technique as compared to the single oil drop technique.

It is seems clear that further chemical, biochemical and physiological studies of lipase inhibitors will improve our knowledge of this important and interesting family of carboxylester hydrolases. Orlistat was the first lipase inhibitor to be launched on the market. It is to be expected that other molecules will be developed as beneficial drugs for human health.

9.5 References

1 Desnuelle, P., Sjöström, H. and Nonén, O. (1986) Molecular and cellular basis of digestion. Elsevier, Amsterdam

2 Carey, M.C., Small, D.M. and Bliss, C.M. (1983) Lipid digestion and absorption of fat. Ann. Rev. Physiol. **45**, 651–677

2a Gargouri, Y., Moreau, H. and Verger, R. (1989) Gastric lipases: Biochemical and physiological studies. Biochim. Biophys. Acta **1006**, 255–271

2b Carrière, F., Gargouri, Y., Moreau, H., Ransac, S., Rogalska, E. and Verger, R. (1994) *in* Lipases: Their structure, biochemistry and application (Wooley, P. and Petersen, S.B., eds.), pp. 181–205. Cambridge University Press, Cambridge, England

2c Hamosh, M. (1990) Lingual and gastric lipases: their role in fat digestion. CRC Press, Boston

3 Borgström, B. (1975) On the interactions between pancreatic lipase and colipase and the substrate, and the importance of bile salts. J. Lipids Res. **16**, 411–417

4 Momsen, W.E. and Brockmann, H.L. (1976) Effect of Colipase and Taurodeoxycholate on the Catalytic and Physical Properties of Pancreatic Lipase B at an Oil-Water Interface. J. Biol. Chem. **251**, 378–383

5 Borgström, B. and Erlanson-Albertson, C. (1971) Pancreatic juice colipase: Physiological importance. Biochim. Biophys. Acta **242**, 509–513

6 Borgström, B. and Erlanson-Albertson, C. (1973) Pancreatic lipase and colipase. Interactions and effects of bile salts and other detergents. Eur. J. Biochem. **37**, 60–68

7 Vandermeers, A., Vandermeers-Piret, M.C., Rathe, J. and Christophe, J. (1974) On human pancreatic triacylglycerol lipase: isolation and some properties. Biochim. Biophys. Acta **370**, 257–268

8 De Caro, A., Figarella, C., Amic, J., Michel, R. and Guy, O. (1977) Human pancreatic lipase: A glycoprotein. Biochim. Biophys. Acta **490**, 411–419

9 Lowe, M.E., Rosenblum, J.L. and Strauss, A.W. (1989) Cloning and characterization of human pancreatic lipase cDNA. J. Biol. Chem. **264**, 20042–20048

10 Thirstrup, K., Carrière, F., Hjorth, S., Rasmussen, P.B., Wöldike, H., Nielsen, P.F. and Thim, L. (1993) One-step purification and characterization of human pancreatic lipase expressed in insect cells. FEBS Lett. **327**, 79–84

11 Bodmer, M.W., Angal, S., Yarranton, G.T., Harris, T.J.R., Lyons, A., King, D.J., Piéroni, G., Rivière, C., Verger,

R. and Lowe, P.A. (1987) Molecular cloning of a human gastric lipase and expression of the enzyme in yeast. Biochim. Biophys. Acta **909**, 237–244

12 Bennet, W. (1987) Dietary treatments of obesity. Ann. N.Y. Acad. Sci. **449**, 250–263

13 Winkler, F.K., d'Arcy, A. and Hunziker, W. (1990) Structure of human pancreatic lipase. Nature **343**, 771–774

14 Kirchgessner, T.G., Chuat, J.C., Heinzmann, C., Etienne, J., Guilhot, S., Svenson, K., Ameis, D., Pilon, C., d'Auriol, L., Andalibi, A., Schotz, M.C., Galibert, F. and Lusis, A.J. (1989) Organization of the human lipoprotein lipase gene and evolution of the lipase gene family. Proc. Natl. Acad. Sci. USA **86**, 9647–9651

15 Persson, B., Bentsson-Olivecrona, G., Enerbäck, S., Olivecrona, T. and Jörnvall, H. (1989) Structural features of lipoprotein lipase. Lipase family relationships, binding interactions, non-equivalence of lipase cofactors, vitellogenin similarities and functional subdivision of lipoprotein lipase. Eur. J. Biochem. **179**, 39–45

16 Derewenda, Z.S. and Cambillau, C. (1991) Effects of gene mutations in lipoprotein and hepatic lipases as interpreted by a molecular model of the pancreatic triglyceride lipase. J. Biol. Chem. **266**, 23112–23119

17 Hide, W.A.C.L. and Li, W.H. (1992) Structure and evolution of the lipase superfamily. J. Lipid Res. **33**, 167–178

18 van Tilbeurgh, H., Roussel, A., Lalouel, J.M. and Cambillau, C. (1994) Lipoprotein lipase. Molecular model based on the pancreatic lipase X-Ray structure: consequences for heparin binding and catalysis. J. Biol. Chem. **269**, 4626–4633

19 Carrière, F., Verger, R., Lookene, A. and Olivecrona, G. (1995) *in* Interface between chemistry and biochemistry (Jörnvall, H. and Jollès, P., eds.). Birkhäuser, Basel

20 Winkler, F.K. and Gubernator, K. (1994) *in* Lipases. Their structures, biochemistry and application (Wooley, P. and Petersen, S.B., eds.), pp. 139–157. Cambridge University Press, Cambridge

21 Ollis, D.L., Cheah, E., Cygler, M., Dijkstra, B., Frolow, F., Franken, S.M., Harel, M., Remington, S.J., Silman, I., Schrag, J., Sussman, J.L., Verschueren, K.H.G. and Goldman, A. (1992) The α/β hydrolase fold. Protein Eng. **5**, 197–211

22 Blow, D. (1991) Lipases reach the surface. Nature **351**, 444–445

23 Derewenda, Z.S. and Derewenda, U. (1991) Relationships among serine hydrolases: evidence for a common structural motif in triacylglyceride lipases and esterases. Biochem. Cell Biol. **69**, 842–851

24 Derewenda, Z.S., Derewenda, U. and Dodson, G.G. (1992) The crystal and molecular structure of the *Rhizomucor miehei* triacylglyceride lipase at 1.9 Å resolution. J. Mol. Biol. **227**, 818–839

25 Anthonsen, H.W., Baptista, A., Drablos, F., Martel, P., Petersen, S.B., Sebastiao, M. and Vaz, L. (1995) *in* Biotechnology Annual Review, Vol 1, (Elgewely, M.R., ed.), pp. 315–371. Elsevier Science, Amsterdam

26 Jaeger, K.-E., Ransac, S., Dijkstra, B.W., Colson, C., Vanheuvel, M. and Misset, O. (1994) Bacterial lipases. FEMS Microbiol. Rev. **15**, 29–63

27 Guidoni, A., Benkouka, F., de Caro, J. and Rovery, M. (1981) Characterization of the serine reacting with diethyl *p*-nitrophenylphosphate in porcine pancreatic lipase. Biochim. Biophys. Acta **660**, 148–150

28 Chapus, C. and Sérémiva, M. (1976) Mechanism of pancreatic lipase action. 2. Catalytic properties of modified lipases. Biochemistry **15**, 4988–4991

29 Lüthi-Peng, Q. and Winkler, F.K. (1992) Large spectral changes accompany the conformational transition of human pancreatic lipase induced by acylation with the inhibitor tetrahydrolipstatin. Eur. J. Biochem. **205**, 383–390

30 Van Tilbeurgh, H., Egloff, M.-P., Martinez, C., Rugani, N., Verger, R. and Cambillau, C. (1993) Interfacial activation of the lipase-procolipase complex by mixed micelles revealed by X-ray crystallography. Nature **362**, 814–820

31 Ferrato, F., Carrière, F., Sarda, L. and Verger, R. (1997) *in* Methods Enzymol., vol. 286 (Dennis, E. and Rubin, B., eds.), pp. 327–347, Academic Press, New York

32 van Tilbeurgh, H., Sarda, L., Verger, R. and Cambillau, C. (1992) Structure of the pancreatic lipase-procolipase complex. Nature **359**, 159–162

33 van Tilbeurgh, H., Egloff, M.-P., Martinez, C., Rugani, N., Verger, R. and Cambillau, C. (1993) Interfacial activation of the lipase-procolipase complex by mixed micelles revealed by X-Ray crystallography. Nature **362**, 814–820

34 Bourne, Y., Martinez, C., Kerfelec, B., Lombardo, D., Chapus, C. and Cambillau, C. (1994) Horse pancreatic lipase – the crystal structure refined at 2.3 Å resolution. J. Mol. Biol. **238**, 709–732

35 Roussel, A., Canaan, S., Egloff, M.P., Riviere, M., Dupuis, L., Verger, R. and Cambillau, C. (1999) Crystal structure of human gastric lipase and model of lysosomal acid lipase, two lipolytic enzymes of medical interest. J. Biol. Chem. **274**, 16995–17002

36 Kraut, J. (1977) Serine proteases: Structure and mechanism of catalysis. Ann. Rev. Biochem. **46**, 331–358

37 Liao, D.I., Breddam, K., Sweet, R.M., Bullock, T. and Remington, S.J. (1992) Refined atomic model of wheat serine Carboxypeptidase-II at 2.2–Angstrom resolution. Biochemistry **31**, 9796–9812

38 Rudenko, G., Bonten, E., d'Azzo, A. and Hol, W. (1995) Three-dimensional structure of the human 'protective protein': structure of the precursor form suggests a complex activation mechanism. Structure **3**, 1249–1259

39 Egloff, M.-P., Marguet, F., Buono, G., Verger, R., Cambillau, C. and van Tilbeurgh, H. (1995) The 2.46 Å resolution structure of the pancreatic lipase-colipase complex inhibited by a C_{11} alkyl phosphonate. Biochemistry **34**, 2751–2762

40 Brzozowski, A.M., Derewenda, U., Derewenda, Z.S., Dodson, G.G., Lawson, D.M., Turkenburg, J.P., Bjorkling, F., Huge-Jensen, B., Patkar, S.A. and Thim, L. (1991) A model for interfacial activation in lipases from the structure of a fungal lipase-inhibitor complex. Nature **351**, 491–494

41 Roussel, A., Miled, N., Berti-Dupuis, L., Riviere, M., Spinelli, S., Berna, P., Gruber, V., Verger, R. and Cambillau, C. (2002) Crystal structure of the open form of dog gastric lipase in complex with a phosphonate inhibitor. J. Biol. Chem. **277**, 2266–2274

42 Verger, R., Mieras, M.C.E. and de Haas, G.H. (1973) Action of phospholipase A at interfaces. J. Biol. Chem. **248**, 4023–4034

43 Ransac, S., Rivière, C., Gancet, C., Verger, R. and de Haas, G.H. (1990) Competitive inhibition of lipolytic enzymes. I. A kinetic model applicable to water-insoluble competitive inhibitors. Biochim. Biophys. Acta **1043**, 57–66

44 Ransac, S., Gargouri, Y., Moreau, H. and Verger, R. (1991) Inactivation of pancreatic and gastric lipases by tetrahydrolipstatin and alkyl-dithio-5-(2-nitrobenzoic acid). A kinetic study with 1,2-didecanoyl-sn-glycerol monolayers. Eur. J. Biochem. **202**, 395–400

45 Verger, R. and de Haas, G.H. (1976) Interfacial enzyme kinetics of lipolysis. Annual Review Biophys. Bioeng. **5**, 77–117

46 Dennis, E.A. (1987) Drug. Dev. Res. **10**, 205–220

47 Verger, R. (1984) *in* Lipases (Borgström, B. and Brockman, H.L., eds.), pp. 83–149. Elsevier Science, Amsterdam

48 Kurganov, B.I., Tsetlin, L.G., Malakhova, E.A., Chebotareva, N.A., Lankin, V.Z., Glebova, G.D., Berezovsky, V.M., Levashov, A.V. and Martinek, K. (1985) A novel approach to study of action of water-insoluble inhibitors of enzyme reactions. J. Biochem. Biophys. Methods **11**, 177–184

49 Piéroni, G., Gargouri, Y., Sarda, L. and Verger, R. (1990) Interactions of lipases with lipid monolayers. Facts and questions. Adv. Colloid Interface Sci. **32**, 341–378

50 Moulin, A., Fourneron, J.-D., Piéroni, G. and Verger, R. (1989) Interface-mediated inactivation of pancreatic lipase by a water-reactive compound: 2-sulfobenzoic cyclic anhydride. Biochemistry **28**, 6340–6346

51 Gargouri, Y., Chahinian, H., Moreau, H., Ransac, S. and Verger, R. (1991) Inactivation of pancreatic and gastric lipases by THL and $C_{12:0}$-TNB: a kinetic study with emulsified tributyrin. Biochim. Biophys. Acta **1085**, 322–328

52 Cudrey, C., van Tilbeurgh, H., Gargouri, Y. and Verger, R. (1993) Inactivation of pancreatic lipases by amphiphilic reagents 5-(Dodecyldithio)-2-nitrobenzoic acid and tetrahydrolipstatin. Dependence upon partitioning between micellar and oil phases. Biochemistry **32**, 13800–13808

53 Ransac, S., Gargouri, Y., Marguet, F., Buono, G., Beglinger, C., Hildebrand, P., Lengsfeld, H., Hadváry, P. and Verger, R. (1997) Covalent inactivation of lipases. Methods Enzymol. **286**, 190–231

54 Gargouri, Y., Ransac, S. and Verger, R. (1997) Covalent inhibition of digestive lipases: an in vitro study. Biochim. Biophys. Acta **1344**, 6–37

55 Dervichian, D.G. (1971) Méthode d'étude des réactions enzymatiques sur une interface. Biochimie **53**, 25–34

56 Zografi, G., Verger, R. and de Haas, G.H. (1971) Kinetic analysis of the hydrolysis of lecithin monolayers by phospholipase A. Chem. Phys. Lipids **7**, 185–206

57 Piéroni, G. and Verger, R. (1979) Hydrolysis of mixed monomolecular films of triglyceride/lecithin by pancreatic lipase. J. Biol. Chem. **254**, 10090–10094

58 Nury, S., Piéroni, G., Rivière, C., Gargouri, Y., Bois, A. and Verger, R. (1987) Lipase kinetics at the triacylglycerol-water interface using surface tension measurements. Chem. Phys. Lipids **45**, 27–37

59 Chen, P., del Rio, O.I. and Neumann, A.W. (2000) Axisymmetric drop shape analysis. *in* Physical Chemistry of Biological Interfaces. Marcel Dekker, New York

60 Labourdenne, S., Gaudry-Rolland, N., Letellier, S., Lin, M., Cagna, A., Esposito, G., Verger, R. and Rivière, C. (1994) The oil-drop tensiometer: potential applications for studying the kinetics of (phospho)lipase action. Chem. Phys. Lipids **71**, 163–173

61 Grimaldi, M., Bois, A., Nury, S., Rivière, C., Verger, R. and Richou, J. (1991) Analyse de la forme du profil d'une goutte pendante par traitement d'images numériques. (Mesure en temps réel de la tension interfaciale). Opto. **91**, 104–110

62 Verger, R., de Haas, G.H., Sarda, L. and Desnuelle, P. (1969) Purification from porcine pancreas of two molecular species with lipase activity. Biochim. Biophys. Acta **188**, 272–282

63 Maylié, M.F., Charles, M. and Desnuelle, P. (1972) Action of organophosphate and sulfonyl halides on porcine pancreatic lipase. Biochim. Biophys. Acta **276**, 162–175

64 Verger, R. (1970) Contribution à l'étude de la lipase pancréatique de porc. PhD thesis, Faculté des Sciences de Marseille

65 Desnuelle, P., Sarda, L. and Ailhaud, G. (1960) Inhibition de la lipase pancréatique par le diéthyl-*p*-nitrophényl phosphate en émulsion. Biochim. Biophys. Acta **37**, 570–571

66 Rouard, M., Sari, H., Nurit, S., Entressangles, B. and Desnuelle, P. (1978) Inhibition of pancreatic lipase by mixed micelles of diethyl *p*-nitrophenyl phosphate and the bile salts. Biochim. Biophys. Acta **530**, 227–235

67 Sikk, P.F., Osa, A.V. and Aaviksaa, A.A. (1985) Irreversible inhibition of pancreatic lipase by bis-*p*-nitrophenyl methylphosphonate. FEBS Lett. **184**, 193–196

68 Desnuelle, P. (1986) *in* Molecular and cellular basis of digestion (Desnuelle, P., Sjöström, H. and Norén, O., eds.), pp. 275–296, Elsevier Science, Amsterdam

69 Maraganore, J.M. and Heinrikson, R.L. (1986) Which class of serine is involved in the lipase mechanism? Trends Biochem. Sci. (Pers. Ed.) **11**, 497–498

70 Komaromy, M.C. and Schotz, M.C. (1987) Cloning of rat hepatic lipase cDNA: Evidence for a lipase gene family. Proc. Natl. Acad. Sci. USA **84**, 1526–1530

71 Chapus, C., Rovery, M., Sarda, L. and Verger, R. (1988) Minireview on pancreatic lipase and colipase. Biochimie **70**, 1223–1234

72 Shimada, Y., Sugihara, A., Tominaga, Y., Lizumi, T. and Tsunasawa, S. (1989)

cDNA cloning of *Geotrichum candidum* lipase. J. Biochem. **106**, 383–388

73 Brady, L., Brzozowski, A. M., Derewenda, Z. S., Dodson, E., Dodson, G., Tolley, S., Turkenburg, J. P., Christiansen, L., Huge-Jensen, B., Norskov, L., Thim, L. and Menge, U. (1990) A serine protease triad forms the catalytic centre of a triacylglycerol lipase. Nature **343**, 767–770

74 Schrag, J. D., Li, Y., Wu, S. and Cygler, M. (1991) Ser-His-Glu triad forms the catalytic site of the lipase from *Geotrichum candidum*. Nature **351**, 761–764

75 Martinez, C., de Geus, P., Lauwereys, M., Matthyssens, G. and Cambillau, C. (1992) *Fusarium solani* cutinase is a lipolytic enzyme with a catalytic serine accessible to solvent. Nature **356**, 615–618

76 Grochulski, P., Li, Y., Schrag, J. D., Bouthillier, F., Smith, P., Harrison, D., Rubin, B. and Cygler, M. (1993) Insights into interfacial activation from an open structure of *Candida rugosa* lipase. J. Biol. Chem. **268**, 12843–12847

77 Noble, M. E. M., Cleasby, A., Johnson, L. N., Egmond, M. R. and Frenken, L. G. J. (1993) The crystal structure of triacylglycerol lipase from *Pseudomonas glumae* reveals a partially redundant catalytic aspartate. FEBS Lett. **331**, 123–128

78 Lang, D., Haalck, L., Hofmann, B., Hecht, H.-J., Spener, F., Schmid, R. D. and Schomburg, D. (1994) Crystallization and preliminary X-ray analysis of a lipase from *Chromobacterium viscosum*. Acta Cryst. **50**, 225–227

79 Lawson, D. M., Brzozowski, A. M., Rety, S., Verma, C. and Dodson, G. G. (1994) Probing the nature of substrate binding in *Humicola lanuginosa* lipase through X-Ray crystallography and intuitive modelling. Protein Eng. **7**, 543–550

80 Uppenberg, J., Hansen, M. T., Patkar, S. and Jones, T. A. (1994) Sequence, crystal structure determination and refinement of two crystal forms of lipase B from *Candida antarctica*. Structure **2**, 293–308

81 Tiruppathi, C. and Balasubramanian, K. A. (1982) Purification and properties of an acid lipase from human gastric juice. Biochim. Biophys. Acta **712**, 692–697

82 Moreau, H., Gargouri, Y., Lecat, D., Junien, J.-L. and Verger, R. (1988) Purification, characterization and kinetic properties of the rabbit gastric lipase. Biochim. Biophys. Acta **960**, 286–293

83 Gargouri, Y., Moreau, H., Piéroni, G. and Verger, R. (1988) Human gastric lipase: A sulfhydryl enzyme. J. Biol. Chem. **263**, 2159–2162

84 Moreau, H., Gargouri, Y., Piéroni, G. and Verger, R. (1988) Importance of sulfhydryl group for rabbit gastric lipase activity. FEBS Lett. **236**, 383–387

85 Moreau, H., Moulin, A., Gargouri, Y., Noël, J.-P. and Verger, R. (1991) Inactivation of gastric and pancreatic lipases by diethyl *p*-nitrophenyl phosphate. Biochemistry **30**, 1037–1041

86 Lowe, M. E. (1992) The catalytic site residues and interfacial binding of human pancreatic lipase. J. Biol. Chem. **267**, 17069–17073

87 Bousset-Risso, M., Bonicel, J. and Rovery, M. (1985) Limited proteolysis of porcine pancreatic lipase. FEBS Lett. **182**, 323–326

88 De Caro, J. D., Rouimi, P. and Rovery, M. (1986) Hydrolysis of *p*-nitrophenyl acetate by the peptide chain fragment (336–449) of porcine pancreatic lipase. Eur. J. Biochem. **158**, 601–607

89 Kerfelec, B., Foglizzo, E., Bonicel, J., Bougis, P. E. and Chapus, C. (1992) Sequence of horse pancreatic lipase as determined by protein and cDNA sequencing. Implications for *p*-nitrophenyl acetate hydrolysis by pancreatic lipases. Eur. J. Biochem. **206**, 279–287

90 Patkar, S. and Björkling, F. (1994) *in* Lipases. Their structure, biochemistry and application (Wooley, P. and Petersen, S. B., eds.), pp. 207–224. Cambridge University Press, Cambridge

91 Marguet, F., Cudrey, C., Verger, R. and Buono, G. (1994) Digestive lipases: Inactivation by phosphonates. Biochim. Biophys. Acta **1210**, 157–166

92 Marguet, F. (1994) Synthèse d'inhibiteurs chiraux organophosphores: Contribution à l'étude du mécanisme d'action des lipases digestives. Thèse d'Université, University of Aix-Marseille II

93 Mannesse, M.L.M., Boots, J.W.P., Dijkman, R., Slotboom, A.J., Vanderhijden, H.T.W.V., Egmond, M.R., Verheij, H.M. and Dehaas, G.H. (1995) Phosphonate analogues of triacylglycerols are potent inhibitors of lipase. Biochim. Biophys. Acta **1259**, 56–64

94 Stadler, P., Zandonella, G., Haalck, L., Spener, F., Hermetter, A. and Paltauf, F. (1996) Inhibition of microbial lipases with stereoisomeric triacylglycerol analog phosphonates. Biochim. Biophys. Acta **1304**, 229–244

95 Stadler, P., Kovac, A., Haalck, L., Spener, F. and Paltauf, F. (1995) Stereoselectivity of microbial lipases. The substitution at position *sn*-2 of triacylglycerol analogs influences the stereoselectivity of different microbial lipases. Eur. J. Biochem. **227**, 335–343

96 Rogalska, E., Cudrey, C., Ferrato, F. and Verger, R. (1993) Stereoselective hydrolysis of triglycerides by animal and microbial lipases. Chirality **5**, 24–30

97 Ransac, S., Rogalska, E., Gargouri, Y., Deveer, A.M.T.J., Paltauf, F., de Haas, G.H. and Verger, R. (1990) Stereoselectivity of lipases. I. Hydrolysis of enantiomeric glyceride analogues by gastric and pancreatic lipases. A kinetic study using the monomolecular film technique. J. Biol. Chem. **265**, 20263–20270

98 Rogalska, E., Nury, S., Douchet, I. and Verger, R. (1995) Lipase stereoselectivity and regioselectivity toward three isomers of dicaprin: A kinetic study by the monomolecular film technique. Chirality **7**, 505–515

99 Egloff, M.-P., Ransac, S., Marguet, F., Rogalska, E., van Tilbeurgh, H., Buono, G., Cambillau, C. and Verger, R. (1995) Enzymes lipolytiques et lipolyse. Les lipases: Cinétiques, spécificités et aspects structuraux. Oléagineux, Corps Gras et Lipides **2**, 52–67

100 Miled, N., Roussel, A., Bussetta, C., Berti-Dupuis, L., Riviere, M., Buono, G., Verger, R., Cambillau, C. and Canaan, S. (2003) Inhibition of dog and human gastric lipases by enantiomeric phosphonate inhibitors: a structure-activity study. Biochemistry **42**, 11587–11593

101 Hogan, S., Fleury, P. and Hadvàry, P. (1987) Studies on the antiobesity activity of tetrahydrolipstatin, a potent and selective inhibitor of pancreatic lipase. Int. J. Obes. **11**, 35–42

102 Hadvàry, P., Lengsfeld, H. and Wolfer, H. (1988) Inhibition of pancreatic lipase *in vitro* by the covalent inhibitor tetrahydrolipstatin. Biochem. J. **256**, 357–361

103 Borgström, B. (1988) Mode of action of tetrahydrolipstatin: A derivative of the naturally occurring lipase inhibitor lipstatin. Biochim. Biophys. Acta **962**, 308–316

104 Hadvàry, P., Sidler, W., Meister, W., Vetter, W. and Wolfer, H. (1991) The lipase inhibitor tetrahydrolipstatin binds covalently to the putative active site serine of pancreatic lipase. J. Biol. Chem. **266**, 2021–2027

105 Lüthi-Peng, Q., Maerki, H.P. and Hadvàry, P. (1992) Identification of the active-site serine in human pancreatic lipase by chemical modification with tetrahydrolipstatin. FEBS Lett. **299**, 111–115

106 Güzelhan, C., Odink, J., Jansen-Zuidema, J.J.N. and Hartmann, D. (1994) Influence of dietary composition on the inhibition of fat absorption by orlistat. J. Int. Med. Res. **22**, 255–265

107 Meier, M.K., Blum-Kählin, D., Bremer, K., Isler, D., Joly, P., Keller-Rupp, P. and Lengsfeld, H. (1991) Preclinical profile of the lipase inhibitor tetrahydrolipstatin (Ro 18-0647, THL), a potential drug for the treatment of obesity. Int. J. Obes. **15**, 31

108 Hauptman, J., Jeunet, F. and Hartmann, D. (1992) Initial studies in humans with the novel gastrointestinal lipase inhibitor, Ro 18-0647 (tetrahydrolipstatin). Am. J. Clin. Nutr. **55**, 309S–313S

109 Zhi, J., Mulligan, T. and Hauptman, J. (1999) Long-term systemic exposure of orlistat, a lipase inhibitor, and its metabolites in obese patients. J. Clin. Pharmacol. **39**, 41–46

110 Tiss, A., Lengsfeld, H., Hadvàry, P., Cagna, A. and Verger, R. (2002) Transfer of orlistat through the oil-water interfaces. Chem. Phys. Lipids **119**, 41–49

111 Lookene, A., Skottova, N. and Olivecrona, G. (1994) Interactions of lipoprotein lipase with the active-site inhibitor

tetrahydrolipstatin (Orlistat) (R). Eur. J. Biochem. **222**, 395–403

112 Tiss, A., Lengsfeld, H., Hadvàry, P. and Verger, R. (2003) Inhibition of Human Pancreatic Lipase by Orlistat: further Kinetic Studies Showing its Reversibility. Eur. J. Biochem. (submitted)

113 Cygler, M., Grochulski, P., Kazlauskas, R. J., Schrag, J. D., Bouthillier, F., Rubin, B., Serreqi, A. N. and Gupta, A. K. (1994) A structural basis for the chiral preferences of lipases. J. Am. Chem. Soc. **116**, 3180–3186

114 Yapoudjian, S., Ivanova, M. G., Brzozowski, A. M., Patkar, S. A., Vind, J., Svendsen, A. and Verger, R. (2002) Binding of Thermomyces (Humicola) lanuginosa lipase to the mixed micelles of cis-parinaric acid/NaTDC. Eur. J. Biochem. **269**, 1613–1621

115 Gargouri, Y., Moreau, H., Piéroni, G. and Verger, R. (1989) Role of a sulfhydryl group in gastric lipases. A binding study using the monomolecular-film technique. Eur. J. Biochem. **180**, 367–371

116 Gargouri, Y., Piéroni, G., Rivière, C., Sarda, L. and Verger, R. (1986) Inhibition of lipases by proteins: A binding study using dicaprin monolayers. Biochemistry **25**, 1733–1738

117 Tiss, A., Carrière, F. and Verger, R. (2001) Effects of gum arabic on lipase interfacial binding and activity. Anal. Biochem. **294**, 36–43.

118 van Tilbeurgh, H., Gargouri, Y., Dezan, C., Egloff, M.-P., Nésa, M.-P., Ruganie, N., Sarda, L., Verger, R. and Cambillau, C. (1993) Crystallization of pancreatic procolipase and of its complex with pancreatic lipase. J. Mol. Biol. **229**, 552–554

119 Piéroni, G. and Verger, R. (1983) Hydrolysis of mixed monomolecular films of phosphatidylcholine/triacylglycerol by pancreatic phospholipase A_2. Eur. J. Biochem. **132**, 639–644.

10
Physiology of Gastrointestinal Lipolysis and Therapeutical Use of Lipases and Digestive Lipase Inhibitors

Hans Lengsfeld, Gabrielle Beaumier-Gallon, Henri Chahinian, Alain De Caro, Robert Verger, René Laugier, and Frédéric Carrière

10.1
Introduction

Digestion of dietary lipids is a complex process which includes mechanical, physicochemical as well as biochemical steps. Human gastric (HGL) and pancreatic (HPL) lipases are the enzymes responsible for the actual processing of triglycerides in the gastrointestinal (GI) tract. They allow the intestinal absorption of fat by converting triglycerides into free fatty acids (FFA) and 2-monoglycerides (MG). These enzymes need an emulsification of their substrate for optimal activity. The emulsification of triglycerides starts in the stomach and is mediated by physical forces through peristalsis and amphiphilic molecules such as lipolysis products, dietary proteins and phospholipids. Preduodenal lipolysis has thus a qualitative importance, but for decades the gastric phase of lipolysis in humans was assumed to be negligible in comparison with the intestinal one. The importance of preduodenal lipase became clear when it was observed that patients in the late stage of pancreatitis or cystic fibrosis (i.e. with no pancreatic lipase secretion) still absorb a large amount of dietary lipids (Lapey et al., 1974). Later on, it has been shown that pancreatic lipase alone does not readily hydrolyse triglyceride droplets covered by phospholipids such as Intralipid (emulsion of soybean TG stabilized by egg yolk lecithins) or native milk fat globules. After pre-incubating these substrates with gastric lipase, the released FFA were found to trigger the action of pancreatic lipase (Bernbäck et al., 1989; Gargouri et al., 1986a). The existence of a gastric lipase in humans was also questioned for many years before its cellular and tissular origins were clearly demonstrated (Moreau et al., 1988b, 1989). Preduodenal lipases have been screened in various species from the tongue to the pylorus (Moreau et al., 1988a). In each species tested, the preduodenal lipase activity was associated mainly with a single tissue, which was located either in the lingual (rat and mouse), pharyngeal (calf, lamb and sheep) or gastric area (humans, cat, rabbit, dog, monkey, horse, hog, guinea-pig). As we will discuss in this chapter, the contribution of HGL to gastrointestinal lipolysis is by now well documented. Quantitatively, HPL remains the major lipase involved in the lipolysis of dietary fat. HPL and its specific protein cofactor, colipase, act in the small intestine in the presence of bile. Contrary to what is often stated, bile salts do not constitute good

Lipases and Phospholipases in Drug Development
Edited by G. Müller and S. Petry

ISBN: 3-527-30677-3

emulsifiers of triglycerides, but they adsorb at the triglyceride–water interface because they are amphiphilic molecules. An essential role of bile salts is, therefore, not the fine emulsification of TG but the solubilization of lipolysis products. As opposed to TG and DG, FFA and MG interact with bile salts and phospholipids to form mixed micelles. This micellar solubilization allows to drive lipolysis products from TG droplets to the intestinal brush border where they are absorbed by enterocytes. During the last fifty years, HPL has been extensively studied in its biochemical, structural and physiological properties (Carrière et al., 1998; Lowe, 2002; Verger, 1984). Surprisingly, some basic data such as the amounts of HPL produced during a meal or its specific activity on meal triglycerides were only obtained in recent years, together with the characterization of HGL. These data were mainly generated during clinical studies performed with the aim of establishing the contribution of HGL (Carrière et al., 1993a, 2000) or the mechanism of action of the lipase inhibitor orlistat (Carrière et al., 2001). Here, we show basic data on gastrointestinal lipolysis (lipase concentrations and outputs, lipolysis levels, specific activities, etc.) to provide a real quantitative approach of lipolysis under physiological conditions. The molecular aspects of HGL and HPL have already been reviewed (Miled et al., 2000) and will not be described in detail here. Also the putative physiological role of bile salt-stimulated "lipase" (or carboxyl ester lipase) will not be discussed here because this enzyme does not display a significant activity on meal triglycerides as compared to HGL and HPL (Carrière et al., 1997, 2000), and its contribution to the lipolysis of other glycerides such as monoglycerides seems physiologically relevant only during the lactation period in humans (Bernbäck et al., 1990). Generally speaking, the physiological role of this enzyme is far from being clearly understood, even in the lipolysis of cholesterol esters (Howles et al., 1996).

10.2 Tissular and Cellular Origins of HGL and HPL

Native HGL is a highly glycosylated 50 kDa protein which is only produced in the fundic mucosa of human stomach (Moreau et al., 1988b). HGL is colocalized with the pepsinogen in chief cells of fundic glands (Fig. 10.1). Pentagastrin stimulates the secretions of HGL, pepsinogen and gastric acid (Moreau et al., 1988c, 1990; Szafran et al., 1978) and this secretagogue is commonly used to collect gastric juice for HGL purification (Ville et al., 2002). Other effective signals for the start of gastric lipase secretion are stomach motility, cholinergic stimuli (Carrière et al., 1992) or test meals (Carrière et al., 1993a). Neither the hormone cholecystokinin (CCK-PZ) nor secretin have an effect on HGL secretion (Moreau et al., 1988c). The lack of effect of CCK-PZ has also been observed in clinical experiments performed with the lipase inhibitor orlistat (Carrière et al., 2001). By inhibiting digestive lipases, orlistat reduces the production of FFA and indirectly the release of CCK-PZ. As a consequence, the CCK-PZ-dependent secretion of pancrea-

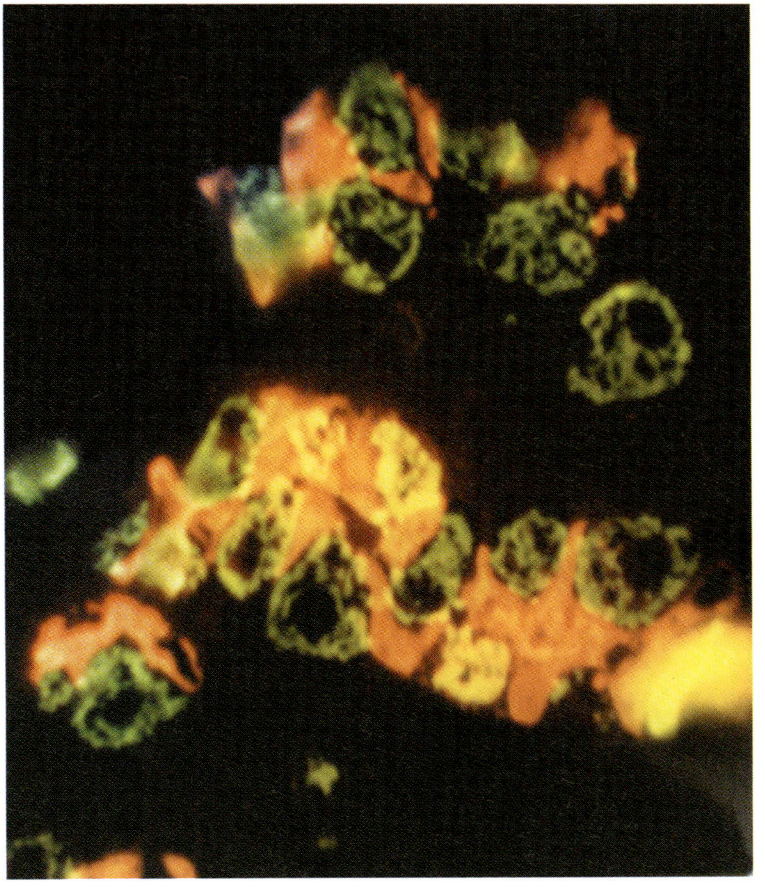

Fig. 10.1 Immunocytolocalization of HGL in chief cells of fundic glands. Chief cells containing HGL were stained by rabbit polyclonal anti-HGL antibodies coupled to rhodamin, and parietal cells producing gastric HCl were stained by a monoclonal antibody specific of the proton pump and coupled to fluorescein.

tic juice and pancreatic lipase output are reduced (Hildebrand et al., 1998). Such an effect is not observed with HGL, the output of which remains unchanged in test meal experiments performed with orlistat (Carrière et al., 2001). No case of HGL deficiency has been reported to date.

Native HPL is a glycosylated 50 kDa protein which is synthesized by acinar cells of the pancreas (Fig. 10.2). In the cell, this lipase is located with many other enzymes into the zymogen granules, and is then secreted through the pancreatic duct into the intestinal lumen together with the bile. Secretion of HPL can be stimulated by intestinal hormones such as CCK-PZ and secretin (Zieve et al., 1966a, b). Lipolysis products (FFAs) are known to stimulate the HPL secretion indirectly through the CCK action (Hildebrand et al., 1998). HPL secretion is drasti-

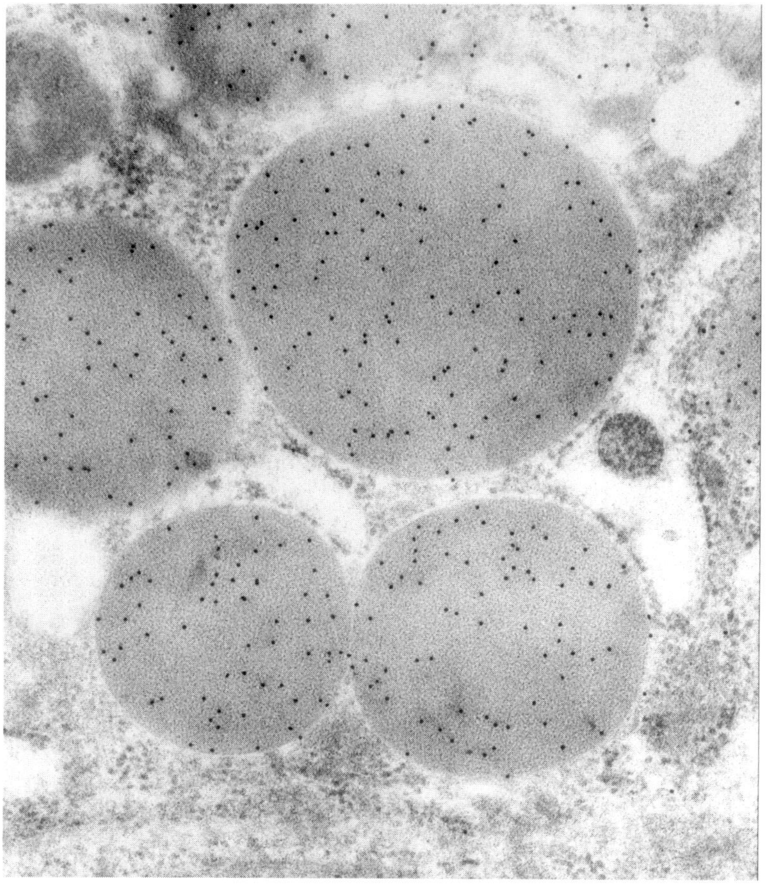

Fig. 10.2 Immunocytolocalization of HPL in the zymogen granules of pancreatic acinar cells. Zymogen granules are specifically stained by rabbit polyclonal anti-HPL antibodies coupled to gold particles and are observed by electron microscopy.

cally reduced in pathological cases of chronic pancreatic insufficiency or cystic fibrosis due to the destruction of acinar cells or an impaired secretion. The human pancreas also produces two pancreatic lipase-related proteins (PLRP1 and PLRP2) which are encoded by distinct genes and share around 70% amino acid identities with classical HPL (Giller et al., 1992). PLRP1 and PLRP2 differ from HPL by their biochemical properties and their physiological roles remain either unknown for PLRP1 (De Caro et al., 1998; Roussel et al., 1998a) or unclear for PLRP2 due to its broader substrate specificity and its various tissular locations (Lowe, 2000; Thirstrup et al., 1994).

10.3 Hydrolysis of Acylglycerols by HGL and HPL

10.3.1 Substrate Specificity

Both HGL and HPL display a high substrate selectivity towards triacylglycerols and do not hydrolyze phospholipids. They differ, however, in their specificity: HGL can hydrolyze the three ester bonds of acylglycerols whereas HPL is a regioselective lipase hydrolyzing only the ester bonds at the *sn*-1 or *sn*-3 position of glycerol (Carrière et al., 1991). When using a prochiral triglyceride substrate such as triolein, HPL does not show a high stereospecificity whereas HGL preferentially hydrolyzes the ester bond at the *sn*-3 position of triglycerides, generating a large enantiomeric excess of 1,2-diglycerides (Rogalska et al., 1990). From this property results the apparent HGL specificity for short acyl chains, because these acyl chains are found at the *sn*-3 position in milk triglycerides. In fact, HGL like HPL can hydrolyze triglycerides with various chain length (Gargouri et al., 1986b). In the subsequent conversion of diglycerides into monoglycerides, gastric lipase shows a reverse stereospecificity with a preference for the hydrolysis of the ester bond at the *sn*-1 position of 1,2-diolein versus the *sn*-3 position of 2,3-diolein (Carrière et al., 1997). Pancreatic lipase does not show a high stereospecificity in this second reaction. Globally, digestive lipase are tailored to produce 2-monoglycerides which are the glycerides absorbed at the intestinal level.

In both HGL and HPL, the catalytic machinery consists of a triad (Ser, His, Asp) and an oxyanion hole as observed in serine proteases (Blow, 1991; Roussel et al., 1999; Winkler et al., 1990). Substrate specificity is mainly due to acyl chain binding sites and to a "cap" or a "lid" controlling the access to the active site, as well as to the ability of digestive lipases to adsorb at an oil–water interface. This later property distinguishes lipases from classical enzymes. HGL and HPL are soluble enzymes which act on an insoluble substrate and therefore carry out an interfacial catalysis. Such a catalytic process can be described by two minimal steps: a step of lipase adsorption at the oil–water interface, followed by the catalytic step. Substrate specificity of interfacial enzymes can result from these two steps. From the crystal structures of HPL and dog gastric lipase obtained in the presence of substrate analogs and amphiphiles, it is expected that digestive lipases become fully active once they bind an oil–water interface together with an opening of the lid (Roussel et al., 2002; van Tilbeurgh et al., 1993). This behaviour might explain why both HGL and HPL are produced and secreted as potentially active enzymes in contrast to most other digestive enzymes which are secreted as proenzymes (pepsinogen, chymotrypsinogen, pro-phospholipase A2, etc.) and further activated in the GI tract by proteolytic cleavage.

10.3.2
Specific Activities of HGL and HPL

In the standard *in vitro* assays of lipase activity using pure triglycerides and optimized conditions, HGL often appeared to be less active than HPL (Tab. 10.1). But assay conditions for measuring maximum activities were quite different and a strict comparison between the two lipases was only obtained using the triglycerides present in a test meal as substrate (Tab. 10.1). Gastric and pancreatic lipases can hydrolyze dietary triglycerides at similar rates but their specific activities on test meal triglycerides are much lower than those recorded *in vitro* under optimized conditions using olive oil–gum arabic emulsion (the standard assay of the European Pharmacopeia), Intralipide™ (soybean oil emulsified by lecithin) or a synthetic short-chain TG, tributyrin (Tab. 10.1). However, these activities are consistent with the activities estimated from *in vivo* experiments in humans (Carrière et al., 2000). Also, these activities are high enough for dietary triglycerides to be completely hydrolyzed during the digestion of a meal, taking into account the amounts of lipases produced during a meal in humans (Carrière et al., 2000).

Why is lipase activity on test meal triglycerides the best reference for comparing digestive lipases, their respective contribution to gastrointestinal lipolysis and their potential benefit? The standard assay of the European Pharmacopea for measuring lipase activity uses an emulsion of olive oil–gum arabic as substrate. According to this assay, one lipase unit (=1 European Pharmacopae Unit=1 FIP unit) is defined as 1 micromole of free fatty acid released per minute. The lipolytic activity of a commercial pancreatic extract such Creon® (Solvay Pharmaceuti-

Tab. 10.1 Maximum specific activities of HGL, DGL, HPL and pancreatic extracts (Creon®) on various triglycerides. Specific activities are expressed in international units (1 U = 1μ mol of free fatty acid released per minute) per mg of pure enzyme, except for Creon® for which specific activities are expressed in U per mg of pancreatic extract (PE). One Creon 25000® capsule contains 500 mg pancreatic extract and 25000 lipase units measured using olive oil-gum arabic as substrate. This lipolytic activity corresponds to around 8–9 mg of active pancreatic lipase in one capsule of Creon 25000®.

Lipase source	*Substrates*			
	Olive oil – Gum arabic	*Intralipid*	*Tributyrin*	*Test meal triglycerides*
HPL	3000 U/mg (pH 7.5–8.0)	0	8000–12000 U/mg (pH 7.5–8.0)	12–15 U/mg (pH 6.25)
HGL	0	600 U/mg (pH 5.4)	1000 U/mg (pH 6)	30 U/mg (pH 5.5)
DGL	0	1000 U/mg (pH 4)	500–600 U/mg (pH 5.5)	n.d.
Creon 25000®	50 U/mg PE (pH 8.0)	n.d.	150 U/mg PE (pH 8.0)	0.5 U/mg PE (pH 6.0)

cals) is defined according to this assay, and for instance, a Creon 25000® capsule contains 25000 Eur. Pharm. U. Gastric lipase cannot be compared with pancreatic extracts on this basis because it simply does not hydrolyze olive oil emulsified by gum arabic. Therefore, according to the European Pharmacopeia, gastric lipase would not be classified as a lipase. Gastric lipase is, however, a true lipase since it

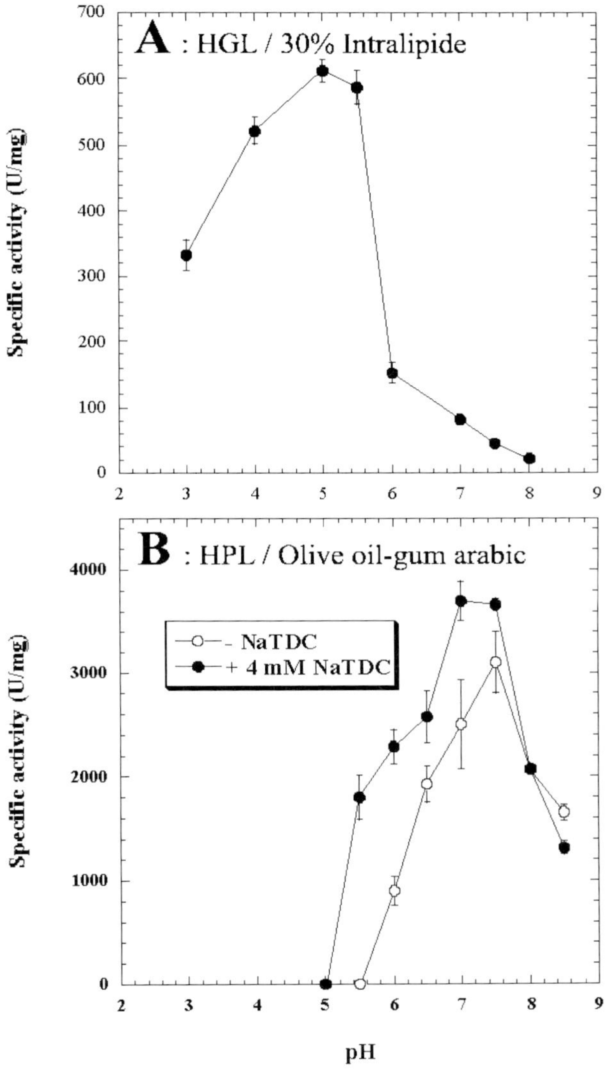

Fig. 10.3 Variations with pH of the specific activities of HGL and HPL. A: Activity of pure HGL on 30% Intralipide™ as substrate. Adapted from Gargouri et al. (1986) with permission. B: Activity of pure HPL (+colipase) on olive oil emulsified with gum arabic, in the presence or absence of 4 mM sodium taurodeoxycholate (NaTDC).

hydrolyzes the long-chain triglycerides from an Intralipide™ emulsion (soybean oil) (Carrière et al., 1991; Gargouri et al., 1986 b). Conversely, pancreatic lipase is not directly active on Intralipide™ (Gargouri et al., 1986 a) or on milk triglycerides (Bernbäck et al., 1989). These lipid emulsions do not represent however the real physicochemical state of triglycerides in a meal or in the GI tract. Using a real meal as substrate allows to measure the lipolytic activities of both HGL and HPL under similar experimental conditions simulating the *in vivo* conditions (Carrière et al., 2000).

10.3.3
Lipase Activity as a Function of pH

Gastric lipase displays its maximum lipolytic activity at acidic pH values (3 to 6, see Fig. 10.3) whereas pancreatic lipase is not significantly active below pH 5.0 and displays its maximum activity at pH 7.0–7.5 (Fig. 10.3). In the presence of bile salts, the maximum activity of HPL is shifted towards lower pH values. HGL and HPL are thus well adapted to the pH conditions recorded *in vivo*, both in the stomach and in the small intestine (Fig. 10.6).

10.3.4
Effects of Bile Salts on the Activity of HGL and HPL

Bile salts are amphiphilic molecules produced by the liver and delivered into the duodenum together with other bile lipids (phospholipids and cholesterol) and proteins forming a bile–lipoprotein complex (Hauton, 1986). Before reaching the duodenum, bile and pancreatic secretions are mixed in the Water ampoula. In the GI tract, bile salts are either found in mixed micelles or adsorbed at the lipid–water interface. Bile salts are competing with other amphiphilic molecules for adsorption at lipid–water interfaces. Since they are highly tensio-active, they can displace most of the proteins or other amphiphilic molecules present at the lipid–water interface in the GI tract. It is often stated that they are "cleaning" the interface. From *in vitro* experiments, it has been shown that HPL alone is desorbed from the lipid–water interface by bile salts and TG hydrolysis is therefore inhibited (Bezzine et al., 1999). HPL needs the presence of a 10 kDa protein also produced by the pancreas, colipase. This specific pancreatic lipase-anchoring protein makes a 1:1 complex with HPL (van Tilbeurgh et al., 1992) and counteracts the effect of bile salts by permitting an interfacial binding of HPL and therefore TG hydrolysis (Bezzine et al., 1999). Contrary to HPL, and contrary to what is often mentioned in the recent literature (Layer et al., 2001; Suzuki et al., 1997), HGL is not inhibited by bile salts (Carrière et al., 1993 a; Gargouri et al., 1986 b). Bile salts are even required to stabilize HGL during *in vitro* lipase assays. In the absence of amphiphiles, HGL is irreversibly inactivated at the oil–water interface because of a high surface tension. Bile salts decrease the surface tension and preserve HGL from inactivation (Fig. 10.4). Maximum activity is reached at bile salts concentrations above the critical micellar concentration (CMC, 1–2 mM). With HPL, a similar stabilizing effect of bile salts is observed *in vitro* at low concentrations. Using sodium taurodeoxycholate (NaTDC), a maximum activity is observed at 0.5 mM NaTDC, but HPL activity decreases rapidly above the CMC due to HPL desorption

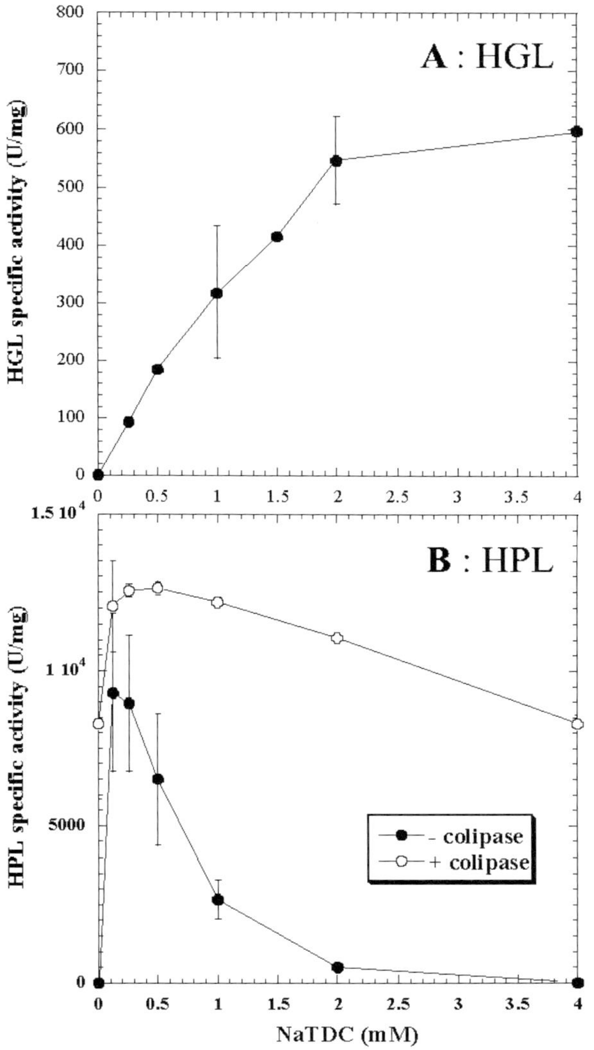

Fig. 10.4 Effects of bile salts on the lipase activities of HGL and HPL. A: Variation with NaTDC concentration in the specific activity of HGL using tributyrin as substrate. Adapted with permission from Gargouri et al. (1986) B: Variation with NaTDC concentration in the specific activity of HPL using tributyrin as substrate, in the presence or absence of colipase. Adapted with permission from Thirstrup et al. (1993).

from the oil–water interface (Fig. 10.4). These data indicate that HGL is a more tensio-active protein than HPL. This high tensioactivity is one of the striking properties of HGL because most lipases are inhibited by bile salts.

All main bile salts (glyco- and taurocholate, deoxycholate and chenodeoxycholate) have the same effects on HGL activity which increases up to a bile salt concentration of around 3 mM for a single species and 4 mM for a mixture of bile salts. However, it is worth noting that synthetic detergents (Triton X-100, Tween, benzalkonium chloride) that dramatically decrease surface tension (<8 mN/m), actually inhibit HGL.

10.4 Gastrointestinal Lipolysis of Test Meals in Healthy Human Volunteers

The relative contributions of HGL and HPL to the *in vivo* lipolysis of dietary triglycerides were established a few years ago (Carrière et al., 1993a). Healthy subjects were intubated with gastric and duodenal tubes as shown in Fig. 10.5, and this experimental device allowed us to measure separately HGL and HPL outputs (lipase units and mass of enzyme) and to monitor the amounts of lipolytic products at various stages of the GI tract. In previous studies, pancreatic lipase outputs were never expressed in term of mass of enzyme but only in terms of enzymatic units (Brunner et al., 1974; DiMagno et al., 1973; Johansson et al., 1972). Because the lipase assays and the definition of one lipase unit varied from one study to another, it was impossible to estimate the real HPL output during a meal or following a stimulation of pancreatic secretion by various secretagogues. The secretion of HGL during a meal was unknown before 1993 and the contribution of preduodenal lipase activity to the lipase activity measured in the GI tract was neglected, except in some studies (Abrams et al., 1987). Because the interference of HPL could not be ruled out during the measurement of HGL activity using standard assays (Gargouri et al., 1986b), especially in duodenal samples, a discriminative assay was developed in our laboratory to specifically measure HGL activity in the presence of HPL. Using tributyrin as substrate, at pH 4.5 and in the presence of a high sodium chloride concentration (0.6 M), HGL still displays a high specific activity of 1300 U/mg (1 U = 1 μmol of free fatty acid released per minute), whereas HPL is inactive under these conditions (Carrière et al., 1993a). Conversely, HPL activity can be assayed at pH 8.0, a pH value at which HGL is not active. The most sensitive pH-stat assay for measuring pancreatic lipase activity is performed with tributyrin as substrate in the presence of 4 mM NaTDC and colipase (Erlanson and Borgström, 1970). Under these conditions, HPL displays a high specific activity of 8000 U/mg. The interference by carboxyl ester lipase is not significant because this enzyme displays a much lower specific activity than HPL under these conditions (less than 1% activity). Based on the known specific activities of HGL and HPL, the mass concentrations of active lipase (μg or mg/mL) can be deduced from the lipase activities (U/mL) measured in gastric and duodenal sam-

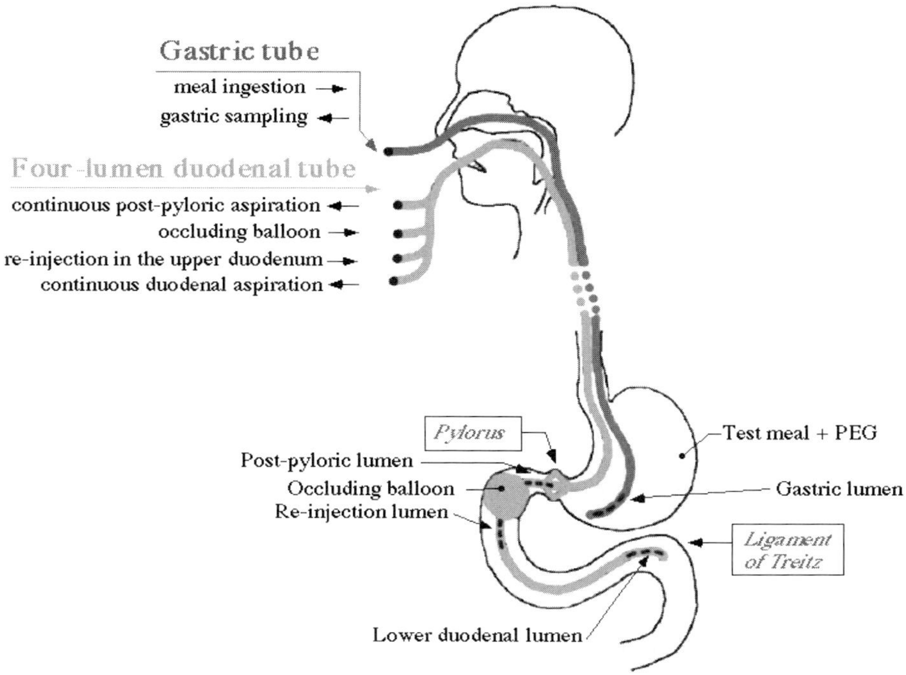

Fig. 10.5 Experimental device for collecting samples *in vivo* during the digestion of a test meal.

ples. The respective mass concentrations of HGL and HPL can also be measured by ELISA using monoclonal antibodies (De Caro et al., 1999). The comparison of mass concentrations obtained by ELISA and those deduced from the enzymatic assays allowed to study the stability of each lipase activity, as well as the effects of lipase inhibitors (see section 10.7). ELISA measurements of HPL mass concentrations in duodenal samples were similar with the mass concentrations deduced from activity measurements. These results not only demonstrated the high stability of HPL in duodenal contents during a meal, but they also confirmed the high specificity for HPL of the tributyrin assay at pH 8.0, in the presence of carboxyl ester lipase (Carrière et al., 2001).

10.4.1
Test Meals

Different complete test meals containing proteins, carbohydrates and lipids have been tested for measuring digestive lipase secretions and lipolysis levels in the GI tract. Shak Iso® (initially from Sopharga Laboratories, later from Nestlé Clinical Nutrition, France) was used as a complete liquid test meal. A standard 375 mL bottle contains 14 g protein (casein, lactoserum proteins, skimmed dry milk, egg yolk powder), 52 g carbohydrate (maltodextrin, saccharose) and 12.5 g lipid. These lipids are mainly triglycerides (11.2 g) from butter oil (62.5%), corn oil (26.9%) and soybean oil (10.6%). The overall TG concentration in the liquid meal was 30 mM. For the gastrointestinal lipolysis experiments in healthy volunteers, one meal consisted of 500 mL Shak Iso® and the total amounts of ingested triglycerides were 15 g (18 mmoles TG) (Carrière et al., 2001). This test meal, like other liquid test meals used in previous clinical studies, is easier to work with than a "true meal" because it can be rapidly delivered in the stomach through a gastric tube and quantitatively recovered by aspiration from the upper part of the small intestine. It is, however, a pre-emulsified meal which contains large amounts of lecithin to stabilize the fine droplets of triglycerides. Moreover, all the meal components are emptied from the stomach at similar rates without significant phase separation between lipids and water-soluble components of the meal (Cortot et al., 1979), whereas a phase separation occurs in the stomach during the digestion of a "normal" liquid-solid meal (Cortot et al., 1981). The extent of the phase separation depends on the rate of gastric emptying.

Since all these parameters have potentially an impact on gastrointestinal lipolysis, mixed solid/liquid test meals were also used to study lipase secretions and lipolysis levels. A typical meal used in our experiments contained 80 g string beans, 90 g beef, 70 g French fries, 10 g butter and water, making a total volume of 700 mL. Before being mixed together, the string beans, grilled beef meat and fries were put into a mincer with 2-mm apertures to allow a subsequent administration of the meal and its recovery by aspiration through the gastric and duodenal tubes. The total amount of fat present in the liquid-solid meal was estimated to be 17 g, including 15 g TG (18 mmol TG). The TG concentration in the solid meal was 25.7 mM. It is worth noting that these amounts of lipids are lower than the average lipid intake in developed countries (100 g per day, 30 g per meal). It was, however, necessary to reduce the solid part of the meal to ensure a high recovery of the meal from the intestine and to allow a real quantification of lipolysis.

In addition, a non-absorbable marker, polyethylene glycol (PEG 4000, 10 g/L) was added to each meal in order to estimate the percent recovery of the meal and to adjust the volumes at the end of the test meal digestion (Borgström et al., 1957). This marker also allowed to estimate the rate of gastric emptying, the volume of gastric contents and the pyloric outputs of HGL and lipolysis products (Carrière et al., 1993 a).

The importance of the physical state of the lipids in the meal was demonstrated by measuring *in vitro* the specific activities of HGL and HPL on the triglycerides of the liquid and the mixed liquid-solid test meals. HPL was found to display a higher activity on the pre-emulsified TG of Shak Iso® (43–47 U/mg) than on the TG found in the mixed liquid-solid meal (12–15 U/mg) (Carriere et al., 2000). As will be shown below (section 10.7), these observations were essential to understand the mechanism of lipolysis inhibition by the lipase inhibitor orlistat®.

10.4.2 Experimental Device for Collecting Samples *in vivo*

Various types of duodenal tubes have been used to estimate gastrointestinal lipolysis (Carrière et al., 1993a, 2001). Fig. 10.5 describes an experimental device in which volunteers were intubated with a four-lumen duodenal tube and with a single lumen gastric tube. An inflatable balloon present on the duodenal tube was inflated to occlude the upper part of the duodenum, close to the pylorus where no HPL is present. Continuous aspiration was performed upstream of the balloon to determine the pyloric outputs of HGL and of the lipolysis products generated into the stomach. After being sampled, the postpyloric aspirates were re-injected downstream of the occluding balloon in order to investigate the duodenal step of digestion. A second continuous system of aspiration was set up at the ligament of Treitz to assess the outputs of HGL and HPL, and of the overall lipolysis products resulting from both intragastric and duodenal lipolysis. The contents of the upper and the lower parts of duodenum were continuously aspirated for 15 min periods. The total volumes of both aspirates were measured and samples were collected for lipase, pH, PEG and lipid analysis. The single lumen gastric tube was used to administer the meal and to collect gastric samples every 15 minutes as well as the total residual gastric contents at the end of the experiment.

10.4.3 Gastric and Duodenal pH Variations

Since the lipase activities of HGL and HPL are highly dependent on pH values, it was particularly important to assess gastric and duodenal pH variations during the digestion period. After ingestion of a meal, the pH values in the stomach reach 6–7 depending on the type of meal ingested (Fig. 10.6A). Then, the gastric pH decreases with time due to the combined effects of gastric acid secretion and gastric emptying, which increase the proportion of gastric secretion in gastric contents and reduce the buffering effects of the meal components. At half gastric emptying (60 min on average), the intragastric pH value is close to 5. At the end of gastric emptying (150 min or more), very low pH values (1–2) corresponding to nearly pure gastric juice are reached.

Duodenal pH variations are less important and usually restricted to the 5–7 pH range (Fig. 10.6B).

10.4.4 Lipase Concentrations and Outputs

The mean basal concentration of HGL in gastric juice is found in the 100–120 μg/mL range, but a high variability is observed because HGL stability is decreased at pH values below 1.5 (Ville et al., 2002). Immediately after the ingestion

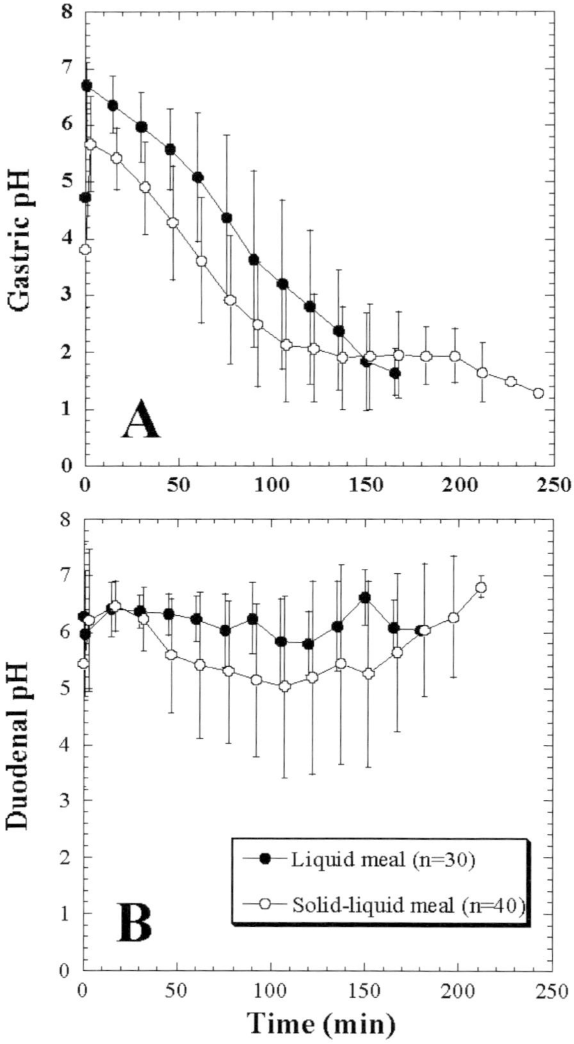

Fig. 10.6 Gastric and duodenal pH variations during test meals in humans. Mean values and standard deviations were estimated from data of 30 experiments performed with a liquid test meal and 40 experiments performed with a solid-liquid test meal.

of a meal, HGL concentration in gastric contents is usually very low due to the large dilution of basal gastric juice by the meal (Fig. 10.7). HGL concentration is then increasing because its secretion is stimulated by the meal (Carrière et al., 1993a). Fig. 10.7A shows a high dispersion of the HGL concentration in gastric contents which is increasing with time (data from ten individual experiments with healthy volunteers). When the same HGL data are plotted as a function of

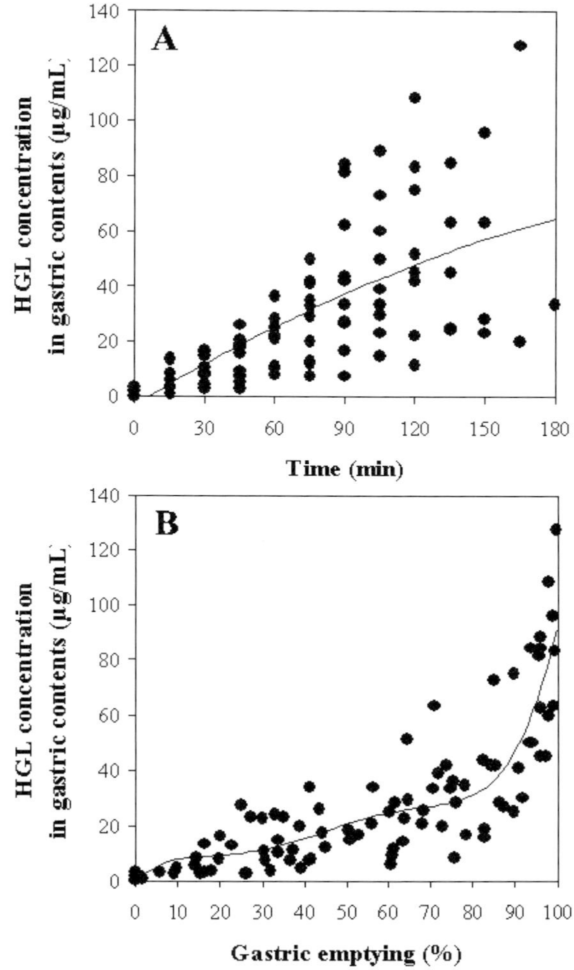

Fig. 10.7 Variations with time (A) and meal gastric emptying (B) of HGL concentration in gastric contents during test meals in humans. Mean values and standard deviations were estimated from data of 10 experiments performed with a liquid test meal (Carrière et al., 2001). Similar values are recorded with solid-liquid test meals.

meal gastric emptying (in %, estimated from PEG emptying) instead of time, dispersion of the values is reduced (Fig. 10.7 B). A sharp increase in HGL concentration is observed after 70–80% gastric emptying. As expected, HGL concentration reaches its basal concentration in pure gastric juice at the end of meal gastric emptying. Gastric emptying is a major cause of variability when studying gastrointestinal parameters as a function of time. Expressing gastric HGL concentration as a function of gastric emptying instead of time makes it possible to reduce the individual variability and the deviation between experiments (Carrière et al., 2001).

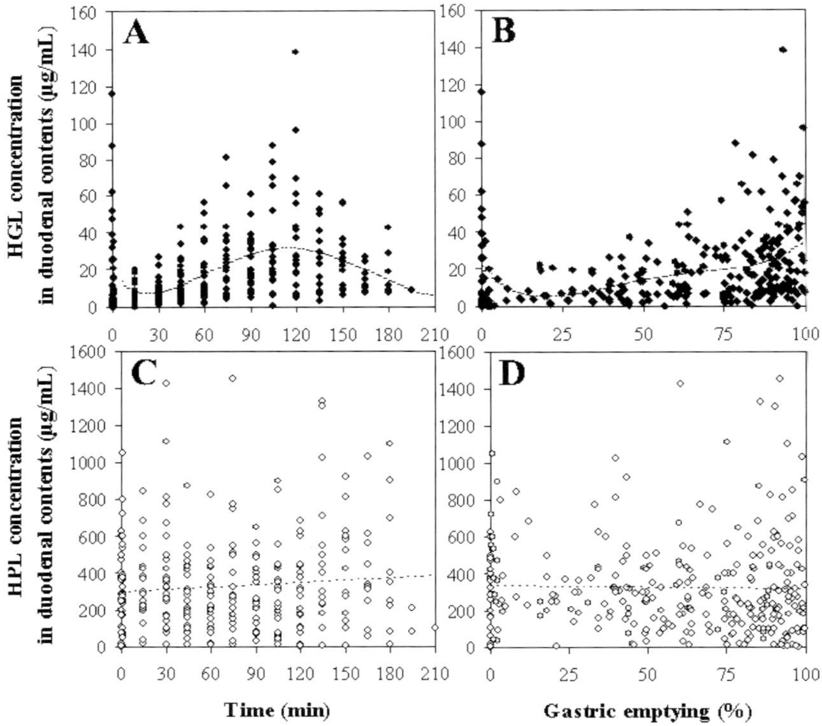

Fig. 10.8 Variations with time (A, C) and meal gastric emptying (B, D) of HGL and HPL concentrations in duodenal contents during test meals in humans. Mean values and standard deviations were estimated from data (268 values for HGL concentration and 296 values for HPL concentration) of 29 experiments performed with both liquid and solid-liquid test meals. Data obtained with both types of meals are not significantly different.

In duodenal contents, HGL concentration shows a Gaussian profile as a function of time, and maximum HGL concentrations are observed after 100–120 minutes (Fig. 10.8 A). When the same data are plotted as a function of meal gastric emptying, one can see that the highest values are measured at the end of gastric emptying (Fig. 10.8 B), like in gastric contents (Fig. 10.7). The mean HGL concentration in duodenal contents is 20 µg/mL, with a maximum value of 140 µg/mL. A high variability of HPL concentration is observed as a function of time (Fig. 10.8 C) and expressing HPL concentration as a function of gastric emptying instead of time (Fig. 10.8 D) does not reveal a specific pattern of variation as previously observed with HGL (Fig. 10.7). The mean HPL concentration in duodenal contents is 330 µg/mL, with a maximum value of 1450 µg/mL.

The HGL output during a meal was estimated from several clinical studies (Tab. 10.2), with mean values ranging from 9 to 25 mg HGL and high standard deviations. Whatever the type of meal and the amounts of fat ingested, the values obtained in the various studies were not found to be significantly different. The

Tab. 10.2 HGL outputs in various clinical studies

Test meals	*Fat amounts*	*HGL outputs (mean ± SD, mg)*	*Number of experiments*	*Reference*
Liquid meal	11.25 g TG	22.6 ± 8.1	n = 15	Carrière et al., 1993
Liquid meal	13.4 g TG	24.7 ± 9.1	n = 6	Borovicka et al., 1997
Liquid meal	15 g TG	21.6 ± 14	n = 3	Carrière et al., 2001
Liquid meal + Xenical	15 g TG	17.6 ± 6.7	n = 3	Carrière et al., 2001
Liquid meal + Orlistat	15 g TG	17.2 ± 11.6	n = 3	Carrière et al., 2001
Mixed solid/liquid meal	15 g TG	15.2 ± 5.1	n = 7	Carrière et al., 2001
Mixed solid/liquid meal + Xenical	15 g TG	9.1 ± 3.2	n = 5	Carrière et al., 2001
Mixed solid/liquid meal + Orlistat	15 g TG	16.3 ± 9.6	n = 5	Carrière et al., 2001
Liquid meal	11.25 g TG	15.4 ± 8.0	n = 6	Renou et al., 2001
Liquid meal + lansoprazole	11.25 g TG	19.0 ± 7.4	n = 6	Renou et al., 2001

Tab. 10.3 HPL outputs in various clinical studies

Test meals	*Fat amounts*	*HPL outputs (mean ± SD, mg)*	*Reference*
500 mL Liquid test meal	11.25 g TG	88.2 ± 25	Carrière et al., 1993
500 mL Liquid test meal	15 g TG	253.5 ± 95.5	Carrière et al., 2000, 2001
700 mL Mixed solid/liquid test meal	15 g TG	202.9 ± 96.1	Carrière et al., 2000, 2001
500 mL of 10% Intralipid	50 g TG	442 ± 87	Schwizer et al., 1997
500 mL Fatty acid meal	40.4 g FFA	645 ± 96	Schwizer et al., 1997
500 mL Liquid test meal	13.4 g TG	245 ± 22	Borovicka et al., 1997

HPL outputs are shown in Tab. 10.3, with mean values ranging from 88 to 645 mg HPL and high standard deviations. Contrary to what is observed with HGL, HPL secretion seems to increase with the amounts of fat present into the meal. This finding is in agreement with the increased pancreatic lipase expression (mRNA) observed in rats when the diet is enriched in fat (Wicker-Planquart and Puigserver, 1993).

10.4.5 Lipolysis Levels

Intragastric lipolysis of meal triglycerides was estimated from the cumulated pyloric outputs of lipolysis products (free fatty acids: FFA; diglycerides: DG; monoglycerides: MG) and residual triglycerides (TG) and was expressed as the percentage

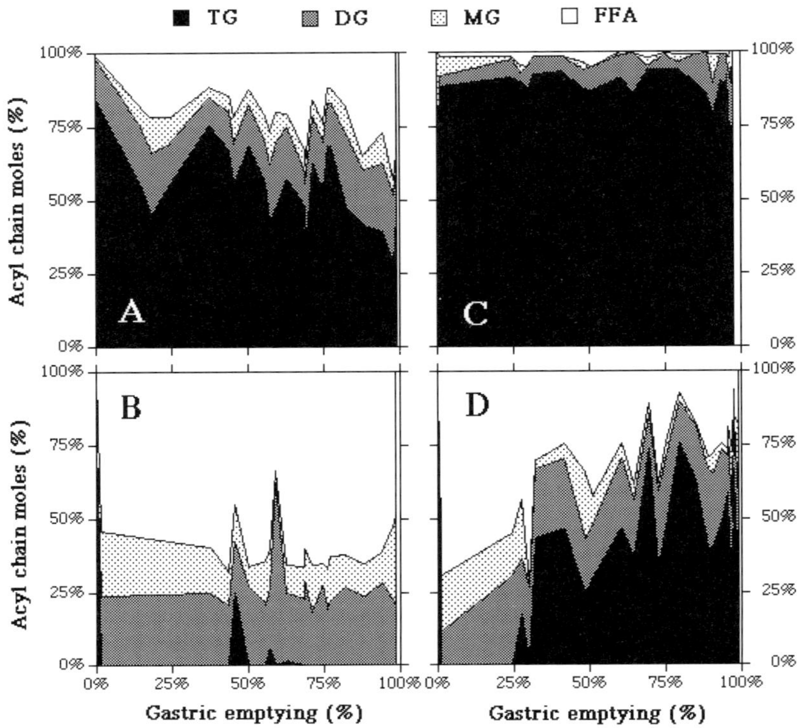

Fig. 10.9 Pyloric and duodenal outputs of lipolysis products and residual meal triglycerides during the digestion of a liquid test meal. The distribution of total meal TG acyl chains (mol%) in lipolysis products (free fatty acids: FFA; diglycerides: DG; monoglycerides: MG) and residual triglycerides (TG) is plotted as a function of meal gastric emptying (%). Panels A and C show the pyloric output of lipolysis products in the absence and in the presence of the lipase inhibitor orlistat, respectively. Panels B and D show the duodenal output at the Angle of Treitz of lipolysis products in the absence and in the presence of the lipase inhibitor orlistat, respectively.

of TG acyl chains converted into absorbable acyl chains, i.e. FFA and MG. Fig. 10.9 A shows the distribution of total acyl chains (mol%) in the lipolysis products passing through the pylorus as a function of meal gastric emptying. One can clearly see that the proportion of acyl chains present in meal triglycerides is decreasing with meal gastric emptying, whereas the proportion of FFA is increasing. These data are consistent with the fact that HGL concentration in gastric contents is also increasing with meal gastric emptying (Fig. 10.7 B). Intragastric lipolysis by HGL was found to be significantly lower ($P<0.02$) with the solid meal ($9.3 \pm 6.3\%$) than with the liquid meal ($24.4 \pm 5.7\%$; see Tab. 10.6). MG concentration in gastric contents was usually very low.

The overall lipolysis occurring in both the stomach and the duodenum was estimated from the cumulated outputs of lipolysis products and residual triglycerides at the Angle of Treitz. Fig. 10.9 B shows the distribution of total acyl chains (mol%) in the lipolysis products passing through the Angle of Treitz as a function of meal gastric emptying. At this stage of the GI tract and with a liquid test meal, around 50% of the meal TG acyl chains have already been converted in FFA. Most of the TG have disappeared but significant amounts of DG remain to be converted into MG and FFA.

Duodenal lipolysis level was deduced by subtracting the gastric lipolysis level from the overall lipolysis level. Duodenal lipolysis was also found to be lower with the solid meal (18.4%) than with the liquid meal (35.0±2.0%; Tab. 10.6). The triglycerides of the liquid test meal therefore provided a better substrate for HGL and HPL, probably because they were pre-emusified and stabilized in the form of a fine emulsion by the large amounts of phospholipids present in the Shak Iso® liquid test meal. In the solid meal, the physicochemical state of the lipids is more heterogeneous and most of the TG have to be emulsified in the course of the digestion process.

Although the lipolysis levels were found to increase in the duodenal contents, it is worth noting that the overall lipolysis levels at the Angle of Treitz reached only 59.4±5.6% with the liquid meal and 27.7±6.8% with the solid meal. The lipolysis could not be complete because the duodenal contents were collected continuously at the Angle of Treitz for 15 minute periods and at the end of each period, the lipolysis of the aspirate was stopped in order to analyze the lipolysis products. The mean residence time of the meal lipids in the small intestine and hence their contact with lipases were therefore shortened artificially in comparison with the situation under normal physiological conditions.

The contribution of HGL to overall lipolysis is difficult to estimate because both HGL and HPL act on the same substrate in duodenal contents and generate similar lipolysis products. From the respective amounts of HGL and HPL produced during a meal and the specific activities of these two lipases on meal triglycerides, it was however estimated that HGL might be responsible for the release of one acyl chain out of four that have to be released to convert two triglyceride molecules into 2-monoglycerides and FFA during the whole digestion process (Carrière et al., 1993 a).

10.5 HGL and HPL Stability

Residual activity of purified HGL or HGL in gastric juice has been tested after the enzyme was incubated for 2 h at various pH values (Ville et al., 2002). In between pH 2 and 6–7, HGL remains highly stable in gastric juice with a recovery of lipase activity as high as 93%. For extreme pH values, the activity decreases rapidly (pH 8) or is definitely lost (pH 0.5, 0.75 and 1) which means that basal HGL secreted in the fasting state is most likely inactivated and that the buffering effect of a meal seems compulsory to preserve HGL activity. Pure HGL is more pH-sensitive and the greatest stability was observed at neutral pH with a recovery of lipase activity of 95% after 1 min and 59% after 2 hours incubation at pH 7. Half inactivation time decreases as pH decreases, except at pH 8 where the loss of activity is extremely fast. Under physiological conditions, HGL is most likely stabilized by other components of the gastric juice such as the mucus. Inactivation of the HGL at very low pH values (0.5–1) and at pH 8.0 is irreversible and probably results from a denaturation of the 3D structure. In addition to this factor of inactivation, HGL is also degraded by pepsin in gastric juice at low pH values. In this case, immunoreactivity of HGL using western blot analysis and polyclonal antibodies is completely lost, thus revealing a complete degradation of the polypeptide. HGL is only sensitive to pepsin at very low pH values (0.5–1), suggesting that only denaturated HGL is sensitive to proteolytic cleavage by pepsin. At the optimum pH value for pepsin activity (pH 3.0), HGL is highly stable in gastric juice.

The role of preduodenal lipase in the overall gastrointestinal lipolysis has often been underestimated because this enzyme was considered to be highly sensitive to proteolytic degradation by pancreatic proteases (Bernbäck et al., 1987; Roberts and Hanel, 1988). The *in vivo* stability of preduodenal lipase in the rat intestine was however studied in the absence of a meal. Pancreatic lipase is also highly sensitive to pancreatic proteases under these conditions, but not in the presence of a meal, the components of which stabilize the lipase (Muller and Ghale, 1982). Similarly, it was observed that preduodenal lipase activity was preserved in the duodenum of CP patients, in the presence of a test meal (Abrams et al., 1987). The presence of active preduodenal lipase was also detected in the intestinal contents of CF patients (Fredrikzon and Blackberg, 1980). When the duodenal levels of HGL were precisely measured during a test meal with both a specific enzymatic assay (pH-stat) and an ELISA, it was deduced that most of the HGL quantified by ELISA remained active. Moreover, the output of active HGL at the Angle of Treitz was similar to the pyloric output of HGL, demonstrating a high stability in the duodenum of the HGL produced in the stomach (Carrière et al., 1993 a). Similar experiments were performed in dogs. In this case, significant levels of active gastric lipase were also measured in the jejunum (Carrière et al., 1993 b).

In order to estimate the stability of HGL and HPL lipase activities in duodenal contents during a test meal in humans, samples of duodenal contents collected at the Angle of Treitz by aspiration were incubated for three hours at 37 °C. Fig. 10.10 shows the residual lipase activities measured after 3 hours, as a function of the pH value of the sample. One can see that HGL activity is highly stable at acidic pH value (more than 80% residual activity at pH 2) whereas HPL activity drastically decreases at pH 4 with less than 20% residual activity recovered. Reverse stabilities are observed at pH values close to neutrality. HPL is highly stable at pH 7 with 95% residual activity recovered whereas HGL residual activity is around 5%.

10.6 Potential Use of Gastric Lipase in the Treatment of Pancreatic Insufficiency

Several observations made with cystic fibrosis (CF) or chronic pancreatitis (CP) patients suggest that gastric lipase can partly replace pancreatic lipase in case of exocrine pancreatic insufficiency (EPI). In CF patients, 20 to 80% of the ingested triglycerides remain absorbed (Fredrikzon and Blackberg, 1980; Lapey *et al.*, 1974; Ross, 1955; Ross and Sammons, 1955). In very rare cases of total congenital pancreatic lipase deficiency, 50% of the ingested triglycerides can be absorbed (Muller et al., 1975). During a test meal, the lipolytic activity of gastric contents is sometimes higher in CF patients than in control subjects (Roulet et al., 1980).

It has been shown that the secretory output of HGL under pentagastrin stimulation is significantly increased in case of severe CP resulting from alcoholism (Moreau et al., 1990). These results suggested an adaptation of HGL in case of EPI. Contradictory observations were however reported (Moreau et al., 1988c). Abrams et al. did not observe an increase in preduodenal lipase levels in CF and CP patients (Abrams *et al.*, 1984, 1987). But they reported that 90% of the lipolytic activity measured at the Angle of Treitz during the digestion of a meal was due to preduodenal lipase. These studies referred to "lingual lipase" because human preduodenal lipase was thought to be of lingual origin at that time. It was demonstrated later on that preduodenal lipase is of gastric origin in humans (Bodmer et al., 1987; Moreau *et al.*, 1988b). According to the data obtained by Abrams, one would expect that the increased role of preduodenal lipase in case of EPI is not due to an increase in the enzyme secretion, but to an increased specific activity. In pathological cases of EPI, the pH of small intestine contents is lower than in normal subjects due to a reduced bicarbonate secretion and a weak neutralization of gastric acid. As already mentioned, gastric lipase displays a higher specific activity at acidic pH values (Fig. 10.3) and its contribution to the overall gastrointestinal lipolysis is expected to be higher in cases of EPI. Moreover, the low levels or the absence of pancreatic proteases are in favour of a higher stability of gastric lipase in the small intestine contents.

The use of gastric lipase, a naturally acid stable and active lipase, to improve fat absorption in cases of EPI is an alternative to the use of porcine pancreatic extracts (Bénicourt et al., 1993; Moreau et al., 1987; Roulet et al., 1980), because pancreatic enzymes are particularly sensitive to the more acidic conditions prevailing in the small intestine in cases of EPI. Pancreatic extracts (PE) containing porcine pancreatic lipase have been the therapeutic standard for several decades for the treatment of EPI. Modern therapeutic concepts recommend administration of 25 000–40 000 units of porcine lipase per meal using pH-sensitive pancreatin microspheres (Layer and Keller, 2003). Still, in most patients, lipid digestion cannot be completely normalized by current standard therapy, and future developments are needed for optimizing treatment. In fact the low efficiency of PE is not really surprising since 25 000–40 000 Eur. Pharm. U. only correspond to 8–13 mg of active pancreatic lipase. We have already mentioned that pancreatic lipase output during a meal is much higher than these values (Tab. 10.3). Moreover, experiments performed with animal models of EPI clearly show that much higher amounts of PE have to be administered for a complete reduction of steatorhrea (Gregory et al., 2002). Novel lipases are now tested for higher efficacy. Fungal lipase is already found in some enzyme supplements but has inferior properties compared with conventional pancreatic extracts (Layer and Keller, 2003). Bacterial lipase products show promising potential and might offer future therapeutic alternatives (Suzuki et al., 1997). The most promising development, however, concerns gastric lipase, a recombinant dog gastric lipase (DGL) being produced in transgenic plants for that purpose (Gruber et al., 2001). Among all lipases, DGL shows the highest specific activity on long-chain triglycerides at pH 4 (Carrière et al., 1991) and it is a good enzyme candidate for acting under the acidic environment of the small intestinal in patients with EPI (Bénicourt et al., 1993). The results of the first study to be carried out on the efficacy of gastric lipase in an animal EPI model indicate that supplementing the diet with exogenous purified rabbit gastric lipase reduces steatorrhea in minipigs (Milano and Oliveira, 1998).

10.7
Inhibition of Gastrointestinal Lipolysis by Orlistat for Obesity Treatment

10.7.1
The Lipase Inhibitor Orlistat

Obesity is a serious disease which predisposes to numerous health problems, including diabetes, hypertension and atherosclerosis. It accounts for 2–7% of the total healthcare costs in Western countries and its prevalence is increasing rapidly in the developing countries, where it is already out of control (Garrow, 1998). Dietary therapy is the first line treatment for obesity, but since it has not proved to be sufficient in many cases, drug therapy also has to be envisaged. There were high hopes for the efficacy of drugs acting on the central nervous system as appetite suppressants. Serious cardiovascular side effects were reported, however, with drugs such as fenfluramine, and their clinical use was therefore severely restricted. As excess fat consumption is widely thought to be one of the main causes of obesity, means of specifically inhibiting triglyceride digestion have been developed as a new approach to reducing fat absorption.

Orlistat (tetrahydrolipstatin) is a covalent inhibitor of digestive lipases (Borgström, 1988; Hadvàry et al., 1988), derived from lipstatin, a natural product of *Streptomyces toxytricini* (Weibel et al., 1987). It is an active site-directed inhibitor which reacts with the nucleophilic serine residue from the catalytic triad of pancreatic lipase (Fig. 10.11) (Hadvàry et al., 1991; Lüthi-Peng et al., 1992). By covalently blocking the lipase active site, orlistat inhibits the hydrolysis of dietary triglycerides and thus reduces the subsequent intestinal absorption of the lipolysis products, monoglycerides and free fatty acids. It also inhibits gastric lipase, cholesterol esterase and various other lipases which are all serine hydrolases (Borgström, 1988; Cudrey et al., 1993; Gargouri et al., 1991; Ransac et al., 1991; Roussel et al., 1998 b), but it does not inactivate serine proteases such as trypsin and chymotrypsin (Hadvàry et al., 1988). When administered with a meal, orlistat acts locally in the gastrointestinal tract since its systemic absorption is negligible (Zhi et al., 1995). The drug therefore acts specifically on digestive lipases *in vivo*, although it can inhibit other human lipases such as lipoprotein lipase (Lookene et al., 1994) and hormone-sensitive lipase (Smith et al., 1996) *in vitro*.

Many clinical trials have been performed on orlistat during the last few years and one formulation of the drug (Xenical®) was licensed in the European Union in 1998 (Brooks, 1998) and in the United States in 1999. The results of clinical studies have indicated that Xenical® reduces fat absorption up to 30% (Zhi et al., 1994) during a meal. Xenical® has dose-dependent effects up to 400 mg per day and a plateau is then observed (Zhi et al., 1994). When Xenical® is administered in therapeutic doses (120 mg with the main meals) in conjunction with a mildly hypocaloric diet, obese patients have lost around 10% of their bodyweight over a 1 year period (Sjoström et al., 1998).

10.7.2
Design of Clinical Studies for Quantification of Lipase and Lipolysis Inhibition

In most clinical studies, the effects of orlistat have been estimated indirectly from the levels of fecal fat excretion. Its direct inhibitory effects on digestive lipases and lipolysis have been quantified more recently. Hildebrand and coworkers observed that a dose-dependent inhibition of human pancreatic lipase (HPL) occurred concomitantly with a decrease in the free fatty acid levels at the Angle of Treitz (Hildebrand *et al.*, 1998). Schwizer and coworkers have established that HPL activity in the duodenal contents was almost completely inhibited by orlistat, whereas the total output of immunoreactive HPL remained unchanged in comparison with that recorded in control experiments without orlistat (Schwizer et al., 1997). Even though HGL is the second most important enzyme involved in dietary triglyceride hydrolysis, HGL was not a main target in the development of orlistat, and the respective effects of orlistat on gastric and duodenal lipolysis were only investigated in a recent study (Carrière et al., 2001). The aim of this study was to quantify the inhibition of HGL and HPL by orlistat, as well as the decrease in the lipolysis occurring in response to orlistat at the post-pyloric and lower duodenal levels, in the course of test meals in healthy human volunteers. Two different formulations of orlistat (Xenical® pellets containing 120 mg orlistat and 120 mg micronized orlistat powder) and placebo (controls) were studied in association with one complete liquid test meal and one mixed solid/liquid test meal. The experimental devices used to measure the concentrations and outputs of digestive lipases and lipolysis products were similar to the device described in section 10.4. In addition, the HGL and HPL inhibition levels were estimated by comparing the amounts of active lipase (pH-stat measurements) with the total amounts of lipase determined by ELISA. Lipolysis levels were estimated from the recovery of FFA, MG, DG and

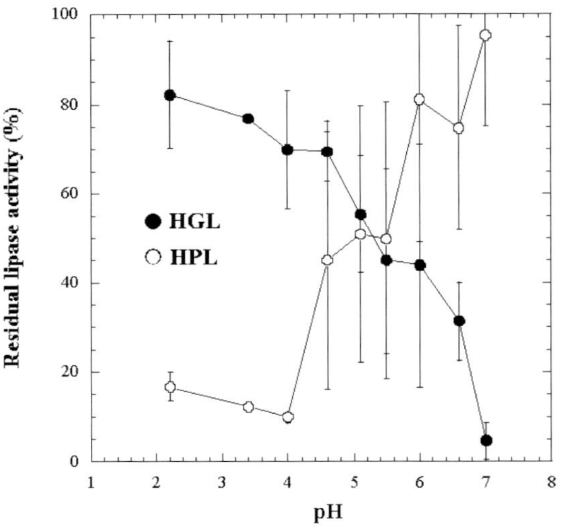

Fig. 10.10 Stability of HGL and HPL lipase activities in duodenal contents during a test meal. Samples of duodenal contents collected at the Angle of Treitz by aspiration were incubated for three more hours at 37 °C, after the initial HGL and HPL activities were measured. The residual activities measured after 3 hours were expressed as percentage of the initial activities and plotted as a function of the pH value of the sample.

residual meal TG at the post-pyloric level (intragastric lipolysis) and at the Angle of Treitz (overall lipolysis), respectively. Duodenal lipolysis level was obtained by subtracting intragastric lipolysis from overall lipolysis.

10.7.3 HGL Inhibition by Orlistat

Some HGL inactivation was observed in the control experiments (29.4% with the liquid meal; 12.3% with the solid meal; Tab. 10.4). Partial inactivation of HGL was found to occur mostly in the samples with the lowest pH values (1–2), which were collected at the end of the meal gastric emptying process.

In the presence of a lipase inhibitor, the HGL inhibition was always very high (Tab. 10.4) whatever the mode of orlistat administration (Xenical® pellets and micronized orlistat powder) or the type of meal ingested (liquid and solid). The differences observed in comparison with the control experiments were significant ($p < 0.05$) for both Xenical® pellets and micronized orlistat powder given with the liquid meal and were highly significant for Xenical® pellets ($p < 0.005$) and for micronized orlistat powder ($p < 0.001$) given with the solid meal . The mean inhibition level of HGL was found to be higher with micronized orlistat powder than with Xenical® pellets (Tab. 10.4), but this difference was not statistically significant. Micronized orlistat powder pre-mixed with the meal fat was therefore as efficient as the orlistat present in the pellets of Xenical®. It was observed that the HGL inhibition was also very high in the duodenal samples.

Tab. 10.4 Effects of Orlistat on HGL secretion and activity during digestion of test meals. Values are means ± SD.

Meal	*Treatment*	*Amounts of HGL secreted (mg)*	*Active HGL*		*HGL inhibition (%)*
			Amounts (mg)	*% control*	
Liquid meal:	Xenical® pellets	17.6 ± 6.7	5.4 ± 4.6	37.2	75.2 ± 17.7
Shak Iso®	Orlistat powder	17.2 ± 11.6	1.4 ± 1.0	10.0	91.4 ± 4.4
	Control	21.6 ± 14.5	14.5 ± 10.8	100.0	29.4 ± 29.6
Solid meal:	Xenical® pellets	9.1 ± 3.2	4.8 ± 1.7	36.4	46.6 ± 10.5
Hamburger,	Orlistat in butter	16.3 ± 9.6	5.4 ± 3.4	41.0	65.2 ± 15.2
fries, butter	Control	15.2 ± 5.1	13.2 ± 4.6	100.0	12.3 ± 14.0

Tab. 10.5 Effects of Orlistat on HPL secretion and activity during digestion of test meals. Values are means ± SD.

Meal	*Treatment*	*Amounts of HPL secreted (mg)*	*Active HPL*		*HPL inhibition (%)*
			Amounts (mg)	*% control*	
Liquid meal:	Xenical® pellets	173.1 ± 46.1	39.9 ± 12.1	16.2	77.2 ± 3.7
Shak Iso®	Orlistat powder	192.4 ± 78.6	34.4 ± 23.7	14.0	77.2 ± 19.8
	Control	253.5 ± 95.5	246.0 ± 73.0	100	2.5 ± 3.9
Solid meal:	Xenical® pellets	127.3 ± 78.9	25.2 ± 22.2	13.2	82.6 ± 5.4
Hamburger,	Orlistat in butter	137.6 ± 67.8	67.2 ± 64.5	35.3	51.2 ± 34.6
fries, butter	Control	202.9 ± 96.1	190.3 ± 85.4	100	6.7 ± 6.3

10.7.4 HPL Inhibition by Orlistat

The results obtained with the liquid and solid test meals are listed in Tab. 10.5. A high level of HPL inhibition was observed in all the groups treated with the two forms of orlistat, and with both type of meals. With the liquid meal, the difference with controls was highly significant in the Xenical® pellet group ($p < 0.001$) and the micronized orlistat powder group ($p < 0.01$). With the solid test meal, the difference with controls was highly significant in the Xenical® pellet group ($p < 0.001$) and significant in the micronized orlistat powder group ($p < 0.05$). The difference between the two treated groups was not significant, whatever the type of meal ingested.

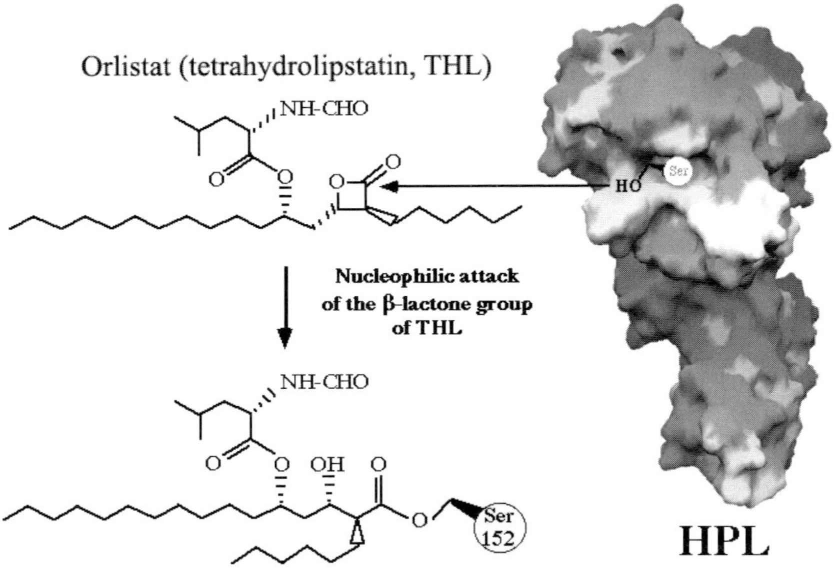

Fig. 10.11 Covalent inhibition of human pancreatic lipase (HPL) by orlistat.

10.7.5 Effects of Orlistat on Gastric Lipolysis

With both types of test meal and all drug formulations, the level of gastric lipolysis was highly and significantly ($p<0.05$) reduced *in vivo* by orlistat (10.9 to 33.4% of controls; Tab. 10.6), and a good correlation was always observed with the high level of HGL inhibition (46.6 to 91.4%; Tab. 10.4).

Tab. 10.6 Effects of Orlistat on gastrointestinal lipolysis during digestion of test meals in healthy volunteers. Values are means ± SD.

Meal	*Treatment*	*Gastric lipolysis (%)*	*Duodenal lipolysis (%)*	*Overall lipolysis (%)*
Liquid meal:	Xenical® pellets	8.1 ± 6.8 (33.4)	40.5 ± 19.2 (115.7)	48.6 ± 12.3 (81.9)
Shak Iso®	Orlistat powder	2.7 ± 1.4 (10.9)	26.1 ± 10.4 (74.5)	28.7 ± 9.4 (48.4)
	Control	24.4 ± 5.7 (100)	35.0 ± 2.0 (100)	59.4 ± 5.6 (100)
Solid meal:	Xenical® pellets	1.9 ± 1.3 (20.7)	6.0 (32.6)	8.0 ± 1.5 (28.8)
Hamburger, fries,	Orlistat in butter	2.3 ± 1.6 (25.1)	6.9 (37.6)	9.2 ± 4.5 (33.4)
butter	Control	9.3 ± 6.3 (100)	18.4 (100)	27.7 ± 6.8 (100)

10.7.6 Effects of Orlistat on Duodenal Lipolysis

With the liquid meal, the duodenal lipolysis (Tab. 10.6) decreased only slightly when micronized orlistat powder was pre-mixed with the meal (74.5% of controls), and did not decrease at all when Xenical® pellets were added in the course of the administration of the meal (115.7% of controls). The differences between treated and control groups were not significant. This lack of effect of orlistat as means of reducing the duodenal lipolysis was not correlated with the high level of HPL inhibition observed in the duodenal contents (Tab. 10.5). These paradoxical results were however supported and explained by the results of further *in vitro* experiments. The rates of HPL inhibition by orlistat were found to be similar with both types of meals (half-inhibition time = 5–6 min), but the finely pre-emulsified TG of the liquid meal were rapidly hydrolyzed by HPL before the enzyme was significantly inhibited by orlistat (Carrière et al., 2001).

With the solid meal, the *in vivo* duodenal lipolysis was highly and significantly reduced by orlistat with both forms of the drug (32.6 to 37.6% of controls; Tab. 10.6), and in this case, a good correlation was observed with the high level of HPL inhibition (51.2 to 82.6%; Tab. 10.5). *In vitro* experiments revealed that with the mixed solid/liquid meal, the rate of hydrolysis of the meal TG by HPL was slower than the rate of HPL inhibition by orlistat (Carrière et al., 2001).

10.7.7 Effects of Orlistat on Overall Lipolysis

Generally speaking, the overall level of lipolysis of the liquid test meal did not decrease significantly in response to Xenical® pellets (81.9% of controls). The decrease in the overall lipolysis was however found to be significant ($p < 0.05$) in the group treated with micronized orlistat powder (48.4% of controls). The difference between the two treated groups was mainly due to the very high level of inhibition of the gastric lipolysis observed in the group treated with micronized orlistat powder (Tab. 10.6).

With the solid test meal, the overall level of lipolysis was highly and significantly reduced by the double action of orlistat on the gastric and duodenal lipolysis (28.8% of controls for Xenical® pellets, $p < 0.005$; 33.4% of controls for micronized orlistat powder in butter, $p < 0.001$; Tab. 10.6). The difference between the two treated groups was not significant.

10.7.8 Effects of Orlistat on Fat Excretion

Similar experiments were performed without gastric and duodenal tubes, and fat absorption was indirectly measured from fecal fat excretion. As predicted from the *in vivo* and *in vitro* gastrointestinal lipolysis experiments, it was observed that Xenical® pellets poorly inhibited the absorption of fat from the finely pre-emulsified liquid test meal. Only 4.2% fat excretion was achieved as compared to 40.5% fat excretion with the solid test meal ingested with Xenical® pellets (Tab. 10.7). These results are in good agreement with the *in vivo* data (Tab. 10.6) showing that the overall level of lipolysis undergone by the liquid meal was not significantly reduced by Xenical® pellets in comparison with what occurred in the control experiments. Since it was observed in the same experiments that the level of gastric lipolysis was significantly reduced, the data on fat excretion indicate that duodenal lipolysis by HPL compensates for the decrease in the gastric lipolysis during the whole digestion process.

When the liquid meal was given with micronized orlistat powder instead of Xenical® pellets, the mean level of fat excretion reached 18.8%, but was very variable and remained significantly lower than the fat excretion level determined after ingestion of the solid meal with micronized orlistat powder (57.4% fat excretion; Tab. 10.7). Here again, the results are in good agreement with the previous *in vivo* data (Tab. 10.6) showing that micronized orlistat powder significantly decreased the overall lipolysis level of the liquid meal versus control experiments. This decrease was mainly due to the drastic drop in the level of gastric lipolysis (10.9% of controls). Interestingly, the level of fat excretion (18.8% of ingested fat) was found to be similar to the decrease in the level of gastric lipolysis (from 24.4% in controls to 2.7% in the presence of orlistat powder). It therefore emerged from the latter experiments that duodenal lipolysis by HPL does not suffice to compensate for the decrease in the gastric lipolysis. One possible explanation for these results focuses on the synergy existing between the action of HGL and that of HPL. It has been established by Gargouri et al. (1986 a) that HPL hydrolyzed a triglyceride emulsion covered by phospholipids only after partial hydrolysis by HGL had occurred. The free fatty acids generated by HGL were found to trigger the action of HPL by changing the "quality" of the oil–water interface. One can therefore envision a further indirect inhibitory effect of orlistat on HPL when the gastric lipolysis is highly reduced.

10.7.9
Weight Management by Orlistat in Obese Patients

A randomised controlled trial to assess the efficacy and tolerability of orlistat in promoting weight loss and preventing weight regain in obese patients was carried out over a 2 year period (Sjoström et al., 1998). Obese patients with body-mass index 28–47 kg/m^2 (n=743) were recruited at 15 European centres, and entered a 4 week, single-blind, placebo lead-in period on a slightly hypocaloric diet (600 kcal/day deficit). The patients who completed the lead-in (n=688) were assigned a double-blind treatment either with orlistat 120 mg (three times a day) or with placebo, for 1 year in conjunction with the hypocaloric diet. In a second 52 week double-blind period, patients were reassigned orlistat or placebo with a weight

Tab. 10.7 Effects of Orlistat on fat excretion after ingestion of liquid and solid meals by healthy volunteers. Values are means ± SD. The number of volunteers is indicated in brackets.

Meal	*Treatment*	*Dose (mg)*	*Fat excretion (% of ingested fat)*
Liquid meal: Shak Iso® (990 mL)	Xenical®	120	4.2 ± 6.8 (n = 5)
	Orlistat powder	120	18.8 ± 31.7 (n = 5)
Solid meal: Hamburger, fries, butter	Xenical®	120	40.5 ± 26.1 (n = 7)
	Orlistat in butter	60	57.4 ± 16.8 (n = 6)

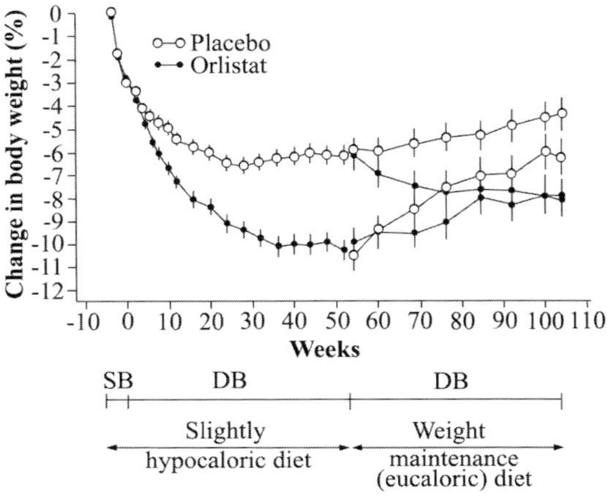

Fig. 10.12 Changes in body weight of obese patients upon treatment with orlistat during two years. The figure shows mean percentage change in bodyweight from start of single-blind lead-in until 2 year examination in orlistat and placebo groups. Initial bodyweight was close to a mean of 100 kg in both groups, percentage change therefore approximately matches kg lost. SB = single-blind lead-in period of 4 weeks; DB = double-blind, placebo-controlled treatment during years 1 and 2. Error bars = SE. Adapted from Sjoström et al. (1998) with permission.

maintenance (eucaloric) diet. From the start of lead-in to the end of year 1, the orlistat group lost, on average, more bodyweight than the placebo group (10.2% [10.3 kg] vs. 6.1% [6.1 kg]; $p < 0.001$; Fig. 10.12). During year 2, patients who continued with orlistat regained, on average, half as much weight as those patients switched to placebo ($p < 0.001$; Fig. 10.12). Patients switched from placebo to orlistat lost an additional 0.9 kg during year 2, compared with a mean regain of 2.5 kg in patients who continued on placebo ($p < 0.001$; Fig. 10.12). Total cholesterol, low-density lipoprotein (LDL) cholesterol, LDL/high-density lipoprotein ratio, and concentrations of glucose and insulin decreased more in the orlistat group than in the placebo group. Gastrointestinal adverse events were more common in the orlistat group. Other adverse symptoms occurred at a similar frequency during both treatments.

From this study and similar studies reported more recently (Broom et al., 2002; Hanefeld and Sachse, 2002; Kelley et al., 2002), it was concluded that orlistat taken together with an appropriate diet promotes clinically significant weight loss and reduces weight regain in obese patients over a 2 year period. The use of orlistat beyond 2 years needs careful monitoring with respect to efficacy and adverse events. These results appeared quite satisfactory from the clinical point of view. The biochemists involved in the orlistat studies were however rather frustrated by the 10% weight loss observed in obese patients after one year treatment, whereas single test meal experiments have shown an inhibition of fat absorption as high as 60% (Tab. 10.7) in healthy volunteers. The current galenic form of orlistat (Xenical) may not be the optimal one since better results are obtained with pure orlistat (Tab. 10.7). Also, during long-term studies, there might be an adaptation of the physiology of obese patients in order to counteract the inhibitory effects of orlistat.

10.7.10 Conclusions

The clinical study by Carrière et al. (2001) provided the first comprehensive overview of test meal lipolysis by HGL and HPL in healthy human volunteers, in the presence and absence of a lipase inhibitor (orlistat). The results obtained showed that orlistat, administered either as Xenical® pellets, micronized powder or melted in butter, is a powerful HGL inhibitor, which strongly reduces the level of gastric lipolysis of liquid and solid test meals. The effects of orlistat on duodenal lipolysis depend on the type of meal ingested. This study highlights the fact that lipase inhibition and lipolysis are two competitive processes. The balance between these processes depends on the physicochemical properties of the dietary TG. Improving the lipase inhibition rate versus the lipolysis rate may be one way in the future development of lipase inhibitors for obesity treatment. Before new inhibitors and formulations are available, an alternative approach to improve the drug efficiency will probably be to associate the prescriptions of Xenical® with more specific diets in which fat is poorly pre-emulsified, and is therefore not a good substrate for gastrointestinal lipases.

10.8 References

Abrams, C. K., Hamosh, M., Dutta, S. K., Hubbard, V. S. and Hamosh, P. (1987) Role of nonpancreatic lipolytic activity in exocrine pancreatic insufficiency. *Gastroenterology* **92**, 125–129.

Abrams, C. K., Hamosh, M., Hubbard, V. S., Dutta, S. K. and Hamosh, P. (1984) Lingual lipase in cystic fibrosis. Quantification of enzyme activity in the upper small intestine of patients with exocrine pancreatic insufficiency. *J. Clin. Invest.* **73**, 374–382.

Bénicourt, C., Blanchard, C., Carrière, F., Verger, R. and Junien, J. L. (1993) Potential use of a recombinant dog gastric lipase as an enzymatic supplement to pancreatic extracts in cystic fibrosis. In *Clinical Ecology of Cystic Fibrosis* (Escobar, H., Baquero, C.F. and Suárez, L., eds.), pp. 291–295. Elsevier Science, Amsterdam.

Bernbäck, S., Blackberg, L. and Hernell, O. (1990) The complete digestion of human milk triacylglycerol in vitro requires gastric lipase, pancreatic colipase-dependent lipase, and bile salt-stimulated lipase. *J. Clin. Invest.* **85**, 1221–1226.

Bernbäck, S., Blackberg, L. and Hernell, O. (1989) Fatty acids generated by gastric lipase promote human milk triacylglycerol digestion by pancreatic colipase-dependent lipase. *Biochim. Biophys. Acta* **1001**, 286–291.

Bernbäck, S., Hernell, O. and Blackberg, L. (1987) Bovine pregastric lipase: a model for the human enzyme with respect to properties relevant to its site of action. *Biochim. Biophys. Acta* **922**, 206–213.

Bezzine, S., Ferrato, F., Ivanova, M.G., Lopez, V., Verger, R. and Carrière, F. (1999) Human pancreatic lipase: colipase dependence and interfacial binding of lid domain mutants. *Biochemistry* **38**, 5499–5510.

Blow, D. (1991) Lipases reach the surface. *Nature* **351**, 444–445.

Bodmer, M.W., Angal, S., Yarranton, G.T., Harris, T.J.R., Lyons, A., King, D.J., Piéroni, G., Rivière, C., Verger, R. and Lowe, P.A. (1987) Molecular cloning of a human gastric lipase and expression of the enzyme in yeast. *Biochim. Biophys. Acta* **909**, 237–244.

Borgström, B. (1988) Mode of action of tetrahydrolipstatin: A derivative of the naturally occurring lipase inhibitor lipstatin. *Biochim. Biophys. Acta* **962**, 308–316.

Borgström, B., Dahlqvist, A., Lundh, G. and Sjövall, J. (1957) Studies of intestinal digestion and absorption in the human, *J. Clin. Invest.* **36**, 1521–1536.

Borovicka, J., Schwizer, W., Mettraux, C., Kreiss, C., Remy, B., Asal, K., Jansen, J.B., Douchet, I., Verger, R. and Fried, M. (1997) Regulation of gastric and pancreatic lipase secretion by CCK and cholinergic mechanisms in humans. *Am.J. Physiol.* **273**, G374–G380.

Brooks, A. (1998) Obesity drug is licensed in the UK. *British Medical Journal* **317**, 835.

Broom, I., Wilding, J., Stott, P. and Myers, N. (2002) Randomised trial of the effect of orlistat on body weight and cardiovascular disease risk profile in obese patients: UK Multimorbidity Study. *Int. J. Clin. Pract.* **56**, 494–499.

Brunner, H., Northfield, T.C., Hofmann, A.F., Go, V.L.W. and Summerskill, W.H.J. (1974) Gastric emptying and secretion of bile acids, cholesterol, and pancreatic enzymes during digestion. Duodenal perfusion studies in healthy subjects. *Mayo Clin. Proc.* **49**, 851–860.

Carrière, F., Barrowman, J.A., Verger, R. and Laugier, R. (1993a) Secretion and contribution to lipolysis of gastric and pancreatic lipases during a test meal in humans. *Gastroenterology* **105**, 876–888.

Carrière, F., Laugier, R., Barrowman, J.A., Douchet, I., Priymenko, N. and Verger, R. (1993b) Gastric and pancreatic lipase levels during a test meal in dogs, *Scand. J. Gastroenterol.* **28**, 443–454.

Carrière, F., Moreau, H., Raphel, V., Laugier, R., Bénicourt, C., Junien, J.-L. and Verger, R. (1991) Purification and biochemical characterization of dog gastric lipase. *Eur. J. Biochem.* **202**, 75–83.

Carrière, F., Raphel, V., Moreau, H., Bernadac, A., Devaux, M.-A., Grimaud, R., Barrowman, J.A., Bénicourt, C., Junien, J.-L., Laugier, R. and Verger, R. (1992) Dog gastric lipase: Stimulation of its secretion in vivo and cytolocalization in mucous pit cells. *Gastroenterology* **102**, 1535–1545.

Carriere, F., Renou, C., Lopez, V., De Caro, J., Ferrato, F., Lengsfeld, H., De Caro, A., Laugier, R. and Verger, R. (2000) The specific activities of human digestive lipases measured from the in vivo and in vitro lipolysis of test meals. *Gastroenterology* **119**, 949–960.

Carrière, F., Renou, C., Ransac, S., Lopez, V., De Caro, J., Ferrato, F., De Caro, A., Fleury, A., Sanwald-Ducray, P., Lengsfeld, H., Beglinger, C., Hadvàry, P., Verger, R. and Laugier, R. (2001) Inhibition of gastrointestinal lipolysis by Orlistat during digestion of test meals in healthy volunteers. *Am. J. Physiol.* **281**, G16–28.

Carrière, F., Renou, C., Ville, E., Grandval, P. and Laugier, R. (2001) Advantage of expressing the variations in some digestive parameters as a function of gastric emptying instead of time. *Digestion* **64**, 46–53.

Carrière, F., Rogalska, E., Cudrey, C., Ferrato, F., Laugier, R. and Verger, R. (1997) In vivo and in vitro studies on the stereoselective hydrolysis of tri- and diglycerides by gastric and pancreatic lipases. *Bioorg. Med. Chem.* **5**, 429–435.

Carrière, F., Withers-Martinez, C., van Tilbeurgh, H., Roussel, A., Cambillau, C. and Verger, R. (1998) Structural basis for the substrate selectivity of pancreatic lipases and some related proteins. *Biochim. Biophys. Acta* **1376**, 417–432.

Cortot, A., Phillips, S.F. and Malagelada, J.R. (1979) Gastric emptying of lipids after ingestion of an homogenized meal, *Gastroenterology* **76**, 939–944.

Cortot, A., Phillips, S.F. and Malagelada, J.R. (1981) Gastric emptying of lipids after ingestion of a solid-liquid meal in humans. *Gastroenterology* **80**, 922–927.

Cudrey, C., van Tilbeurgh, H., Gargouri, Y. and Verger, R. (1993) Inactivation of pancreatic lipases by amphiphilic reagents 5-(Dodecyldithio)-2-nitrobenzoic acid and tetrahydrolipstatin. Dependence upon partitioning between micellar and oil phases. *Biochemistry* **32**, 13800–13808.

De Caro, A., Bezzine, S., Lopez, V., Aoubala, M., Daniel, C., Verger, R. and Carrière, F. (1999) Immunological Characterization of Digestive Lipases. In *Lipase and Phospholipase Protocols* (Doolitle, M.H. and Reue, K., eds.), Vol. 109, pp. 239–256. Humana Press, Totowa.

De Caro, J., Carrière, F., Barboni, P., Giller, T., Verger, R. and de Caro, A. (1998) Pancreatic lipase-related protein 1 (PLRP1) is present in the pancreatic juice of several species. *Biochim. Biophys. Acta* **1387**, 331–341.

DiMagno, E.P., Go, V.L.W. and Summerskill, W.H.J. (1973) Relations between pancreatic enzyme outputs and malabsorption in severe pancreatic insufficiency. *N. Engl. J. Med.* **288**, 813–815.

Erlanson, C. and Borgström, B. (1970) Tributyrin as a substrate for determination of lipase activity of pancreatic juice and small intestinal content. *Scand. J. Gastroenterol.* **5**, 293–295.

Fredrikzon, B. and Blackberg, L. (1980) Lingual lipase: an important lipase in the digestion of dietary lipids in cystic fibrosis? *Pediatr. Res.* **14**, 1387–1390.

Gargouri, Y., Chahinian, H., Moreau, H., Ransac, S. and Verger, R. (1991) Inactivation of pancreatic and gastric lipases by THL and $C_{12:0}$-TNB: a kinetic study with emulsified tributyrin. *Biochim. Biophys. Acta* **1085**, 322–328.

Gargouri, Y., Piéroni, G., Rivière, C., Lowe, P.A., Saunière, J.-F., Sarda, L. and Verger, R. (1986a) Importance of human gastric lipase for intestinal lipolysis: An *in vitro* study. *Biochim. Biophys. Acta* **879**, 419–423.

Gargouri, Y., Piéroni, G., Rivière, C., Saunière, J.-F., Lowe, P.A., Sarda, L. and Verger, R. (1986b) Kinetic assay of human gastric lipase on short- and long-chain triacylglycerol emulsions. *Gastroenterology* **91**, 919–925.

Garrow, J. (1998) Flushing away the fat. *British Medical Journal* **317**, 830–831.

Giller, T., Buchwald, P., Blum-Kaelin, D. and Hunziker, W. (1992) Two novel human pancreatic lipase related proteins, hPLRP1 and hPLRP2. Differences in colipase dependence and in lipase activity. *J. Biol. Chem.* **267**, 16509–16516.

Gregory, P.C., Tabeling, R., Fassmann, C. and Kamphues, J. (2002) Therapy of pancreatic exocrine insufficiency: new experimental data. In *Chronic pancreatitis: Novel concepts in biology and therapy* (Buchler, M., Friess, H., Uhl, W. and Malfertheiner, P., eds.), pp. 445–456. Blackwell Wissenschafts-Verlag, Berlin.

Gruber, V., Berna, P., Arnaud, T., Bournat, P., Clément, C., Mison, D., Olagnier, B., Philippe, L., Theisen, M., Baudino, S., Bénicourt, C., Cudrey, C., Blois, C., Duchateau, N., Dufour, S., Gueguen, C., Jacquet, S., Ollivo, C., Poncetta, C., Zorn, N., Ludevid, D., Van Dorsselaer, A., Verger, R., Doherty, A., Mérot, B. and Danzin, C. (2001) Large-scale production of a therapeutic protein in transgenic tobacco plants: effect of subcellular targeting on quality of a recombinant dog gastric lipase. *Molecular Breeding* **7**, 329–340.

Hadvàry, P., Lengsfeld, H. and Wolfer, H. (1988) Inhibition of pancreatic lipase *in vi-*

tro by the covalent inhibitor tetrahydrolipstatin. *Biochem. J.* **256**, 357–361.

Hadvàry, P., Sidler, W., Meister, W., Vetter, W. and Wolfer, H. (1991) The lipase inhibitor tetrahydrolipstatin binds covalently to the putative active site serine of pancreatic lipase. *J. Biol. Chem.* **266**, 2021–2027.

Hanefeld, M. and Sachse, G. (2002) The effects of orlistat on body weight and glycaemic control in overweight patients with type 2 diabetes: a randomized, placebo-controlled trial. *Diabetes Obes. Metab.* **4**, 415–423.

Hauton, J.C. (1986) A quantitative dynamic concept on the role of bile in fat digestion. In *Molecular and cellular basis of digestion* (Desnuelle, P., Sjoström, H. and Norén, O., eds.), pp. 147–170. Elsevier Science, Amsterdam.

Hildebrand, P., Petrig, C., Burckhardt, B., Ketterer, S., Lengsfeld, H., Fleury, A., Hadvàry, P. and Beglinger, C. (1998) Hydrolysis of dietary fat by pancreatic lipase stimulates cholecystokinin release. *Gastroenterology* **114**, 123–129.

Howles, P.N., Carter, C.P. and Hui, D.Y. (1996) Dietary free and esterified cholesterol absorption in cholesterol esterase (bile salt-stimulated lipase) gene-targeted mice. *J. Biol. Chem.* **271**, 7196–7202.

Johansson, C., Lagerlöf, H.O., Ekelund, K., Kulsdom, N., Larsson, I. and Nylind, B. (1972) Studies of gastrointestinal interactions. III. Determination of gastric secretion and evacuation, biliary and pancreatic secretion, intestinal absorption, intestinal transit time and flow of water in man. *Scand. J. Gastroenterol.* **7**, 489–499.

Kelley, D.E., Bray, G.A., Pi-Sunyer, F.X., Klein, S., Hill, J., Miles, J. and Hollander, P. (2002) Clinical efficacy of orlistat therapy in overweight and obese patients with insulin-treated type 2 diabetes: A 1-year randomized controlled trial. *Diabetes Care* **25**, 1033–1041.

Lapey, A., Kattwinkel, J., di Sant Agnese, P.A. and Laster, L. (1974) Steatorrhea and azotorrhea and their relation to growth and nutrition in adolescents and young adults with cystic fibrosis. *J. Pediatr.* **84**, 328–334.

Layer, P. and Keller, J. (2003) Lipase supplementation therapy: standards, alternatives, and perspectives. *Pancreas* **26**, 1–7.

Layer, P., Keller, J. and Lankisch, A.P. (2001) Pancreatic enzyme replacement therapy. *Curr. Gastroenterol. Rep.* **3**, 101–108.

Lookene, A., Skottova, N. and Olivecrona, G. (1994) Interactions of lipoprotein lipase with the active-site inhibitor tetrahydrolipstatin (Orlistat)(R). *Eur. J. Biochem.* **222**, 395–403.

Lowe, M.E. (2000) Properties and function of pancreatic lipase related protein 2. *Biochimie* **82**, 997–1004.

Lowe, M.E. (2002) The triglyceride lipases of the pancreas. *J. Lipid Res.* **43**, 2007–2016.

Lüthi-Peng, Q., Maerki, H.P. and Hadvàry, P. (1992) Identification of the active-site serine in human pancreatic lipase by chemical modification with tetrahydrolipstatin. *FEBS Lett.* **299**, 111–115.

Milano, S. and Oliveira, M. (1998) Effet d'une lipase gastrique dans un modèle d'insuffisance pancréatique exocrine chronique chez le miniporc. *Gastroenterol. Clin. Biol.* **22**.

Miled, N., Canaan, S., Dupuis, L., Roussel, A., Rivière, M., Carrière, F., de Caro, A., Cambillau, C. and Verger, R. (2000) Digestive lipases: from three-dimensional structure to physiology. *Biochimie* **82**, 973–986.

Moreau, H., Bernadac, A., Gargouri, Y., Benkouka, F., Laugier, R. and Verger, R. (1989) Immunocytolocalisation of human gastric lipase in chief cells of the fundic mucosa. *Histochemistry* **91**, 419–423.

Moreau, H., Gargouri, Y., Lecat, D., Junien, J.-L. and Verger, R. (1988 a) Screening of preduodenal lipases in several mammals. *Biochim. Biophys. Acta* **959**, 247–252.

Moreau, H., Laugier, R., Gargouri, Y., Ferrato, F. and Verger, R. (1988 b) Human preduodenal lipase is entirely of gastric fundic origin. *Gastroenterology* **95**, 1221–1226.

Moreau, H., Sauniere, J.F., Gargouri, Y., Pieroni, G., Verger, R. and Sarles, H. (1988 c) Human gastric lipase: variations induced by gastrointestinal hormones and by pathology. *Scand. J. Gastroenterol.* **23**, 1044–1048.

Moreau, H., Verger, R., Lecat, D. and Junien, J.L. (1987) Lipases et extraits lipasiques, leur procédé de préparation et leur application notamment en thérapeutique. *European Patent* **87401984.7**.

Moreau, J., Bouisson, M., Balas, D., Ravaud, A., Stupnik, S., Buscail, L., Vaysse, N. and Ribet, A. (1990) Gastric lipase in alcoholic pancreatitis. Comparison of secretive profiles following pentagastrin stimulation in normal adults and patients with pancreatic insufficiency. *Gastroenterology* **99**, 175–180.

Muller, D.P. and Ghale, G.K. (1982) Stability of pancreatic enzyme activities in duodenal juice after pancreatic stimulation by a test meal or exogenous hormones. *Ann. Clin. Biochem.* **19**, 89–93.

Muller, D.P.R., McCollum, J.P.K., Tompeter, R.S. and Harries, J.T. (1975) Studies on the mechanism of fat absorption in congenital isolated lipase deficiency. *Gut* **16**, 838.

Ransac, S., Gargouri, Y., Moreau, H. and Verger, R. (1991) Inactivation of pancreatic and gastric lipases by tetrahydrolipstatin and alkyl-dithio-5-(2-nitrobenzoic acid). A kinetic study with 1,2-didecanoyl-sn-glycerol monolayers. *Eur. J. Biochem.* **202**, 395–400.

Renou, C., Carrière, F., Ville, E., Grandval, P., Joubert-Collin, M. and Laugier, R. (2001) Effects of Lansoprazole on Human Gastric Lipase Secretion and Intragastric Lipolysis in Healthy Human Volunteers. *Digestion* **63**, 207–213.

Roberts, I.M. and Hanel, S.I. (1988) Stability of lingual lipase in vivo: studies of the iodinated enzyme in the rat stomach and duodenum. *Biochim. Biophys. Acta* **960**, 107–110.

Rogalska, E., Ransac, S. and Verger, R. (1990) Stereoselectivity of lipases. II. Stereoselective hydrolysis of triglycerides by gastric and pancreatic lipases. *J. Biol. Chem.* **265**, 20271–20276.

Ross, C.A.C. (1955) Fat absorption studies in the diagnosis and treatment of pancreatic fibrosis. *Arch. Dis. Child* **30**, 316–320.

Ross, C.A.C. and Sammons, H.C. (1955) Non-pancreatic lipase in children with pancreatic fibrosis. *Arch. Dis. Child* **30**, 428–431.

Roulet, M., Weber, A.M., Paradis, Y., Roy, C.C., Chartraud, L., Lasalle, R. and Morin, C.L. (1980) Gastric emptying and lingual lipase activity in cystic fibrosis. *Pediatr. Res.* **14**, 1360–1362.

Roussel, A., Canaan, S., Egloff, M.P., Rivière, M., Dupuis, L., Verger, R. and Cambillau, C. (1999) Crystal structure of human gastric lipase and model of lysosomal acid lipase, two lipolytic enzymes of medical interest. *J. Biol. Chem.* **274**, 16995–17002.

Roussel, A., de Caro, J., Bezzine, S., Gastinel, L., de Caro, A., Carrière, F., Leydier, S., Verger, R. and Cambillau, C. (1998a) Reactivation of the totally inactive pancreatic lipase RP1 by structure-predicted point mutations. *Proteins* **32**, 523–531.

Roussel, A., Miled, N., Berti-Dupuis, L., Rivière, M., Spinelli, S., Berna, P., Gruber, V., Verger, R. and Cambillau, C. (2002) Crystal structure of the open form of dog gastric lipase in complex with a phosphonate inhibitor. *J. Biol. Chem.* **277**, 2266–2274.

Roussel, A., Yang, Y., Ferrato, F., Verger, R., Cambillau, C. and Lowe, M. (1998b) Structure and activity of rat pancreatic lipase-related protein 2. *J. Biol. Chem.* **273**, 32121–32128.

Schwizer, W., Asal, K., Freiss, C., Mettraux, C., Borovicka, J., Rémy, B., Güzelhan, C., Hartman, D. and Fried, M. (1997) Role of lipase in the regulation of gastrointestinal function in humans. *Am. J. Physiol.* **273**, G612–G620.

Sjoström, L., Rissanen, A., Andersen, T., Boldrin, M., Golay, A., Koppeschaar, H.P. and Krempf, M. (1998) Randomised placebo-controlled trial of orlistat for weight loss and prevention of weight regain in obese patients. European multicentre orlistat study group. *Lancet* **352**, 167–172.

Smith, G.M., Garton, A.J., Aitken, A. and Yeaman, S.J. (1996) Evidence for a multidomain structure for hormone-sensitive lipase. *FEBS Lett.* **396**, 90–94.

Suzuki, A., Mizumoto, A., Sarr, M.G. and Dimagno, E.P. (1997) Bacterial lipase and high-fat diets in canine exocrine pancreatic insufficiency: a new therapy of steatorrhea? *Gastroenterology* **112**, 2048–2055.

Szafran, Z., Szafran, H., Popiela, T. and Trompeter, G. (1978) Coupled secretion of gastric lipase and pepsin in man following pentagastrin stimulation. *Digestion* **18**, 310–318.

Thirstrup, K., Verger, R. and Carrière, F. (1994) Evidence for a pancreatic lipase subfamily with new kinetic properties. *Biochemistry* **33**, 2748–2756.

van Tilbeurgh, H., Egloff, M.-P., Martinez, C., Rugani, N., Verger, R. and Cambillau, C. (1993) Interfacial activation of the lipase-procolipase complex by mixed micelles revealed by X-Ray crystallography. *Nature* **362**, 814–820.

van Tilbeurgh, H., Sarda, L., Verger, R. and Cambillau, C. (1992) Structure of the pancreatic lipase-procolipase complex. *Nature* **359**, 159–162.

Verger, R. (1984) Pancreatic lipases. In *Lipases* (Borgström, B. and Brockman, H. L., eds.), pp. 83–149. Elsevier Science, Amsterdam.

Ville, E., Carrière, F., Renou, C. and Laugier, R. (2002) Physiological study of pH stability and sensitivity to pepsin of human gastric lipase. *Digestion* **65**, 73–81.

Weibel, E. K., Hadvàry, P., Hochuli, E., Kupfer, E. and Lengsfeld, H. (1987) Lipstatin, an inhibitor of pancreatic lipase, produced by *Streptomyces toxytricini*. I. Producing organism, fermentation, isolation and biological activity. *J. Antibiotics* **40**, 1081–1085.

Wicker-Planquart, C. and Puigserver, A. (1993) Regulation of gastrointestinal lipase. Gene expression by dietary lipids. In *Nutrition and gene expression* (Berdanier and Hargrove, eds.), pp. 56–89. CRC Press.

Winkler, F. K., d'Arcy, A. and Hunziker, W. (1990) Structure of human pancreatic lipase. *Nature* **343**, 771–774.

Zhi, J., Melia, A. T., Guerciolini, R., Chung, J., Kinberg, J., Hauptman, J. B. and Patel, I. H. (1994) Retrospective population-based analysis of the dose-response (fecal fat excretion) relationship of orlistat in normal and obese volunteers. *Clin. Pharmacol. Ther.* **56**, 82–85.

Zhi, J. G., Melia, A. T., Eggers, H., Joly, R. and Patel, I. H. (1995) Review of limited systemic absorption of orlistat, a lipase inhibitor, in healthy human volunteers. *J. Clin. Pharmacol.* **35**, 1103–1108.

Zieve, L., Mulford, B. and McHale, A. (1966a) Secretion of pancreatic enzymes. II. Comparative response following test meal or injection of secretin and pancreozymin. *Am. J. Dig. Dis.* **11**, 685–694.

Zieve, L., Silvis, S. E., Mulford, B. and Blackwood, W. D. (1966b) Secretion of pancreatic enzymes. I. Response to secretin and pancreozymin. *Am. J. Dig. Dis.* **11**, 671–684.

11
Physiological and Pharmacological Regulation of Triacylglycerol Storage and Mobilization

Günter Müller and Stefan Petry

Abbreviations

aa, amino acid(s); AC, adenylate cyclase; AR, adenosine receptor; α/β-AR, α/β-adrenergic receptor; (D/T)AG, (di/tri)acylglycerol; DGAT, acyl-CoA:diacylglycerol acyltransferase; DIGs, detergent-insoluble glycolipid-enriched lipid raft microdomains; ER, endoplasmic reticulum; FA, fatty acids; H/LDL, high/low density lipoprotein; HSL, hormone-sensitive lipase; LD, lipid droplets; LPL, lipoprotein lipase; NA, nicotinic acid; NEFA, non-esterified fatty acids; NIDDM, non-insulin-dependent diabetes mellitus; PDE, phosphodiesterase; PI3K, phosphatidylinositol-3-kinase; PKA/B/C, protein kinase A/B/C; PP, protein phosphatases; TNF-α, tumor necrosis factor-α; VLDL, very low density lipoproteins; WAT, white adipose tissue

11.1
Metabolic Role of Triacylglycerol

11.1.1
Triacylglycerol and Energy Storage

Triacylglycerol (TAG) is the most concentrated form of energy available to biological tissues which emerged during biological evolution. For instance, the work covered by birds during their non-stop long-distance migrations is powered almost exclusively by fat reserves. The extra weight of carbohydrate required to produce the same calories would prevent the birds from ever taking off. In terms of human survival, the first unaided crossing of the polar ice cap was made possible by the very high butter-fat content of the 220 kg of food reserves transported by the sledges that were man-powered over the frozen wastes. The extra weight of a similar energy content as carbohydrate would have made this pioneering journey impossible in the short and medium terms. TAG also exhibits relatively low "biological toxicity" compared to FA (see below) and is well tolerated even at high concentrations in the blood plasma. TAG therefore provides a benign form of FA storage and transport. Before its use as an energy source, however, it must undergo lipolytic mobilization to provide FA. The potent "toxicity" of FA in the long term re-

Lipases and Phospholipases in Drug Development
Edited by G. Müller and S. Petry

ISBN: 3-527-30677-3

quires a finely tuned regulatory mechanism that ensures that their release from TAG stores is correctly balanced by uptake and utilization.

Although the cells of most tissues synthesize TAG [1, 2], only a few are adapted to store TAG in considerable quantity. Some, such as heart and skeletal muscle, store small quantities of TAG for their own use (albeit the heart has been recently shown to have low capacity for lipoprotein export [3]). In other tissues, for instance in the pancreatic β-cells, TAG synthesis, storage and lipolysis may be involved in the regulation of the production of molecules, such as DAG and FA(-derivatives), which participate in various signal transduction processes [4] or in providing a source of FA committed to membrane phospholipid synthesis [5]. In mammals, of course, adipose tissue is the most important storage pool of whole-body TAG and, consequently, the body's largest energy reservoir; for example, an adult with 15 kg of body weight has >460 MJ of TAG fuel stores, which could provide 8.37 MJ daily for 2 months [6]. The primary role of adipocytes is to store TAG during caloric excess and to mobilize this reserve when expenditure exceeds intake. Mature adipocytes are uniquely equipped to perform these functions. They possess the full complement of enzymes and regulatory proteins needed to carry out both TAG storage and mobilization (see below). In adipocytes, the regulation of these processes is exquisitely responsive to hormones, cytokines and other factors that are involved in energy homeostasis (Tab. 11.1 and below).

Nevertheless, the liver also has an important TAG storage function. These two tissues export a proportion of their mobilized TAG. FA are released from adipose tissue, and TAG is exported as VLDL from the liver. Both organs are connected via a metabolic cycle in which FA released from adipose tissue are the major substrates for hepatic VLDL production and, in turn, FA from TAG contained in VLDL are returned to adipose tissue by the action of LPL. In addition, TAG in adipose tissue represents the major source of plasma glycerol, which accounts for about 90% of the substrates for hepatic gluconeogenesis at fasted condition in rodents [7]. Central to the control of this inter-tissue cycle between WAT and liver is the regulation of TAG mobilization and release by both organs. The different cellular and molecular mechanisms involved in the regulation of TAG mobilization from the storage pools in each of these tissues reflect the importance of the different roles of the lipids and FA which are exported.

FA stored within the TAG of adipose cells constitute the vast majority of energy reserves in animals. While the TAG pool in adipocytes provides a reservoir of energy for times of prolonged restricted caloric intake, there is also a normal "acute" flux of FA into and out of adipose tissue in response to meals and other factors, such as insulin [8, 9]. TAG hydrolysis in adipocytes, termed lipolysis, contributes to the increased levels of plasma NEFA by two mechanisms, one direct and the other indirect. In the post-absorptive state or during starvation FA are normally released from adipocytes directly into the plasma and transported to various tissues where they serve as a source of energy. In the insulin-resistant state this direct release of FA is exacerbated, because the reduction in the anti-lipolytic action of insulin (see below) leads to increased lipolysis and a consequent increase in the release of FA from adipocytes into plasma. In contrast to the postabsorptive direct FA release, there is

Tab. 11.1 Natural and synthetic ligands known to affect TAG storage and/or mobilization in WAT via receptor-mediated mechanisms. Meanwhile, the adipocyte has gained the status of an endocrine cell with the discovery of regulated production of several secreted products of various nature (e.g. lipid metabolites, cytokines and other proteins) that are involved in autocrine, paracrine and endocrine effects on TAG storage and mobilization. Adapted from Ref. [559] with modifications.

Ligand	*Effect on TAG mobilization*
Insulin	Inhibition and stimulation of TAG storage
Insulin-like growth factor I and II	Inhibition
Thyroid-stimulating hormone	Regulation
Prostaglandin E2	Inhibition
Adenosine	Inhibition
Neuropeptide Y-Y1	Inhibition
Epidermal growth factor	Antagonism of lipolytic catecholamine effect
α_2 catecholamine	Inhibition
Nicotinic acid/acipimox	Inhibition
Gastric inhibitory peptide	Stimulation of TAG storage and amplification of insulin effects
Muscarinic receptor agonists	Inhibition
ASP	Inhibition and stimulation of TAG storage
Phosphoinositolglycan(-peptides)	Inhibition of catecholamine-induced TAG mobilization and stimulation of TAG storage
Glimepiride	Inhibition of catecholamine-induced TAG mobilization and stimulation of TAG storage
Glucagon	Stimulation
Leptin	Stimulation and antagonism of anti-lipolytic insulin effect
Growth hormone	Stimulation
TNF-α	Stimulation
Interleukin-6	Stimulation and inhibition of LPL
Angiotensin II	Stimulation
$\beta_1/\beta_2/\beta_3/\beta_4$ catecholamines	Stimulation
Triiodothyronine T_3	Stimulation

an indirect effect of adipocyte lipolysis on FA release in the postprandial state. There is a normal postprandial influx into adipocytes of FA produced by LPL action on VLDL and chylomicrons in the adipose endothelium, and this influx depends on a favorable downhill FA concentration gradient across the adipocyte plasma membrane. Again, this indirect effect is exacerbated in the insulin-resistant state, since the increased lipolysis and, in consequence, the increased intracellular FA concentration in adipocytes abrogate the FA concentration gradient required for influx of FA. Thus, in an indirect fashion, adipocyte lipolysis can increase plasma FA by reducing the postprandial influx of FA into adipocytes [8, 9]. Given the apparent interrelated contribution of the cellular and LPL-mediated lipolytic reactions in adipose tissue to the plasma NEFA pool, it is important to understand the molecular basis of TAG hydrolysis by these mechanisms.

The two key enzymes involved in TAG mobilization within adipose tissue are LPL and HSL. The function of LPL in mobilising FA from TAG contained in VLDL and chylomicrons in the capillary endothelium of adipose tissue is not so clear as originally thought. It is now becoming evident that a variable (and regulated) proportion of NEFA released by LPL from TAG-rich lipoproteins does not enter adipose tissue but "leaks" into the plasma compartment [10] and that this proportion may be under metabolic control dependent upon the nutritional state [11]. For instance, in the post-absorptive state and during starvation, very little, if any, NEFA derived from TAG of VLDL enter the adipose tissue. The major portion enters the blood plasma and supplements the HSL-mediated provision of NEFA to the periphery during this period [8]. Any excess NEFA produced which override the immediate oxidative capacity of energy-requiring tissues are re-esterified by the liver and stored as TAG or VLDL. The potential hazards of hyperlipidemia for the cardiovascular system in the postprandial state are now well established and accepted [12–15]. The liver may, thus, play a major protective role during this period by removing excessive NEFA from the circulation, temporarily storing them as a "benign" derivative, TAG, and secreting them later as VLDL when the period of maximum danger has passed. This protective function may prove to be yet another facet of the liver's more general role as a front-line defence to protect body tissues from the potentially dangerous but inevitable consequences of major and abrupt physiological transitions. In this way, the liver may act as an essential (storage) buffer by which the supply of and demand for adipose tissue FA are matched in a way which does not result in or neutralize the potential FA-induced "lipotoxicity" to peripheral tissues (see below; for a review see Ref. [16]).

11.1.2
Lipolysis and Re-esterification

Adipocytes are highly differentiated cells specialized in handling large quantities of long-chain FA (Fig. 11.1). During lipid storage in the fed state, FA are released by LPL bound to the endothelial capillary wall into the blood from chylomicrons and lipoproteins assembled and secreted by intestinal and liver cells, respectively (step 1), or from albumin, move through the endothelium via passage through tight junctions or endothelial cells (step 2), bind to the outer leaflet of the plasma membrane of the target cells, and cross the membrane bilayer by simple diffusion or protein-mediated transport (step 3). While the direct involvement of FA not bound to albumin in FA uptake by adipocytes is generally accepted [17], it is still debated whether, and if so to what extent, the albumin-bound FA pool is directly engaged in this process [18]. Using the hindlimb perfusion technique it was demonstrated recently that FA incorporation into intramuscular TAG depends not only on the unbound to albumin FA concentration but also, to some extent, on the total FA concentration [19]. This supports the previous speculation that albumin-bound FA could be directly involved in the uptake process mediated by an interaction of the FA-albumin com-

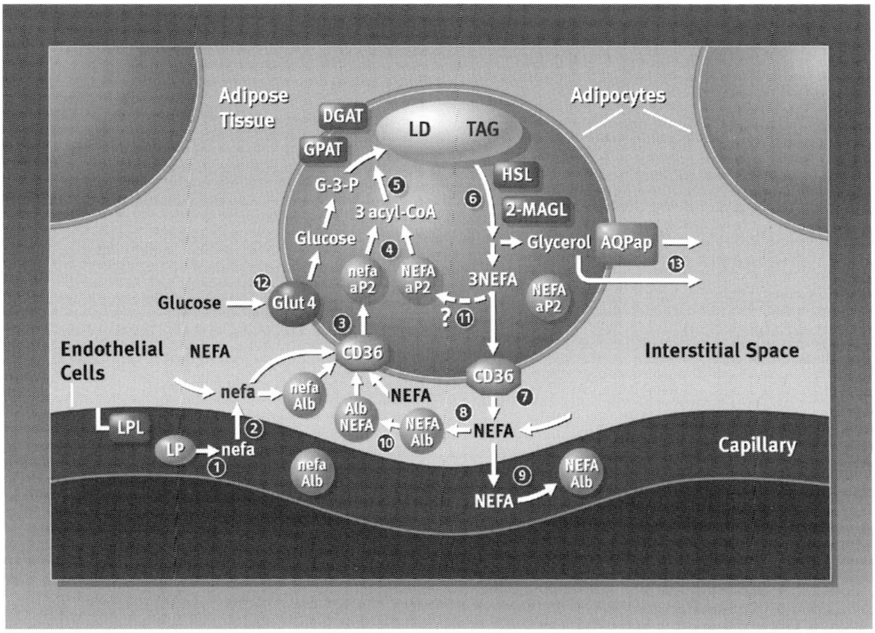

Fig. 11.1 Working model for FA re-esterification and its control by the efficacy of removal of newly released NEFA from the interstitial space into the circulation. Elevation/reduction of the blood flow would decrease/increase the accumulation of NEFA in the interstitial space in the vicinity of adipocytes where the endogenously released NEFA intermingle with and are diluted by the nefa generated by LPL action from lipoproteins (LP) in the endothelial capillaries, and thereby decrease/increase the probability of re-uptake and re-esterification of NEFA released from adipocytes by HSL and 2-monoacylglycerol lipase (2-MAGL) action. This allows rapid adaptation of the supply of NEFA from adipose tissue according to the acute requirements of the organism (e.g. during starvation), which, to a certain degree, operates independent of the actual (insulin-controlled) lipolytic/anti-lipolytic state of the adipocytes. The major portion of NEFA/nefa in the interstitial space and capillaries is bound to serum albumin (Alb). The transport of glucose (as precursor for glycerol-3-phosphate, G-3-P) and the bidirectional flux of NEFA/nefa across the adipocyte plasma membrane is facilitated by GLUT4 and CD36/FAT, respectively, whereas simple diffusion and/or specific transport via AQPap is involved in the release of glycerol from adipocytes. See text for details. Adapted from Ref. [28] with modifications.

plex with an albumin receptor located at the cell surface [20], putatives candidate for which are the albumin-binding proteins (ABP) [21]. They could either promote a direct transfer of FA from the complex into the plasma membrane or facilitate dissociation of FA molecules from albumin, and thus locally increase the number of FA molecules available for uptake. Subsequently, FA transported across the plasma membrane are trapped in the cytoplasm by conversion into mobilised acyl-CoA (step 4), (re-)esterified and stored primarily as TAG in LD (step 5). During lipolysis of stored TAG in the fasting state, FA are released from intracellular LD primarily by action of HSL and 2-monoacylglycerol lipase (step 6) [22], move to and cross the plasma membrane (step 7), and are released into the interstitial space, where they bind to albumin (step 8). Subsequently, the FA pass through the endothelial cells via transcytosis or diffuse through the spaces between them (tight junctions) to reach the blood (step 9).

The temporal and spatial relationship between breakdown (lipolysis) and re-synthesis (re-esterification) of TAG (TAG-FA cycling) has to be tightly controlled to either provoke or avoid generation of heat and energy expenditure through the utilization of ATP during substrate cycling. A substrate cycle exists when opposing, non-equilibrium reactions catalyzed by different enzymes are operating simultaneously [23, 24]. At least one of the reactions must involve the hydrolysis of ATP. Thus, a substrate cycle both liberates heat and increases energy expenditure in the absence of net conversion of substrate into product. Some of the NEFA formed during lipolysis can be re-esterified to TAG [25, 26], whereas little or no glycerol is re-utilized by fat cells [27]. This pathway of lipolysis and NEFA re-esterification forms an important cycle for energy turnover, allowing the fat cell to respond rapidly to changes in peripheral requirements for NEFA (Fig. 11.1). This is reflected in the FA concentration gradient between plasma and adipose tissue (mainly fat cell cytosol) which is determined by both the blood stream and uptake of NEFA by peripheral tissues. Two types of NEFA re-esterification in fat cells can be recognized [28]. Primary re-esterification is the total amount of NEFA that is re-esterified during a given situation and reflects the TAG synthesis capacity of the fat cells. Fractional re-esterification is the proportion of NEFA re-esterified in relation to the amount of NEFA formed by lipolysis in fat cells [28]. The latter constitutes a futile cycle, energy-rich NEFA first being formed by lipolysis of TAG and then re-synthesized to TAG. There is experimental evidence that NEFA have to be released from adipocytes into the interstitial space, where they form a common pool with NEFA generated by LPL action, and subsequently be taken up by the adipocytes for efficient re-esterification rather than that NEFA accumulating in the cytoplasm can be used for a "short-circuit" re-esterification [28, 29]. Fat cell re-esterification can be determined *in vitro* by simultaneous measurement of the release of glycerol and NEFA using fluorescence, luminescence or dual radioisotope techniques or a combination thereof [24].

11.1.3 TAG Storage/Mobilization and Disease

11.1.3.1 Diabetes Mellitus and Metabolic Syndrome

During the last decade the traditional glucose-insulin axis with respect to examining glucose tolerance and as a basic pathophysiological mechanism operating in diabetes, although remaining diagnostically useful, has been criticized [30–33]. The glucose-insulin axis has been overemphasized, while alterations in the insulin to NEFA ratio and "NEFA" metabolism received much less attention [33]. The major change in the appreciation of the nature of molecules supplying body energy came with the discoveries by Dole [34] and Gordon and Cherkes [35] that plasma NEFA are the major "lipid vehicles" in the mammalian organism. Furthermore, NEFA are the only form of TAG mobilization from adipose tissue, subject to hormonal regulation. The very rapid turnover of plasma NEFA indicates that they are an important fuel even in non-fasting conditions and represent an energy source to body tissues equivalent to or greater than glucose. Early data on the role of NEFA in glucose homeostasis were reviewed in 1969 by Ruderman and coworkers [36]. The importance of NEFA in the regulation of gluconeogenesis was extensively dealt with, but no clear pathogenic conclusion with relation to major diabetes perturbations emerged.

Significantly, Zierler and Rabinowitz [37] showed, in 1964, that infusion of a very small amount of insulin through the brachial artery in human forearm leads to a significant decrease in plasma NEFA concentrations without any concomitant effect on glucose uptake. Consequently, the restraint of NEFA release from adipose tissue is the most sensitive action of insulin compared with other insulin effects. The lack of restraint of NEFA mobilization in hypoinsulinemic states during pathogenesis of NIDDM leads to a marked plasma NEFA increase, ectopic deposition of TAG and insulin resistance with regard to non-oxidative as well as oxidative glucose metabolism. The latter fact stresses the negative effect of excessive availability of NEFA in the short term on insulin-mediated glucose utilization in muscle, as shown by Randle and coworkers [38, 39]. It involves a reduction of muscle glycolysis by inhibiting phosphofructokinase in course of elevated cytosolic concentrations of citrate and consequent accumulation of glucose-6-phosphate. Pyruvate dehydrogenase is also inhibited by rising mitochondrial concentrations of acetyl-CoA and NADH. In the liver the same changes in redox ratio and activation of pyruvate carboxylase by acetyl-CoA result in stimulation of gluconeogenesis. This mechanism links acute NEFA oversupply to impaired glucose tolerance or hyperglycemia in the developing or frank NIDDM, respectively.

Moreover, the increased plasma NEFA concentration in NIDDM patients [40] provides substrate for VLDL production by the liver (see above), which is manifested as hypertriglyceridemia [41]. Thus, drugs with the potential of lowering NEFA levels [42] may reduce hepatic VLDL production, improve glucose tolerance and counteract lipotoxicity and therefore be useful for anti-diabetic therapy. The generation and careful analysis of transgenic and knockout animals has aided insight into the recognised but partly misunderstood participation of cellular lipid metabolism in energy homeostasis. Numerous mouse models with disruptions in pathways relevant to FA as well as TAG storage and mobilization have been generated with dysregulated circulation of FA in the bloodstream as non-esterified molecules, bound to albumin, or in the core of lipoproteins, such as VLDL (see below).

Because of its role in FA mobilization, HSL is among the candidates for such dysregulation, and, furthermore, HSL has been suggested as a target for the development of new antidiabetic drugs. In adipocytes from elderly male subjects with several manifestations of the metabolic syndrome, a marked change in the lipolytic response toward noradrenaline due to dysregulation in the number of βAR was observed [43]. In addition, the maximum lipolytic effects of isoproterenol, forskolin and dibutyryl cyclic AMP were also considerably altered, suggesting modifications at the level of the cAMP-dependent regulation of HSL activity [43]. However, more recently, HSL activity measured *in vitro* was found not to differ significantly between patients suffering from the metabolic syndrome and healthy subjects [44].

With regard to genetic studies, a polymorphism in exon 4 of the HSL gene, changing an arginine to a cysteine, showed no difference in allele frequency between NIDDM patients and healthy subjects [45]. In another study, a polymorphic marker in intron 7 of the human HSL gene showed a significant difference in allele frequency distribution between NIDDM patients, in particular abdominally obese NIDDM patients with the metabolic syndrome, and control subjects [46]. Although in agreement with another study, showing positive association between a different polymorphic marker in the HSL gene and NIDDM [47], transmission disequilibrium tests suggested linkage disequilibrium between the HSL gene marker and an allele or gene that increases susceptibility to abdominal obesity and thereby possibly NIDDM [46]. In conclusion, although most available data seem to argue against a direct involvement of genetic HSL aberrations in the development of the metabolic syndrome/NIDDM, clarification requires more studies.

11.1.3.2 **Lipotoxicity**

In the last decade the deleterious effects of the deposition of TAG in non-adipose tissues during long-term elevations of plasma NEFA have been widely acknowledged. Whereas adipocytes have a unique capacity to store excess FA in the form of TAG in LD, non-adipose tissues, such as cardiac and skeletal muscle myocytes and pancreatic β-cells, have a limited capacity for storage of TAG. In hyperhyperlipidemic states, accumulation of excess TAG in non-adipose tissues leads to cell dysfunction and/or cell death, a phenomenon known as lipotoxicity and lipoapoptosis, respectively. Excess lipid accumulation in skeletal muscle is associated with the development of insulin resistance [48].

11.1.3.2.1 **β-Cells**

The "lipotoxic" consequences of excessive FA release from adipose tissue for pancreatic β-cells have been described in a provocative report by Unger and colleagues [49]. Abnormal β-cell function and insulin resistance are the major characteristics in the pathogenesis of NIDDM. Lipotoxicity was proposed as an attractive concept to explain parallel developments in the impairment of insulin signaling and β-cell function [49–52]. Before the onset of overt NIDDM, insulin resistance is closely associated with β-cell dysfunction characterized by β-cell hyperplasia and insulin hypersecretion. During further progression of the disease, however, the β-cell can no longer compensate for peripheral insulin resistance and declines in its function. From animal models of obesity-linked NIDDM, such as the obese and insulin-resistant Zucker rats, as well as *Psammomys obesus*, both impairment of insulin secretion linked to intracellular fat deposition [53] and loss of cell mass by FA-induced apoptotic cell death contribute to the development of relative insulinopenia [54]. Exposure of

NEFA has cytostatic and proapoptotic effects on human pancreatic β-cells [55]. Lipid overload in pancreatic β-cells also leads to dysregulated insulin secretion [56, 57] followed by apoptotic cell death [58]. Furthermore, human autopsy studies show a relative reduction of β-cell mass in patients with NIDDM compared with weight-matched non-diabetic subjects [59, 60]. These findings support the idea that, in genetically predisposed human subjects, prolonged exposure to elevated NEFA may contribute to β-cell death and development/progression of NIDDM. The intracellular signaling pathways for FA-induced β-cell apoptosis are incompletely understood. Different mechanisms for lipoapoptosis, such as increased ceramide formation, mitochondrial cytochrome c release, and nitric oxide generation, have been suggested [54, 61, 62]. Furthermore, in different cell types the FA-dependent signaling process might involve activation of different protein kinases C isoforms [63, 64]. However, the exact role of PKC isoforms in FA-induced apoptosis has not been defined (see below).

TAG storage and lipolysis in the pancreatic β-cell may play a dual role. First, by way of storage, as a protective mechanism against the toxic effects of excess NEFA, and second by way of lipolysis to provide a signal (FA, DAG) that participates in the stimulus/secretion-coupling which regulates acute insulin release [65, 66]. With regard to the latter function, islets depleted of TAG show an abrogated stimulation of insulin release by glucose [67]. HSL may balance TAG storage and TAG mobilization in pancreatic β-cells, and, thus, via PKA, provide a critical link in the stimulus/secretion-coupling response [4]. HSL is predominantly expressed in pancreatic islets as a 3.1 kb mRNA encoding an $\sim$89 kDa isoform containing an additional 43 aa N-terminal to the adipocyte form of the protein, as well as the 2.8 kB mRNA encoding the 84 kDa adipocyte form [4]. HSL immunoreactivity in islets is detected primarily in β-cells, but some HSL is also observed in α-cells [4, 68]. Long-term incubation of β-cells or islets with high concentrations of glucose increases HSL mRNA and protein expression, as well as HSL activity [69]. This induction of HSL expression appears to be regulated transcriptionally and to depend on the metabolism of glucose. Treatment of ob/ob mice with leptin increased HSL expression in islets, along with a decrease in islet TAG content and an improvement in insulin secretion [70]. HSL is definitely functional in islets, since islet TAG content is increased 2–2.5-fold in HSL null mice compared with wild-type mice [69]. These observations together with studies on insulin release in HSL null mice upon glucose challenge [71] suggest that HSL might be important in glucose-induced insulin secretion and are consistent with the current view that FA metabolites are critical for insulin secretion, but that excessive NEFA can lead to lipotoxicity and dysfunction of β-cells [33]. The exact function of HSL in insulin secretion from β-cells awaits further study.

11.1.3.2.2 **Cardiac Myocytes**

Endogenous myocardial TAG can provide a substantial proportion of the FA oxidative demands of the working rat heart [72], especially in acute NIDDM [73]. However, in these situations the myocardial TAG content does not seem to decline, suggesting continuous synthesis from plasma NEFA [74]. These metabolic requirements are met by an extremely high lipolytic turnover of myocardial TAG, amounting to approximately 1 pool h^{-1} [75]. By contrast, hepatic and adipose tissue TAG have turnover rates of approximately 0.04 and 0.004 pool h^{-1}, respectively [76, 77]. Similar to the liver, this extensive recycling between myocardial TAG and its lipolytic products might occur in a pool that is physically separated from that involved in *de novo* TAG synthesis from exogenous FA [69]. This compartmentation implies two separate intracellular sites of DGAT activity in heart muscle. Whether these are the products of distinct genes remains an open question.

Cardiomyocytes are known to express HSL [68, 78, 79; see below] and the hormone-responsiveness of myocardial lipolysis suggests that HSL is probably responsible. In addition, the heart can synthesize and secrete apoB-containing lipoproteins [3]. If, as apparently with the liver, endogenous TAG is mobilized for lipoprotein assembly via a lipolytic event the question arises: can HSL fulfil this function in the heart or is another lipase involved in a mechanism identical to that in the liver? It has been proposed that mobilization of endogenous cardiac TAG for lipoprotein secretion provides the heart with a safety valve to dispose of excess lipid [3] and thus protects against lipotoxicity [65]. Lipoapoptosis observed in the heart leads to the development of heart failure [80, 81]. If, as apparently with skeletal muscle [82], excess TAG accumulation plays a role in insulin resistance, lipoprotein secretion may enable cardiac muscle insulin sensitivity in the face of an excessive flux of FA.

11.1.3.2.3 Molecular Mechanisms

In various experimental systems, saturated and unsaturated FA differ in their contributions to lipotoxicity. Previous studies in CHO cells [83], cardiac myocytes [84], pancreatic β-cells [85, 86], breast cancer cell lines [87], and hematopoietic precursor cell lines [88] suggest that lipotoxicity from the accumulation of long-chain FA is specific for saturated FA. This selectivity has been attributed to the generation of specific proapoptotic lipid species or signaling molecules in response to saturated but not unsaturated FA. The nature of such signals may differ across cell types, but includes reactive oxygen species generation [83], *de novo* ceramide synthesis [89], nitric oxide generation [90], decreases in PI3K [87], primary effects on mitochondrial structure and function [91] and suppression of anti-apoptotic factors, such as *Bcl*II [92].

A recent study [93] provided evidence that the differential toxicity of these FA is directly related to their ability to promote TAG accumulation. It was demonstrated that exogenous or endogenously generated unsaturated FA rescue palmitate-induced apoptosis by promoting palmitate incorporation into TAG. Moreover, unsaturated as well as saturated long-chain FA were toxic in cells with impaired TAG synthetic capacity (e.g. lack of DGAT activity) [91]. *In vivo*, TAG accumulation and storage in non-adipose tissues may occur in the setting of mismatch between cellular FA influx and FA utilization. TAG accumulation in non-adipose cells provoked by either enhanced synthesis or impaired mobilization in response to acute FA overload may represent an initial cellular defense mechanism against lipotoxicity. Consequently, accumulation of TAG in non-adipose tissues probably serves as a barometer of the

FA overload state in human disorders such as hyperlipidemia and lipodystrophies, and in animal models such as the Zucker (diabetic) fatty rat or the ob/ob mouse. Accumulation of cellular TAG *per se* does not seem to be toxic, at least initially. Rather, storage of excess FA in TAG pools of "inert" LD probably diverts these molecules from pathways that lead to cellular dysfunction and/or cytotoxicity and may thus serve as a buffer against lipotoxicity. In pathologic states, lipotoxicity may occur over time, despite TAG accumulation and storage, when either the cellular capacity for TAG storage is exceeded or when TAG pools are hydrolyzed (during HSL action), resulting in increased cellular NEFA levels. Thus, the duration and extent of TAG overload in conjunction with (regulation of) TAG storage and mobilization may determine whether a cell is protected or damaged.

This mechanism may also underlie the FA-induced apoptosis in rat and human pancreatic β-cells. Chronic treatment with high physiological levels of saturated FA, but not with mono- or polyunsaturated FA, triggers apoptosis in about 20% of cultured β-cells [94]. Apoptosis restricted to saturated FA was also observed in primary cultured human β-cells, suggesting that this mechanism is potentially relevant *in vivo* in humans. Apoptosis was accompanied by a rapid nuclear translocation of protein kinase C-δ (PKC-δ) and subsequent lamin B1 disassembly. This translocation was impaired by inhibition of phospholipase C, which also considerably reduced apoptosis. Furthermore, lamin B1 disassembly and apoptosis were decreased by cell transfection with a dominant-negative mutant form of PKC-δ [94]. These data suggest that nuclear translocation and kinase activity of PKC-δ are both necessary for saturated FA-induced lipoapoptosis. In analogy to recent findings with CHO cells [93], it is conceivable that in β-cells accumulation of DAG generated by phospholipase C action and the resulting activation of PKC-δ can be prevented by conversion of DAG into and storage as TAG via an ill-defined regulatory process of DAG acylation that is stimulated by unsaturated FA.

11.1.3.3 **Inborn Errors of TAG Storage and Metabolism**

Several distinct abnormalities of intracellular TAG mobilization and transfer occur which manifest themselves in rare inherited metabolic disorders, such as Wolman's disease (WD) and neutral lipid storage disease (NLSD). The former arises from an inherited deficiency of lysosomal acid lipase (LAL) and causes a massive increase in neutral lipids due to an inability of tissues to degrade exogenous lipoprotein-derived TAG and cholesterylester. However, the metabolism of endogenously derived TAG is unaffected [95, 96]. The disease is characterized by hepatomegaly, maladsorption, steatorrhea and abdominal distension [96]. The human enzyme amino acid sequence has been deduced from cDNA [97, 98], and the mutation has been identified in a rat model of WD through cloning of rat lysosomal acid lipase cDNA [99]. In a sense, WD is a consequence of defective lipolytic transfer of exogenous TAG into normal storage pools.

NLSD is characterized by muscular weakness, ichthyosis, neuropathy and multisystemic TAG storage [100, 101]. Unlike WD, this disease is accompanied by massive accumulation of TAG in the cytosol of many cell types, including hepatocytes [102, 103]. LAL is unaffected. The defect results from an abnormality in the processing of cytosolic TAG derived from both endogenous synthesis and exogenous sources [95, 100]. Despite some unexplained observations, it had long been assumed that the abnormality was caused by a defective cytosolic TAG lipase system. However, since DAG derived from stored TAG may serve as a precursor for phospholipid biosynthesis, it has since been suggested that lipase activity is normal in NLSD, but that transfer of the FA and AG products into phospholipid is impaired. Thus, the accumulating AG is reacylated and cycled back to the cytosolic TAG pool [104, 105]. If true, this suggests that, under normal conditions, FA flux from TAG to phospholipids is very large indeed and that this route may provide a large proportion of total cellular phospholipid. However, phospholipid synthesis *de novo* from newly acquired FA is not impaired in NLSD [106]. Of the many reports of TAG-phospholipid interactions, including those which may be involved in the mobilization of lipids for hepatic synthesis of VLDL [77, 106, 107], of particular interest here is the recently proposed role of phosphatidic acid, a product of the phospholipase D-mediated lipolysis of phosphatidylcholine, in the bulk lipid addition step of VLDL assembly [108]. Phosphatidic acid is, of course, a precursor of TAG and this may provide a metabolic route for the transfer of phospholipid-derived FA into TAG of VLDL [107].

On the above view, NLSD is primarily a disease of defective phospholipid metabolism and the TAG-derived route of phospholipid biosynthesis is essential for the normal functioning of skin, muscle, liver, and the central nervous system. Further studies on the fatty liver dystrophic mouse which bears a close resemblance to human NLSD [103] may identify the defective gene. Defective neutral phospholipid metabolism in this mouse model is reflected by a decreased phospholipid content of peripheral nerve myelin [109]. The nature of the lipase(s) involved in normal TAG-to-phospholipid acyltransfer has not been characterized, but may be similar to a previously described microsomal neutral lipase [110] or to a carboxylesterase [111].

11.2 Components for TAG Storage and Mobilization

11.2.1 TAG in Lipoproteins

Homozygous VLDL receptor knockout mice are resistant to both genetic and diet-induced obesity caused by a decreased peripheral and whole-body uptake of NEFA with no alterations in either food intake or fat absorption [112]. The reduction in adipocyte TAG storage as shown by a decreased average fat cell size in VLDL receptor-deficient rodents implies an impaired FA delivery to adipose tissue in the absence of this lipoprotein receptor. Analogously, hepatic overexpression of human apolipoprotein C-I leads to hyperlipidemia accompanied by decreased visceral fat depots and lack of subcutaneous WAT in mice [113]. Conversely, overexpression of apoA-II, the second most abundant HDL component, increases adiposity and insulin resistance in relation to decreased skeletal muscle glucose utilization [114]. In the capillaries of skeletal muscle and adipose tissue at the luminal face of endothelial cells, LPL catalyses the rate-limiting step in the hydrolysis of TAG from circulating VLDL and chylomicrons. Thus LPL plays an important role in directing fat partitioning. In fact, complete LPL deficiency in mice results in minimal amounts of tissue lipids, leading to neonatal death due to marked hypoglycemia and hypertriglyceridemia [115]. In heterozygotes only mild hypertriglyceridemia with impaired LDL clearance and mild hyperinsulinemia accompanied by an approximately 20% decrease in fasting glucose concentrations were observed. Exclusive LPL deficiency in adipose tissue on a standard genetic background leads to normal growth and body composition. This implies that, although LPL controls FA entry into adipose tissue, fat mass is preserved by endogenous FA synthesis [116]. When the same lack of LPL in adipose tissue is generated on an ob/ob background a diminished weight gain is attained due to an impaired lipid accumulation in adipocytes. Conversely, targeted overexpression of LPL in skeletal muscle or liver has no effect on body weight, but produces an increase in FA uptake into the respective tissue that adversely affects glucose metabolism [117].

11.2.2
TAG in Adipose Cells

Adipocytes, which vary enormously in size (20–200 μm in diameter), are embedded in a connective tissue matrix and are uniquely adapted to store and mobilize energy as TAG and NEFA/glycerol, respectively. Surplus energy is assimilated by fat cells and stored as TAG in LD. To accommodate the TAG, adipocytes can change their diameter up to 20-fold and their volumes by several hundred-fold. Because ~90% of the cell volume is constituted by LD(s), the nucleus and the thin cytoplasm rim are pushed to the periphery of the adipocytes. WAT is actively involved in cell function regulation through a complex network of endocrine, paracrine and autocrine signals that influence the response of many tissues, including hypothalamus, pancreas, liver, skeletal muscle and others. Until recently, adipocytes were seen as passive tissues for the storage of excess energy as TAG [118]. However, they clearly act as endocrine secretory cells [118, 119], releasing in constitutive or regulated fashion several hormones (e.g. leptin), growth factors (TGFβ), cytokines (e.g. TNF-α) and other factors (e.g. ASP) which act as feedback signals for the adipose tissue itself or other tissues. Adipose tissue signaling pathways, arranged in a hierarchical fashion, constitute one of the voices of the body that enable the organism to adapt to a range of metabolic challenges, such as starvation, stress, and infection, as well as periods of energy excess.

11.2.2.1 Enzymes of TAG Synthesis

Accumulating evidence for important functions of enzymes involved in lipid synthesis and its control has been obtained through targeted disruption in rodent models of the corresponding genes (Fig. 11.2). However, direct links between abnormal expression or genetic variants and human disorders, such as obesity, hyperlipidemia, insulin resistance, and NIDDM await further clarification. Glycerol-3-phosphate dehydrogenase (GPDH) is a ubiquitously expressed enzyme that participates in TAG synthesis and in shuttling NADH into mitochondria for oxidative metabolism. Transgenic mice overexpressing GPDH show a normal body weight but with an increased interscapular brown fat depot and virtually no WAT [120]. Conversely, animals lacking mitochondrial GPDH show a decreased body weight compared with wild-type mice with no reported effect on adipose tissue depots but a marked insulin release defect when the malate-aspartate shuttle is blocked. This indicates an important role of the NADH shuttle or glycerol phosphate system in glucose-induced activation of mitochondrial metabolism and insulin secretion [121]. Therefore, the two mouse models with altered expression of GPDH provide evidence for an involvement of this enzyme in lipogenesis and WAT development as well as in the regulation of the glycolytic pathway.

Malonyl-CoA is a key metabolite generated by acetyl-CoA carboxylases, ACC1 and ACC2, which plays a pivotal role in linking FA and carbohydrate metabolism through regulation of FA synthesis and oxidation, respectively, in response to hormonal and dietary influences. ACC1 is highly expressed in the cytosol of adipocytes and hepatocytes (Fig. 11.2), while ACC2 is attached to the outer face of mitochondria and is predominantly expressed in skeletal muscle and heart. The roles of the carboxylases in energy homeostasis in lipogenic and non-lipogenic tissues have deciphered by the generation of ACC2$^{-/-}$ mice [122]. In comparison with wild-type mice controls, mutant mice had a 30% higher fatty oxidation rate and accumulated 50% less lipid in their adipose depots. Mutants, however, had a normal growth rate despite consuming 20 to 30% more food than the control mice [122].

The synthesis of TAG and glycerophospholipids starts with the acylation of glycerol-3-phosphate by glycerol-3-phosphate acyltransferase (GPAT) to form lysophosphatidic acid (Fig. 11.2). Mammals have two isoforms of the enzyme, located in the outer mitochondrial membrane (mtGPAT) and the ER (microsomal GPAT). mtGPAT$^{-/-}$ mice have reportedly reduced weight and fat pad mass accompanied by lower liver and plasma TAG, together with lower VLDL secretion rate and an altered glycerolipid FA composition [123].

The final step in TAG synthesis occurs by the action of acyl-CoA:diacylglycerol acyltransferase (DGAT) with DAG and fatty acyl-CoA as substrates (Fig. 11.2). DGAT is assumed to be a microsomal enzyme. The membrane-bound enzyme has been only partially purified from rat liver microsomes [124]. Cases and coworkers [125] have identified a mouse cDNA encoding DGAT and, in addition, an expressed sequence tag clone that shared regions of similarity with acyl-CoA:cholesterol acyltransferase, an enzyme that also uses fatty acyl-CoA as a sub-

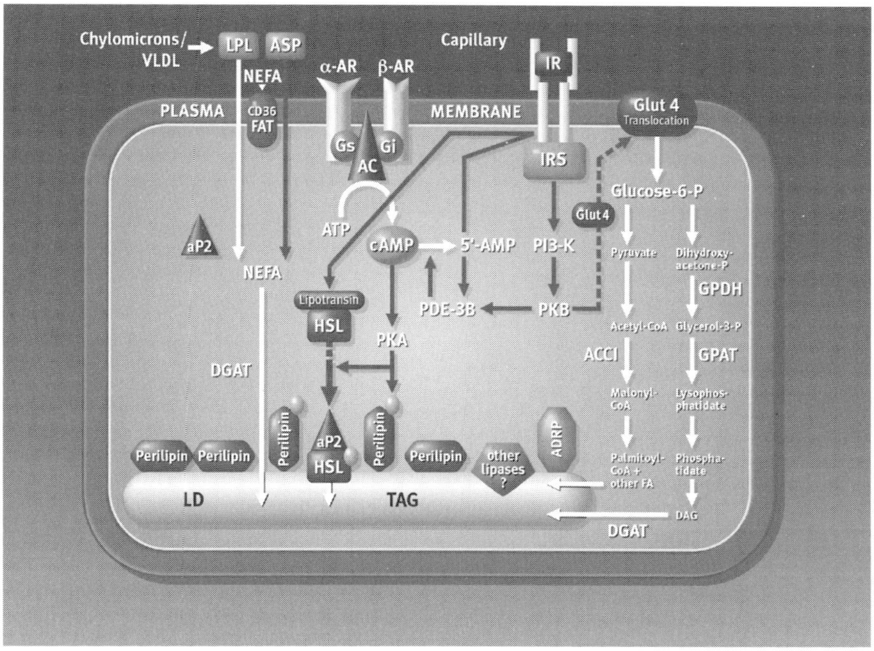

Fig. 11.2 Schematic model for the main signaling and metabolic pathways coupling hormonal stimuli and NEFA to (regulated) TAG synthesis, storage and mobilization in adipocytes. PKA-dependent phosphorylation is depicted as filled circles. See text for details. Adapted from Ref. [119] with modifications.

strate [126]. Expression of this mouse cDNA in insect cells resulted in high levels of DGAT activity in cell membranes whereas the activity of other acyltransferases was unaffected. During differentiation of NIH 3T3-L1 cells into adipocytes, the expression of the gene, now referred to as DGAT2, increased markedly in parallel with an increase in DGAT activity [127]. Smith and coworkers reported that DGAT1-deficient mice exhibit altered TAG metabolism but are still viable and reach a normal body weight on a standard chow diet although with an approximately 50% reduction in fat mass, adipocyte size and leptinemia [127]. In addition, after a glucose load DGAT1-deficient rodents have a decreased plasma glucose response, consistent with an increased insulin sensitivity that correlates with a diminished muscular and hepatic TAG content [117]. Surprisingly, DGAT1 deficiency has no effect on plasma TAG and NEFA, probably due to compensation by DGAT2 or alternative pathways for the conversion of DAG into TAG [121]. DGAT1$^{-/-}$ mice, however, are resistant to diet-induced obesity, which is related to an increased energy expenditure of increased lean body mass or changes in cold-induced thermogenesis [127].

11.2.2.2 Lipid Droplets

11.2.2.2.1 Morphology and Lipid Composition

In different types of mammalian cells, intracellular lipids are stored in discrete LD. White adipocytes have only one or a few large TAG-rich LD (10–100 μm diameter) per cell [128]. Brown adipocytes, which occur in some hibernating animals and the fetuses/neonates of many mammals, contain many smaller storage LD (2–10 μm diameter). Small LD (<1 μm) are also present in various other cells such as liver, muscle, kidney, intestine and mammary gland [129, 130], cultured lines such as 3T3-L1 fibroblasts, CHO cells [131], melanocytes, cultured pulmonary epithelial cells, various animal tissues [132] and macrophages [133]. Other cells, such as cultured HepG2 hepatoma cells [131] and primary lung lipofibroblasts may contain LD up to 2–3 μm in diameter. Cholesterylester-rich LD are contained in steroidogenic cells of adrenal cortex, testes and ovary [134–136].

Among the first reports on the composition of LD in mammalian cells, of the 1960s. Jensen and Kofod [137] described LD from bovine heart with a high content of TAG and a small amount of protein. Membrane-bound LD of rat liver, which consist mainly of neutral lipids [138], and a lipid-rich fraction of pig liver, which mainly contains TAG [139], were isolated and characterized. Angel [140] described liposome particles that were regarded as the metabolically active lipid pool in adipose tissue. Further investigations revealed that the composition of cellular LD varies depending on the cell type. LD from bovine heart have similarities to the fat globule of bovine milk in structure and, to a certain extent, in chemical composition [141]. Under electron microscopy, bovine heart LD appear as irregular bodies 0.5–2 μm in diameter. Occasionally, a few larger particles (5–8 μm in diameter) were detected as distinct structures without indication of fusion. They consist of a core of TAG surrounded by a phospholipid monolayer with some proteins attached to the surface.

Approximately 95% of total lipids of LD from bovine heart are constituted by TAG. The amount of protein present in these LD is about 5% of total mass, and the amount of phospholipids varies from 3 to 7% of total lipid. The major phospholipids forming the monolayer of these LD are phosphatidylcholine (ca. 50%) and phosphatidylethanolamine (30–40%). The NEFA content is very low. The chemical composition of LD in beef heart with their high TAG content and the rather small amount of phospholipids resembles the composition of chylomicrons. A striking difference, however, is the lack of cholesterol and cholesterylesters in LD from beef heart whereas these lipid species occur at approximately 1–2% of total lipid in chylomicrons. Furthermore, the protein content of LD is two to three times higher than of chylomicrons. In contrast to bovine heart LD, those of stellate cells from the rat liver consist of retinylesters, TAG, free cholesterol and a small amount of phospholipids [142]. A general characteristic of LD regardless of the cell type appears to be the content of approximately 5% phospholipids of the total mass (reviewed in Ref. [143]).

However, electron microscopy of conventional resin-embedded ultrathin sections cannot visualize any membranous structure around the LD. In the ultrathin section of specimens fixed by aldehydes and then by osmium tetroxide there appears to be no LD, and its periphery is usually seen as a thin intermittent line. In many diagrams, the LD surface has been depicted as a phospholipid monolayer with the hydrophilic head-group facing the cytoplasm and the hydrophobic acyl chains extending into the LD content [144, 145]. It has also been assumed that LD form by accumulation of TAG between the two leaflets of the ER membrane because hydrophobic TAG molecules may exist stably in the aqueous cytoplasm only when covered with amphiphilic molecules with the hydrophilic moiety facing outward. However, experimental evidence to support the above assumptions is scarce. Only freeze-fracture electron microscopy showed that the fracture plane along the LD surface is occasionally continuous with the cytoplasmic leaflet of the ER membrane [146, 147]. Because it is difficult to retain lipid-rich structures by conventional morphological methods, cryoelectron microscopy is of advantage since it observes biological specimens at atomic resolution without fixation or staining. The microscopy showed that the LD surface is indeed a hemi-membrane, or a phospholipid monolayer [147].

A phospholipid monolayer in the surface is consistent with the current model that LD are formed by TAG deposition between the two leaflets of the ER membrane and may remain connected to it [144, 145; see below]. Distribution of acyl-CoA:cholesterol acyltransferase-1, a major enzyme that synthesizes cholesterylester, in the entire ER [148] seems to indicate that LD may bud anywhere along the membrane. However, Cap-LC/ESI mass spectrometry showed that FA moieties of phosphatidylcholine and lyso-phosphatidylcholine in LD are distinct from those in the rough ER [149]. The results do rule out the generation of the LD surface generated from the ER membrane but indicate that the former is a highly differentiated domain. Mature LD might be independent of the ER. Alternatively, the LD may be connected to the ER, but some molecular mechanism may demarcate the LD surface from the bulk ER membrane as postulated for other ER domains [150]. Whatever is true, TAG synthesized in wide areas of the ER do not deposit indiscriminately but are concentrated to loci specialized to make LD. ADRP or other LD-associated proteins may be involved (see below).

Cap-LC/ESI mass spectrometry has also shown that the LD surface differs from the detergent-insoluble glycolipid- and sphingolipid/cholesterol-rich microdomains (DIGs) in FA composition [149]. Interestingly, DIGs contain a higher proportion of long saturated FA than rough ER, whereas LD do not. This indicates that not only interaction between sphingolipids and cholesterol but also interaction between saturated phospholipids and cholesterol are important in forming DIGs, as predicted from model membrane studies [151]. The discovery of caveolin sequestration (see below) led to speculation that the LD surface might be similar to DIGs [152]. However, the present results do not support this in terms of the FA composition. Furthermore, thin-layer chromatography showed that the relative amount of sphingomyelin and free cholesterol in comparison with phosphatidylcholine is far less in LD than in DIGs. In addition to the ER, the LD of mammalian cells seem to be closely associated with other organelles. Christiansen and Jensen [141] described a close association of LD of beef heart with mitochondria, suggesting a role of the LD as a source for FA utilized for energy production by mitochondrial β-oxidation. Franke and coworkers [153] detected cytoskeletal intermediate filament proteins, particularly vimentin, forming a network in the vicinity of LD. The interaction of intermediate filaments and LD was confirmed by transmission electron microscopy [154] and field-emission scanning electron microscopy [155].

11.2.2.2.2 Protein Composition

Perilipins Little is known about proteins on the surface of mammalian intracellular LD. Two groups of proteins whose association with LD in animal cells is well documented are the perilipins and adipocyte differentiation related protein (ADRP) (reviewed in Ref. [156]). The human ortholog of ADRP has been termed adipophilon [157]. Other proteins at or near the surface of the LD of adipocytes include the "capsular" proteins [158–160] and vimentin [153, 161]. Furthermore, certain enzymes, e.g. eicosanoid-forming enzymes, associate with LD in various types of cells [162–164].

Perilipins belong to a family of polypeptides that are expressed in adipocytes [165–168] and steroidogenic cells of adrenal cortex, testes and ovary [166]. Greenberg and coworkers [165] isolated two related classes of full-length cDNAs that encode perilipins A and B from rat adipocyte cDNA expression library. Perilipin A and B, 56 kDa (517 aa) and 46 kDa, respectively, share a common amino-terminal sequence of 406 aa. Similar to adipocytes, steroidogenic cells express perilipin A, but these cells also contain relatively high amounts of perilipin C, a protein of ca. 42 kDa that is not detectable in adipocytes [167]. Perilipins are encoded by a single copy gene that gives rise to the three protein isoforms by alternative mRNA splicing with perilipin A being the most abundant. In murine adipocytes a forth perilipin, named perilipin D, occurs at the mRNA level. The perilipin D polypeptide, however, has not yet been reported [169]. Perilipins exhibit sequence similarity to a few other proteins, one of which is ADRP. The amino-termini of perilipin and ADRP are highly homologous at a region of ca. 100 aa. ADRP was first identified during a search for genes being expressed in an early stage of differentiation of murine adipocytes in culture [170]. The mRNA of ADRP is present in various tissues and cultured cell lines. Immunocytochemical examination revealed that ADRP localizes to LD in murine 3T3-L1 adipocytes, murine MA-10 Leydig cells, CHO fibroblasts and human HepG2 hepatoma cells [167]. In MA-10 Leydig cells ADRP and perilipins co-localize to LD. In 3T3-L1 adipocytes the two proteins were found on small LD although at different stages of differentiation [163]. ADRP is present on these LD in early differentiated adipocytes and is then replaced by perilipins, which can only be found on the surface of the LD at a later growth stage. It was suggested that ADRP is involved in the management of TAG stores in non-adipose tissues. Interestingly, recently, ADRP was found to co-localize with insulin and at the LD surface in human islet cells [171]. Incubation with glucose or FA alone and a combination of glucose and FA increased ADRP expression by 2- and 10-fold, respectively. Oleate induced ADRP expression more potently than palmitate, which correlated with higher levels of neutral lipid accumulation by islets. Consistent with increased ADRP expression the initial uptake rate of FA in islet cells was elevated in islet cells treated with FA *plus* glucose compared with control cells. In addition, ADRP expression is considerably increased in ob/ob mice islets and its upregulation is tightly correlated with TAG accumulation in islets, demonstrating an inverse relationship with impairment of β-cell function [171]. Thus, ADRP may play a role in regulating TAG metabolism also in β-cells and, in particular, be involved in the loss of β-cell function during the pathogenesis of NIDDM.

A cDNA of another protein, named TIP47 or pp17, which has similarity to ADRP has also been described [157, 172]. TIP47/pp17 is a cargo protein involved in the trafficking of the mannose-6-phosphate receptor between endosomes and the Golgi network. Perilipin A, ADRP and TIP47/pp17 exhibit strong sequence homology at their amino-terminal regions. Starting from the amino terminus, perilipin A contains a sequence of 105 aa similar to adipophilin and TIP47 (aa 17–121), five 10-aa domains with amphipathic β-pleated sheet character (between aa 11 and 182), three sequences of moderate hydrophobicity (aa 243–260, 320–342, 349–364), a highly acidic region (aa 291–318), and six consensus sites for PKA phosphorylation. Two such hydrophobic regions and three areas of amphipathic β-pleated sheet structure were detected in ADRP [157]. A protein named S3-12 is also highly homologous to ADRP in a 33-aa repeated sequence [173]. The mRNAs of perilipins, ADRP and S3-12 are expressed much higher in adipose than in other tissues. It was suggested that lipids regulate transcription of perilipin mRNA, because the amount of perilipin mRNA and the corresponding protein increase in the presence of exogenous lipids. In analogy, formation of the mRNA of ADRP and protein expression in pre-adipocytes was also shown to be stimulated by FA in a time- and dose-dependent way [174]. Stimulation could be only achieved by long-chain FA whereas the degree of saturation had no influence. Treatment of such stimulated cells with actinomycin D inhibited the expression of ADRP, suggesting that FA stimulate the gene expression at the transcriptional level. Recently, Henry and coworkers studied the protein expression of perilipin and ADRP in human muscle in greater detail [175]. Perilipin was detectable in biopsies of vastus lateralis muscle (VLM) only at levels consistent with adipocyte infiltration. ADRP was highly expressed in VLM, irrespective of whether derived from obese non-diabetic or diabetic subjects. Weight loss of the latter led to increased whole-body insulin sensitivity and ADRP content in VLM. Treatment of the obese non-diabetic subjects with either troglitazone or metformin in combination with glibenclamide considerably improved the glycemic control, as reflected in lowering of HbA1c and plasma fasting TAG levels, which was well correlated to post-treatment increases in ADRP expression in VLM [175]. These results demonstrate that ADRP is the predominant LD-associated protein in VLM and that its expression is up-regulated under conditions of improved glucose tolerance. It is tempting to speculate that increased expression of ADRP may act to sequester FA in TAG of LD contained in skeletal muscle and thereby protects the muscle from the inhibitory ("lipotoxic", see above) effects of FA (or derivatives thereof) on insulin signaling and action. In consequence, this study provided additional evidence that TAG in LD represents the "benign" storage form of FA and accumulates during overfeeding of myocytes and β-cells with FA, which is compatible rather than interferes with insulin action and secretion, at least up to a certain extent of LD deposition.

Perilipins remain tightly bound to LD through alkaline carbonate or urea wash conditions, and require detergents to remove them. Stable expression of perlipin A in 3T3-L1 fibroblasts and subsequent immunofluorescence microscopy and immunoblotting revealed that neither the amino- nor the carboxyl-terminus is required to target perilipin A to LD [176]. Full-length perilipin A is associated with LD via hydrophobic interactions, as shown by the persistence of perilipins on LD after centrifugation through alkaline carbonate solution. Mutagenesis studies indicate that the signals responsible for tight anchoring of perilipin A to LD are, most probably, multiple, partially redundant targeting sequences of moderately hydrophobic amino acids located within the 25% of the central domain. These signals seem to be, at least in part, constituted by the three regions of moderate hydrophobic character and the five domains with characteristics of amphipathic β-pleated sheets [176]. Cytochemical analysis by immunogold labeling suggests that the perlipins are located on or within the limiting surface of the LD [146].

Adipophilin Adipophilin (ADPH) is a prominent LD-associated protein found in many mammalian cell types, including hepatocytes, adipocytes, muscle cells, and mammary epithelial cells, either during development [131] or in the mature functioning cell [131, 132, 157]. ADPH has proposed functions in LD formation [177], FA transport [177], and milk lipid secretion [178, 179]. Although it can be detected at the plasma membrane in mammary epithelial cells as a result of milk globule secretion [179], the weight of evidence suggests that ADPH specifically targets LD. However, little is known about the structural and biochemical properties involved in its association with LD or how it accomplishes its biological functions. Genetic analysis [180, 181] has demonstrated that ADPH is related to the family of LD-associated proteins that include perilipin (see above) and the mannose-6-phosphate receptor targeting protein TIP47, through a conserved region in their amino-termini, called the "PAT" domain (*p*erilipin, *A*DPH, *T*IP47). The lack of significant sequence conservation outside this region indicated that it is an important determinant in the association of these proteins with LD [180]. However, a recent study revealed that neither the PAT domain nor the carboxy-terminal half of ADPH is required for efficient targeting to or biogenesis of LD [182], but that, nevertheless, specific regions located elsewhere within ADPH are responsible for LD targeting. Previous work indicates that secretion of LD from mammary epithelial cells during lactation involves the formation of a tripartite complex between ADPH, the mammary-specific transmembrane protein, butyrophilin, and xanthine oxidoreductase [179]. As formation of such a complex probably involves regions of ADPH distinct from those involved in LD targeting, the PAT domain or the carboxy-terminal half of ADPH may represent candidate regions for interaction with butyrophilin and/or xanthine oxidoreductase to trigger LD secretion.

Caveolins In primary and cultured adipose cells, caveolin-1 and caveolin-2 constitute a framework of caveolae that represent bulb-shaped plasma membrane invaginations (50–100 nm in size) enriched in cholesterol, (glyco)sphingolipids (sphingomyelin) as well as glycolipids (gangliosides, GPI lipids) and harboring a special set of (dually acylated) signaling proteins, glycosyl-phosphatidylinositol-anchored (GPI) proteins and the caveolae structural and marker proteins, caveolin 1–3 [183–186]. Based purely on ultrastructural comparisons and tissue expression profiles, the adipocyte seems to have the highest concentrations of caveolae and the highest levels of caveolin-1 and -2, i.e. more than any other cell type [187]. Indeed, electron micrographs of adipocytes dating back to 1963 show that caveolae account for up to 30% of the surface area of the adipocyte plasma membrane [188, 189]. Furthermore, in 3T3-L1 adipocytes, a widely used model system for studying adipogenesis, the number of caveolae increases 9-fold, and caveolin-1 and -2 expression increases 20-fold during differentiation from fibroblasts to the adipocyte state [187, 189].

Caveolins are supposed to anchor to the plasma, Golgi and ER membranes by the central hydrophobic domain with both amino- and carboxy-termini exposed to the cytoplasm [190]. Caveolae fulfil important functions in protein and lipid/cholesterol transport (endocytosis, potocytosis, transcytosis, polarized transport) as well as in signal transduction [185, 186, 190–192]. Fujimoto and coworkers showed that caveolin-2, especially its β-isoform, is targeted to the surface of LD, by immunofluorescence and immunoelectron microscopy, and by subcellular fractionation [193]. Brefeldin A-treatment induced further accumulation of caveolin-2 along with caveolin-1 in LD. Analysis of mouse caveolin-2 deletion mutants revealed that the central hydrophobic domain (residues 87–119) and the amino-terminal (residues 70–86) and carboxy-terminal (residues 120–150) hydrophilic domains are all necessary for the localization in LD. The amino- and carboxy-terminal domains appeared to be related to membrane binding to and exit from ER, respectively. This implies that caveolin-2, which is synthesized on cytosolic ribosomes and then incorporated into the ER membrane in post-translational fashion, is transported to LD as a membrane protein [193]. Thus caveolin-2 seems to differ from the hitherto reported LD proteins, such as the perilipins, in that it exists as an integral membrane protein in the ER and plasma membrane. As anti-caveolin-2 antibody, recognizing the N-terminal segment [194, 195], decorated only the P face of LD in a freeze-fracture replica, caveolin-2 may take the same orientation on the LD surface as in the membrane [193]. This is consistent with the hypothesis that the LD surface is a half-membrane, or a phospholipid monolayer, with amphiphilic lipid molecules with their hydrophilic and hydrophobic portions oriented toward the cytoplasm and the LD content, respectively [143]. No labeling for caveolin-2 was seen in the LD surface, but was found in small clusters both in cryosections and in freeze-fracture replicas. The labeling for adipophilin shows the same pattern at the LD surface. These results suggest that the whole LD surface may not be homogenous and that some segregation may occur in the half-membrane [193]. Interestingly, a recent study revealed unesterified cholesterol in LD [196]. Furthermore, LD of eosinophils contain proteins typical for DIGs related to intracellular signaling, such as MAPK and Lyn [163, 197]. Together, these results suggest, intriguingly, that LD might be a novel membrane domain where caveolins may functionally regulate signaling proteins [184, 185, 191].

The recruitment of caveolin-1 to LD in BFA-treated cells [198] may also be due to the hydrophobic domain shared by all caveolins. However, caveolin-1 does not appear to have the same degree of affinity to LD as caveolin-2. First, in cells expressing caveolin-1 alone, its redistribution to LD after BFA-treatment occurred slowly and only in some cells. Second, without the BFA-treatment, caveolin-1 was not found in LD even when overexpressed. The LD localization signal of caveolin-1 is likely to be less effective than that of caveolin-2. Endogenous caveolin-2 was not detected in LD under normal culture conditions. This may be because caveolin-2α, expressed predominantly in most cells, is less efficient than caveolin-2β in LD targeting. But the accumulation of caveolin-2 in LD is transient, so that its distribution in LD can only be detected when the protein is expressed abundantly or when its trafficking is perturbed. The apparent morphological continuity of ER and LD ([146], see above) and the persistence of LD labeling in BFA-treated cells indicate that caveolin-2 goes directly from ER to LD. Whether LD exists independently from the ER-Golgi pathway, or LD is an intermediate compartment between ER and Golgi, is not known. The results of BFA experiments may simply indicate that an excess of caveolin-2 in ER overflowed to LD. However, BFA might also block other pathways (e.g. from LD to the Golgi) and have caused LD retention. The two pathways are not mutually exclusive and need further studies for definition. Trafficking of caveolins to LD is important because it may be linked to that of TAG molecules. A testable question is whether the TAG composition of LD is changed by the presence of caveolins and/or by BFA-treatment. Also, if and how is LD related to the reported trafficking routes of caveolin-1 [199, 200]? Interestingly, overexpression of a mutant caveolin-3 in BHK cells leads to its association with the surface of large LD containing TAG and endogenous caveolins 1–3 [201]. Fluorescence, electron, and video microscopy observations are consistent with formation of the LD by maturation of subdomains of the ER. The mutant caveolin caused the intracellular accumulation of free cholesterol in late endosomes, a decrease in surface cholesterol and a decrease in cholesterol efflux and synthesis. Incubation of the cells with oleic acid induced a significant accumulation of wild-type caveolins in the enlarged LD [201]. Taken together, the available data hint at a role of caveolins in LD biogenesis.

11.2.2.2.3 **Biogenesis**

Role of Caveolins LD biogenesis is incompletely understood. The observations presented above suggest that although the caveolins are not normally associated with internal lipid compartments and particles, such as LD, they can certainly traffic to these locations under certain conditions. As such the caveolins represent the first known integral membrane protein components of LD. This characteristic may also explain the initially surprising finding that caveolin-1 can behave as a soluble protein. Immunogold labeling first detected soluble caveolin-1 in the lumen of the ER after cells were exposed to cholesterol oxidase for conversion of cholesterol at the outer leaflet of the plasma membrane into cholesterone [202]. However, it remained a mystery how caveolin-1 as an integral membrane protein firmly embedded in the cytoplasmic leaflet of the (ER) membrane can move to the luminal leaflet for subsequent release into the lumen. Subsequently, a small pool of soluble caveolin-1 was found in the cytosol of fibroblasts in a complex with chaperones [200, 203]. Finally, the analysis of caveolin-1 distribution in different cell types revealed targeting of caveolin-1 to the cytosol (skeletal muscle cells and keratinocytes), and to the lumen of secretory vesicles (serous cells of the pancreas, fundic stomach, and salivary gland) [204, 205]. Both the secreted and the cytosolic caveolin-1 exhibited characteristics for incorporation in lipoprotein-like particles, which may explain their apparent solubility. Several groups provided additional evidence that the cytosol of many cells contains a pool of apparently "soluble" and "lipoprotein-like" caveolin-1. This caveolin-1 is associated with cholesterol and behaves like a protein that is embedded in a particle with the size [206] and buoyant density [204] of HDL. Interestingly, caveolin-1 is secreted by exocrine secretory cells in HDL-like particles that contain apoA-1, raising the possibility that caveolin-rich lipid particles in the cytoplasm are involved in the assembly of secreted lipoproteins.

Based on the structure of caveolins, with their long stretch of hydrophobic amino acids acting as membrane-anchoring domain, it is unlikely that they can switch the "membraneous" into a true "soluble" conformation compatible with the aqueous milieu of the cytosol. Rather the involvement of caveolins in the biogenesis of various lipid particles by budding from either the cytoplasmic or the luminal leaflets of the ER membrane may explain their cell type-specific localization either in the cytosol (as LD or as precursor particles for milk lipid granules destined for budding from the plasma membrane) or in the lumen of the ER and secretory vesicles (as nascent lipoproteins destined for movement along and release by the secretory pathway), respectively. This unifying concept for the biogenesis of apparently "soluble" and "membraneous" caveolin-containing structures requires a "flip-flop" mechanism for translocation of caveolins from the cytoplasmic ER membrane leaflet (with amino- and carboxy-termini exposed to the cytoplasm) to the luminal ER membrane leaflet (both termini exposed to the lumen), which may be induced during cholesterol depletion (e.g. by cholesterol oxidase). In conclusion, caveolin-1 can be integrated in (one leaflet of) various cellular membranes (plasma membrane, Golgi, ER, caveolae) or in (the phospholipid monolayer of) various cytosolic and secreted lipoprotein-like particles (LD, milk lipid granules, lipoproteins) and in each case may regulate the accessibility of these structures towards degradation by (phospho)lipases in a way similar to that now commonly accepted for perilipins.

Taken together, the prevailing model for LD biogenesis rationalizes that caveolins (but not most other membrane proteins) might be able to enter the LD (Fig. 11.3). Since LD are surrounded by a phospholipid monolayer and are thought to be derived from the ER, where TAG and cholesterylesters are synthesized (see above), it is tempting to speculate that TAG accumulate in the hydrophobic core of the bilayer, forming a bulge that eventually buds from the ER membrane to form free LD (Fig. 11.3 A). Accumulation of TAG in the bilayer core would initially force opposite leaflets of the bilayer apart, increasing bilayer thick-

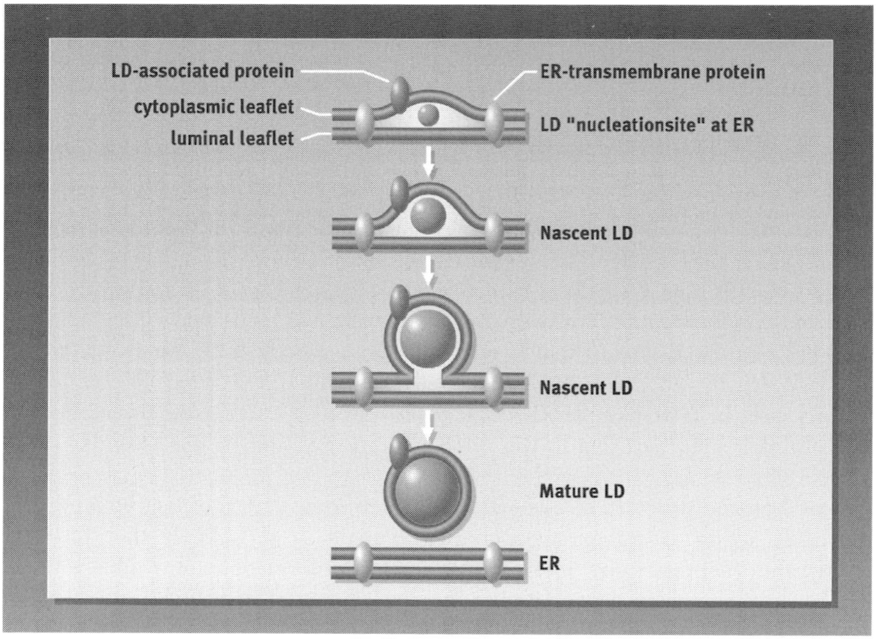
LD-associated protein
cytoplasmic leaflet
luminal leaflet
ER-transmembrane protein
LD "nucleationsite" at ER
Nascent LD
Nascent LD
Mature LD
ER

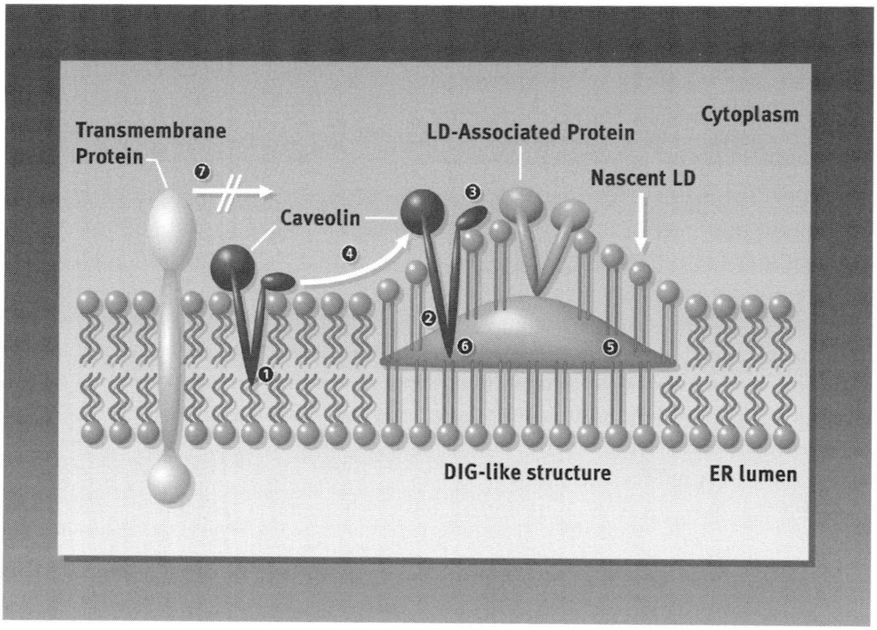
Transmembrane Protein
LD-Associated Protein
Cytoplasm
Nascent LD
Caveolin
DIG-like structure
ER lumen

ness. Such a thickened bilayer could not accommodate transmembrane proteins that have hydrophilic domains on both sides of the membrane and these proteins would be excluded from the forming LD (Fig. 11.3 B). Caveolins, by contrast, lack luminal hydrophilic domains, and could diffuse freely between the ER membrane and the monolayer surrounding the nascent LD. Thus caveolins would not need to dissociate from the ER membrane and expose their hydrophobic domains to the cytosol to enter the LD (Fig. 11.4, steps 1 and 2). This model suggests that caveolins can enter LD only while they are forming and are still in contact with the ER. Consequently, accumulation of caveolins in the ER would trigger their targeting to LD and concomitantly LD biogenesis [145, 193]. Alternatively, transport of caveolins from the plasma membrane and/or Golgi apparatus to the ER via a retrograde pathway might somehow lead to LD targeting. If caveolins normally pass through LD during biosynthetic transport or intracellular cycling, they must exit the LD rapidly as they are not normally (i.e. absence of BFA, wild-type caveolin) detected there.

Although it is unclear how ER accumulation of caveolins might lead to an apparent concentration of the proteins in LD, several possible explanations can be imagined. First, caveolins might have a high affinity for LD, either through specific binding to a protein or (phospho)lipid of the LD, or through preference of caveolin for the physical state of the LD phospholipid monolayer or TAG core leading to its spontaneous partitioning. Efficient packaging of caveolins into transport vesicles might normally prevent entry into the LD. Accumulation in the ER might saturate some component of the transport machinery, possibly a specific cargo receptor, giving the protein time to access the LD. Alternatively, accumulation of caveolins in the ER might change the properties of the ER membrane. The altered properties might increase the affinity of caveolins for LD, either by changing the physical properties of the membrane or by inducing a conformational change in caveolins.

Role of Perilipins In LD, accumulation of perilipins and TAG occurs roughly in parallel. When FA synthesis was inhibited, the concentration of TAG decreased by 90% and correspondingly the amount of perilipin dropped [153]. The availability of only a small amount of TAG may cause formation of small LD with insufficient surface space for the association of perilipin. Decreased synthesis of FA resulting in libera-

Fig. 11.3 "Budding" model for the biogenesis of intracellular LD. (A) LD formation is initiated by accumulation of TAG between the leaflets of the ER membrane. During budding of the forming LD, typical transmembrane proteins with hydrophilic domains protruding into the ER lumen and cytoplasm are segregated from membrane proteins, anchored in the outer leaflet, only, such as caveolin. After detachment of the matured LD from the ER membrane, the TAG core of the LD is surrounded by a phospholipid monolayer with embedded caveolin having both its amino- and carboxy-terminus exposed to the cytoplasm. (B) The topology and segregation of transmembrane proteins and caveolin are shown in greater detail. See text for details. Adapted from Refs. [145 and 152] with modifications.

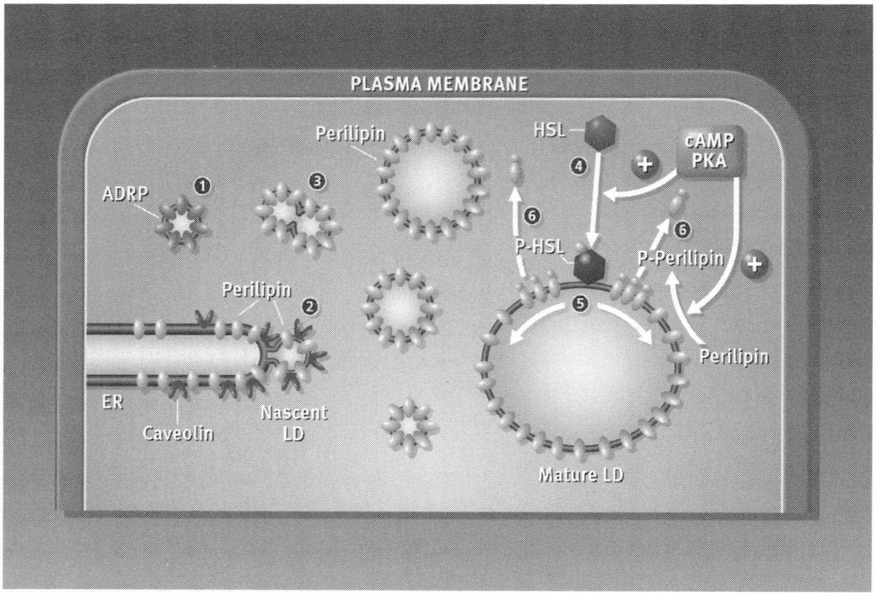

Fig. 11.4 Hypothetical model for integration of the major mechanisms of TAG storage and mobilization: (1) conversion of fibroblasts/preadipocytes into adipocytes, (2) biogenesis of forming LD at the ER membrane, (3) fusion of smaller LD to mature LD, (4) translocation of HSL from the cytosol to the surface of LD, (5) disassembly and reorganization at the LD surface of perilipin and (6) its translocation from LD to the cytosol. Both processes may contribute to relieve the barrier that prevents HSL from gaining access to the LD. PKA-dependent phosphorylation is depicted as filled circles. See text for details. Adapted from Ref. [144] with modifications.

tion of FA from TAG may lead to subsequent detaching of perilipin from the surface of the shrinking LD and subsequent degradation of the protein. Londos and co-workers [157] showed that undifferentiated 3T3-L1 fibroblasts transfected with constructs that encode perilipin A accumulate numerous small LD that are associated with perilipin A. These results indicate that appropriate targeting of perilipins is not restricted to adipocytes or steroidogenic cells, and further suggests that perilipins may serve as nucleation sites for TAG deposition (Fig. 11.2). As nearly all cells can synthesize TAG the formation of LD in the transfected fibroblasts may reflect trapping of TAG in perilipin-bound structures. This is compatible with the observed consequences of introducing perilipin A into other cells that do not normally express this protein. It remains to be demonstrated, however, whether perilipins are directed toward LD, or *vice versa* LD are transported to perilipin aggregates, or association of both components occurs in parallel (Fig. 11.4).

Growth Adipocyte differentiation in WAT involves the coalescence of small nascent LD to form the one or more large LD found in mature cells. By contrast, most other cell types contain numerous smaller LD (see above). Franke and coworkers [153] have proposed that, in adipocytes, cytoskeletal intermediate-filament proteins, particularly vimentin, form cages around small LD and thereby prevent the latter from fusing (Fig. 11.4, step 3). Indeed, a vimentin network is seen near LD in newly differentiating adipocytes. As the cells mature, a decrease in the number of vimentin fibers coincides with, and might play a role in, LD coalescence [153]. In fact, in 3T3-L1 fibroblasts, expression of a dominant-negative vimentin reduces the formation of LD [207]. However, the involvement of vimentin in LD biogenesis came under debate when knockout mice lacking vimentin were shown to exhibit normal biosynthesis of adipocyte LD, even though no compensatory formation of other intermediate-filament proteins could be demonstrated [208]. Nevertheless, coalescence of adipocyte LD must be highly regulated – given the thousand-fold difference in the volume of LD in white and brown adipose tissue – but the mechanism of adipocyte LD maturation remains to be elucidated.

Cell-free System Very recently, a microsome-based, cell-free system containing substrates for TAG synthesis and a phospholipase D-activating protein was introduced that assembles newly formed TAG into spherical LD [209]. These LD were recovered in the low-density fraction by gradient ultracentrifugation and were similar in size and appearance to those isolated from rat adipocytes and 3T3-L1 fibroblasts. Caveolin-1 and -2, vimentin, ADRP, and the 78 kDa glucose-regulated protein, GRP-78, were associated with LD from this cell-free system. The caveolin was soluble in 1% TX-100, as was the caveolin on LD from 3T3-L1 cells [209]. These data suggest that caveolin is released from the microsomes along with LD, indicating that caveolin may help to sort newly formed TAG from the ER into a cytosolic form in a budding process of LD from the ER membrane. The ADRP on the LD was also derived from the microsome-associated pool, suggesting that it can be targeted to the ER membrane. ADRP can be acylated [150, 210], and these acyl moieties may participate in the interaction between ADRP and the membrane, since covalently linked FA are involved in the membrane targeting of other proteins (e.g. ADP ribosylation factor-1) [211, 212]. Thus, ADRP binds to the microsomal membrane and participates in the budding of the small LD. Perilipins were not expressed in the 3T3-L1 fibroblasts used to prepare the microsomes and consequently were not present in the partially purified cytosolic phospholipase D activator preparation. Therefore perilipins were not present on the LD assembled in the cell-free system [209]. Consequently, small LD can be assembled in the absence of perilipins. GRP-78 is a chaperon protein that contains a KDEL sequence and is therefore restricted to the ER/Golgi region of the secretory pathway. GRP-78 appeared to be present on the cytosolic side of the surface layer of LD [213], suggesting that the surface layer is derived from the membrane of the ER. This agrees with the model proposed for the assembly of LD (see above), according to which the TAG will, during the biosynthesis, "oil-out" between the leaflets of the ER bilayers (Fig. 11.3). The LD formed by complete budding from the ER would be covered by a phospholipid monolayer and proteins originally bound to both the luminal and cytosolic surface of the ER membrane.

The LD from the cell-free system, like those from 3T3-L1 cells, contained TAG, DAG, phosphatidylcholine, phosphatidylethanolamine, and phosphatidylserine

[209]. The assembly of these LD-like structures depended on the rate of TAG biosynthesis and required an activator present in the high-speed supernatant from homogenized rat adipocytes. The activator-induced phospholipase D activity, and its effect on the release of the LD from the microsomes, was inhibited by butan-1-ol or 2,3-diphosphoglycerate. The activator could be replaced by a constitutively active phospholipase D or its product, phosphatidic acid. In conclusion, LD seem to be formed from regions of the microsomal membrane that contain caveolin, vimentin, and ADRP in a process that depends on DGAT activity and is driven by the activity of a phospholipase D and the formation of phosphatidic acid. Phosphatidic acid acts as an intracellular messenger or as cone-shaped lipid that alters the curvature of the membrane [214, 215 a]. Its formation seems to be important for the fission of transport vesicles [215 b, 216]. Recent data [209] suggest that it plays a role in (regulation of) the budding step within the biogenesis of LD, as has previously been proposed for the assembly of VLDL in the ER lumen [217, 218].

Targeting Mechanism Three studies have addressed the question of how LD-associated proteins accumulate at the monolayer surface of LD during their biogenesis. Pol and coworkers [201] demonstrated that amino-terminal truncation mutants of caveolin-1, -2, -3, among them Cav-3DGV, are targeted to LD (and the early Golgi). Fujimoto and coworkers [193] found overexpressed caveolin-2 to be associated with LD (and the Golgi). Brown and coworkers [198] showed that caveolin-1 with an ER retention signal fused to its carboxy terminus, such as Cav-1KKSL, as well as an internal-deletion mutant of caveolin-1, are highly concentrated in LD (and the ER and Golgi, respectively). Furthermore, the same group reported that blockade of the anterograde transport along the secretory pathway by brefeldin A leads to increased expression of endogenous caveolin at the surface of LD [198]. Subsequently, Brown suggested [145] that (transient) accumulation of caveolins at the ER as a result of overexpression or impaired release into the secretory pathway during a folding defect caused by the mutation or by brefeldin A-treatment may be sufficient for their efficient partitioning from the cytoplasmic membrane leaflet of the ER to the phospholipid monolayer of LD. According to this view, the association of caveolins with LD under physiological conditions, rather than the unusual experimental conditions of inefficient transport, has still to be demonstrated.

The LD-associated proteins described so far can be classified on basis of their biosynthesis and topology: ADRP, perilipins and HSL are translated on free ribosomes [157] and recruited to the LD from the cytosol as a type of peripheral membrane protein, the latter two shuttling between the LD and the cytosol in reversible fashion in response to the nutritional state of the adipocyte (see below). Caveolins are co-translationally inserted into the ER in a signal-recognition-particle-dependent fashion [195], with their central hydrophobic domain penetrating into the ER cytoplasmic membrane leaflet and the two flanking hydrophilic domains extending into the cytoplasm (Fig. 11.3 B, step 1). Maintaining this topology the caveolins would diffuse laterally from the bulk ER to the surface phospholipid monolayer of nascent LD still attached to the ER without exposing their hydrophobic domains to the cytosol, whereas transmembrane proteins would be excluded from the LD (Fig. 11.3 B, step 4). However, the apparent concentration of proteins at LD requires additional specific targeting information and deciphering mechanisms [128, 163], albeit a consensus sequence for LD targeting has not been identified. The proteins could interact directly with phospholipids located at the LD surface (Fig. 11.3 B, step 2) or with TAG deposited in the LD core (step 6) or with other LD proteins (step 3). Since LD probably arise (also) *de novo* (in contrast to the biogenesis of membranes) in the absence of pre-formed structures which could function as nucleation or docking sites for the incorporation of newly synthesized LD proteins, some LD proteins have to come into direct contact with phospholipids/TAG of the LD and discriminate them from those of the ER. Caveolins represent candidates for LD proteins triggering the initial biogenetic steps of LD assembly on basis of their high affinity for DIGs, which are characterized by tight packing and a high degree of order of their (saturated) lipid acyl chains (see above). The apparent affinity of caveolins for LD may well rely on their affinity for DIGs, which is supported by the finding that the LD-targeted Cav-3DGV mutant resists solubilization by cold non-ionic detergent, as is true for typical DIG-associated proteins [201]. Consequently, a physical state of the LD surface monolayer distinct from the ER cytoplasmic leaflet and more "DIG-like" may attract typical LD proteins, such as caveolins (Fig. 11.3 B, step 7). However, putative targeting information encoded by the "DIG milieu" of nascent LD is not compatible with a recent more detailed analysis of the phospholipids contained in LD and DIGs, which did not reveal close similarity with regard to length and saturation of the constituent FA (see above). Alternatively, contact of the TAG molecules in the LD core with the phospholipids of the LD surface (Fig. 11.3 B, step 5) might trigger a tight packing of the latter, creating a special lipidic environment that causes the selective recruitment of certain phospholipids and proteins to nascent LD.

11.3 Mechanism and Regulation of TAG Mobilization

11.3.1 cAMP

TAG mobilization by lipolysis as a strictly regulated process has been known since the early 1960s, when it was established that fast-acting hormones such as ACTH and epinephrine increased lipolysis [219, 220], and that insulin counteracted this activation [221–223]. It was soon recognized that cAMP was involved in the regulation of the catecholamine-sensitive lipolytic activity in adipose tissue [224]. Catecholamines can either stimulate lipolysis via three subtypes of β-AR, which are positively coupled to cAMP-synthesizing AC by G_s proteins, or inhibit lipolysis via α_2-AR, negatively coupled to the AC by G_i proteins (Fig. 11.2). Concerning the functional significance of adrenoceptor-mediated increases in cAMP levels, two points should be emphasized. First, lipolytic agents generally increase the levels of cAMP far above the concentration required for maximal activation of PKA [225, 226]. Second, considerable species and tissue specificity exists with regard to the distribution of adrenoceptor subtypes. In humans, for instance, the interplay between α_2- and β-AR is important in modulating cAMP levels in adipocytes. It is now generally accepted that lipolytic response of fat cells is controlled mainly by the sympathetic nervous system and by plasma insulin levels and depends on the balance between stimulatory and inhibitory pathways. cAMP activates PKA and this kinase then phosphorylates the two main targets involved in the control of lipolysis in the adipocyte, HSL and the perilipins (Fig. 11.2). Phosphorylation of these proteins dramatically increases lipolysis (see below). Insulin is the physiologically most important antilipolytic hormone. The ability of insulin to antagonize hormone-induced lipolysis can largely be explained by its ability to lower cAMP levels and thereby PKA activity [227], leading to a decrease in the phosphorylation state of HSL and perilipin.

11.3.2
Phosphorylation of HSL

Phosphorylation of partially purified HSL by PKA, leading to a moderate activation of its activity *in vitro*, was described in the early 1970s [228, 229] and later confirmed [225, 230, 231]. Subsequent phosphopeptide mapping and phosphoamino acid analysis suggested that HSL was phosphorylated on a single serine residue, named the regulatory site [232]. Partial amino acid sequencing of phosphopeptides generated from bovine HSL [233], together with determination of the primary structure of rat HSL [234], allowed identification of Ser-563 as the regulatory site [235]. However, more recent data have dramatically challenged this view on the short-term regulation of HSL. The finding that mutation of Ser-563 did not abolish PKA-induced activation of HSL led to the identification of two novel PKA sites, Ser-659 and Ser-660, that seem to be responsible in intact primary adipocytes for the activation of HSL in response to isoproterenol stimulation [236]. The role of phosphorylation of Ser-563 remains elusive (Fig. 11.5).

Apart from three PKA phosphorylation sites described above, HSL is phosphorylated *in vivo* in hormonally quiescent cells at a site named the basal site [235], corresponding to Ser-565 in rat HSL [232, 234], i.e. two residues carboxy-terminal to Ser-563. Glycogen synthase kinase-4 [237], Ca^{2+}/calmodulin-dependent kinase II, and AMP-activated protein kinase (AMPK) [238] phosphorylate Ser-565 *in vitro* without any direct effect on enzyme activity (Fig. 11.5). AMPK has been proposed to be the physiologically relevant kinase [238], based on its involvement in other aspects of lipid metabolism and on its proposed role as fuel gauge [239]. Furthermore, because phosphorylation by AMPK prevented subsequent phosphorylation of Ser-563, and *vice versa*, it was proposed that phosphorylation of Ser-565 exerts an antilipolytic role [238]. This proposal is supported by experiments showing that preincu-

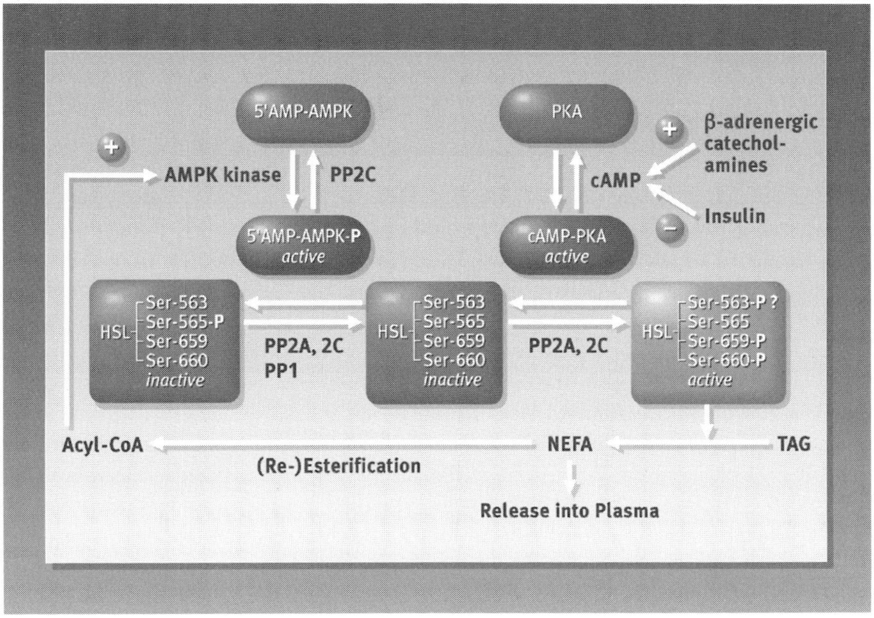

Fig. 11.5 Working model for the interplay between basal and regulatory phosphorylation sites at HSL, including the kinases and phosphatases involved. See text for details. Adapted from Ref. [557] with modifications.

bation with AICAR (5-aminoimidazole-4-carboxamide 1-β-D-ribofuranoside) – an activator of AMPK – causes a moderate reduction of the lipolytic response to catecholamines in primary adipocytes [240, 241]. The data indicating mutually exclusive phosphorylation at the basal and regulatory sites imply that activation of HSL by PKA has to be preceded by dephosphorylation of the basal site. Whether this dephosphorylation is an important regulatory step or merely a result of a high constitutive protein phosphatase activity in the cell is not clear. Moreover, the recent finding that Ser-563 is not essential for HSL activation [236] raises some questions regarding this antilipolytic role of Ser-565 phosphorylation. Also, the moderate antilipolytic effect of AICAR should be interpreted with caution because this substance has many cellular effects and has been questioned as a specific AMPK activator [242].

In fact, recent studies have elucidated completely novel aspects of AMPK physiology as it may be directly involved in the β-adrenergic regulation of lipolysis. Much experimental evidence has promoted the idea that AMPK acts as an intracellular energy sensor, stimulated by the increased intracellular AMP/ATP ratio when cells are stressed by conditions such as hypoxia/ischemia in the heart and excessive contraction in skeletal muscle. Activated AMPK accelerates ATP-producing pathways, such as fatty acid and glucose oxidation, while reducing ATP consumption, ultimately leading to the preservation or restoration of adequate high energy phosphates [243]. Whereas the importance of AMPK to control lipid metabolism in liver and muscle is well established, its role in regulating lipolysis in adipose tissue has remained controversial. Previous observations led to the hypothesis that AMPK antagonizes lipolysis in adipocytes (see above), presumably to prevent futile cycling and depletion of ATP during unrestrained simultaneous lipolysis and re-esterification. However, subsequent mutational analysis of the PKA phosphorylation sites of HSL raised serious doubts on the negative effect of AMPK phosphorylation on PKA-induced HSL activation [236], which were then strengthened by the finding that isoproterenol stimulated AMPK phosphorylation and activity in isolated rat adipocytes, in apparent contradiction to an anti-lipolytic role for AMPK [244]. Confirming and extending the latter study, it was demonstrated recently, that treatment of murine cultured 3T3-L1 adipocytes with isoproterenol or forskolin promoted the phosphorylation of AMPK at a critical activating Thr172 residue in a concentration- and time-dependent fashion [245]. This correlated well with stimulation of AMPK activity as measured in the immune complex. Analogs of cAMP mimicked the effect of isoproterenol and forskolin on AMPK phosphorylation. Treatment of the adipocytes with insulin reduced both basal and forskolin-induced AMPK phosphorylation via a pathway dependent on PI-3′K. Overexpression of a dominant-inhibitory mutant of AMPK blocked isoproterenol-induced lipolysis by about 50% [245]. These data indicate that in adipocytes a novel pathway operates through which cAMP can lead to the activation of AMPK, and that this pathway is required for maximal stimulation of lipolysis. However, overexpression of a constitutively active AMPK version had a minimal effect on both basal and isoproterenol-induced lipolysis only, which may indicate that AMPK is required but not sufficient for lipolysis induction (in the absence of elevated cAMP). Moreover, it remains unclear from this study how agents that increase cAMP or the nucleotide itself activate AMPK in intact cells. Thr172, the phosphorylation site on AMPK responsible for most of the increase in AMPK activity, is not a consensus PKA phosphorylation site, suggesting that PKA regulates AMPK activity through an indirect mechanism, such as phosphorylating and activating an unknown upstream AMPK kinase. Alternatively, the effect of cAMP could be independent of PKA. The other open question is how AMPK regulates lipolysis. As already mentioned, phosphorylation of HSL at Ser563, Ser659, and Ser 660 by PKA alone does not account for the maximal activity of HSL as *in vitro* PKA phosphorylation causes a 1.5- to 2.2-fold increase in HSL lipase activity only, while isoproterenol stimulates FA release from adipocytes by more than 50-fold. AMPK phosphorylates Ser565, which is located in the regulatory domain of HSL, as are the PKA phosphorylation sites, *in vitro*. Phosphorylation of Ser565 by AMPK might, speculatively, be involved in regulation of the translocation of HSL to the LD (see below). This would explain the recent finding that mutation of this residue abrogated the ability of HSL to translocate to LD [246]. In addition, or alternatively, several other proteins, participation of which in lipolysis has been documented, such as perilipin and lipotransin (see below), may represent direct or indirect targets of AMPK. Taken together, the available experimental evidence hints at a positive modulatory role of AMPK in β-adrenergic stimulation of lipolysis, during which it undergoes phosphorylation and activation and then possibly phosphorylates HSL at Ser565, leading to its translocation to LD. Physiologically, this putative mechanism nicely fits the commonly accepted role of AMPK as intracellular energy sensor since it guarantees that ATP depletion in adipocytes and in other tissues coordinated via the β-adrenergic response leads to enhanced lipolysis, resulting in the release of FA that can be oxidized by target tissues to compensate for the increased demand for high energy phosphates.

11.3.3 Dephosphorylation of HSL

Two major studies devoted to clarifying the action of PP on HSL [247, 248] have demonstrated a several-fold higher activity of PP2A and PP2C than PP1 toward PKA sites, even though the source of both HSL and PP differed in each study. The same was true for the basal site, i.e. Ser-565 [247]. Because of their higher abundance in adipose tissue, PP2A and PP1 are the main PP acting on HSL *in vivo*. Wood and coworkers [248] suggested a role for PP1 in selective dephosphorylation of the basal site. This, however, was not supported by another study, which showed a higher preference of PP1 for the PKA sites than for the basal site [247]. The dephosphorylation of the novel PKA sites, Ser-659 and Ser-660, as well as the possible existence of specific HSL phosphatases, remains to be investigated (Fig. 11.4).

11.3.4 Intrinsic HSL Activity

Historically, the major consequence of HSL phosphorylation was thought to be an increase in the catalytic activity of the lipase toward TAG, which thereby triggers lipolysis, but more recent observations indicate that activation of lipolysis involves a more complex and elegant series of reactions. Phosphorylation of HSL *in vitro* leads to a moderate (two-fold) increase in activity against emulsified TAG substrates, whereas β-adrenergic stimulation of adipocytes leads to a >50-fold increase in the hydrolysis of TAG housed within LD [249, 250]. A reasonable explanation for this discrepancy is that (i) cytosolic HSL in unstimulated adipocytes is already highly active but has limited access to its substrate contained within intracellular LD and (ii) the major consequence of lipolytic stimulation is the migration of HSL toward the LD [251, 252] and a change in the LD surface that provides access of the lipase to its TAG substrate in the LD core. That HSL is present in an active state in unstimulated adipocytes is supported by comparisons of lipolysis in intact control and stimulated cells with the lipolytic activities of homogenates or sonicates derived from these cells [249, 253]. Whereas lipolysis in intact stimulated cells is far greater than in intact control cells, there is little between the measurable lipolytic activities in homogenates or sonicates of these control and stimulated cells. Importantly, high HSL activity is found in sonicates or homogenates of control cells that exhibit only minimal lipolysis before disruption of the cells [249, 253]. The conditions of cell disruption in these studies are unlikely to foster phosphorylation of HSL because both PKA and ATP are diluted, and chelators in the homogenization medium reduce the effective concentration of Mg^{2+}, which is required for PKA activity. Hence, it appears that the non-PKA-phosphorylated HSL in homogenates or sonicates of control adipocytes can hydrolyze TAG at a rate nearly equal to that exhibited by PKA-phosphorylated HSL when acting on exogenous emulsified TAG substrates. Strålfors and coworkers [250] proposed that the great increase in the lipolytic activity in sonicates of unstimulated adipocytes results from the large increase in surface area of TAG arising from dispersion of the LD. Okuda and coworkers [253] speculated that the increase was due to destruction of the integrity of the LD surface. Implicit in both explanations is the idea that, in the intact adipocyte, HSL is active but has restricted access to the TAG substrate. As reviewed below, this is compatible with data showing that lipolytic stimulation is accompanied by a migration of HSL to the LD *plus* a dramatic change in the structure of the LD surface.

11.3.5 Translocation of HSL

11.3.5.1 Mechanism

In a seminal paper Hirsch and Rosen reported that cAMP-stimulated lipolysis in cultured adipocytes is accompanied by redistribution of HSL from the aqueous supernatant of homogenates of unstimulated cells to the particulate fraction and the floating fat cake of homogenates of stimulated cells [254]. These findings were later confirmed by the direct immunological demonstration of HSL movement from the cytosol to the LD [251], leading to the hypothesis that a major consequence of PKA activation is the translocation of the phosphorylated HSL from the cytosol to the LD [251]. Subsequently, this hypothesis was confirmed by examining the location of HSL in 3T3-L1 adipocytes with the use of immunofluorescence microscopy [252]. In unstimulated cells, the enzyme is disturbed diffusely in the cytosol, but migrates to the LD surface upon stimulation of lipolysis by the β-AR agonist, isoproterenol. This translocation is reversible since upon addition of the β-AR antagonist, propranolol, to stimulated cells HSL resumes a diffuse distribution. Furthermore, HSL leaks from adipocytes into the medium when digitonin is used to permeabilize unstimulated, but not stimulated cells. These data suggest that HSL is freely soluble in the cytosol of unstimulated adipocytes and that its binding to the LD surface in stimulated cells is reasonably avid [252]. Furthermore, HSL remains bound to LD following fractionation of stimulated cells. The temporal kinetics of HSL translocation to and away from the LD parallels the temporal kinetics of the onset and termination of lipolysis. Because the effects of isoproterenol are mimicked by non-hydrolysable cAMP analogs, the translocation reaction is downstream of PKA activation. There appears to be no cytoskeletal involvement in the translocation reaction, since translocation is unimpeded by cytoskeletal poisons such as nocodazole, cytochalasin B and the combination of these two agents [252]. Thus, an important consequence of lipolytic activation of adipocytes is the translocation of HSL from the cytosol to its substrate, i.e. the TAG in the core of LD (Fig. 11.4, step 4).

Meanwhile, HSL translocation seems to be more complex and may be accompanied by and/or trigger a conformational change of HSL, leading to higher affinity for TAG and/or phospholipids of the LD. *In vitro*, even non-phosphorylated HSL binds readily to phospholipid vesicles [255] and mixed phospholipid/DAG monolayers [256]. However, in intact cells, the interaction with the LD only takes place upon lipolytic stimulation and phosphorylation of HSL. Remarkably, a mutant variant in which the newly described phosphorylation sites (Ser-659, Ser-660) have been replaced by alanines was unable to translocate. Of interest is whether the different sites for PKA phosphorylation on HSL have different roles in promoting cellular lipolysis, i.e. in triggering HSL translocation to LD versus increasing intrinsic HSL activity toward TAG in the LD core. Unpublished work by Londos and coworkers revealed that HSL translocation proceeds normally when either of these two sites is mutated singly, but is abolished on simultaneous mutation of both sites. By analogy with other lipases, HSL might exist in two conformational states, an active form – corresponding to the "open" form exposing a large hydrophobic lid structure as observed for other lipases – and an inactive "closed" form. In intact cells, phosphorylation of HSL would be required to trigger the transition from the "closed" to the "open" form, exposing the hydrophobic lid structure for interaction with TAG. *In vitro*, HSL may be permanently stabilized in the "open" form by the non-ionic detergent required to keep the enzyme soluble and active in its purified form [256], thus making phosphorylation unnecessary. This mechanism of stabilization of the "open" form by a micelle of tensioactive molecules has been elegantly demonstrated for pancreatic lipase [257].

Recently, the involvement of multiple mechanisms for hormonal regulation of lipolysis by HSL has convincingly been demonstrated with transgenic mice expressing human HSL in WAT [258]. HSL mRNA levels were increased in WAT by 3-fold vs. endogenous HSL. The lipolytic activity against TAG was elevated in WAT of transgenic compared with wild-type mice which may explain the observed more pronounced reduction of body weight and fat mass in transgenic compared with wild-type mice after a 4-day calorie restriction. However, and surprisingly, the cAMP-inducible lipolytic response was lower in adipocytes prepared from transgenic mice [258]. Thus these data suggest that mechanisms other than phosphorylation, translocation and intrinsic activity of HSL may become rate-limiting for lipolytic degradation of TAG in LD of intact adipocytes, at least under conditions of HSL overexpression, some of which and the components putatively involved are discussed below.

11.3.5.2 Involvement of Perilipins

Initially it has been proposed that the function of perilipin could be to anchor HSL to its TAG substrate as a "docking protein" at the surface of LD when phosphorylated [251]. However, attempts to coimmunoprecipitate (PKA-phosphorylated) HSL and perilipin offer no evidence of any direct interaction between these two proteins, nor does a yeast two-hybrid screen [259]. A salient feature of perilipins A and B is their phosphorylation by PKA. The A isoform has six predicted sites for PKA phosphorylation, three of which are in perilipin B. Most, if not all, of these sites in perilipin A appear to be phosphorylated upon lipolytic activation of adipocytes [168]. The PKA-mediated phosphorylations occur on serine residues. By contrast, in unstimulated cells one or more threonine residues, as yet not identified, are phosphorylated. Because perilipin A is polyphosphorylated by PKA during lipolytic activation of adipocytes, a role for perilipin in lipolysis has been hypothesized. As cytosolic HSL seems to be in an activated state in the absence of PKA phosphorylation, one must ask why the lipase does not bind to and hydrolyze TAG in unstimulated adipocytes. There are two possible explanations. First, despite the evidence that in intact cells non-phosphorylated HSL behaves like a freely soluble cytosolic protein, this HSL form might be constrained from reaching the LD due to its association with another protein (see below). Secondly, perilipins could serve to block HSL action in unstimulated cells. During the last decade, the common hypothesis has been that non-PKA-phosphorylated perilipins suppress lipolysis by denying HSL access to TAG in the core of the LD (Fig. 11.4, step 5). Several lines of evidence support this:

Mutational analysis of the three consensus PKA sites in perilipin A that are common to both perilipin A and B (Ser-81, Ser-222, Ser-276) was performed to analyze the role of PKA-mediated hyperphosphorylation of perilipin in regulating lipolysis [250]. When the mutant protein (delta3) was expressed in cells overexpressing acyl-CoA synthase-1 ACS1 and fatty acid-binding protein-1 FATP1, like wild-type perilipin A, the mutant one targeted itself to the surface of LD, increased TAG accumulation, and decreased basal lipolysis. However, unlike perilipin A, perilipin-delta3 blocked PKA-stimulated lipolysis in ACS1/FATP1 cells overexpressing HSL [260]. Thus, PKA-stimulated phosphorylation of perilipin A is necessary to abrogate perilipin's A inhibitory action on lipolysis. Because perilipin A and B share the same PKA phosphorylation sites that were mutated, it can be speculated that perilipin B may also modulate PKA-stimulated lipolysis. If perilipins A and B assume similar extended conformations at the LD surface, the greater protection against hydrolysis of the longer splice variant, perilipin A, may be attributed to its greater length and thus greater coverage of the LD surface, i.e. it is the extended unique carboxy-terminal tail ($\sim$12 kDa) of perilipin A that is important in protecting against lipolysis in the unstimulated state. Since perilipins are major substrates of PKA, which polyphosphorylates perilipins on lipolytic stimulation, i.e. in parallel to lipid hydrolysis [146, 157], it follows that phosphorylation of PKA sites in the amino-terminal region induces a conformational change in the carboxy-terminal region that exposes portions of the LD surface to HSL action.

Furthermore, expression of native and mutated forms of perilipin A and B in CHO fibroblasts revealed that perilipin A inhibits TAG hydrolysis by 87% when PKA is quiescent but that activation of PKA and phosphorylation of perilipin A provokes a 7-fold lipolytic activation [261]. No HSL was detected by immunoblotting these cells nor could any PKA-stimulated lipolytic activity in CHO cell homogenates be measured. Thus, the lipase responsible for TAG hydrolysis in CHO cells appears unresponsive to PKA, and PKA regulation of lipolysis in CHO cells expressing perilipin A is concluded to result solely from the PKA phosphorylation of perilipin, which is supported by the finding that mutation of selected PKA sites eliminated this regulated response [262]. Unlike perilipins, ADRP has no consensus sequences for PKA phosphorylation and thus is unlikely to be phosphorylated in lipolytically stimulated cells.

In agreement with the above findings, perilipin A acts cooperatively with HSL in lipolysis when both proteins are introduced into CHO cells. Thus, the ability of perilipin to regulate lipolysis of TAG in the core of LD appears to be manifested with any lipase that has access to the LD surface. The enhancement of lipolytic activity by perilipin A and the endogenous lipase of CHO cells is clearly time-dependent, requiring at least 30 min after stimulation to become manifest. In contrast, when both HSL and perilipin are introduced into CHO cells, the lipolytic activation occurs with no detectable lag. Phosphorylation of HSL is assumed to foster this immediate reaction, in contrast to the endogenous lipase, which is apparently not a PKA substrate.

Additional support for the barrier hypothesis for perilipin is based on analyses of the stimulation of adipose cell lipolysis with the cytokine, TNF-α. Several cytokines, including TNF-α, increase lipolysis in adipocytes [262, 263]. Unlike the fast-acting lipolytic hormones, which activate PKA and increase lipolysis within minutes, TNF-α requires at least 6 to 8 h to initiate a lipolytic response. Moreover, TNF-α does not increase PKA activity. However, the cytokine does terminate expression of the perilipin gene, and within 24 h of exposure of the cells to TNF-α the perilipin content of cells is reduced by up to ca. 50%. That this reduction in perilipin accounts for the increased lipolysis is supported by Sousa and coworkers [264] report that adenovirus-mediated expression of perilipins A or B prevents TNF-α-stimulated lipolysis. In contrast to the endogenous perilipin, that produced by the adenovirus construct is not attenuated by the cytokine. Thus, the data indicate that perilipins not phosphorylated by PKA inhibit lipolysis and that phosphorylation of perilipins abrogates this barrier (Fig. 11.4, step 5).

Most interestingly, the increased PKA activity in 3T3-L1 adipocytes is correlated with dramatic and reversible changes at the LD surface [264]. Immunofluorescence staining for perilipin decreased at the periphery of large LD, suggesting that PKA activation lowers the density of the perilipin coat at the LD surface, either by an increase of the total LD surface area leading to "dilution" of perilipin or, more likely, loss of perilipin from the LD due to its translocation into the cytosol. The apparent reciprocal translocation of HSL (Fig. 11.4, step 4) and perilipin (Fig. 11.4, step 6) between LD and cytosol seems to depend on the "age" of the adipocytes [265]. In young male rats (160–200 g) the translocation of HSL in response to isoproterenol closely paralleled the stimulation of lipolysis, but there was no translocation of perilipin from LD to the cytosol. By contrast, upon lipolytic stimulation and perilipin phosphorylation in adipocytes from older rats, there was no translocation of HSL but a significant translocation of perilipin away from the LD. Furthermore, the maximum rate of lipolysis induced by isoproterenol was markedly lower in cells from older than in those from younger rats [265]. A similar redistribution of perilipin has also been reported for 3T3-L1 adipocytes [157, 253]. This "reverse" perilipin translocation is unlikely to involve a conformational change that allows perilipin to become freely soluble in the cytosol as it is extremely hydrophobic [165, 168]. Instead, phosphorylation of perilipin and/or some other mechanism may trigger an alteration in the LD surface, rendering LD more susceptible to disruption by fractionation procedures, resulting in the release of apparently "free" perilipin in the cytosolic fraction. In conclusion, the results with older rats emphasize that lipolytic stimulation is not due to HSL translocation alone. The current speculation is that this diminution of the perilipin coating may represent one mechanism whereby perilipin contributes to lipolysis in adipocytes. However, it appears that the translocation of perilipin is not essential for the stimulation of lipolysis in all cells, or may not be the only or predominant mechanism as it is not observed in adipocytes from young male rats in which translocation of HSL was apparent upon lipolytic stimulation.

11.3.5.3 Involvement of Lipotransin

Translocation of HSL to LD might be prevented in unstimulated cells by a cytosolic inhibitor protein which interacts with HSL or by a LD-associated protein which prevents HSL gaining access to the LD surface, as has been proposed for perilipin (see above). Alternatively, HSL might interact with a LD-associated "docking" protein upon activation, facilitating its translocation to the LD. Saltiel and coworkers have identified a novel binding-protein for HSL, called lipotransin, that fulfils several criteria for an HSL "docking" protein [266]. It can bind directly to HSL, as demonstrated in two-hybrid assays, and in *in vitro* pulldown experiments. The two proteins form a complex in coprecipitation experiments from cells transfected to overexpress the proteins, as well as in fat cells expressing endogenous levels. HSL and lipotransin are detected in overlapping but not identical adipocyte cellular fractions. However, in response to lipolytic agents, HSL translocates to a lipid/membrane fraction that is highly enriched in lipotransin. Additionally, the lipotransin and HSL genes exhibit a similar pattern of expression in both human and mouse tissues [266]. A human homolog of lipotransin, called p60 katanin [267], was tentatively identified by 5′ RACE and 3′ RACE methods from two clones based on their homology to sea urchin katanin; 87% identity was revealed between lipotransin and human p60 katanin in the amino-terminal 185 aa.

The binding of lipotransin to HSL appears to be controlled by several factors. The interaction was significantly increased after phosphorylation of HSL by PKA [266]. Perhaps Ser-659 and Ser-660 control the binding of HSL to lipotransin, providing a mechanism by which the lipolytic hormones can induce the translocation of the enzyme to LD. In addition to the phosphorylation of HSL, the ATPase activity of lipotransin probably plays a role in controlling the interaction of the proteins. Lipotransin belongs to the AAA family of proteins, characterized by a signature motif of 230 aa containing an ATP-binding site consensus, and a catalytic center for ATPase [268]. A BLAST search revealed sequence similarity to several AAA family members involved in diverse cellular functions. The AAA module in lipotransin is located from aa 190 to the 3′ end of the sequence. One Walker motif A (P loop) was found at aa 249–256. The potential catalytic center for ATPase is at aa 353–371 [266]. Proteins containing these domains play a role in recognizing and fusing specific membrane populations [269], priming the assembly or breaking apart of protein complexes and unfolding stable protein conformations [270]. Although ATPase activity is typically difficult to detect in these proteins, blockade of the activity with non-hydrolyzable analogs of ATP can inhibit their effects [271]. In this regard, the interaction of lipotransin with HSL appears to be sensitive to the MgATP binding or ATPase activity of lipotransin. Inhibition of the activity with EDTA to chelate magnesium, or AMP-PNP to block ATP hydrolysis, promotes the formation of a stable complex [266]. These data suggest that, after HSL interacts with lipotransin, the complex undergoes dissociation due to ATP binding and hydrolysis.

Interestingly, fat cells expressing endogenous amounts of the proteins require prior exposure to insulin to capture a lipotransin: HSL complex, as seen by precipitating with antibodies to either protein, and reproduced in pulldown experiments from insulin-treated cells [266]. These data are consistent with the insulin-promoted translocation of HSL to the LD seen in fractionation experiments, even though the hormone blocks lipolysis under these conditions. These findings suggest that insulin might prevent the translocation of HSL to the LD by "freezing" HSL with lipotransin near or even on the cytoplasmic face of the LD, thus preventing access to TAG in the LD core. The following molecular model for the hormonal regulation of lipolysis was suggested by Syu and Saltiel [266; Fig. 11.6]: Upon phosphorylation, HSL can interact directly with lipotransin, which docks the protein close to the LD, which may be formed by surrounding ER cisternae or, more likely, phospholipid monolayer structures directly and continuously connecting LD and ER. Once bound, lipotransin can undergo a cycle of ATP hydrolysis, permitting the dissociation of HSL and its direct association with the LD. When cells are exposed to insulin, the complex is frozen, blocking the dissociation of HSL and its association with the LD. Although the mechanism by which insulin regulates this activity remains unknown, one interesting possibility is that the ATPase activity of lipotransin is negatively influenced by insulin, perhaps through a phosphorylation mechanism. However, phosphorylation of lipotransin in response to insulin has not been reported.

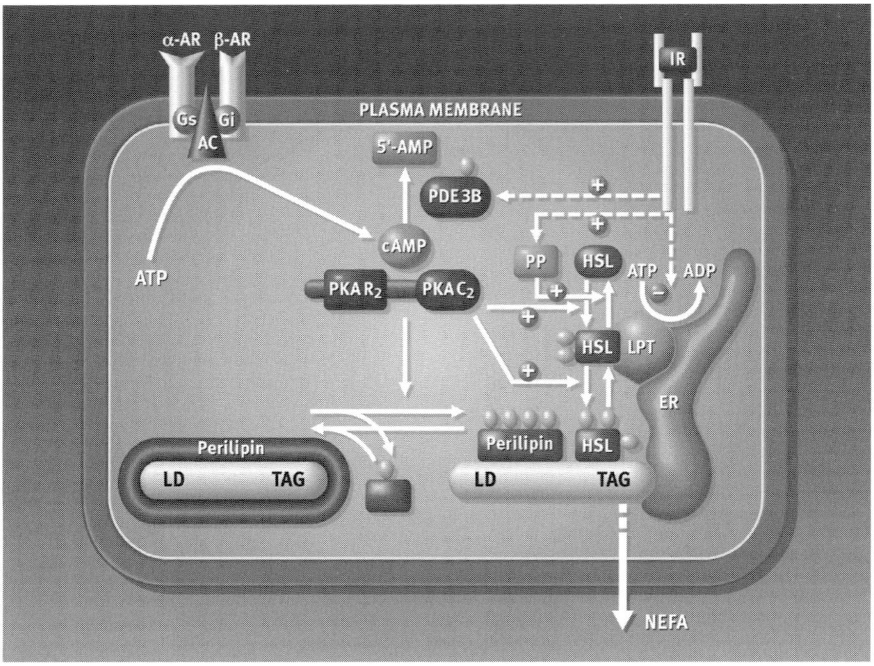

Fig. 11.6 (Hormonal) regulation of TAG mobilization in adipocytes by translocation of HSL and perilipin as well as the molecular mechanisms involved (phosphorylation by PKA, dephosphorylation by protein phosphatases PP and interaction with lipotransin LPT near the ER and LD). Phosphorylation is depicted as filled circles. See text for details. Adapted from Ref. [558] with modifications.

11.3.6 Intrinsic Activity of HSL

11.3.6.1 Feedback Inhibition

HSL is inhibited in a non-competitive manner by oleoyl-CoA, oleic acid and 2-monopalmitoylglycerol, with 50% inhibition at 0.1, 0.5 and 500 μM, respectively [272]. Feedback inhibition of HSL by oleoyl-CoA and oleic acid may therefore prevent accumulation of free FA and cholesterol in the cell, whereas 2-monoacylglycerol may act as a feedback inhibitor if the capacity of 2-monoacylglycerol lipase is exceeded. The inhibition of HSL by oleoyl-CoA and oleic acid at concentrations well below their critical micellar concentrations (reported for fatty acyl CoAs, 30–60 μM; for FA 0.8–1 mM) rules out any detergent effects, implying that inhibition is via binding to a specific site on the enzyme protein [272; Fig. 11.7, step 1]. FA inhibit AC and therefore cAMP accumulation in the adipocyte [273; Fig. 11.7, step 2]. Furthermore, PKA can phosphorylate and thereby stimulate phosphodiesterase, PDE3, directly (Fig. 11.7, step 3) or possibly indirectly by phosphorylating and acti-

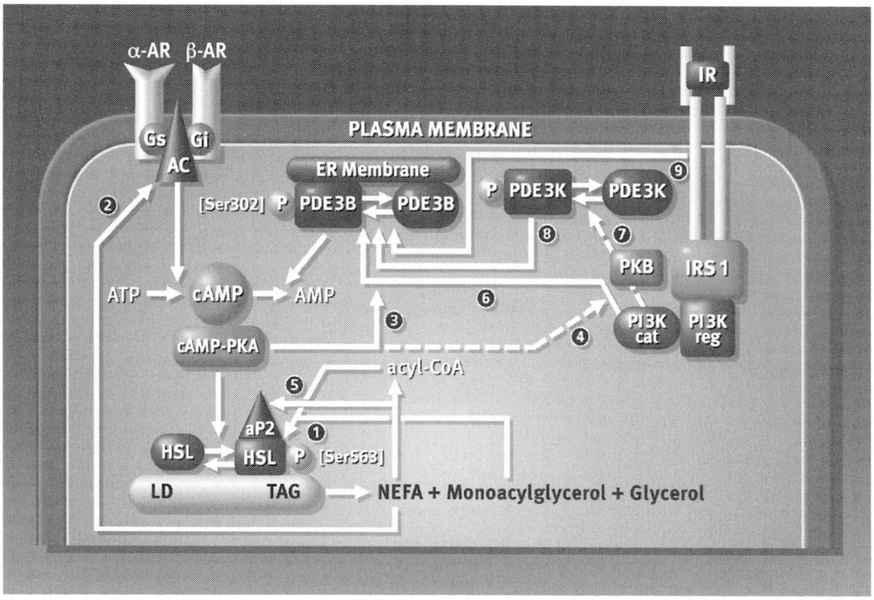

Fig. 11.7 Working model for the short-term inhibition of HSL activity/phosphorylation by insulin, including mechanisms of feedback inhibition by lipolytic end products and positive anti-lipolytic cross-talk to the insulin signaling cascade in adipocytes. Operation of a putative PDE3K downstream of PKB as well as direct phosphorylation/activation of PDE3B by the insulin receptor (IR) and/or PI3K is indicated. See text for details. Adapted from Ref. [557] with modifications.

vating the postulated PDE3 kinase (PDE3K), which finally leads to cAMP degradation [274]. These mechanisms would reduce HSL activation by phosphorylation. The four inhibitory feedback loops may therefore act together to prevent further accumulation of FA released from lipolysis and therefore possible cell damage via detergent properties inherent in high concentrations of FA (Fig. 11.7).

The observed inhibition of HSL by 2-monopalmitoylglycerol may also be of significance as a feedback control of lipolysis. As a result of preferential cleavage of 1(3)-ester bonds by HSL, the hydrolysis of TAG and subsequently of DAG yields 2-monoacylglycerol. This is a poor substrate for HSL and therefore a distinct 2-monoacylglycerol lipase is required for the final step of lipolysis to yield free NEFA and glycerol [275]. If the rate of production of 2-monoacylglycerol by HSL exceeds its rate of hydrolysis by the 2-monoacylglycerol lipase, the resultant accumulation of this lipid may prevent its further generation. However, the physiological conditions under which this might occur are unknown as 2-monoacylglycerol lipase is reported to be present in adipocytes in sufficient quantity and is not rate-limiting, even during hormonally-stimulated lipolysis [275].

The two key enzymes in FA synthesis and cholesterol synthesis, ACC1 and HMG-CoA reductase, are also inhibited by oleic acid and oleoyl-CoA in a non-competitive manner and over similar concentration ranges as for HSL [276, 277]. All three enzymes are phosphorylated by the AMPK. Phosphorylation of ACC1 and HMG-CoA reductase results in direct inhibition of these enzymes [278], whereas phosphorylation of HSL by AMPK blocks subsequent phosphorylation and activation of HSL by PKA [238; see above]. AMPK is itself activated by nanomolar concentrations of fatty acyl CoA [278]. Thus a co-ordinated feedback mechanism may exist whereby elevated levels of FA or fatty acyl CoAs regulate the levels of free cholesterol and FA in the cell via inhibition of their *de novo* synthesis or mobilization from their stores through lipolysis.

11.3.6.2 **Adipocyte Lipid-binding Protein**

Utilizing a yeast two-hybrid screen of a rat adipose tissue library, it was demonstrated that HSL specifically interacts with adipocyte lipid-binding protein (ALBP or aP2; see Ref. [279]) and the amino-terminal 300 aa of HSL has been identified as the region responsible for this interaction. This adipose-specific fatty acid-binding protein is expressed during adipocyte differentiation and makes up to 6% of cytosolic protein in mature fat cells [280]. Fatty acid-binding proteins (FABP) are abundant low-molecular-weight cytoplasmic proteins that are involved in intracellular trafficking and targeting as well as metabolism of FA [281]. ALBP binds FA, retinoids and other hydrophobic ligands [282] and is postulated to shuttle FA within the aqueous cytosol toward the membranes of the relevant intracellular organelles that are involved in TAG synthesis and oxidation. This transfer is believed to occur via a collision mechanism that depends on the interaction of organelle membrane phospholipids with basic residues of ALBP, resulting in partitioning of FA to organelles for metabolic utilization. It has been proposed that intracellular FABP function to sequester FA, thus serving as an intracellular buffer or participating in facilitating the movement of FA within cells. Because HSL and ALBP interact, it was also hypothesized that ALBP might prevent direct feedback inhibition of HSL by high local concentrations of NEFA released at the site of hydrolysis (see above). Consistent with this view, adipocytes from ALBP-null mice exhibit markedly reduced basal and stimulated lipolysis both *in situ* and *in vivo*, but show few phenotypic changes when fed a standard low-fat diet, with the exception of slightly lower plasma glucose and TAG concentrations in the fasted state [283, 284]. This may be based on increased expression of keratinocyte fatty acid-binding protein in ALBP-null mice, normally expressed at low levels in adipose tissue [285]. However, in contrast to wild-type controls, high-fat feeding did not lead to hyperinsulinemia and insulin resistance in ALBP null mice, despite similarly elevated body weights. No major abnormalities in adipocyte TAG synthesis were observed in ALBP null mice with compensation by keratinocyte FABP. Therefore, deficiency of ALBP was associated with partial protection from mild insulin resistance resulting from short-term high-fat feeding, possibly by relieving HSL from feedback inhibition by NEFA. In the opposite concept for the function of ALBP it delivers FA to HSL, thereby mediating the inhibition of its catalytic activity by FA.

Evidence has accumulated that the interaction of ALBP and HSL constitutes an additional mechanism whereby the hydrolytic activity of HSL is regulated (Fig. 11.7, step 5). Thus incubation of ALBP with purified recombinant HSL *in vitro* led to an increase in TAG hydrolysis [286]. This ALBP-induced increase in hydrolytic activity was due to at least two components. First, a small non-specific effect of added ALBP, perhaps altering the surface tension of the TAG substrate and thereby leading to interfacial activation of HSL, and, second, a specific effect. The specific effect of the ALPB-induced increase in HSL hydrolytic activity appeared to be based primarily on the ability of ALBP to interact with HSL. The ability of ALBP to bind FA did not seem to be required for this effect since a FA-binding mutant of ALBP, which binds normally to HSL, displayed a similar capacity to increase activity as wild-type ALBP [286]. There was no evidence that the ALBP-induced increase in HSL activity was due to changes in the dimerization of HSL, which has been reported previously [287]. Furthermore, the ability of ALBP to increase HSL hydrolytic activity was also demonstrated *in situ* in cells co-transfected with HSL and ALBP as compared with HSL and vector alone. Importantly, the ability of FA to inhibit HSL hydrolytic activity was attenuated by co-incubation with ALBP [286]. Here, the specific effect of ALBP to protect HSL from FA-induced inhibition appeared to depend in part on the ability of ALBP both to interact with HSL and to bind FA since the ALBP FA-binding mutant did not preserve HSL activity as effectively as native ALBP when exposed to higher concentrations of FA. The level of ALBP *in vivo* is, notably, extremely high (estimated to be 250–400 μM), suggesting that, within the cellular context, ALBP can relieve product inhibition by sequestration of FA to a much greater degree than measured *in vitro* or *in situ*. Based upon the amount of FA released from isolated adipocytes and the estimated water volume of an adipocyte, the intracellular concentration of FA is estimated to be 600 ± 74 μM [288]. The near 1:1 stoichiometry of the total FA pool with ALBP and the high affinity of FA for this protein, measured by a combination of titration calorimetry and fluorescence displacement assays, suggest that FA in the adipose cell are largely, if not exclusively, protein-bound.

By analysis of a series of deletional mutants of HSL in GST pull-down experiments, the region of HSL interacting with ALBP was localized to aa 192–200 [286]. In particular, site-directed mutagenesis of His-194 or Glu-199 eliminated the ability of HSL to interact with ALBP. Thus, these two residues, either directly or indirectly, are critical for mediating the interaction of HSL with ALBP. Interestingly, HSL mutants H194L and E199A, each of which retained normal basal hydrolytic activity but were unable to bind ALBP, failed to display an increase in hydrolytic activity when cotransfected with wild-type ALBP. The retention of normal basal hydrolytic activity by these mutants argues against mutations at these sites affecting HSL dimerization, since monomeric HSL displays only 1/40 the activity of the dimer [287]. Moreover, this is consistent with the finding that no apparent increase in HSL dimerization occurred upon incubation with ALBP [286]. Consequently, the capacity of ALBP to increase the hydrolytic activity of HSL is, apparently, not due solely to the ability of ALBP to bind and sequester FA, but also appears to depend on the physical interaction of ALBP with HSL. This suggests that the binding of ALBP to HSL might either alter its conformation, allowing TAG to access the catalytic site more efficiently or preventing FA from inhibiting HSL through a conformational change or steric inhibition. Therefore, ALBP can, apparently, increase the hydrolytic activity of HSL through its ability to bind and sequester FA, and through its specific interaction with HSL. Experiments with ALBP-null mice indicate that these functional consequences of the interaction of ALBP with HSL occur *in vivo* [283, 284]. Thus, the current observations are consistent with the proposal that ALBP and HSL constitute a lipolytic complex (Fig. 11.7). This complex functionally results in an increase in the hydrolytic activity of HSL brought about by the physical interaction of HSL with ALBP and by the ability of ALBP to bind and sequester FA. Thereby, lipolysis and intracellular trafficking of FA are controlled in an organized fashion.

11.3.7 Expression of HSL

Existing adipocyte-like cell lines, such as 3T3-L1, 3T3-F442A and BFC-1, together with primary adipocyte cultures, have been used to investigate the effects of different nutritional and hormonal factors on HSL expression. cAMP and phorbol esters decrease HSL expression in mature 3T3-F442A and BFC-1 adipocytes via two apparently distinct mechanisms [289]. Insulin, growth hormone, and retinoic acid [289] and oleic acid [290, 291] reportedly had no effect on HSL expression in mature 3T3-F442A adipocytes. In a study on isolated rat adipocytes in culture, dexamethasone increased HSL mRNA fourfold, whereas adrenaline and growth hormone had no effect [292]. The effect of dexamethasone contrasts with the report by Plee-Gautier and coworkers [289] that dexamethasone alone had no effect, but a potentiation of the cAMP effect. The reason for the discrepancy is not known but may be attributed to the different experimental systems.

The expression of HSL protein and mRNA levels are lower in subcutaneous fat stores compared with internal (visceral, omental) fat depots in the rat [288], suggesting a possible basis for the differences in the rate of lipolysis among various fat depots. In contrast, subcutaneous fat in humans has a higher HSL mRNA expression and HSL activity than omental fat [293]. Human subcutaneous fat cells are larger and there is a positive correlation between fat cell size and HSL expression [293]. When controlled for adipocyte cell size, the amount of HSL protein and HSL mRNA levels in subcutaneous adipocytes show a strong correlation with maximum lipolytic activity [294].

Laurell and coworkers have identified an alternatively spliced shorter (80 kDa) form of human HSL that is exclusively expressed in WAT and devoid of both esterase and lipase activities, and is presumably generated by skipping of exon 6, which encodes the serine residue of the catalytic triad [295]. Subsequently, two HSL immunoreactive bands (88 kDa = L form and 84 kDa = S form) in homogenates from human obese adipose tissue were reported [296]. Immunodetection experiments with an antibody specifically recognizing the domain encoded by exon 6 suggested that the 80 kDa and S forms correspond to the polypeptide translated from the transcript generated through the human-specific skipping of exon 6 [297]. The maximal lipolytic response of adipocytes from obese subjects expressing both the L and S form was 35% lower than those expressing the L form, exclusively, irrespective of the lipolytic stimulus used (βAR agonists, cAMP). The reduction in lipolytic responsiveness correlated well with a 20% decrease in HSL enzymatic activity, whereas no difference in the quantity of L form protein was found between the two groups. This indicates that down-regulation of total HSL enzymatic activity in the presence of the S form is not compensated by overexpression of the L form. One possible mechanism as to how expression of the S form decreases the lipolytic responsiveness may be based on the dimerization of HSL as being essential for its function (see above). Thus, hetero-oligomerization of the L and S forms could render dimers that lack enzymatic activity. Another possibility is that the S form, which can be phosphorylated by PKA, translocates to the LD upon lipolytic stimulation of the adipocytes and thereby competes with the L form as a substrate for the translocation machinery and/or subsequent binding to the LD (see above). The consequences of the concomitant expression of the S and L form on plasma NEFA levels and insulin-stimulated glucose disposal rates in obese subjects have not been reported. Therefore it remains unclear whether the moderate reduction of HSL activity in adipose tissue exerts beneficial or detrimental effects on whole-body insulin sensitivity, possibly based on counteracting or supporting lipotoxic mechanisms (see above).

11.3.8
Release of Lipolytic Products

11.3.8.1 FA Transport

While liberated FA from adipocytes was believed to be transported by simple diffusion (for a review see Ref. [298]) recent studies have identified several membrane proteins that transport NEFA across the adipocyte plasma membrane. These include the FA transport protein, FATP [298, 299], plasma membrane fatty acid-binding protein [300] and fatty acid translocase [298, 301], which are expressed in tissues with a high metabolic capacity for FA, such as WAT, skeletal muscle and heart. Protein-mediated FA transport is assumed to occur via facilitated diffusion along the concentration gradient formed between the intracellular and plasma NEFA concentrations. Thus, during starvation the release of NEFA from adipose tissue into the plasma is determined by the adipocyte lipolytic state, the coronary flow and the peripheral removal of NEFA. CD36 is a class B scavenger receptor that binds multiple ligands, including FA, and has been proposed to function as the predominant transporter of long-chain NEFA, acting as a gatekeeper. Adipose cells of CD36 null mice show a decreased capacity to incorporate long-chain NEFA into TAG [302]. Although neither the effects of CD36 deficiency on adipose mass, adipocyte size or insulinemia have been reported, a ca. 30% reduction in plasma glucose concentrations has been observed. Transgenic mice overexpressing CD36 in skeletal muscle showed increased glucose and insulin concentrations together with decreased plasma concentrations of NEFA and TAG at the same time as having increased FA oxidation in muscle [303]. FATP expression is augmented by fasting and damped by insulin [304, 305].

11.3.8.2 Glycerol Transport

Similar to FA transport, a specific protein is thought to participate in facilitating glycerol release from adipocytes, which can be blocked by mercury – arguing against simple diffusion. Using the cDNA library from human adipose tissue and testis, two groups have identified a novel member of the aqua(glycerol)porin (AQP) family [306–308]. AQPs, which are channel-forming integral proteins with at least 10 family members in mammalian tissues, function as water and/or glycerol channels [309, 310]. This novel AQP, termed AQPap or AQP7, was expressed most abundantly and almost exclusively in adipose tissue and is the only aquaglyceroporin expressed there [306]. Subsequently it was demonstrated that fasting dramatically enhanced and refeeding suppressed its mRNA expression in mice and that mRNA expression of AQPap was overexpressed in obese mice with elevated adipose glycerol concentration [308]. Administration of insulin reduced the expression of AQPap in concentration- and time-dependent manners in 3T3-L1 adipocytes. Interestingly, in non-stimulated 3T3-L1 adipocytes, AQPap was detected in the perinuclear cytoplasm with a scattered distribution, whereas after short-term stimulation with epinephrine AQPap was recovered with the plasma membrane fraction [308]. This suggests translocation of AQPap from intracellular vesicles to the plasma membrane, as has been described for the kidney-specific aquaporin [311], possibly by a PKA-dependent mechanism, although phosphorylation of AQPap has not yet been described.

The coordinated regulation of the expression and activity of HSL, FATP and AQPap (during fasting, see above) will benefit the effective release of FA and glycerol from adipocytes. Physiologically coordinated regulation of the AQPap and FATP1 genes by insulin should be efficient for supplying glycerol and NEFA in accordance with nutritional conditions. However, the suppression of AQPap (as well as FATP1) mRNA expression by feeding was impaired in obese mice with insulin resistance despite high concentrations of plasma insulin leading to higher glycerol and NEFA plasma levels [308]. Consistent with this finding, using microdialysis the interstitial glycerol concentration in subcutaneous fat tissue was found to be elevated in obese mice and increased glycerol production in fat tissue was reported for obese humans [312]. Hyperglycerolemia and increased glycerol turnover rate in combination with increased NEFA influx into the liver in obesity contributes to elevated hepatic glucose production, resulting in hyperglycemia [313–315]. Impaired suppression of AQPap in the insulin-resistant state may be involved in increased hepatic glucose output through hyperglycerolemia, since plasma glycerol is a major substrate for hepatic gluconeogenesis and is derived predominantly from lipolysis rather than from plasma glucose, even under basal, i.e. post-prandial, conditions [316].

Interestingly, a mutation in the conserved GXXXGXXXG motif in the sixth transmembrane domain of aquaporins has been identified for human AQPap (G264V) which renders it incapable of transporting glycerol (and water) [317]. In homozygous subjects this mutation leads to lack of the response of plasma glycerol toward exercise, whereas in control subjects with wild-type AQPap gene plasma glycerol increased in parallel with exercise-induced changes in noradrenaline [318]. These findings indicate that AQPap is responsible for the exercise-induced increase in plasma glycerol in humans. However, normal adiposity and a normal plasma glycerol in G264V homozygous subjects lacking functional AQPap suggest there is another pathway to maintain plasma glycerol in the sedative condition.

11.3.8.3 Cholesterol Transport

Two distinct sources and responsible mechanisms for free cholesterol released from adipocytes during lipolysis can be envisioned, cholesterylesters embedded in the TAG core of LD and cleaved by activated HSL, which exerts pronounced cholesterylesterase activity, and unesterified cholesterol intercalated into the phospholipid monolayer surrounding the TAG core of LD, which may be liberated simply during reduction of the LD surface upon TAG hydrolysis by activated HSL. Thus, for both mechanisms HSL may be rate-limiting for cholesterol release from adipocytes, which consequently would proceed in parallel to mobilization of FA and glycerol from TAG. However, when evaluating the concomitant release of glycerol and cholesterol from 3T3-L1 adipocytes treated with isoproterenol, cholesterol efflux in the medium did not occur during the first 4 h of lipolytic stimulation [319]. A more prolonged lipolytic stimulus (24 h treatment with monobutyryl-cAMP) released significant amounts of cholesterol into the medium. This suggests that prolonged lipolysis (i.e. long-term fasting period) is required for adipocyte cholesterol to re-enter the plasma *in vivo*, and that release of cholesterol from adipocytes is regulated, most likely, by its transport across the plasma membrane in addition to HSL activity. The question of the link between the process of TAG and cholesterol mobilization from adipocytes during lipolysis has received fresh attention since the identification of the ABCA1 protein pathway, which is a key mediator of cellular cholesterol efflux [320]. The reported expression of ABCA1 in adipocytes may be controlled by LXR, which is highly expressed and acts as key transcriptional regulator of ABCA1 in adipocytes [321, 322]. Intracellular cAMP stimulates cholesterol efflux in macrophages by acting through ABCA1 [323]. Thus, TAG mobilization and cholesterol efflux through ABCA1 in adipocytes might be linked.

11.4 Physiological, Pharmacological and Genetic Modulation of TAG Mobilization

11.4.1 Muscle Contraction

Acute muscular contraction increases HSL activity in mouse soleus [324]. In agreement with this finding, muscle contraction, independent of hormonal influence, increases intramuscular TAG utilization [325, 326]. The mechanism underlying the contraction-induced increase in HSL activity is unknown. Furthermore, whether intramuscular HSL activity can be hormonally regulated during muscular contraction is also unclear. The regulation of HSL in skeletal muscle by external hormonal signals during exercise is thought to be unlikely, inasmuch as calcium and metabolites related to the energy status of the cell are expected to be the dominant regulators [327]. However, this may not be the case, since, surprisingly, insulin has been found to play a pivotal role in the regulation of FA oxidation and TAG utilization during intense muscular contraction, when the role of calcium and metabolites would be expected to exert a strong influence [328]. This suggests that the well-known early decline in insulin during exercise may be a major factor in permitting increased TAG mobilization and utilization by muscle. The role of epinephrine in regulating intramuscular lipolysis during contraction is also relatively unexamined. Infusion of various physiological concentrations of epinephrine during low-intensity cycling (25% peak oxygen uptake) enhances whole body lipolysis but actually reduces whole body lipid oxidation [329]. Furthermore, increasing the exercise intensity to 45% peak oxygen uptake blunts the lipolytic action of epinephrine [329]. Thus the most important hormonal signal early in exercise that permits enhanced lipid utilization by muscle may be the decline in insulin, rather than the increase in epinephrine.

The contraction-induced activation of HSL suggests the involvement of calcium in lipolysis control, possibly via Ca^{2+}/calmodulin-dependent kinase(s). Although early observations pointed to a role of calcium at different levels of the adipocyte lipolytic cascade [330–334], this has been contradicted by recent reports suggesting that an increase in intracellular calcium concentrations inhibits lipolysis [333, 334]. However, tissue-specific mechanisms may well exist for lipolysis control.

HSL of 84 kDa size has been reported to be expressed in cardiac muscle [335] and skeletal muscle [336, 337]. In skeletal muscle the expression of HSL is higher in oxidative than glycolytic muscle [336, 337], and HSL expression is reduced in muscle of 24-month-old rats [338], perhaps contributing to the increase in muscle TAG content observed with aging. HSL activity in muscle is stimulated by catecholamines [336, 337] acting via β-AR and cAMP and by contraction acting independently of sympathetic tone or catecholamines [339]. Exercise training does not affect the expression of HSL protein in muscle, but decreases the sensitivity of stimulation of muscle HSL activity by epinephrine [340]. Heart-specific transgenic overexpression of HSL prevents the accumulation of cardiac TAG normally seen in fasted rodents [341]. In addition, heart-specific overexpression of HSL alters the expression of cardiac genes for FA oxidation, transcription factors, signaling molecules, cytoskeletal proteins, and histocompatibility antigens. Thus, HSL in cardiac and skeletal muscle plays a role in controlling the accumulation of LD as well as the mobilization of TAG dependent on the working load.

At present there is no experimental evidence for a direct inhibition of HSL by insulin in skeletal muscle.

11.4.2 Nutritional State

Although HSL is primarily regulated by post-translational mechanisms (see above), pre-translational mechanisms are effective during changes in the nutritional state. High fat-feeding of rats was associated with an increase in both basal and stimulated HSL activity, with a positive relationship between fat cell size and HSL expression [342]. In contrast, following several days of food deprivation in the rat, there is a 2-fold increase in HSL activity, immunoreactive protein, and mRNA levels in adipose tissue that is not observed with short-term fasting and is associated with a reduction in fat cell size [343]. HSL mRNA levels have also been shown to be increased in adipose tissue in hibernating marmots during their time of fasting [344].

Glucose deprivation of primary cultured adipocytes results in a slight decline in HSL expression, whereas incubation with high glucose and insulin maintains HSL expression and leads to an increase of basal and stimulated lipolysis [345]. Recently, a glucose-responsive region was mapped within the proximal promoter of human HSL and the involvement of upstream stimulatory factor-1 and -2 binding to a consensus E-box within this region was shown to be responsible for transcriptional regulation in response to glucose metabolism between glucose-6-phosphate and triose phosphates [346]. Glucose deprivation also reduces HSL mRNA levels by a factor of 2.5–3 in 3T3-F442A cells [291]. The physiological significance of the glucose/insulin effects on HSL expression remains to be determined, but it may reflect an adaptation to nutritional changes.

11.4.3 Hormones and Cytokines

11.4.3.1 Insulin

11.4.3.1.1 Molecular Mechanisms

Insulin, the most important physiological inhibitor of catecholamine-induced lipolysis, induces phosphorylation and activation of a PDE, leading to a decrease in cAMP levels and concomitant decrease of PKA activity (Fig. 11.7). The decrease in cAMP is mainly the result of an insulin-mediated activation of PDE3 [347, 348]. By catalyzing the hydrolysis of cAMP and cGMP, PDEs regulate intracellular concentrations and biological responses of these second messengers. PDEs constitute a group of structurally related enzymes that belong to at least nine related gene families (PDE1-9), which differ in their primary structures, affinities for cAMP and cGMP, responses to specific effectors, and sensitivity for inhibitors and regulatory mechanisms [349, 350]. Two PDEs (PDE3 and PDE4) have high affinity for cAMP and are present in fat cells [351]. PDE3 can be distinguished from PDE4 by its high affinity for both cAMP (low K_m) and cGMP (which down-regulates PDE3 activity) and is associated with the ER membrane [352]. Activation of PDE3 but not PDE4 plays a role in the antilipolytic effect of insulin *in vivo* in human fat tissue [353, 354]. Two distinct PDE3 subfamilies, PDE3A and PDE3B, products of distinct but related genes, have been identified, of which PDE3B is expressed in adipocytes [347, 348, 352]. The importance of PDE3 as an upstream regulator of HSL activity in insulin's antilipolytic pathway has been shown with the use of specific and cell-permeable PDE3 inhibitors [350–357]. Specific inhibition of PDE3 in intact cells completely blocks the antilipolytic action of insulin. Activation of PDE3 is associated with phosphorylation of serine-302 in intact cells [270]. This site is also phosphorylated in response to cAMP-increasing hormones, leading to feedback regulation of the cAMP level [274; see above]. In the presence of both hormones, i.e. the physiological condition during which insulin exerts its antilipolytic action, there is more than additive phosphorylation of serine-302 associated with a more than additive activation of PDE3, suggesting positive cross-talk between the pathways upstream of PDE3 [358, 359]. Thus, the regulation of cAMP content, PKA activity, and thereby the activity of HSL is very complex and includes feedback loops as well as cross-talk between different pathways.

Upstream regulation of PDE3B can be summarized as follows (Fig. 11.7): The insulin-stimulated PDE3 kinase (PDE3K), apparently a serine kinase activity, may be identical to that intrinsically associated with the insulin receptor [360, 361]. PDE3B may thus represent the first substrate of the insulin receptor kinase which is phosphorylated at serine leading to its activation (Fig. 11.7, step 5). However, other experimental evidence argues for the dual specificity protein/lipid kinase, PI3K, consisting of a regulatory and catalytic subunit, as the major upstream regulator of PDE3B and, consequently HSL, since treatment of the cells with the PI3K inhibitor wortmannin completely blocks activation of PDE3K, phosphorylation/activation of PDE3B, and finally the antilipolytic action of insulin [362, 363; Fig. 11.7, step 6]. Upon insulin stimulation PI3K binds to specific tyrosine-phosphorylated motifs on the insulin receptor substrate protein-1 (IRS-1), the major substrate for the insulin receptor tyrosine kinase (IR). The signal transduction between PI3K and PDE3K is not yet known, but probably involves protein kinase B (PKB; Refs. [360, 361]; Fig. 11.7, step 7). The well-known insulin-stimulated serine kinases, MAPK or p70S6 kinases, apparently are not involved in the antilipolytic pathway of insulin [345].

In addition to the short-term insulin regulation of cAMP degradation, long-term effects of insulin on HSL have been reported. HSL activity, immunoreactive protein, and mRNA levels in adipose tissue were increased in response to streptozotocin-induced insulin deficiency in the rat, whereas acute treatment with insulin returned HSL activity to normal without altering the increased amounts of HSL immunoreactive protein and mRNA [366]. Increased HSL activity after prolonged exposure of isolated adipocytes to insulin has been reported in one study [367], whereas others did not reveal an effect of insulin at the mRNA expression level of HSL [368]. Thus, long-term insulin regulation of HSL and the involvement of pre-translational mechanisms remain under debate, while short-term treatment with insulin controls HSL by post-translational mechanisms.

11.4.3.1.2 **Desensitization**

Long-term exposure of 3T3-L1 adipocytes to insulin at or near physiological concentrations results in increased basal glycerol release in the medium without a concomitant rise in NEFA release [368]. These conditions also altered β-adrenergic-stimulated lipolysis, exaggerating glycerol but preventing NEFA release. Stimulation of lipolysis by long-term insulin treatment, as measured by glycerol release is an unexpected finding given the short-term anti-lipolytic action of insulin. This finding, however, agrees with previous studies that showed increased glycerol release after prolonged insulin treatment of isolated rat adipocytes [345, 367]. One of these studies showed that the effect of long-term insulin depends on the presence of high glucose concentrations in the medium [367]. Yet, the role of the hexosamine pathway in mediating this effect of long-term insulin treatment seems unlikely as increased glycerol release occurred irrespective of the presence of glutamine in the medium. In this study, long-term exposure to insulin was done in culture medium, which contained 25 mM glucose. Yet, at the end of the incubation period, glycerol was collected for 1 h in a buffer without glucose [345]. Thus it seems that extracellular glucose could be necessary as a priming factor, particularly under high glucose conditions, but is not required for the actual metabolic flux responsible for glycerol generation. Collectively, these studies demonstrate the ability of insulin, even at sub-nanomolar concentrations, to stimulate glycerol release upon prolonged exposure as opposed to its short-term inhibitory effect on lipolysis.

In contrast to the effect on glycerol release, long-term exposure to insulin was not associated with increased NEFA concentrations in the medium [368], a finding consistent with a previous study [369]. Despite being the two end-products of TAG breakdown, glycerol and NEFA released from adipocytes rarely reach the expected 1 : 3 molar ratio. Short-term insulin stimulation rapidly inhibits NEFA release from adipose tissue, through a combined inhibitory action on HSL and a stimulatory effect on NEFA re-esterification [370, 371]. Recent findings support the possibility that although long-term insulin stimulates glycerol release, the short-term insulin stimulatory effect on NEFA re-esterification is maintained under prolonged treatment. Because the anti-lipolytic effect of insulin and the stimulation of NEFA re-esterification have been shown to involve PI3K [371, 372], the two pathways seem to diverge downstream of PI3K. These seemingly independent pathways to affect glycerol and NEFA release from adipocytes also respond to leptin [373], which was shown to stimulate glycerol but not NEFA release from isolated adipocytes (see below). A role for leptin in mediating the effects of long-term insulin on glycerol and NEFA release from adipose tissue is an intriguing possibility. In conclusion, long-term treatment with insulin apparently leads to desensitization of the signaling pathway down-regulating HSL without concomitant negative interference with stimulation of the re-esterification pathway. The physiological role of long-term insulin regulation of HSL activity remains unclear.

11.4.3.2 **Leptin**

Studies with rodents have demonstrated that leptin induces lipolysis both *in vitro* [373, 374] and *in vivo* [375]. The effect was absent in animals lacking a functional leptin receptor, i.e. db/db mice [374, 375] and in Zucker rats [373]. Intriguingly, FA release does not accompany glycerol release in leptin-induced lipolysis, which suggests simultaneous induction of FA oxidation [369]. In these studies, effects on HSL expression were not investigated. Thus it is not known whether the lipolytic effect of leptin is mediated via phosphorylation-induced activation of HSL, an induction of HSL expression, a combination of both, or some other mechanism. In another study, however, long-term treatment of mice with leptin increased HSL mRNA expression by 30% in WAT, whereas no effect was seen in brown adipose tissue [376]. The leptin signal is presumably transduced via the JAK/STAT pathway. Leptin was also shown to directly activate the TAG/FA substrate cycle, lipolysis and FA oxidation, shifting fuel preference from carbohydrate to FA oxidation. In particular, studies with isolated rat adipocytes in primary culture demonstrated that leptin impairs insulin inhibition of isoproterenol-induced lipolysis and PKA activation [377]. These insulin effects were reduced by leptin (2 nM) with a half-life of 8 h. Leptin concentrations below 1 nM led to a right-ward shift of the corresponding insulin concentration–response curves. At leptin concentrations above 30 nM the responsiveness was diminished, and also resulted in nearly complete relief of lipolysis from insulin control. The IC_{50} for leptin was about 3 nM after 15 h of preincubation of the primary adipocytes. The natural splice variant des-Gln49-leptin exhibited a significantly lower potency. Adipocytes regained full sensitivity for insulin's antilipolytic action within a few hours after leptin removal [377]. Consequently, in ob/ob mice, the loss of leptin stimulation of uncoupling and TAG/FA substrate cycling triggered, at least in part, by leptin blockade of the protein-1 antilipolytic activity of insulin will simultaneously contribute to the decrease in TAG mobilization and utilization as well as to the increase in TAG synthesis and storage. In conclusion, accelerated TAG/FA substrate cycling in adipose tissue provides a new mechanism by which leptin triggers increased metabolic rate and energy expenditure.

11.4.3.3 Growth Hormone

Growth hormone is another hormone signaling via the JAK/STAT pathway and with documented lipolytic effects, albeit with variable efficiency, depending on species and experimental system [378]. Using primary rat adipocytes, growth hormone was found to increase HSL activity, i.e. lipolysis, without causing an increase in HSL mRNA levels [379], supporting a role for post-translational mechanisms in controlling lipolysis. In agreement, cGMP levels appear to be elevated moderately upon challenge with growth hormone. cGMP has been shown to mimic cAMP as an activator of lipolysis [224]. This is supported by *in vitro* studies showing that cGMP-dependent protein kinase phosphorylates and activates HSL [380, 381]. cGMP-dependent protein kinase is present at very low levels in adipose tissue [382], and the physiological role of this kinase and cGMP for activation of lipolysis, if any, is not known. Recently, nitric oxide was shown to exert modulatory actions on lipolysis in adipose tissue via cGMP-independent mechanisms [383]. The effects were complex, with different redox forms acting as either stimulatory or inhibitory.

11.4.3.4 Glucose-dependent Insulinotropic Polypeptide

A recent investigation has demonstrated lipolytic action of glucose-dependent insulinotropic polypeptide (GIP) on 3T3-L1 adipocytes [384]. The effect appeared to be cAMP-mediated and could be blocked by insulin in a wortmannin-sensitive manner.

11.4.3.5 TNF-α

Of the various factors suggested to play a part in impaired adipocyte function in insulin resistance and NIDDM, increased TNF-α expression and production have attracted interest. TNF-α is produced and secreted from adipose tissue in obesity and thus acts in an autocrine fashion to alter adipocyte function during obesity-linked insulin resistance [385–387]. Long-term exposure (>6 h) to TNF-α has been shown to stimulate lipolysis in adipocytes [368, 388], despite inducing reduced HSL expression [389, 390]. An early study showed a down-regulation of HSL gene expression upon TNF-α treatment of 3T3-L1 cells, as measured by Northern blot analysis and enzyme activity measurements [391]. A similar effect, although much more moderate, was seen more recently at the protein level [264]. In a study with primary rat adipocytes, however, no alteration of the levels of HSL protein occurred upon treatment with TNF-α [392].

Studies using human adipose tissue have now shown that TNF-α inhibits LPL activity by down-regulating its protein expression [388]. Increased TNF-α mRNA levels are correlated with decreased LPL activity in human subcutaneous adipose tissue [393]. In addition, TNF-α reduces the expression of FA transporters in adipose tissue of the Syrian hamster, and decreases the expression of enzymes involved in lipogenesis, such as acetyl-CoA carboxylase, FA synthase and acyl-CoA synthase [394]. TNF-α could thus impair the synthesis and entry of FA into the adipocyte, curtailing an increase in the intracellular TAG pool size.

TNF-α has also been found to promote lipolysis. Indirectly, the down-regulation of HSL observed in 3T3-L1 cells may reflect the known dedifferentiating effect of TNF-α in these cells [395], whereas, directly, the lipolytic effect of TNF-α could be explained by the decrease in LPL activity and intracellular reesterification [396] and/or down-regulation of G_i [365; see below] and perilipin expression [264, 397, 398]. The latter effect of TNF-α, as well as the activation of lipolysis, was prevented by thiazolidinediones [390], a class of insulin-sensitizing anti-diabetic agents [399]. The mechanisms involved in regulation of lipolysis by TNF-α remain unclear. A study with human subcutaneous adipocytes [400] demonstrated that, concomitant with an increase in lipolysis induced by TNF-α, there is activation of mitogen-activated protein kinases (MAPK) and ERK1/2, leading to a decrease in the expression of PDE3B, together with an elevation in intracellular cAMP and hence activation of PKA [401, 402]. Since these two pathways are classically not coupled, the significance of changes in both is unclear in relation to an increase in lipolysis. Interestingly, TNF-α stimulated the three mammalian MAPK, p42/44 (ERK2/1), JNK and p38 in time- and concentration-dependent manners in parallel to induction of lipolysis (3-fold at maximal dose). Lipolysis was completely abrogated by inhibitors specific for p42/44 and JNK but was not affected by a p38 inhibitor [403]. These data suggest involvement of p42/44 and JNK in the TNF-α up-regulation of lipolysis in human preadipocytes. The close association between the expression of enzymes involved in lipolysis and lipogenesis with adipogenesis has made it difficult to tease out process-specific actions of TNF-α [404]. TNF-α can thus act on the adipocyte to shift lipid metabolism away from lipid accumulation and towards a reduction in the TAG pool, which could represent one mechanism by which TNF-α can reduce total adipose tissue weight.

Severe weight loss or cachexia is a detrimental end-point of several diseases, including cancer, infection and congestive heart failure. Unlike starvation, cachexia results from a severe loss of lean body mass (mostly skeletal muscle) and the depletion of the fat depots [405]. The pathogenesis of this weight loss is multifactorial, but evidence suggests that cytokines are key players. The predominant reason for the loss of adipose tissue is the fall in the activity of LPL and an increase in the activity of HSL. Since TNF-α promotes lipolysis and inhibits lipogenesis, it represents an ideal player in the depletion of adipose tissue mass seen with cachexia. An elevation in plasma levels of TNF-α, as opposed to adipose tissue-derived TNF-α, is suggested to be responsible for the metabolic alterations in adipose tissue with cachexia [406, 407].

11.4.4 ASP

Acylation-stimulating protein (ASP) is identical to C3adesArg, a cleavage product of complement C3. Cleavage of complement C3 is mediated through the alternate complement pathway via the interaction of C3, factor B, and adipsin that generates C3a. Rapid cleavage of the carboxy-terminal arginine of C3a by carboxypeptidase N generates ASP [403, 408]. Adipocytes are one of the few cells capable of producing all three factors (factor B, adipsin, and C3) that are required for the production of ASP [403, 408, 409]. ASP production increases consequent to adipocyte differentiation [410], and plasma ASP levels are elevated in obesity [411, 412]. Chylomicrons *in vitro* stimulate ASP production by adipocytes [413, 414]. *In vivo* arterial-venous gradients across a subcutaneous adipose tissue bed in humans demonstrate direct postprandial production of ASP [415]. The postprandial increase in ASP is adipose tissue-specific and is not observed in the general circulation [416]. Altogether, these data suggest that ASP and TAG storage are metabolically intertwined.

ASP acts as an adipocyte autocrine factor and has been proposed to play a central role in the metabolism of adipose tissue by increasing the efficiency of TAG storage in adipocytes, an action that results in more rapid postprandial TAG clearance [417]. This is based on stimulation of both glucose uptake and NEFA storage in human adipose tissue [418, 419] via translocation of glucose transporters (GLUT1, GLUT3 and GLUT4) from intracellular sites to the cell surface [420, 421] and an increase in DGAT activity [419]. In addition, ASP has been shown to inhibit HSL in adipocytes, independently and additively to insulin [422].

The pathways distal to PI3K that are involved in the effect of insulin on FFA re-esterification and in the action of ASP on FFA re-esterification and lipolysis were unknown until now. In the study mentioned above, PKC blockers could not counteract the effects of ASP or insulin on fractional NEFA re-esterification (see above) or lipolysis, suggesting that the actions of the two hormones on NEFA release are not mediated by PKC [422]. This is in contrast to earlier results on the action of ASP on TAG synthesis and glucose transport in human fibroblasts [423], which is counteracted by PKC blockers. The PI3K blocker wortmannin counteracted the effect of insulin on NEFA release by reversing the effect of insulin on both fractional NEFA re-esterification and lipolysis. The effects of ASP, however, were not affected by wortmannin [422]. Thus the pathway mediating the effect of ASP on NEFA release does not involve PI3K. This strongly supports the hypothesis that ASP and insulin follow different pathways in their action on NEFA release. Because the ASP receptor protein(s) has not been identified, further research in this area is difficult at present. The question was addressed whether a more distal pathway could be common for the actions of insulin and ASP on NEFA metabolism. Interest was focused on PDE, which previously was found to mediate not only the antilipolytic effect of insulin (see above) but also contributes to the stimulatory effect of the hormone on glucose transport in human fat cells [424]. A non-selective PDE inhibitor that does not interact with the antilipolytic effects of adenosine, counteracted the actions of both ASP and insulin on NEFA release by reversing their effects on fractional NEFA re-esterification and lipolysis. The selective PDE3 blocker OPC3911 could also counteract these effects of ASP and insulin. The selective PDE4 blocker rolipram could not counteract the effect of insulin on fractional NEFA re-esterification and lipolysis. Rolipram did not alter the effect of ASP on lipolysis either, but it counteracted slightly the effect of ASP on fractional NEFA re-esterification. These results agree with those obtained with direct measurements of PDE activity. PDE3 activity in fat cells was stimulated with both ASP and insulin in a concentration-dependent manner, whereas PDE4 activity was only slightly stimulated with ASP. Thus, the effects of both ASP and insulin on NEFA release appear to be mediated predominantly by PDE3. PDE4 seems to be involved in the action of ASP as well, although to a lesser extent than PDE3.

11.4.5
Acipimox and Nicotinic Acid

Acipimox (5-methylpyrazine carboxylic acid-4-oxide) is a newer, well-tolerated derivative of nicotinic acid (NA) which shows lipid-lowering properties in patients with types II and IV hyperlipidemia [425]. The fall in plasma TAG is associated with a decrease in total and LDL cholesterol and an increase in HDL cholesterol [426–429]. NA has long been used in the treatment of dyslipidemia, producing a very desirable modification of multiple cardiovascular risk factors, increasing HDL, TAG, and lipoprotein(a), which results in a reduction in mortality [430]. NA is an effective therapeutic agent. However, it has to be administered at high doses and has a characteristic side-effect profile defined by intense, but transient, prostaglandin-mediated cutaneous vasodilation ("flushing") that affects patient compliance [431, 432]. Despite their long history of clinical use, the precise mechanism of action of acipimox and NA was unknown until recently, although inhibition of adipocyte lipolysis via the activation of a G_i-coupled receptor is believed to contribute [433, 434]. It has been postulated that a reduction in NEFA liberated from adipose tissue results in reduced influx of NEFA into the liver and hepatic re-esterification of NEFA and thereby diminishes the production rate of hepatic TAG available for VLDL and LDL synthesis, which in part explains the hypolipidemic effects observed during NA and acipimox therapy.

Whereas most studies have focused on the effects of NA on lipid metabolism, the action of NA on carbohydrate metabolism is less well understood. After acute NA administration, glucose concentrations have been reported to decrease [435], rise [436] or remain unaltered [437] in rats and humans. Results of glucose tolerance tests after acute NA intake have also been inconsistent [438, 439]. Chronic administration of NA has consistently resulted in deterioration of glucose tolerance and elevation of fasting blood glucose concentrations in normal humans [439–441] and impairment of glycemic control in NIDDM patients [440]. These effects are contrary to expectations based on the glucose-FA cycle of Randle and coworkers [38, 39]. If reduction of lipolysis and NEFA availability reduces oxidation of fat in tissues and availability of gluconeogenic precursors in the liver, insulin sensitivity should improve and glucose concentrations fall, as observed with inhibitors of mitochondrial FA oxidation [442]. It is unclear whether this failure of NA to act as an anti-diabetic drug is caused by its unfavorable pharmacokinetic profile (short half-life time) and/or its mode of action as inhibitor of adipocyte lipolysis.

11.4.5.1 Mode of Action

The effect of acipimox on adipose tissue lipolysis has been examined under three experimental conditions [443]. Firstly, in the presence of high levels of the non-selective β-AR agonist isoproterenol, secondly in the presence of low levels of adenosine deaminase (ADA), which when used alone have a minimal effect on basal lipolysis, yet when combined with sub-maximal concentrations of isoproterenol, substantially facilitate the lipolytic response. Thirdly, acipimox was examined in the presence of high levels of ADA. Fat cells produce adenosine from cAMP via consecutive action of PDE and 5′-nucleotidase which upon release into the medium act at specific AR to inhibit AC [444], this inhibitory effect being mediated by the GTP-dependent [445], pertussis toxin-sensitive [446] regulatory protein, G_i. High levels of ADA, by converting the endogenous adenosine into inosine, relieve the inhibition of AC, which results in the observed increase in basal lipolysis. The data show that acipimox has only a minor, if any, anti-lipolytic effect on lipolysis stimulated by isoproterenol alone [443]. In contrast, when the lipolytic rate is increased by a lower concentration of isoproterenol in combination with ADA, acipimox becomes a potent anti-lipolytic agent. Wieser and Fain reported similar results for NA, with the addition of ADA increasing the sensitivity to catecholamine-induced lipolysis towards inhibition by NA [447]. The dependence of acipimox upon removal of adenosine suggests that the anti-lipolytic mechanisms downstream of the AR and the putative acipimox receptor (see below) and upstream of AC are identical for adenosine and acipimox, respectively.

These results show that acipimox acts to inhibit cAMP-stimulated lipolysis through suppression of intracellular cAMP levels, with a subsequent decrease in PKA activity. The anti-lipolytic activity of acipimox is associated with a quantitative shift of HSL from the LD to the cytosol in rat adipocytes, suggesting that, by decreasing intracellular cAMP levels, acipimox can reduce the association of HSL with TAG substrate of the LD. Pharmacokinetic studies show that the plasma concentration of acipimox in man after a single 250 mg oral dose reaches 33 μM, and remains above 7 μM for 6–8 h [448]. In comparison, acipimox is a potent anti-lipolytic agent at a concentration of 10 μM and above, in human adipose tissue stimulated maximally and sub-maximally by isoproterenol [449]. Thus, plasma acipimox levels correlate well with its *in vitro* effect on lipolysis and cAMP synthesis in rat adipocytes.

Work on NA showed that it also decreased intracellular cAMP levels [450] through inhibition of AC [447]. The conditions were identical to those required to inhibit the enzyme by the anti-lipolytic hormonal factors, prostaglandin E_2, adrenaline, and the adenosine analogue, PIA [445, 452]. These findings indicated that

NA inhibits AC via a receptor-mediated process in a hormone-like manner, which indicates the involvement of the regulatory protein, G_i. This was confirmed by the discovery that the anti-lipolytic effects of prostaglandin E_2, PIA and NA were abolished by the pretreatment of adipocytes with pertussis toxin, to inactivate G_i [453]. Using the AR antagonist, 3-isobutyl-1-methylxanthine, Londos and coworkers [454] demonstrated that NA and adenosine act through separate receptors to inhibit adipocyte AC activity. Since NA shares its ability to inhibit lipolysis via a G_i-dependent mechanism with acipimox, and since the drugs are structurally related, it seemed possible that acipimox suppresses intracellular cAMP levels through binding to the same G_i-linked receptor (see below).

11.4.5.2 **Molecular Mechanism**

G protein activation by NA and derivatives was recently assessed as stimulation of guanosine 5′-(γ-[^{35}S]-thio)triphosphate binding, and [^{3}H]NA was used to specifically label binding sites [455]. NA ($EC_{50} \sim 1$ μM) stimulated [^{35}S]GTPγS binding in membranes from rat adipocytes and spleen, but not from other tissues. G protein activation in adipocyte membranes in the presence of maximally activating concentrations of a selective A_1 AR agonist, 2-chloro-N^6-cyclopentyl-adenosine, and NA was almost additive, indicating that G proteins of mostly distinct pools were activated by these agonists. G protein activation by NA and related substances in spleen and adipocytes revealed identical pharmacological profiles. [^{3}H]NA specifically detected guanine nucleotide-sensitive binding sites of identical pharmacology in adipocyte and spleen membranes [455]. These data indicate that NA most probably acts on a specific G protein-coupled receptor (GPCR).

To identify the G_i-linked GPCR for NA, orphan receptors were selected based on their tissue expression profiles for a rational screening [456]. Subsequently, a subset of 10 orphan GPCR, which by mRNA distribution analysis exhibited significant expression levels in both adipose tissue and spleen, were expressed in an appropriate mammalian cell line to allow measurement of functional GTPγS binding following NA treatment. Thereby HM74 was identified as a low-affinity receptor for NA which represents an orphan GPCR previously cloned from a cDNA library derived from human monocytes [457]. The half-maximal effector concentration for NA with 1 mM at HM74 was 1000-fold higher than that previously reported in membranes produced from rat adipose tissue or spleen [455]. Subsequent identification of a novel paralogue of HM74, termed HM74A, by a molecular biology approach and its expression in various test systems revealed that HM74A acts as a high affinity receptor for NA [456]. The activity and affinity of NA was in good agreement with that previously reported [455]. In the [^{3}H]NA displacement assay, both the absolute potency and the rank order of potency of various HM74A ligands was the same, whether tested against the stable CHO cell line expressing recombinant human HM74A or human adipocytes [456]. These data strongly suggest that HM74A is the G_i-linked GPCR for NA on human adipocytes. Compatible with this conclusion, nicotinamide, which unlike NA produces no alteration in lipoprotein profiles [432], acted only as a very weak agonist at HM74A with 1000-fold lower potency than NA. In contrast, acipimox has also been identified as a full high-affinity agonist for HM74A. Thus activation of HM74A seems to account for the inhibition of lipolysis observed with NA and acipimox. Therefore, of the NA derivatives tested in man, it appears that potency at HM74A is linked with their efficacy in normalizing lipoprotein profiles [456].

11.4.5.3 **Desensitization**

The established phenomenon that exposure of a cell to an agonist can cause desensitization, so that a second exposure to the agonist is less effective than the first, has been studied extensively and various mechanisms have been described. First, short-term exposure to a ligand (minutes) can induce a change in the receptor. For example, β-AR can become phosphorylated by a specific kinase (β-AR kinase) after exposure to an agonist [458, 459]. Second, after more prolonged exposure to an agonist, the number of receptors on the cell surface can decrease by a process termed down-regulation. Down-regulation is generally considered to occur in two steps. Initial rapid sequestration of receptors from the cell surface through endocytosis is followed by a slower loss of total cellular receptors involving intracellular degradation, although at least a portion is often recycled back to the cell surface, as has been demonstrated for the insulin receptor [460, 461]. In addition to increased degradation of receptors, recent studies for β-AR have suggested that alterations in receptor mRNA turnover, i.e. decreased transcription and/or increased mRNA degradation, may also play a role [462, 463]. These phenomena can readily explain homologous desensitization, where exposure of a cell to a hormone results in subsequent insensitivity to that hormone. Another commonly reported phenomenon is known as heterologous desensitization, in which treatment of a cell with a hormone subsequently can make the cell less sensitive to another hormone that operates through a different receptor. Heterologous desensitization has been described for many signaling systems, including GPCRs coupled to AC [464, 465] and to phospholipase C [466, 467]. Mechanisms of heterologous desensitization, however, are relatively poorly understood.

A_1 AR can, reportedly, be down-regulated in rat adipocytes following prolonged incubation with the non-metabolizable adenosine analog and agonist, PIA [468]. Although there are two major classes of AR, known as A_1 and A_2 receptors [469], adipocytes possess primarily receptors of the A_1 class [470, 471]. These receptors are inhibitory toward AC, and their action is thought to be mediated via a pertussis toxin-sensitive G-protein known as G_i, of which three subtypes 1–3 have been identified in rat adipocytes. Prolonged incubation of adipocytes with PIA leads to a marked down-regulation of A_1 AR [472]. In addition, an approximately 50% decrease in G_i, as measured by pertussis-catalyzed ADP-ribosylation, has been found. This observation suggested that G_i was down-regulated along with the AR [472]. Subsequently, a series of specific antisera was used to demonstrate that PIA causes a marked (ca. 90%) loss of G_i1 and G_i3 with a more modest (50%) loss of

G_i2 [473]. When PIA is washed away, G_i down-regulation was reversed, suggesting that it is a real regulatory phenomenon rather than a simple toxic effect [473]. Furthermore, Stiles and co-workers [474, 475] reported similar findings in adipocytes derived from rats after chronic infusion of PIA. More recent studies suggest that G-protein down-regulation is a common phenomenon resulting from chronic exposure to agonists. Since several agonists can presumably couple through a single G-protein, it is clear that this phenomenon of G-protein down-regulation could account for heterologous desensitization.

In addition to impairment of signal transduction activity at the level of G-proteins, heterologous desensitization of GPCR regulation of lipolysis can be caused by changes at the level of cell surface expression of GPCRs or signaling activity of signaling components, such as AC, PDE and PKA (catalytic activity), as well as function of the lipolytic end effectors (e.g. HSL/perilipin translocation apparatus, lipotransin action). Interestingly, prolonged incubation of rat adipocytes in primary culture with NA at concentrations up to 1 mM had no effect on the levels of G_i1-3 as well as AC [476]. However, pretreatment of the cells with 1 mM NA for 4 days decreased their sensitivity for subsequent acute inhibition of β-adrenergic stimulation of lipolysis and AC activity by NA by one order of magnitude (but had no effect on the sensitivity of the cells toward PIA). The molecular mechanism of the homologous desensitization of lipolysis by NA, which may form the molecular basis for this observed loss of efficacy of NA and acipimox treatment during chronic anti-hyperlipidemic therapy, remains to be elucidated.

11.4.6
Glimepiride and Phosphoinositolglycans

Recent studies have identified two completely different exogenous stimuli which mimic the anti-lipolytic activity of insulin in insulin target cells (for a review see Ref. [477]). They bypass activation of the insulin receptor, but rather act by triggering redistribution of signaling components between DIGs of different lipid and protein composition at the target cell plasma membrane. (i) Phosphoinositolglycan(-peptides) (PIG[-P]) prepared by chemical synthesis [478] or isolation from natural sources [475] reduce isoproterenol-stimulated lipolysis in concentration-dependent fashion to almost basal levels in primary and cultured adipocytes. PIG(-P) are derived from the polar core glycan head group of glycosylphosphatidylinositol (GPI) anchored proteins [480] consisting of 2′,3′-cyclic phospho-*myo*-inositol coupled to five sugar moieties in typical glycosidic linkages (PIG portion) and, optionally, contain an additional terminal ethanolamine-linked tripeptide residue (P portion) [478, 479]. (ii) The hypoglycemic sulfonylurea, glimepiride, which in NIDDM patients exerts an insulin-independent blood glucose-decreasing activity in addition to its insulin-releasing potency [481] blocks isoproterenol-stimulated lipolysis in adipocytes from insulin-resistant Zucker fatty rats or in adipocytes made insulin-resistant *in vitro* by incubation in primary culture in the presence of high concentrations of insulin and glucose [482, 483].

Interestingly, hydrophilic PIG-P and lipophilic glimepiride, albeit differing completely in structure, have in common part of the signaling pathway that cross-talks to the insulin signal transduction cascade and thereby negatively regulates HSL activity and lipolysis in adipocytes [480]. The available data support the following model for their molecular mode of action. PIG(-P) bind as agonists/antagonists to the receptor, p115, a transmembrane protein that is anchored at the cell surface within DIGs [484]. Upon binding, PIG(-P) displace endogenous GPI proteins from binding to p115 by competition with their GPI head-groups and thereby reduce their concentration at DIGs [485]. In contrast, glimepiride is not recognized by a proteinaceous receptor of the adipocyte plasma membrane but spontaneously inserts into DIGs via direct hydrophobic interactions in non-saturatable fashion with GPI lipids thereby altering the structural organization of DIGs [486, 487]. Both events induce the redistribution of lipid-modified signaling protein, which are known to reside within DIGs on basis of their acylation/glypiation and/or direct binding to caveolin (for reviews see Refs. [477, 488, 489]). In particular, GPI proteins and palmitoylated non-receptor tyrosine kinases of the src class, such as pp^{59}Lyn, translocate from DIGs of lower buoyant density (higher cholesterol content) to those of higher buoyant density (lower cholesterol content) accompanied by dissociation of pp^{59}Lyn from the DIG resident protein, caveolin (see above; Refs. [485, 488]). This relieves pp59^{Lyn} from inhibition by caveolin [485, 488], interaction of which with various signaling proteins down-regulates their signaling activity (for review see Refs. [489, 490]). In turn activated pp^{59}Lyn phosphorylates IRS-1 at tyrosine residues, presumably at the same sites used by the insulin receptor [491]. In consequence, activation of the PI3K pathway leads to upregulation of PDE3B and down-regulation of HSL activity via the same series of events engaged by insulin (see above). Very recent experiments revealed a possible explanation for the molecular link between the interaction of glimepiride with DIGs and the resulting reorganization and redistribution of structural (GPI proteins) and signaling (non-receptor tyrosine kinases) components within the DIGs [492]. The binding of glimepiride to GPI lipids triggers the activation of a DIG-associated GPI-specific phospholipase which cleaves GPI proteins and GPI lipids [492]. PIG(-P) thereby generated may bind to the p115 at the DIGs and subsequently induce redistribution of signaling proteins from DIGs and the same downstream series of events, as recently demonstrated for exogenously added synthetic or natural PIG(-P) [485]. Alternatively, GPI cleavage may directly affect the structure of DIGs and thereby cause the redistribution of DIG components at both the outer (GPI proteins) and inner (non-receptor tyrosine kinases) leaflets of the DIG membrane, accumulation of which critically depends on the presence of glycolipids [493, 494]. The latter possibility seems more likely since p115 is apparently not involved in mediating the anti-lipolytic activity of glimepiride [495]. It seems plausible that the loss of long-chain FA from the outer leaflet of DIGs caused by redistribution of GPI proteins or cleavage of GPI lipids within DIGs simultaneously triggers the alteration of the FA composition of the DIG inner leaflet. This "adjustment to a common FA milieu" across the leaflets of DIGs may facilitate the redistribution of pp59^{Lyn}. Interestingly, the GPI-specific phospholipase stimulated by glimepiride in cultured and primary adipocytes [496, 497] is reportedly activated also by insulin in the presence of glucose [496–498].

11.4.7
Differences in Regulation of TAG Storage and Mobilization between Visceral and Subcutaneous Adipocytes

Individuals with peripheral obesity possess fat distribution subcutaneously in gluteofemoral areas and the lower part of the abdomen, and are at little risk of metabolic complications, such as NIDDM. Conversely, individuals with upper-body obesity accumulate fat in subcutaneous and visceral deposits and are more susceptible to metabolic problems, in particular when visceral fat deposits are abundant. Visceral fat deposits, located in the body cavity, are composed of the omental and mesenteric fat, and comprise the minor component of total body fat, representing ~20% and ~5–8% of total body fat in men and women, respectively. In upper-body obesity, fat excess is present in the visceral abdominal regions but also in the subcutaneous abdominal regions. There are several explanations but no clear proof why upper-body obesity is more at risk of developing metabolic disease than lower-body obesity [499]. Since the visceral deposit is in direct contact with the liver through the portal circulation and considering the alterations of hormonal control of lipolysis of its adipocytes (insulin vs. catecholamines, see below), the "portal paradigm" was postulated on the basis that visceral adipocytes, through enhancement of lipolysis due to both reduced insulin-induced anti-lipolysis and enhanced catecholamine-induced lipolysis, will release portal NEFA that disturb liver metabolism [500]. Chronic NEFA excess will in turn lead to glucose intolerance, hyperinsulinemia, insulin resistance and dyslipidemia. However, experimental evidence is accumulating that non-visceral upper-body fat rather than visceral fat deposits is the predominant source of excess postprandial systemic NEFA availability in upper-body obese women and NIDDM patients [501, 502]. This raises the question as to whether the extent of visceral fat just represents another marker of ectopic fat distribution (i.e. fat present at the wrong place in tissues such as muscle, heart, liver and pancreatic β-cells). Alternatively, visceral adipocytes may possess properties that could be the origin of metabolic disturbances. In fact, striking adipose-location-related differences have been reported in adipocyte responsiveness to insulin and catecholamines. Moreover, differences have been described in the characteristics of various biochemical pathways regulating TAG storage and mobilization between subcutaneous and visceral fat deposits (Tab. 11.2).

In detail, adipocytes from visceral adipose tissue are more resistant to insulin-induced anti-lipolysis and re-esterification of NEFA than those from leg and non-visceral body fat both *in vitro* [503, 504] and *in vivo* [505]. Various functional differences in these cells have been identified at the level of the insulin receptor and the post-receptor insulin signaling cascade [503, 504]. PDE3B involved in lipolysis regulation by insulin (see above, Fig. 11.7) and protein tyrosine phosphatases dephosphorylating the insulin receptor, such as PTP1b, could be affected in differen-

Tab. 11.2 Differences in the regulation of TAG mobilization and storage between fat cells from visceral (vAT) and subcutaneous adipose tissue (scAT). The extent of some of these differences depends on the adipocyte hyperplasie (total fat mass) and hypertrophy as well as on the sex. Adapted from Ref. [499] with modifications (see [449] for references).

Effects and factors	*Regional differences*
TAG mobilization	
$\beta_{1,2}$-Adrenoceptor number	vAT > scAT
β-Adrenoceptor-dependent stimulation of TAG mobilization	vAT > scAT
Isoproterenol-induced stimulation of AC	scAT = vAT
Basal rat of lipolysis	scAT > vAT
HSL protein level and activity	scAT > vAT
ALBP mRNA and protein levels	scAT > vAT
Leptin mRNA and protein secretion	scAT > vAT
TNFα secretion	vAT = scAT
TAG Storage	
α_2-Adrenoceptor number	scAT > vAT
α_2-Adrenoceptor-dependent inhibition of TAG storage	scAT > vAT
Insulin receptor expression (exon 11 deleted)	vAT > scAT
IRS-1 mRNA and protein levels	scAT > vAT
Insulin-induced insulin receptor tyrosine phosphorylation	scAT > vAT
Insulin-induced IRS-1 tyrosine phosphorylation	scAT > vAT
Insulin-induced PI3K activation	scAT > vAT
PTP1b activity	vAT > scAT
ASP mRNA levels	scAT > vAT
LPL mRNA and protein levels	scAT > vAT
FA Transport by human preadipocytes	scAT > vAT

tial fashion in visceral vs. subcutaneous adipocytes. This concurs with the endogenous PTP1b activity found to be elevated in visceral adipose tissue and might contribute to the relative insulin resistance of this fat deposit [506]. Exacerbating the impairment of insulin inhibition of lipolysis, the lipolytic response toward catecholamines is more pronounced in isolated adipocytes from visceral adipose tissue than from subcutaneous gluteal, femoral and abdominal adipose tissue [507]. This higher lipolytic activity can be explained by altered expression or function of HSL (increase) and/or proteins interacting with either HSL, such as ALBP (increase), or the LD, such as perilipin (decrease) (see above, Fig. 11.6). An enhanced α_2-adrenoceptor responsiveness associated with a concomitant decrease in β-adrenoceptor responsiveness explains the lower lipolytic effect of catecholamines in gluteal and femoral adipocytes of normal and obese women and abdominal adipocytes of obese men compared with visceral adipocytes. Conversely, visceral adipocytes exhibit the highest $\beta_{1,2}$-adrenoceptor-mediated and the weakest α_2-adrenoceptor-mediated lipolytic and anti-lipolytic responsiveness, respectively, to catecholamines, which seems to correlate with decreased expression of α_2-adrenoceptors and concomitantly increased expression of β-adrenoceptors [508, 509]. In obese subjects, unrestrained lipolysis leads to excessive NEFA release from visceral hypertrophied adipocytes and may prevent their further enlargement in contrast to subcutaneous adipocytes, which actually represent the largest ones. The search for adipose tissue-specific local differences in the expression of genes that regulate the differentiation and expansion of adipose tissue as well as the regulation of TAG storage and mobilization is underway using genomic and proteomic approaches.

11.4.8 Up-/Down-regulation of Components of TAG Storage and Mobilization

11.4.8.1 HSL

Recently, the functional significance of HSL in adipose tissue metabolism has begun to be clarified in studies using HSL null mice [510–512]. These mice showed normal growth rates and body weights. While the epididymal retroperitoneal and femoral WAT depots of HSL null mice remained unchanged, they displayed a 65% increase in brown adipose tissue mass compared with wild-type control mice. Inactivation of HSL resulted in the complete absence of neutral cholesterylester hydrolase activity in adipose tissue (both white and brown). However, TAG lipase activity in WAT was reduced by only 40% and TAG lipase activity in brown adipose tissue was similar to wild-type mice [510]. Basal lipolysis, i.e. glycerol release, was reduced in isolated adipose cells from HSL null mice in one study [498], but was unaffected and seemingly increased in another [511]. Nonetheless, there was a marked defect or complete absence of catecholamine-stimulated glycerol release in adipose cells from HSL null mice, whereas catecholamine-stimulated NEFA release was still observed, but attenuated [510, 511]. This apparent discrepancy in the release of glycerol and NEFA from adipose cells of HSL null mice has been clarified by the observation that the basal and catecholamine-induced DAG content increased markedly in white and brown adipose tissue of HSL null mice [512]. Therefore, the studies with HSL null mice appear to substantiate that HSL is the rate-limiting enzyme for DAG hydrolysis in adipose tissue and is essential for hormone-stimulated lipolysis. Remarkably, the absence of HSL was not associated with the development of obesity. However, adipose cells from HSL null mice, while displaying size heterogeneity, tended to be hypertrophic [510, 511]. Moreover, due to defective lipolysis during fasting, there was a reduction in circulating NEFA and a decreased hepatic production of VLDL secondary to the diminished release of NEFA from adipose tissue [513]. This was associated with an induction of LPL in WAT, as well as in skeletal and cardiac muscle, but a decrease of LPL in brown adipose tissue [513]. It remains to be determined whether operation of the TAG lipase in HSL null mice is a consequence of its compensatory upregulation in HSL-deficient adipocytes, only, or is indicative for expression of a second TAG-specific HSL-like lipase in wild-type adipocytes, too. Taken together, the findings on the lack of obesity and mild adipocyte hypertrophy observed in HSL null mice suggest that other lipases could also play a role in TAG mobilization.

Interestingly, in HSL null mice after prolonged fasting, plasma NEFA and TAG levels as well as hepatic TAG stores were reported to be significantly lower than with wild-type mice [513, 514]. This low hepatic TAG content was associated with improved hepatic insulin sensitivity since insulin caused a greater reduction in endogenous hepatic glucose production in HSL null mice (by ~71%) than in wild-type mice (by ~31%). This increase in hepatic insulin sensitivity was associated with elevated insulin receptor protein levels and activation of components of the insulin signaling cascade, such as phosphorylation of PKB and activity of PI3K [514]. The low hepatic TAG content in HSL null mice can be explained by their low plasma NEFA levels, since liver-specific NEFA uptake is commonly regarded to be a concentration-driven process facilitated by specific membrane transporters (see above). The inverse relationship between insulin sensitivity and hepatic TAG content may be explained by alterations in gene expression during activation of nuclear transcription factors such as peroxisome proliferator-activated receptors by intracellular TAG and/or fatty acyl derivatives. Furthermore, the diminished plasma NEFA levels *per se* cannot be excluded from being, at least partly, responsible for the increased hepatic insulin sensitivity, since plasma NEFA concentrations are inversely correlated with insulin sensitivity (see above). Strikingly, in HSL null mice, no differences were observed in insulin-mediated whole-body glucose uptake compared with wild-type mice as revealed during hyperinsulinemic euglycemic clamp studies [514]. The lack of improvement of whole-body insulin sensitivity may be explained by the reported absence of significant differences in muscle TAG content in combination with unchanged insulin-induced PKB phosphorylation in HSL null compared with wild-type mice despite lower plasma NEFA levels.

Recently, HSL null mice have been shown to have impaired insulin secretion, a 2- to 2.5-fold increase in islet TAG content, and elevated basal insulin secretion from isolated islets that fails to rise further upon challenge with glucose [515]. Comparison of the gene expression profile between islets isolated from wild-type and HSL null mice using microarray chip technology and Taqman quantitative PCR for selected genes revealed changes in genes that are involved in lipid metabolism (methyl sterol oxidase, short chain acyl-CoA dehydrogenase), insulin response (insulin-induced growth response protein) and cytoskeleton (profilin, cofilin). In particular, the mRNAs for UCP-2, SREBP-1c, and PPARγ were up-regulated in islets from HSL null mice as compared with wild-type mice. The transcriptional up-regulation of UCP-2, which is regulated by SREBP-1c and PPAR-γ, was associated with elevated NEFA and correlated with impaired glucose-stimulated insulin secretion. Finally, C/EBPα, which is involved in cytokine-regulated apoptosis of pancreatic β-cells, was also up-regulated. Collectively, these findings suggest that HSL is important in maintaining lipid homeostasis in β-cells by directly and indirectly controlling gene expressing for multiple metabolic and signaling pathways and hence in the regulation of glucose-stimulated insulin secretion.

A recent study using transgenic mice overexpressing HSL specifically in β-cells provided novel and important insights with regard to the role of HSL in the development of lipotoxicity [516]. These transgenic mice were characterized by impaired glucose and severely blunted glucose-stimulated insulin secretion upon challenge with a high-fat diet. Their islets displayed both elevated HSL activity and forskolin-induced lipolysis compared with wild-type islets, which resulted in significantly reduced TAG levels in transgenic compared with wild-type islets provided the mice have been fed a high-fat diet. Thus, the rate of influx of NEFA into TAG of the islet LD and the capacity to mobilize this TAG pool seem to determine the emergence of islet lipotoxicity. This is in accord with the recently reported inverse correlation between apoptosis and TAG accumulation in cultured β-cells, suggesting a cytoprotective function of TAG in cytoplasmic LD formed by normal β-cells against NEFA-induced islet dysfunction [517]. Moreover, prolonged high-fat feeding of mice is accompanied by down-regulation of the expression of islet HSL [518]. Consequently, physiological down-regulation of HSL or pharmacological inhibition of HSL and possibly of other as yet unknown TAG lipases in β-cells could be interpreted in terms of the operation of a defense mechanism against the emergence of NEFA-induced islet cell dysfunction.

In conclusion, HSL degrading TAG of LD in islets seems to exert at least three different functions in the regulation of insulin secretion and its coupling to external stimuli. (i) In the short-term, upon challenge of β-cells with an insulin secretagogue, HSL is phosphorylated and activated via (glucagon-like peptide 1 and gastric inhibitory polypeptide) receptor/cAMP-dependent mechanisms lipolytically releasing NEFA and/or NEFA derivatives from islet TAG. These may act as acute stimulus (glucose)-secretion coupling factors for the exocytosis of insulin-containing secretory vesicles. (ii) In the long-term diet-induced NIDDM and the increased influx of NEFA derived from plasma lipoproteins through the action of LPL and plasma NEFA into β-cells lead to the accumulation of TAG in LD of β-cells which apparently exceeds the degradation capacity of the basal HSL. Thus the low activity state of HSL under these conditions guarantees the storage of NEFA (derivatives) which potentially compromise the insulin secretory mechanism and induce β-cell apoptosis. (iii) Long-lasting aberrations in the regulation of lipolytic activity, such as overexpression of HSL, failure to down-regulate HSL during prolonged high-fat feeding or the mass effect of TAG overstorage *per se* leading to incremental lipolytic release of NEFA even at low constitutive HSL activity, may result in exceeding a certain threshold level of NEFA flux. This triggers the lipotoxic pathway, presumably involving binding of NEFA (derivatives) as ligands to transcription factors, such as PPAR, and other mechanisms. The resulting alterations in the expression of genes regulating insulin secretory vesicle biogenesis and exocytosis finally lead to dampening of glucose-induced insulin secretion.

11.4.8.2 **ALBP**

In the absence of ALBP, intracellular concentrations of NEFA unbound to protein were increased in adipose cells, and basal as well as isoproterenol-stimulated lipolysis is decreased by 40% [519, 520]. These ALBP-deficient rodents showed a normal growth rate, body weight and body composition compared with wild-type littermates due to functional compensation by the keratinocyte-type FABP [521, 522; see above]. However, when exposed to a high-fat diet, ALBP null mice developed diet-induced obesity, reaching a greater total weight gain than the control mice as a consequence of an increased fat pad weight. Both lean and obese ALBP null mice displayed normoglycemia and normoinsulinemia, providing evidence for an uncoupling of obesity from insulin resistance through ALBP deficiency [519, 520].

11.4.8.3 **Perilipin**

Recently, Arner and coworkers investigated the amount of perilipin protein and lipolysis rates in human subcutaneous adipocytes of non-obese and obese women [523]. A polymorphism in intron 6 (rs891460 A/G) in the perilipin gene seems to be associated with a 50% decrease in perilipin expression and increased basal lipolysis (AA vs. GG carriers) in adipocytes of the obese women, which in subjects matched for body mass index and adipocyte volume was inversely correlated with the plasma concentrations of NEFA and glycerol and thus may contribute to the development of insulin resistance. Furthermore, they found two- to four-fold increases in noradrenaline-induced lipolysis [523]. These data also reinforce the role of perilipin for the regulation of lipolysis in human fat cells.

The participation of perilipin in the regulation of TAG storage and mobilization has been addressed by knockout approaches from two different experimental groups [524, 525]. Perilipin-deficient mice showed an increased metabolic rate together with an increased basal lipolytic rate in adipocytes, which confer resistance to diet-induced obesity. Nonetheless, they were severely impaired in their ability to respond to β-adrenergic stimulation by an appropriate elevation in circulating NEFA. Although absence of perilipin resulted in leanness and reversal of obesity with a ca. 50% decrease in adipose mass and adipocyte size, which was not reflected in the lean body mass (that even increased slightly), it did not affect glucose tolerance [524], even with perilipin ablation being able to worsen peripheral glucose disposal [525]. Perhaps most surprising is that mice lacking perilipin are more drastically affected in regard to regulation of FA mobilization than those deficient in HSL (see above). These observations could be explained if perilipin is not only required to maintain adipose cells in the quiescent state but is also involved in the functional lipolytic activation in adipose cells. To test this hypothesis experimentally, Londos and coworkers have recently investigated HSL translocation in adipocytes differentiated *in vitro* from embryonic fibroblasts of wild-type and perilipin-null mice [526]. Strikingly, they found that HSL fails to translocate to LD in cells lacking perilipin. Furthermore, introduction of normal perilipin A but not of an unstimulatable mutant variant (lacking the amino-terminal PKA phosphorylation sites) into CHO fibroblasts significantly stimulated the PKA-induced translocation of HSL. CHO cells expressing both perilipin and HSL were able to elicit a greater and more rapid response to a lipolytic stimulus than when either of these proteins was expressed separately [526]. Thus, the perilipin-null mice emphasize the two major actions of perilipin: (i) to protect TAG against hydrolysis in the basal state and thus permit TAG storage; and (ii) to facilitate HSL translocation to LD, leading to hydrolysis of TAG and release of FA in the stimulated state. Interestingly, the perilipin-mediated translocation process does not seem to depend on additional factors unique to adipocytes since it can be reconstituted by expressing both perilipin A and HSL in CHO fibroblasts. Moreover, phosphorylation of perilipin A at its three most amino-terminal PKA sites is apparently required to promote HSL translocation to LD [526]. Recently, the same group demonstrated that expression of wild-type perilipin alone, but not of a mutant lacking these PKA sites, in CHO cells increases the sensitivity of lipolysis towards β-adrenergic stimulation [261]. In contrast, LD coated with ADRP from cells which do not express perilipin (i.e. all with the exception of adipocytes and steroidogenic cells) neither facilitate HSL binding to the LD surface nor respond to elevated PKA activity nor are capable to restrain lipolysis to a minimum. The accelerated lipolytic activity exhibited by perilipin A-coated LD may result from increased accessibility of HSL (or some unknown neutral lipase) to TAG within LD when perilipin is phosphorylated at its three most amino-terminal PKA sites. Importantly, the binding of HSL to LD seems to be based on its direct interaction with the TAG substrate since translocation of HSL is blocked upon mutating the serine residue in its catalytic triad [526].

Two hypotheses have been proposed that reconcile the present data. Phosphorylated perilipin could directly recruit HSL to the LD surface by direct binding to the enzyme. However, pull-down and two-hybrid experiments did not provide experimental evidence for a direct interaction between perilipin and HSL so far. Alternatively, phosphorylated perilipin may modify the LD surface to indirectly facilitate interaction of HSL with the TAG core within the LD. HSL translocation and initiation of lipolysis occur within minutes [252]. Initially, HSL and perilipin co-localize on LD but then, within a much longer period, perilipin slowly departs from large LD, and is subsequently recovered from dispersed, much smaller, LD (see above). Londos and coworkers recently stated that even after the departure of perilipin, HSL remains bound to the larger LD which now lack perilipin, where it continues to cleave TAG [526]. Taken together, the present data favor the indirect mechanism, although the two hypotheses are not mutually exclusive. In any case, the mechanism underlying the action of phosphorylated perilipin seems to modify the TAG of LD such that it is more effectively presented as a substrate, e.g. by triggering dispersion of LD upon contact with phosphorylated perilipin leading to a larger total LD surface area and thereby facilitated access of HSL. It is now generally accepted that protein–phospholipid/cholesterol interactions can modify the shape of the plasma membrane [527] and this mechanism seems to form the molecular basis for the biogenesis of caveolae (see above), for instance. The operation of similar mechanisms for neutral lipids seems to be more speculative.

11.4.8.4 PKA

PKA is a heterotrimer consisting of two regulatory cAMP-binding and two catalytic subunits [528]. Four regulatory isoforms (RIα, RIβ, RIIα, and RIIβ) and two catalytic isoforms (Cα and Cβ) are expressed in the mouse, and each is encoded by a separate gene. The RIIβ subunit is expressed principally in three tissues known to regulate energy homeostasis: brown adipose tissue, WAT, and brain [528, 529]. It has been suggested that the induction of PKA in certain tissues may decrease obesity. For example, activation of the adipose-specific β-AR [530], which signals via PKA, decreased obesity in both genetically obese (ob/ob) [531, 532] and diet-induced obese mice [533], suggesting that signaling mechanisms through this pathway are important in preventing obesity.

Studies in mice lacking RIIβ have revealed an unexpected role for this protein in regulating energy balance [530]. These animals remained remarkably lean even when challenged with a high-fat diet and had increased metabolic activity, manifested by increases in body temperature, uncoupling protein-1 concentration, and lipolysis. Biochemical studies [534] have shown that loss of RIIβ was compensated by increased expression of RIα, which is more sensitive to cAMP activation, resulting in a net increase in basal PKA activity, but markedly impaired cAMP-induced activation of HSL in mutant WAT, yet left cAMP-regulated gene expression relatively unperturbed. The disruption of lipolytic stimulation was seen both *in vitro* and *in vivo* and occurred equally for signaling from the β_3-receptor-specific agonist CL316.243 and a non-specific β-agonist, isoproterenol [531]. It is unlikely that the signaling defect in RIIβ mutant WAT was caused by decreased expression of either β-AR or HSL, given that mRNA levels for these genes were unaffected. Conceivably, impairment of lipolytic hormonal response could be a consequence of the chronic stimulation of basal lipolysis seen in RIIβ null mice due to increased basal PKA activity. However, adipocytes from knockout mice deficient in the G-protein subunit, $G_i\alpha_2$, showed a 3-fold increase in basal cAMP levels and an elevated basal rate of lipolysis but retained normal maximal response to βAR agonists [535]. These studies suggest that increasing basal PKA activity in adipose tissue and brain ameliorates obesity.

In a recent study, wild-type and RIIβ null mice maintained on the C57BL/6 genetic background strain were fed a high-fat, high-carbohydrate diet [536], which is known to induce obesity and diabetes in C57BL/6 mice [537, 538]. RIIβ null mice were resistant to weight gain and hyperinsulinemia [537]. *In vivo* insulin sensitivity and glucose disposal were dramatically improved in the RIIβ null mice, as were plasma lipid profiles. When mice were corrected for differences in body weight, improved insulin-mediated glucose disposal was still observed in the RIIβ null mice, suggesting an obesity-independent effect of RIIβ on promoting insulin resistance [537]. This suggests that PKA activity in both adipose tissue and brain is important for determining body composition, food intake and diabetogenic parameters.

It has been speculated that at least some of the elements mediating lipolytic stimulation (e.g. β-AR, AC, PKA, HSL) be co-localized within adipocytes to facilitate efficient signal transduction, and that lipolytic stimulation is impaired in

RIIβ-deficient rodents because RIIβ participates specifically in the formation of this complex. Previous studies suggest a compartmentalized apparatus that mediates lipolytic stimulation as (i) At any given intracellular cAMP concentration, the lipolytic response toward catecholamines is greater than toward forskolin, a nonspecific AC activator [539–541]. Hence low concentrations of isoproterenol (~10 nM) can stimulate lipolysis without measurably altering overall cAMP levels, whereas low concentrations of forskolin (0.1–1.0 μM) increase intracellular cAMP without affecting lipolysis [541]. (ii) The concentration of isoproterenol or β_3-specific agonists required for half-maximal activation of AC activity in adipocyte membranes is ~80-fold greater than the concentration required to activate lipolysis in intact adipocytes [542–544]. (iii) A signaling complex including β_2-AR, PKA, and phosphatases has been isolated that appears to be assembled by the scaffold protein, gravin [545]. In summary, catecholamines stimulate lipolysis more potently than they increase overall intracellular cAMP, suggesting a preferential association of β-AR, PKA, and perhaps its substrate, HSL, in a "lipolytic" complex.

In numerous cell types, co-localization of key components involved in PKA signaling is accomplished by PKA-anchoring proteins (AKAPs), multivalent binding proteins that serve as platforms for the assembly of signal transduction modules [546, 547]. These targeting proteins bind simultaneously to specific sites on the amino-terminus of *R* subunits and to discrete subcellular structures. By tethering PKA at precise intracellular sites, AKAPs ensure that the kinase is exposed to localized changes in cAMP adjacent to appropriate substrates, thus preventing cross-talk between functionally unrelated PKA signaling units within the same cell. Although AKAPs have not yet been described in WAT, they have been found in virtually all other tissues investigated and it seems reasonable to speculate that AKAPs may facilitate phosphorylation of HSL by PKA in compartment-specific fashion in fat cells of WAT. Since most mammalian AKAPs identified to date are strongly RII-specific, PKA-AKAP binding might be disrupted in RIIβ mutant WAT due to the RIIβ-to-RIα isoform switch. This alteration could displace the PKA holoenzyme from local waves of cAMP generated by β-AR agonists and/or from HSL, which would explain the blunted lipolytic response to adrenergic stimuli in mutant WAT. Indeed, disruption of RII-AKAP binding has resulted in diverse functional consequences in other model systems [548, 549].

11.4.8.5 **ASP**

Since mice lacking complement C3 can not produce ASP, C3 gene knockout mice are consequently ASP-null animals. ASP null mice show a reduced WAT weight, distributed evenly through all depots in females while primarily affecting gonadal and perirenal localizations in males, as well as reduced body fat [550, 551]. In addition, male mice have delayed TAG clearance [551]. The reduced adipose mass was accompanied by hypoleptinemia with a modest increase in food intake. When fed on either a low-fat or high-fat diet male ASP null mice showed no differences in body weight compared with wild-type animals. Female ASP null mice, however, had a decreased body weight compared with their wild-type counterparts [550, 551]. The relatively mild phenotype of these mice suggests a minor role of ASP in TAG storage and mobilization control or the existence of compensatory mechanisms rescuing part of the physiological function of ASP.

Very recently, the effect of ASP deficiency has been examined in ob/ob mice, which are leptin-deficient. The ob/ob mice have been used to test for protection from obesity in several double knockout models (for a review see Ref. [119]). In most cases, the decrease in body weight, which was evident in the single knockout, was enhanced when examined on the background of the ob/ob obese mouse model. Compared with age-matched ob/ob mice, ob/ob C3 null mice had delayed postprandial TAG and FA clearance, associated with decreased body weight and according to the homeostasis model assessment index increased insulin sensitivity. By contrast, food intake in the double-knockout mice was slightly elevated compared with ob/ob mice. The hyperphagia/leanness was balanced by a significant increase in energy expenditure and oxygen consumption. These results suggest that the ASP regulation of energy storage may influence energy expenditure and dynamic metabolic balance.

Since ASP increases TAG synthesis and re-esterification [409, 422] and decreases intracellular TAG mobilization, such as norepinephrine-stimulated NEFA release from fat cells as well [422], the effect of ASP, at least in adipose tissue, is to effectively decrease substrate cycling and increase TAG storage. ASP deficiency could possibly release the brake on this cycling process, allowing for increased substrate cycling and augmenting energy expenditure. This pattern is different to that observed for insulin, which had marked effects on both lipolysis and fractional NEFA re-esterification. When fat cells were incubated with maximally effective concentrations of each protein, ASP inhibited NEFA release during lipolysis to a lesser degree than insulin, whereas it stimulated fractional NEFA re-esterification to the same, or even greater, extent as insulin [422].

11.4.8.6 Caveolin

Caveolin null mice shed additional light onto the proposed new roles of caveolin and caveolae in TAG storage and mobilization (see above). Caveolin-1 null mice were found to store less TAG in WAT than control mice [552]. This defect led to resistance to diet-induced obesity and to increased blood concentrations of NEFA and TAG. Because caveolin can be recovered with LD, at least under certain experimental conditions (see above), it is tempting to speculate that the lack of caveolin interferes with normal LD biogenesis or enables HSL to gain access to TAG in the LD core. Alternatively, in the absence of caveolin-1 adipocytes lack functional caveolae and this could lead to diminished FA uptake. As the flux of FA into primary adipocytes and 3T3-L1 adipocytes follows saturatable kinetics [553], facilitated membrane transport has been proposed as the uptake mechanism. Interestingly, labeling of membrane proteins with photoreactive long-chain FA identified caveolin-1 as the major FA-binding protein in adipocytes [554, 555].

Together with the demonstrated association of the caveolins with LD (see above), it is possible that caveolae and the caveolins act as portals for the uptake and transport of FA to LD. However, it cannot be excluded that the negative effect of missing caveolin-1 on TAG storage is secondary to impairment of insulin signaling to the glucose transport system and the resulting reduction of glucose uptake and FA esterification. Caveolae and lipid rafts act as scaffolding platform for the so-called Cbl-CAP pathway that is required for insulin stimulation of glucose transport in addition to the IRS-PI3K pathway (for a review see Refs. [495, 556]).

11.5 Concluding Remarks

The experimental evidence for involvement of aberrant (with regard to amount and/or localization) intracellular storage of TAG in LD of various tissues (muscle, liver, heart and pancreas) and dysregulated mobilization of FA from TAG in LD of adipose tissue in the development of the metabolic syndrome and further on to frank NIDDM is increasing steadily. Consequently, the appropriate down-regulation of unrestrained TAG mobilization seems to represent a promising mode of action for future antidiabetic drugs which encompass three different levels of molecular mechanisms, each engaging several potentially interesting molecular targets as discussed in the preceding chapters:

1. Direct inhibition of TAG degradation by blockade of the catalytic activity of HSL or its physical interaction with TAG or by increasing the sensitivity of HSL for feedback inhibition by FA or by down-regulation of HSL expression.
2. Inhibition of the translocation of HSL from the cytosol to LD, for instance, by upregulation of perilipin expression or blockade of PKA-dependent phosphorylation of HSL and perilipin, or enhancement of the HSL-lipotransin interaction during inhibition of the lipotransin ATPase.
3. Positive or negative interference with signaling cascades mediating hormonal control of lipolysis at the receptor or post-receptor level, for instance, by triggering insulin receptor-independent tyrosine phosphorylation of insulin receptor substrates and downstream signaling along the PI3K pathway during direct (glimepiride) or receptor-mediated (phosphoinositolglycans) interaction with DIGs, or by blockade or activation, respectively, of stimulatory or inhibitory GPCRs or by upregulation/down-regulation of the expression of corresponding $G_{i/s}$ proteins.

Each of these mechanisms and targets offers putative advantages or raises potential issues for drug discovery with regard to

1. higher mechanistic selectivity (receptors, HSL activity and translocation, lipotransin ATPase vs. signaling components, transcription factors);
2. higher tissue specificity (receptors vs. HSL activity, transcription factors, signaling components);

3. higher probability for the identification of small drug-like chemical entities (receptors, HSL activity, lipotransin ATPase vs. binding of HSL to LD, interaction between signaling components, transcription factors);
4. better technical feasibility of a screening approach (receptor, HSL activity, lipotransin ATPase, transcription factors/gene expression vs. HSL translocation and interaction with lipotransin, interaction between signaling components);
5. less side effects concerning long-term treatment and desensitization (HSL activity, translocation and down-regulation vs. receptors, signaling components).

Not surprisingly, agonists/antagonists for receptors and enzyme inhibitors comprise by far the major portion of drugs currently on the market. But a consequent and careful analysis of the criteria mentioned above for each putative molecular target/mechanism will be helpful to identify the most attractive projects for the discovery of drugs modulating TAG storage and mobilization. The modern and powerful methods for generation and optimization of lead compounds by a combination of random, i.e. high-throughput-screening based, and rationale, i.e. structure-based, approaches (Chapter 3) will undoubtedly support the finding of drugs that operate at the different known levels of regulation of TAG storage and mobilization.

11.6 References

1 H. S. Sul and D. Wang, Nutritional and hormonal regulation of enzymes in fat synthesis: Studies on fatty acid synthase and mitochondrial glycerol-3-phosphate acyltransferase gene transcription, *Annu. Rev. Nutr.*, **1998**, *18*, 331–351.

2 R. Lehner and A. Kuksis, Biosynthesis of triacylglycerols, *Prog. Lipid Res.*, **1996**, *35*, 169–201.

3 L. B. Nielsen, M. Veniant, and J. Boren, et al., Genes for apolipoprotein B and microsomal triglyceride transfer protein are expressed in the heart: evidence that the heart has the capacity to synthesize and secrete lipoproteins, *Circulation*, **1998**, *98*, 13–16.

4 H. Mulder, L. S. Holst, H. Svensson, E. Degerman, F. Sundler, B. Ahren, P. Rorsman, and C. Holm, Hormone-sensitive lipase, the rate-limiting enzyme in triglyceride hydrolysis, is expressed and active in beta-cell, *Diabetes*, **1999**, *48*, 228–232.

5 R. H. Unger, Y.-T. Zhou, and L. Orci, Regulation of fatty acid homeostasis in cells: novel role of leptin, *Proc. Natl. Acad. Sci. USA*, **1999**, *96*, 2327–2332.

6 M. D. Jensen, Lipolysis: contribution from regional fat, *Annu. Rev. Nutr.*, **1997**, *17*, 127–139.

7 O. Peroni, V. Large, and M. Beylot, Measuring gluconeogenesis with [2-^{13}C]glycerol and mass isotopomer distribution analysis of glucose, *Am. J. Physiol.*, **1995**, *269*, 516–523.

8 K. N. Frayn, Non-esterified fatty acid metabolism and postprandial lipaemia, *Atherosclerosis*, **1998**, *141*, 41–46.

9 K. N. Frayn, Regulation of fatty acid delivery in vivo, *Adv. Exp. Med. Biol.*, **1998**, *441*, 171–179.

10 K. N. Frayn, S. W. Coppack, B. A. Fielding, and S. M. Humphreys, Coordinated regulation of hormone-sensitive lipase and lipoprotein lipase in human adipose tissue in vivo: implications for the control of fat storage and fat mobilization, *Adv. Enzyme Regul.*, **1995**, *35*, 163–178.

11 J. Saleh, L. K. Summers, K. Cianflone, B. A. Fielding, A. D. Sniderman, and

K. N. Frayn, Coordinated release of acylation stimulating protein (ASP) and triacylglycerol clearance by human adipose tissue in vivo in the postprandial period, *J. Lipid Res.*, **1998**, *39*, 884–891.

12 D. B. Zilversmit, Atherogenesis: a postprandial phenomonen, *Circulation*, **1979**, *60*, 473–485.

13 P. H. Groot, W. A. van Stiphout, and X. H. Krauss et al., Postprandial lipoprotein metabolism in normolipidemic men with and without coronary artery disease, *Arterioscler. Thromb.*, **1991**, *11*, 653–662.

14 F. Karpe, G. Steiner, T. Olivecrona, L. A. Carlson, and A. Hamsten, Metabolism of triglyceride-rich lipoproteins during alimentary lipemia, *J. Clin. Invest.*, **1993**, *91*, 748–758.

15 R. J. Havel, Chylomicron remnants: hepatic receptors and metabolism, *Curr. Opin. Lipidol*, **1995**, *6*, 312–316.

16 G. F. Gibbons, K. Islam, and R. J. Pease, Mobilisation of triacylglycerol stores, *Biochim. Biophys. Acta*, **2000**, *1483*, 37–57.

17 P. D. Berk and D. D. Stump, Mechanisms of cellular uptake of long chain free fatty acids, *Mol. Cell. Biochem.*, **1999**, *192*, 17–31.

18 M. J. Arthur, B. P. Atshaves, A. Frolov, W. D. Foxworth, A. B. Kier, and F. Schroeder, Cellular uptake and intracellular trafficking of long chain fatty acids, *J. Lipid Res.*, **1999**, *40*, 1371–1383.

19 M. Synak, M. Gorecka, J. Langfort, E. Smol, and E. Zernicka, Palmitic acid incorporation into intramuscular acylglycerols depends on both total and unbound to albumin palmitic acid concentration, *Biochem. Cell. Biol.*, **2003**, *81*, 35–41.

20 R. Weisinger, J. Gollan, and R. Ockner, Receptor for albumin on the liver cell surface may mediate uptake of fatty acids and other albumin-bound substances, *Science*, **1981**, *211*, 13395–13401.

21 J. E. Schnitzer and P. Oh, Albondin-mediated capillary permeability to albumin. Differential role of receptors in endothelial transcytosis and endocytosis of native and modified albumins, *J. Biol. Chem.*, **1994**, *269*, 6072–6082.

22 F. B. Kraemer and W.-J. Shen, Hormone-sensitive lipase: control of intracellular tri(di)acylglycerol and cholesteryl ester hydrolysis, *J. Lipid Res.*, **2002**, *43*, 1585–1594.

23 E. A. Newsholme, Substrate cycles: their metabolic, energetic and thermic consequences in man, *Biochem. Soc. Symp.*, **1978**, *43*, 183–205.

24 E. A. Newsholme and B. Crabtree, Substrate cycles in metabolic regulation and in heat generation, *Biochem. Soc. Symp.*, **1976**, *41*, 61–109.

25 D. Steinberg, Fatty acid mobilization – mechanisms of regulation and metabolic consequence, *Biochem. Soc. Symp.*, **1963**, *24*, 111–138.

26 M. Vaughan, The production and release of glycerol by adipose tissue incubated in vitro, *J. Biol. Chem.*, **1962**, *237*, 3354–3358.

27 D. Steinberg, M. Vaughan, and S. Margolis, Studies of triglyceride biosynthesis in homogenates of adipose tissue, *J. Biol. Chem.*, **1961**, *236*, 1631–1637.

28 N. K. Edens, J. Hirsch, and J. J. Hirsch, Mechanism of free fatty acid re-esterification in human adipocytes in vitro, *Lipid Res.*, **1990**, *31*, 1423–1431.

29 R. L. Leibel and J. Hirsch, A radioisotopic technique for analysis of free fatty acid reesterification in human adipose tissue, *Am. J. Physiol.*, **1985**, *248*, 140–147.

30 K. N. Frayn, C. M. Williams, and P. Arner, Are increased plasma non-esterified fatty acid concentrations a risk marker for coronary heart disease and other chronic diseases?, *Clin. Sci.*, **1996**, *90*, 243–253.

31 G. M. Reaven, The fourth musketeer – from Alexandre Dumas to Claude Bernard, *Diabetologia*, **1995**, *38*, 3–13.

32 E. Shafrir and I. Raz, Diabetes: mellitus or lipidus, *Diabetologia*, **2003**, *46*, 433–440.

33 J. D. McGarry, Banting Lecture 2001: Dysregulation of fatty acid metabolism in the etiology of type 2 diabetes, *Diabetes*, **2002**, *51*, 7–18.

34 V. P. Dole, A relation between nonesterified fatty acids in plasma and the metabolism of glucose, *J. Clin. Invest.*, **1956**, *35*, 150–154.

35 R. S. Gordon and A. Cherkes, Unesterified fatty acid in human blood plasma, *J. Clin. Invest.*, **1956**, *35*, 206–212.

36 N. B. Ruderman, C. J. Toews, and E. Shafrir, Role of free fatty acids in glucose homeostasis, *Arch. Int. Med.*, **1969**, *123*, 299–313.

37 K. L. Zierler and D. Rabinowitz, Effect of very small concentrations of insulin on forearm metabolism. Persistence of its action on potassium and free fatty acids without its effect on glucose, *J. Clin. Invest.*, **1964**, *43*, 950–962.

38 P. J. Randle, P. B. Garland, C. N. Hales, and E. A. Newsholme, The glucose fatty-acid cycle: Its role in insulin sensitivity and the metabolic disturbances of diabetes mellitus, *Lancet*, **1963**, *1*, 785–789.

39 P. J. Randle, A. L. Kerbey, and J. Espinal, Mechanisms decreasing glucose oxidation in diabetes and starvation: role of lipid fuels and hormones, *Diabetes Metab. Rev.*, **1988**, *4*, 623–638.

40 G. M. Reaven, C. Hollenbeck, C.-Y. Jeng, M. S. Wu, and Y.-D. I. Chen, Measurement of plasma free fatty acid, lactate and insulin for 24 h in patients with NIDDM, *Diabetes*, **1988**, *37*, 1020–1024.

41 B. V. Howard, Lipoprotein metabolism in diabetes mellitus, *J. Lipid Res.*, **1987**, *28*, 613–628.

42 G. R. Fulcher, M. Walker, C. Catalano, I. Agius, and K. G. M. M. Alberti, Metabolic effects of suppression of non-esterified fatty acid levels with acipimox in obese NIDDM subjects, *Diabetes*, **1992**, *41*, 1400–1408.

43 S. Reynisdottir, K. Ellerfeldt, H. Wahrenberg, H. Lithell, and P. Arner, Multiple lipolysis defects in the insulin resistance (metabolic) syndrome, *J. Clin. Invest.*, **1994**, *93*, 2590–2599.

44 S. Reynisdottir, B. Angelin, and D. Langin, et al., Adipose tissue lipoprotein lipase and hormone-sensitive lipase. Contrasting findings in familial combined hyperlipidemia and insulin resistance syndrome, *Arterioscler. Thromb. Vasc. Biol.*, **1997**, *17*, 2287–2292.

45 F. Shimada, H. Makino, and N. Hashimoto, et al., Detection of an amino acid polymorphism in hormone-sensitive lipase in Japanese subjects, *Metabolism*, **1996**, *45*, 862–864.

46 M. Klannemark, M. Orho, and D. Langin, et al., The putative role of the hormone-sensitive lipase gene in the pathogenesis of type II diabetes mellitus and abdominal obesity, *Diabetologia*, **1998**, *41*, 1516–1522.

47 J. Magre, H. Laurell, and C. Fizames, et al., Human hormone-sensitive lipase: genetic mapping, identification of a new dinucleotide repeat, and association with obesity and NIDDM, *Diabetes*, **1998**, *47*, 284–286.

48 G. I. Shulman, Cellular mechanisms of insulin resistance, *J. Clin. Invest.*, **2000**, *106*, 171–176.

49 R. H. Unger, Lipotoxicity in the pathogenesis of obesity-dependent NIDDM. Genetic and clinical implications, *Diabetes*, **1995**, *44*, 863–870.

50 R. H. Unger, Lipotoxic diseases, *Annu. Rev. Med.*, **2002**, *53*, 319–336.

51 M. E. Griffin, M. J. Marcucci, G. W. Cline, K. Bell, N. Barucci, D. Lee, L. J. Goodyear, E. W. Kraegen, M. F. White, and G. I. Shulman, Free fatty acid-induced insulin resistance is associated with activation of protein kinase Cδ and alterations in the insulin signaling cascade, *Diabetes*, **1999**, *48*, 1270–1274.

52 G. Boden, Role of fatty acids in the pathogenesis of insulin resistance and NIDDM, *Diabetes*, **1997**, *46*, 3–10.

53 R. H. Unger, Lipotoxicity in the pathogenesis of obesity-dependent NIDDM: genetic and clinical implications, *Diabetes*, **1995**, *44*, 863–870.

54 M. Shimabukuro, Y. T. Zhou, M. Levi, and R. H. Unger, Fatty acid-induced beta cell apoptosis: a link between obesity and diabetes, *Proc. Natl. Acad. Sci. USA*, **1998**, *95*, 2498–2502.

55 R. Lupi, F. Dotta, and L. Marselli, et al., Prolonged exposure to free fatty acids has cytostatic and pro-apoptotic effects on human pancreatic islets: evidence that β-cell death is caspase mediated, partially dependent on ceramide pathway, and Bcl-2, *Diabetes*, **2002**, *51*, 1437–1442.

56 Y. P. Zhou and V. Grill, Long term exposure to fatty acids and ketones inhibits B-cell functions in human pancreatic islets of Langerhans, *J. Clin. Endocrinol. Metab.*, **1995**, *80*, 1584–1590.

57 M. Prentki, S. Vischer, M. C. Glennon, R. Regazzi, J. T. Deeney, and B. E. Cor-

KEY, Malonyl-CoA and long chain acyl-CoA esters as metabolic coupling factors in nutrient-induced insulin secretion, *J. Biol. Chem.*, **1992**, *267*, 5802–5810.

58 M. SHIMABUKURO, Y.T. ZHOU, M. LEVI, and R.H. UNGER, Fatty acid-induced β-cell apoptosis: A link between obesity and diabetes, *Proc. Natl. Acad. Sci. USA*, **1998**, *95*, 2498–2502.

59 W. GEPTS and P.M. LECOMPTE, The pancreatic islets in diabetes, *Am. J. Med.*, **1981**, *70*, 105–115.

60 G. KLOPPEL, M. LOHR, K. HABICH, M. OBERHOLZER, and P.U. HEITZ, Islet pathology and the pathogenesis of type 1 and type 2 diabetes mellitus revisited, *Surv. Synth. Pathol. Res.*, **1985**, *4*, 110–125.

61 M. SHIMABUKURO, M. OHNEDA, Y. LEE, and R.H. UNGER, Role of nitric oxide in obesity-induced beta cell disease, *J. Clin. Invest.*, **1997**, *100*, 290–295.

62 K. MAEDLER, G.A. SPINAS, D. DYNTAR, W. MORITZ, N. KAISER, and M.Y. DONATH, Distinct effects of saturated and monounsaturated fatty acids on β-cell turnover and function, *Diabetes*, **2001**, *50*, 69–76.

63 H.Y. YU, T. INOGUCHI, M. KAKIMOTO, N. NAKASHIMA, M. IMAMURA, T. HASHIMOTO, F. UMEDA, and H. NAWATA, Saturated non-esterified fatty acids stimulate de novo diacylglycerol synthesis and protein kinase C activity in cultured aortic smooth muscle cells, *Diabetologia*, **2001**, *44*, 614–620.

64 S.P. COUSIN, S.R. HUGL, C.E. WREDE, H. KAJIO, M.G. MYERS, and C.J. RHODES, Free fatty acid-induced deoxyribonucleic acid synthesis in the pancreatic beta-cell line INS-1, *Endocrinology*, **2001**, *142*, 229–240.

65 J.D. MCGARRY and R.L. DOBBINS, Fatty acids, lipotoxicity and insulin secretion, *Diabetologia*, **1999**, *42*, 128–138.

66 M. PRENTKI and B.E. CORKEY, Are the beta-cell signaling molecules malonyl-CoA and cytosolic long-chain acyl-CoA implicated in multiple tissue defects of obesity and NIDDM?, *Diabetes*, **1996**, *45*, 273–283.

67 K. KOYAMA, G. CHEN, M.Y. WANG, Y. LEE, M. SHIMABUKURO, C.B. NEWGARD, and R.H. UNGER, Beta-cell function in normal rats made chronically hyperleptinemic by adenovirus-leptin gene therapy, *Diabetes*, **1997**, *46*, 1276–1280.

68 C. HOLM, P. BELFRAGE, and G. FREDRIKSON, Immunological evidence for the presence of hormone-sensitive lipase in rat tissues other than adipose tissue, *Biochem. Biophys. Res. Commun.*, **1987**, *148*, 99–105.

69 M.S. WINZELL, H. SVENSSON, P. ARNER, B. AHREN, and C. HOLM, The expression of hormone-sensitive lipase in clonal β-cells and rat islets is induced by long-term exposure to high glucose, *Diabetes*, **2001**, *50*, 2225–2230.

70 A. KHAN, S. NARANGODA, B. AHREN, C. HOLM, F. SUNDLER, and S. EFENDIC, Long-term leptin treatment of ob/ob mice improves glucose-induced insulin secretion, *Int. J. Obes. Relat. Metab. Disord.*, **2001**, *25*, 816–821.

71 R. RODUIT, P. MASIELLO, S.P. WANG, H. LI, G.A. MITCHELL, and M. PRENTKI, A role for hormone-sensitive lipase in glucose-stimulated insulin secretion: a study in hormone-sensitive lipase deficient mice, *Diabetes*, **2001**, *50*, 1970–1975.

72 M. SADDIK and G.D. LOPASCHUK, Myocardial triglyceride turnover and contribution to energy substrate utilization in isolated working hearts, *J. Biol. Chem.*, **1991**, *266*, 8162–8170.

73 M. SADDIK and G.D. LOPASCHUK, Triacylglycerol turnover in isolated working hearts of acutely diabetic rats, *Can. J. Physiol. Pharmacol.*, **1994**, *72*, 1110–1119.

74 E.M. SWANTON and E.D. SAGGERSON, Effects of adrenaline on triacylglycerol synthesis and turnover in ventricular myocytes from adult rats, *Biochem. J.*, **1997**, *328*, 913–922.

75 M.J. GROOT, P.H. WILLEMSEN, W.A. COUMANS, M. VAN BILSEN, and G. J. VAN DER VUSSE, Lactate-induced stimulation of myocardial triacylglycerol turnover, *Biochem. Biophys. Acta*, **1989**, *1006*, 111–115.

76 D. WIGGINS and G.F. GIBBONS, The lipolysis/esterification cycle of hepatic triacylglycerol. Its role in the secretion of very-low-density lipoprotein and its response to hormones and sulphonylureas, *Biochem. J.*, **1992**, *284*, 457–462.

77 G.F. GIBBONS and D. WIGGINS, Intracellular triacylglycerol lipase: its role in the assembly of hepatic very-low-density lipoprotein (VLDL), *Adv. Enzyme Regul.*, **1995**, *35*, 179–198.

78 C.A. Small, A.J. Garton, and S.J. Yeaman, The presence and role of hormone-sensitive lipase in heart muscle, *Biochem. J.*, **1989**, *258*, 67–72.

79 C. Holm, T.G. Kirchgessner, and K.L. Svenson, et al., Hormone-sensitive lipase: sequence, expression, and chromosomal localization to 19cent-q13.3, *Science*, **1988**, *241*, 1503–1506.

80 H.C. Chiu, A. Kovacs, D.A. Ford, F.F. Hsu, R. Garcia, P. Herrero, J.E. Saffitz, and J.E. Schaffer, A novel mouse model of lipotoxic cardiomyopathy, *J. Clin. Invest.*, **2000**, *107*, 813–822.

81 Y.-T. Zhou, P. Grayburn, A. Karim, M. Shimabukuro, M. Higa, D. Baetens, L. Orci, and R.H. Unger, Lipotoxic heart disease in obese rats: Implications for human obesity, *Proc. Natl. Acad. Sci. USA*, **2000**, *97*, 1784–1789.

82 D.A. Pan, S. Lillioja, A.D. Kriketos, M.R. Milner, L.A. Baur, C. Bogardus, A.B. Jenkins, and L.H. Storlien, Skeletal muscle triglyceride levels are inversely related to insulin action, *Diabetes*, **1997**, *46*, 983–988.

83 L.L. Listenberger, D.S. Ory, and J.E. Schaffer, Palmitate-induced apoptosis can occur through a ceramide-independent pathway, *J. Biol. Chem.*, **2001**, *276*, 14890–14895.

84 J.E. deVries, M.M. Vork, T.H.M. Roemen, Y.F. deJong, J.P.M. Cleutjens, G.J. van der Vusse, and M. van Bilsen, Saturated, but not mono-unsaturated fatty acids induce apoptotic cell death in neonatal rat ventricular myocyctes, *J. Lipid Res.*, **1997**, *38*, 1384–1394.

85 K. Maedler, G.A. Spinas, D. Dyntar, W. Moritz, N. Kaiser, and M.Y. Donath, Distinct effects of saturated and monounsaturated fatty acids on β-cell turnover and function, *Diabetes*, **2001**, *50*, 69–76.

86 M. Cnop, J.C. Hannaert, A. Hoorens, D.L. Eizirik, and D.G. Pipeleers, Inverse relationship between cytotoxicity of free fatty acids in pancreatic islet cells and cellular triglyceride accumulation, *Diabetes*, **2001**, *50*, 1771–1777.

87 S. Hardy, Y. Langelier, and M. Prentki, Oleate activates phosphatidylinositol 3-kinase and promotes proliferation and reduces apoptosis of MDA-MB-231 breast cancer cells, whereas palmitate has opposite effects, *Cancer Res.*, **2000**, *60*, 6353–6358.

88 M.B. Paumen, Y. Ishida, M. Muramatsu, M. Yamamoto, and T. Honjo, Inhibition of carnitine palmitoyltransferase I augments sphingolipid synthesis and palmitate-induced apoptosis, *J. Biol. Chem.*, **1997**, *272*, 3324–3329.

89 M. Shimabukuro, M. Higa, Y.T. Zhou, M.Y. Wang, C.B. Newgard, and R.H. Unger, Lipoapoptosis in beta-cells of obese prediabetic fa/fa rats. Role of serine palmitoyltransferase overexpression, *J. Biol. Chem.*, **1998**, *273*, 32487–32490.

90 M. Shimabukuro, M. Ohneda, Y. Lee, and R.H. Unger, Role of nitric oxide in obesity-induced β-cell disease, *J. Clin. Invest.*, **1997**, *100*, 290–295.

91 M. Shimabukuro, K. Koyama, Y. Lee, and R.H. Unger, Leptin- or troglitazone-induced lipopenia protects islets from interleukin 1β cytotoxicity, *J. Clin. Invest.*, **1997**, *100*, 1750–1754.

92 D.B. Ostrander, G.C. Sparagna, A.A. Amoscato, J.B. McMillan, and W. Dowhan, Decreased cardiolipin synthesis corresponds with cytochrome c release in palmitate-induced cardiomyocyte apoptosis, *J. Biol. Chem.*, **2001**, *276*, 38061–38067.

93 L.L. Listenberger, X. Han, S.E. Lewis, S. Cases, R.V. Farese, D.S. Ory, and J.E. Schaffer, Triglyceride accumulation protects against fatty acid-induced lipotoxicity, *Proc. Natl. Acad. Sci. USA*, **2003**, *100*, 3077–3082.

94 K. Eitel, H. Staiger, J. Rieger, H. Mischak, H. Brandhorst, M.D. Brendel, R.G. Bretzel, H.-U. Häring, and M. Kellerer, Protein kinase C δ activation and translocation to the nucleus are required for fatty acid-induced apoptosis of insulin-secreting cells, *Diabetes*, **2003**, *52*, 991–997.

95 N. Hilaire, A. Negre-Salvayre, and R. Salvayre, Cellular uptake and catabolism of high-density-lipoprotein triacylglycerols in human cultured fibroblasts: degradation block in neutral lipid storage disease, *Biochem. J.*, **1994**, *297*, 467–473.

96 G. Schmitz and G. Assman, *Metabolic Basis of Inherited Disease*, 6th edn, **1989**, vol. 2, 1623–1644.

97 D. Ameis, M. Merkel, C. Eckerskorn, and H. Greten, Purification, characterization and molecular cloning of human hepatic lysosomal acid lipase, *Eur. J. Biochem.*, **1994**, *219*, 905–914.

98 R.A. Anderson and G.N. Sando, Cloning and expression of cDNA encoding human lysosomal acid lipase/cholesteryl-ester hydrolase. Similarities to gastric and lingual lipases, *J. Biol. Chem.*, **1991**, *266*, 22479–22484.

99 H. Nakagawa, S. Matsubara, M. Kuriyama, H. Yoshidome, J. Fujiyama, H. Yoshida, and M. Osame, Cloning of rat lysosomal acid lipase cDNA and identification of the mutation in the rat model of Wolman's disease, *J. Lipid Res.*, **1995**, *36*, 2212–2218.

100 S. Di-Donato, B. Garavaglia, P. Strisciuglio, C. Borrone, and G. Andria, Multisystem triglyceride storage disease is due to a specific defect in the degradation of endocellularly synthesized triglycerides, *Neurology*, **1988**, *38*, 1107–1110.

101 M.L. Williams, D.J. Monger, S.L. Rutherford, M. Hincenbergs, S.J. Rehfeld, and C. Grunfeld, Neutral lipid storage disease with ichthyosis: lipid content and metabolism of fibroblasts, *J. Inherit. Metab. Dis.*, **1988**, *11*, 131–143.

102 J. Radom, R. Salvayre, A. Negre, A. Maret, and L. Douste-Blazy, Metabolism of neutral lipids in cultured fibroblasts from multisystemic (or type 3) lipid storage myopathy, *Eur. J. Biochem.*, **1987**, *164*, 703–708.

103 K. Reue and M.H. Doolittle, Naturally occurring mutations in mice affecting lipid transport and metabolism, *J. Lipid Res.*, **1996**, *37*, 1387–1405.

104 R.A. Igal and R.A. Coleman, Acylglycerol recycling from triacylglycerol to phospholipid, not lipase activity, is defective in neutral lipid storage disease fibroblasts, *J. Biol. Chem.*, **1996**, *271*, 16644–16651.

105 R.A. Igel and R.A. Coleman, Neutral lipid storage disease: a genetic disorder with abnormalities in the regulation of phospholipid metabolism, *J. Lipid Res.*, **1998**, *39*, 31–43.

106 H. Bar-On, P.S. Roheim, O. Stein, and S. Stein, Contribution of floating fat triglyceride and of lecithin towards formation of secretory triglyceride in perfused rat liver, *Biochim. Biophys. Acta*, **1971**, *248*, 1–11.

107 D. Wiggins and G.F. Gibbons, Origin of hepatic very-low-density lipoprotein triacylglycerol: the contribution of cellular phospholipid, *Biochem. J.*, **1996**, *320*, 673–679.

108 S.-O. Olofsson, L. Asp, and J. Boren, The assembly and secretion of apolipoprotein B-containing lipoproteins, *Curr. Opin. Lipidol*, **1999**, *10*, 341–346.

109 C.A. Langner, E.H. Birkenmeier, O. Ben-Zeev, M.C. Schotz, H.O. Sweet, M.T. Davisson, and J.I. Gordon, The fatty liver dystrophy (fld) mutation. A new mutant mouse with a developmental abnormality in triglyceride metabolism and associated tissue-specific defects in lipoprotein lipase and hepatic lipase activities, *J. Biol. Chem.*, **1989**, *264*, 7994–8003.

110 E. Levy, R. Goldstein, S. Freier, and E. Shafrir, Characterization of gastric lipolytic activity, *Biochim. Biophys. Acta*, **1981**, *664*, 316–326.

111 A. Becker, A. Böttcher, K.J. Lackner, P. Fehringer, F. Notka, C. Aslanidis, and G. Schmitz, Purification, cloning, and expression of a human enzyme with acyl coenzyme A: cholesterol acyltransferase activity, which is identical to liver carboxylesterase, *Arterioscler. Thromb. Vasc. Biol.*, **1994**, *14*, 1346–1455.

112 J.P. Goudriann, P.J. Tacken, and V.E. Dahlmans, et al., Protection from obesity in mice lacking the VLDL receptor, *Arterioscler. Thromb. Vasc. Biol.*, **2001**, *21*, 1488–1493.

113 M.C. Long, M.J. Gijbels, and V.E. Dahlmans, et al., Hyperlipidemia and cutaneous abnormalities in transgenic mice overexpressing human apolipoprotein C1, *J. Clin. Invest.*, **1998**, *101*, 145–152.

114 L.W. Castellani, A.M. Goto, and A.J. Lusis, Studies with apolipoprotein A-II transgenic mice indicate a role for HDLs in adiposity and insulin resistance, *Diabetes*, **2001**, *50*, 643–651.

115 P.H. Weinstock, C.L. Bisgaier, T. Hayek, and Aalto-Setälä, et al., Decreased HDL cholesterol levels, but normal lipid absorption, growth and feeding behavior in apolipoprotein A-IV knockout

mice, *J. Clin. Invest.*, **1997**, *38*, 1782–1794.

116 P.H. Weinstock, S. Levak-Frank, and L.C. Hudgins, et al., Lipoprotein lipase controls fatty acid entry into adipose tissue, but fat mass is preserved by endogenous synthesis in mice deficient in adipose tissue lipoprotein lipase, *Proc. Natl. Acad. Sci. USA*, **1997**, *94*, 10261–10266.

117 H.C. Chen and R.V. Farese, Fatty acids, triglycerides, and glucose metabolism: recent insights from knockout mice, *Curr. Opin. Clin. Nutr. Metab. Care*, **2002**, *5*, 359–363.

118 J.S. Flier, The adipocyte: storage depot or node on the energy information superhighway?, *Cell*, **1995**, *80*, 15–18.

119 G. Frühbeck and J. Gomez-Ambrosi, Control of body weight: a physiologic and transgenic perspective, *Diabetologia*, **2003**, *46*, 143–172.

120 L.P. Kozak, U.C. Kozak, and C.T. Clarke, Abnormal brown and white fat development in transgenic mice overexpressing glycerol 3-phosphate dehydrogenase, *Genes Dev.*, **1991**, *5*, 2256–2264.

121 K. Eto, Y. Tsubamoto, and Y. Terauchi, et al., Role of NADH shuttle system in glucose-induced activation of mitochondrial metabolism and insulin secretion, *Science*, **1999**, *283*, 981–985.

122 L. Abu-Elheiga, M.M. Matzuk, K.A. Abo-Hashema, and S.J. Wakil, Continuous fatty acid oxidation and reduced fat storage in mice lacking acetyl-CoA carboxylase 2, *Science*, **2001**, *291*, 2613–2616.

123 L.E. Hammond, P.A. Gallagher, and S. Wang, et al., Mitochondrial glycerol-3-phosphate acyltransferase-deficient mice have reduced weight and liver triacylglycerol content and altered glycerolipid fatty acid composition, *Mol. Cell. Biol.*, **2002**, *22*, 8204–8214.

124 M.A. Polokoff and R.M. Bell, Solubilization, partial purification and characterization of rat liver microsomal diacylglycerol acyltransferase, *Biochim. Biophys. Acta*, **1980**, *618*, 129–142.

125 S. Cases, S.J. Smith, and Y.W. Zheng, et al., Identification of a gene encoding an acyl CoA:diacylglycerol acyltransferase, a key enzyme in triacylglycerol synthesis, *Proc. Natl. Acad. Sci. USA*, **1998**, *95*, 13018–13023.

126 S. Cases, S. Stones, and P. Zhou, et al., Cloning of DGAT2, a second mammalian diacylglycerol acyl transferase, and related family members, *J. Biol. Chem.*, **2001**, *276*, 38870–38876.

127 S.J. Smith, S. Cases, and D.R. Jensen, et al., Obesity resistance and multiple mechanisms of triglyceride synthesis in mice lacking Dgat, *Nat. Gen.*, **2000**, *25*, 87–90.

128 D.J. Murphy and J. Vance, Mechanism of lipid-body formation, *Trends Biochem. Sci.*, **1999**, *24*, 109–115.

129 F. Hammerson, *Histology: Color Atlas of Microscopic Anatomy*, 2nd edn, Urban & Schwarzenberg, Baltimore, **1985**.

130 P.R. Weather, H.G. Burkiott, and V.G. Daniels, *Functional Histology: A text and colour atlas*, 2nd edn, Churchill Livingstone, Edinburgh, **1987**.

131 D.L. Brasaemle, T. Barber, N.E. Wolins, G. Serrero, E.J. Blanchette-Mackie, and C. Londos, Adipose differentiation-related protein is an ubiquitously expressed lipid storage droplet-associated protein, *J. Lipid Res.*, **1997**, *38*, 2249–2263.

132 H.W. Heid, R. Moll, I. Schwetlick, H.R. Rackwitz, and T.W. Keenan, Adipophilin is a specific marker of lipid accumulation in diverse cell types and diseases, *Cell Tissue Res.*, **1998**, *294*, 309–321.

133 A.A. Spector, S.N. Mathur, T.L. Kaduce, and B.T. Hydman, Lipid nutrition and metabolism of cultured mammalian cells, *Prog. Lipid Res.*, **1981**, *19*, 155–186.

134 J.T. Gwynne and J.F. Strauss, The role of lipoproteins in steroidogenesis and cholesterol metabolism in steroidogenic glands, *Endocrinol. Rev.*, **1982**, *3*, 299–329.

135 D.A. Freeman and M. Ascoli, Studies on the source of cholesterol used for steroid biosynthesis in cultured Leydig tumor cells, *J. Biol. Chem.*, **1982**, *257*, 14231–14238.

136 C.L. Bisgaier, R. Chanderbhan, R. W. Hinds, and G.V. Vahouny, Adrenal cholesterol esters as substrate source for steroidogenesis, *J. Steroid Biochem.*, **1985**, *23*, 967–974.

137 P. K. Jensen and B. Kofod, Isolation from beef-heart homogenate of a particulate lipoprotein containing beta-carotene, *Biochim. Biophys. Acta*, **1966**, *116*, 579–582.
138 F. F. Schlunk and B. Lombardi, Liver liposomes. I. Isolation and chemical characterization, *Lab. Invest.*, **1967**, *17*, 30–38.
139 J. Stadler and H. Kleining, Nuclear membranes from mammalian liver. III. Fatty acids, *Biochim. Biophys. Acta*, **1971**, *233*, 315.
140 A. Angel, Studies on the compartmentation of lipid in adipose cells. I: Subcellular distribution, composition, and transport of newly synthesized lipid: liposomes, *J. Lipid Res.*, **1970**, *11*, 420.
141 K. Christiansen and P. K. Jensen, Membrane-bound lipid particles from beef heart chemical composition and structure, *Biochim. Biophys. Acta*, **1972**, *260*, 449–459.
142 M. Yamada, W. S. Blaner, D. R. Soprano, J. L. Dixon, H. M. Kjieldbye, and D. S. Goodman, Biochemical characteristics of isolated rat liver stellate cells, *Hepatology*, **1987**, *7*, 1224–1229.
143 D. Zweytick, K. Athenstaedt, and G. Daum, Intracellular lipid particles of eukaryotic cells, *Biochim. Biophys. Acta*, **2000**, *1469*, 101–120.
144 D. J. Murphy and J. Vance, Mechanisms of lipid-body formation, *Trends Biochem. Sci.*, **1999**, *24*, 109–115.
145 D. A. Brown, Lipid droplets: Proteins floating on a pool of fat, *Curr. Biol.*, **2001**, *11*, 446–449.
146 E. J. Blanchette Mackie, N. K. Dwyer, T. Barber, R. A. Coxey, T. Takeda, C. M. Rondinone, J. L. Theodorakis, A. S. Greenberg, and C. Londos, Perilipin is located on the surface layer of intracellular lipid droplets in adipocytes, *J. Lipid Res.*, **1995**, *36*, 1211–1226.
147 A. Peixoto Menezes de and P. Pinto Silva da, Fat droplet formation in rat lactating mammary gland and mammary carcinomas viewed by freeze-fracture, *Lab. Invest.*, **1979**, *40*, 545–553.
148 C. C. Chang, J. Chen, M. A. Thomas, D. Cheng, V. A. Priore Del, R. S. Newton, M. E. Pape, and T. Y. Chang, Regulation and immunolocalization of acyl-coenzyme A: cholesterol acyltransferase in mammalian cells as studied with specific antibodies, *J. Biol. Chem.*, **1995**, *270*, 29532–29540.
149 K. Tauchi-Sato, S. Ozeki, T. Houjou, R. Taguchi, and T. Fujimoto, The surface of lipid droplets is a phospholipid monolayer with a unique fatty acid composition, *J. Biol. Chem.*, **2002**, *277*, 44507–44512.
150 S. J. Stone and J. E. Vance, Phosphatidylserine synthase-1 and -2 are localized to mitochondria-associated membranes, *J. Biol. Chem.*, **2000**, *275*, 34534–34540.
151 R. Schroeder, E. London, and D. Brown, Interactions between saturated acyl chains confer detergent resistance on lipids and glycosylphosphatidylinositol (GPI)-anchored proteins: GPI-anchored proteins in liposomes and cells show similar behavior, *Proc. Natl. Acad. Sci. USA*, **1994**, *91*, 12130–12134.
152 T. Fujimoto, H. Kogo, K. Ishiguro, K. Tauchi, and R. Nomura, Caveolin-2 is targeted to lipid droplets, a new "membrane domain" in the cell, *J. Cell. Biol.*, **2001**, *152*, 1079–1085.
153 W. W. Franke, M. Hergt, and C. Grund, Rearrangement of the vimentin cytoskeleton during adipose conversion: formation of an intermediate filament cage around lipid globules, *Cell*, **1987**, *49*, 131–141.
154 G. Almahbobi, L. J. Williams, X.-G. Han, and P. F. Hall, Binding of lipid droplets and mitochondria to intermediate filaments in rat Leydig cells, *J. Reprod. Fertil.*, **1993**, *98*, 209–217.
155 G. Almahbobi, Adhesion of intermediate filaments and lipid droplets in adrenal cells studied by field emission scanning electron microscopy, *Cell. Tissue Res.*, **1995**, *281*, 387–390.
156 C. Londos, D. L. Brasaemle, C. J. Schultz, D. C. Adler-Wailes, D. M. Levin, A. R. Kimmel, and C. M. Rondinone, On the control of lipolysis in adipocytes, *Ann. New York Acad. Sci.*, **1999**, *892*, 155–168
157 C. Londos, D. L. Brasaemle, C. J. Schultz, J. P. Segrest, and A. R. Kimmel, Perilipins, ADRP, and other proteins that associate with intracellular neutral lipid droplets in animal cells, *Semin. Cell. Dev. Biol.*, **1999**, *10*, 51–58.

158 S. M. Wang and T. H. Fong, A lipid droplet-specific capsule is present in rat adrenal-cells: Evidence from a monoclonal antibody, *Biochem. Biophys. Res. Commun.*, **1995**, *217*, 81–88.

159 T. H. Fong, S. M. Wang, and H. S. Lin, Immunocytochemical demonstration of a lipid droplet-specific capsule in cultured Leydig cells of the golden hamsters, *J. Cell. Biochem.*, **1996**, *63*, 366–373.

160 S. M. Wang, T. H. Fong, S. Y. Hsu, C. L. Chien, and J. C. Wu, Dissection of the signaling mechanism for capsule detachment of lipid droplets in rat adrenocortical cells, *J. Cell. Biochem.*, **1997**, *67*, 84–91.

161 J. G. Lieber and R. M. Evans, Disruption of the vimentin intermediate filament system during adipose conversion of 3T3-L1 cells inhibits lipid droplet accumulation, *J. Cell. Sci.*, **1996**, *109*, 3047–3058.

162 P. F. Weller and A. M. Dvorak, Lipid bodies: Intracellular sites for eicosanoid formation, *J. Allergy Clin. Immunol.*, **1994**, *94*, 1152–1156.

163 W. Yu, P. T. Bozza, D. M. Tzizik, J. P. Gray, J. Cassara, A. M. Dvorak, and P. F. Weller, Co-compartmentalization of MAP kinases and cytosolic phospholipase A2 at cytoplasmic arachidonate-rich lipid bodies, *Am. J. Pathol.*, **1998**, *152*, 759–769.

164 P. T. Bozza, W. Yu, J. F. Penrose, E. S. Morgan, A. M. Dvorak, and P. F. Weller, Eosinophil lipid bodies: Specific, inducible intracellular sites for enhanced eicosanoid formation, *J. Exp. Med.*, **1997**, *186*, 909–920.

165 A. S. Greenberg, J. J. Egan, S. A. Wek, M. C. Moos, C. Londos, and A. R. Kimmel, Isolation of cDNAs for perilipins A and B: Sequence and expression of lipid droplet-associated proteins of adipocytes, *Proc. Natl. Acad. Sci. USA*, **1993**, *90*, 12035–12039.

166 D. A. Servetnick, D. L. Brasaemle, J. Gruia-Gray, A. R. Kimmel, J. Wolff, and C. Londos, Perilipins are associated with cholesteryl ester droplets in steroidogenic adrenal cortical and leydig cells, *J. Biol. Chem.*, **1995**, *270*, 16970–16973.

167 D. L. Brasaemle, T. Barber, N. E. Wolins, G. Serrero, E. J. Blanchette-Mackie, and C. Londos, Adipose differentiation-related protein is an ubiquitously expressed lipid storage droplet-associated protein, *J. Lipid Res.*, **1997**, *38*, 2249–2263.

168 A. S. Greenberg, J. J. Egan, S. A. Wek, N. B. Garty, E. J. Blanchette-Mackie, and C. Londos, Perilipin, a major hormonally regulated adipocyte-specific phosphoprotein associated with the periphery of lipid storage droplets, *J. Biol. Chem.*, **1991**, *266*, 11341–11346.

169 C. Londos, D. L. Brasaemle, J. Gruia-Gray, D. A. Servetnick, C. J. Schultz, D. M. Levin, and A. R. Kimmel, Perilipin: unique proteins associated with intracellular neutral lipid droplets in adipocytes and steroidogenic cells, *Biochem. Soc. Trans.*, **1995**, *23*, 611–615.

170 H. P. Jiang, S. E. Harris, and G. Serrero, Molecular cloning of a differentiation-related mRNA in the adipogenic cell line 1246, *Cell. Growth Differ.*, **1992**, *3*, 21–30.

171 G. Kwon, W. S. Cruz, C. A. Marshall, K. L. Pappan, and H. Imachi, et al., Role of ADRP in lipid metabolism by isolated rat islets, *Diabetes*, **2003**, *52* (Suppl), A378.

172 E. Diaz and S. R. Pfeffer, TIP47: A cargo selection device for mannose 6-phosphate receptor trafficking, *Cell*, **1998**, *93*, 433–443.

173 P. E. Scherer, P. E. Bickel, M. Kotler, and H. F. Lodish, Cloning of cell-specific secreted and surface proteins by subtractive antibody screening, *Nat. Biotechnol.*, **1998**, *16*, 581–586.

174 J. Gao, H. Ye, and G. Serrero, Stimulation of adipose differentiation related protein (ADRP) expression in adipocyte precursors by long-chain fatty acids, *J. Cell. Physiol.*, **2000**, *182*, 297–302.

175 C. Choe, T. P. Ciaraldi, A. S. Greenberg, and A. P. S. Kong, et al., Expression and regulation of perilipin and ADRP in human skeletal muscle, *Diabetes*, **2003**, *52*, A51.

176 A. Garcia, A. Sekowski, V. Subramanian, and D. L. Brasaemle, The central domain is required to target and anchor perilipin A to lipid droplets, *J. Biol. Chem.*, **2003**, *278*, 625–635.

177 J. Gao and G. Serrero, Adipose differentiation related protein (ADRP) expressed in transfected COS-7 cells selectively stimulates long chain fatty acid uptake, *J. Biol. Chem.*, **1999**, *274*, 16825–16830.

178 I.H. Mather and T.W. Keenan, Origin and secretion of milk lipids, *J. Mammary Gland Biol. Neoplasia*, **1998**, *3*, 259–273.

179 J.L. McManaman, W. Zabaronick, J. Schaack, and D.J. Orlicky, Lipid droplet targeting domains of adipophilin, *J. Lipid Res.*, **2003**, *44*, 668–673.

180 X. Lu, J. Gruia-Gray, and N.G. Copeland, et al., The murine perilipin gene: the lipid droplet-associated perilipins derive from tissue-specific mRNA slice variants and define a gene family of ancient origin, *Mamm. Genome*, **2001**, *12*, 741–749.

181 S.J. Miura, W. Gan, and J. Brzostowski, et al., Functional conservation for lipid storage droplet association among Perilipin, ADRP, and TIP47 (PAT)-related proteins in mammals, Drosophila, and Dictyostelium, *J. Biol. Chem.*, **2002**, *277*, 32253–32257.

182 J.L. McManaman, C.A. Palmer, R.M. Wright, and M.C. Neville, Functional regulation of xanthine oxidoreductase expression in the mouse mammary gland: Evidence of a role in lipid secretion, *J. Physiol.*, **2002**, *545*, 567–579.

183 R.G. Parton, Caveolae and caveolins, *Curr. Opin. Cell. Biol.*, **1996**, *8*, 542–548.

184 B. Razani, S.E. Woodman, and M.P. Lisanti, Caveolae: From cell biology to animal physiology, *Pharmacol. Rev.*, **2002**, *54*, 431–467.

185 M.P. Lisanti, P. Scherer, and M. Sargiacomo, Caveolae, caveolin and caveolin-rich membrane domains: A signalling hypothesis, *Trends Cell. Biol.*, **1994**, *4*, 231–235.

186 R.G.W. Anderson, The caveolae membrane system, *Annu. Rev. Biochem.*, **1998**, *67*, 199–225.

187 P.E. Scherer, M.P. Lisanti, and G. Baldini, et al., Induction of caveolin during adipogenesis and association of GLUT4 with caveolin-rich vesicles, *J. Cell. Biol.*, **1994**, *127*, 1233–1243.

188 L.M. Napolitano, The differentiation of white adipose cells. An electron microscope study, *J. Cell. Biol.*, **1963**, *18*, 663–679.

189 J.Y. Fan, J.-L. Carpentier, and E. Obberghen van, et al., Morphological changes of the 3T3-L1 fibroblast plasma membrane upon differentiation to the adipocyte form, *J. Cell. Sci.*, **1983**, *61*, 219–230.

190 S. Monier, R.G. Parton, F. Vogel, J. Behlke, A. Henske, and T.V. Kurzchalia, VIP21-caveolin, a membrane protein constituent of the caveolar coat, oligomerizes in vivo and in vitro, *Mol. Biol. Cell.*, **1995**, *6*, 911–927.

191 C.J. Fielding and P.E. Fielding, Cholesterol and caveolae: structural and functional relationships, *Biochim. Biophys. Acta*, **2000**, *1529*, 210–222.

192 E.J. Smart, G.A. Graf, and M.A. McNiven, et al., Caveolins, liquid-ordered domains, and signal transduction, *Mol. Cell. Biol.*, **1999**, *19*, 7289–7304.

193 T. Fujimoto, H. Kogo, K. Ishiguro, K. Tauchi, and R. Nomura, Caveolin-2 is targeted to lipid droplets, a new "membrane domain" in the cell, *J. Cell. Biol.*, **2001**, *152*, 1079–1086.

194 K. Das, R.Y. Lewis, P.E. Scherer, and M.P. Lisanti, The membrane-spanning domains of caveolins-1 and -2 mediate the formation of caveolin hetero-oligomers. Implications for the assembly of caveolae membranes in vivo, *J. Biol. Chem.*, **1999**, *274*, 18721–18728.

195 S. Monier, R.G. Parton, F. Vogel, J. Behlke, A. Henske, and T.V. Kurzchalia, VIP21-caveolin, a membrane protein constituent of the caveolar coat, oligomerizes in vivo and in vitro, *Mol. Biol. Cell*, **1995**, *6*, 911–927.

196 S. Prattes, G. Horl, A. Hammer, A. Blaschnitz, W.F. Graier, W. Sattler, R. Zechner, and E. Steyrer, Intracellular distribution and mobilization of unesterified cholesterol in adipocytes: triglyceride droplets are surrounded by cholesterol-rich ER-like surface layer structures, *J. Cell. Sci.*, **2000**, *113*, 2977–2989.

197 W. Yu, J. Cassara, and P.F. Weller, Phosphatidylinositide 3-kinase localizes to cytoplasmic lipid bodies in human polymorphonuclear leukocytes and other myeloid-derived cells, *Blood*, **2000**, *95*, 1078–1085.

198 A.G. Ostermeyer, J.M. Paci, Y. Zeng, D.M. Lublin, S. Munro, and D.A.

Brown, Accumulation of caveolin in the endoplasmic reticulum redirects the protein to lipid storage droplets, *J. Cell. Biol.*, **2001**, *152*, 1071–1078.

199 P.A. Conrad, E.J. Uittenbogaard, Y.S. Ying, R.G.W. Anderson, and G.S. Bloom, Caveolin cycles between plasma membrane caveolae and the Golgi complex by microtubule-dependent and microtubule-independent steps, *J. Cell. Biol.*, **1995**, *131*, 1421–1433.

200 A. Uittenbogaard, Y. Ying, and E.J. Smart, Characterization of a cytosolic heat-shock protein-caveolin chaperone complex. Involvement in cholesterol trafficking, *J. Biol. Chem.*, **1998**, *273*, 6525–6532.

201 A. Pol, R. Luetterforst, M. Lindsay, S. Heino, E. Ikonen, and R.G. Parton, A caveolin dominant negative mutant associates with lipid bodies and induces intracellular cholesterol imbalance, *J. Cell. Biol.*, **2001**, *152*, 1057–1070.

202 E.J. Smart, Y.S. Ying, P.A. Conrad, and R.G.W. Anderson, Caveolin moves from caveolae to the Golgi apparatus in response to cholesterol oxidation, *J. Cell. Biol.*, **1994**, *127*, 1185–1197.

203 A. Uittenbogaard and E.J. Smart, Palmitoylation of caveolin-1 is required for cholesterol binding, chaperone complex formation, and rapid transport of cholesterol to caveolae, *J. Biol. Chem.*, **2000**, *275*, 25595–25599.

204 W.P. Li, P. Liu, B.K. Pilcher, and R.G.W. Anderson, Cell-specific targeting of caveolin-1 to caveolae, secretory vesicles, cytoplasm or mitochondria, *J. Cell. Sci.*, **2001**, *114*, 1397–1408.

205 P. Liu, W.P. Li, T. Machleidt, and R.G.W. Anderson, Identification of caveolin-1 in lipoprotein particles secreted by exocrine cells, *Nat. Cell. Biol.*, **1999**, *1*, 369–375.

206 J. Ito, Y. Nagayasu, K. Kato, R. Sato, and S. Yokoyama, Apolipoprotein A-I induces translocation of cholesterol, phospholipid, and caveolin-1 to cytosol in rat astrocytes, *J. Biol. Chem.*, **2002**, *277*, 7929–7935.

207 J.G. Lieber and R.M. Evans, Disruption of the vimentin intermediate filament system during adipose conversion of 3T3-L1 cells inhibits lipid droplet accumulation, *J. Cell. Sci.*, **1996**, *109*, 3047–3058.

208 E. Colucci-Guyon, M.-M. Portier, I. Dunia, D. Paulin, S. Pournin, and C. Babinet, Mice lacking vimentin develop and reproduce without an obvious phenotype, *Cell*, **1994**, *79*, 679–694.

209 D. Marchesan, M. Rutberg, L. Andersson, L. Asp, T. Larsson, J. Boren, B.R. Johansson, and S.-O. Olofsson, A phospholipase D-dependent process forms lipid droplets containing caveolin, adipocyte differentiation-related protein (ADRP), and vimentin in a cell-free system, *J. Biol. Chem.*, **2003**, *278*, 27 293–27 300.

210 H.W. Heid, M. Schnolzer, and T.W. Keenan, Adipocyte differentiation-related protein is secreted into milk as a constituent of milk lipid globule membrane, *Biochem. J.*, **1996**, *320*, 1025–1030.

211 J. Moss and M. Vaughan, Molecules in the ARF orbit, *J. Biol. Chem.*, **1998**, *273*, 21431–21434.

212 M.G. Roth, Lipid regulators of membrane traffic through the Golgi complex, *Trends Cell. Biol.*, **1999**, *9*, 174–179.

213 S. Prattes, G. Horl, A. Hammer, A. Blaschnitz, W.F. Graier, W. Sattler, R. Zechner, and E. Steyrer, Intracellular distribution and mobilization of unesterified cholesterol in adipocytes: triglyceride droplets are surrounded by cholesterol-rich ER-like surface layer structures, *J. Cell. Sci.*, **2000**, *113*, 2977–2989.

214 D. Jones, C. Morgan, and S. Cockcroft, Phospholipase D and membrane traffic: Potential roles in regulated exocytosis, membrane delivery and vesicle budding, *Biochim. Biophys. Acta*, **1999**, *1439*, 229–244.

215 M. Liscovitch, M. Czarny, G. Fiucci, and X. Tang, Phospholipase D: molecular and cell biology of a novel gene family, *Biochem. J.*, **2000**, *345*, 401–415.

215 A. Schmidt, M. Wolde, C. Thiele, W. Fest, H. Kratzin, A.V. Podtelejnikov, W. Witke, W.B. Huttner, and H.D. Soling, Endophilin I mediates synaptic vesicle formation by transfer of arachidonate to lysophosphatidic acid, *Nature*, **1999**, *401*, 133–141.

216 R. Weigert, M.G. Silletta, and S. Spano, et al., CtBP/BARS induces fission of

Golgi membranes by acylating lysophosphatidic acid, *Nature*, **1999**, *402*, 429–433.

217 S.-O. Olofsson, P. Stillemark-Billton, and L. Asp, Intracellular assembly of VLDL: two major steps in separate cell compartments, *Trends Cardiovasc. Med.*, **2000**, *10*, 338–345.

218 C.A. Alexander, R.L. Hamilton, and R.J. Havel, Subcellular localization of B apoprotein of plasma lipoproteins in rat liver, *J. Cell. Biol.*, **1976**, *69*, 241–263.

219 C.H. Hollenberg, M.S. Raben, and E.B. Astwood, The lipolytic response to corticotropin, *Endocrinology*, **1961**, *68*, 589–598.

220 M.A. Rizack, An epinephrine-sensitive lipolytic activity in adipose tissue, *J. Biol. Chem.*, **1961**, *236*, 657–662.

221 E.R. Froesch, H. Burgi, P. Bally, and A. Labhart, Insulin inhibition of spontaneous adipose lipolysis and effects upon fructose and glucose metabolism, *Mol. Pharmacol.*, **1965**, *1*, 280–296.

222 A.G. Goldberg and E.G. Ball, Studies on the metabolism of adipose tissue. 18. In vitro effects of insulin, epinephrine and glucagon on lipolysis and glycolysis in pigeon adipose tissue, *Comp. Biochem. Physiol.*, **1965**, *16*, 367–381.

223 M. Rodbell and A.B. Jones, Metabolism of isolated fat cells. 3. The similar inhibitory action of phospholipase C and insulin on lipolysis stimulated by lipolytic hormones and theophylline, *J. Biol. Chem.*, **1966**, *241*, 140–142.

224 M.A. Rizack, Activation of an epinephrine-sensitive lipolytic activity from adipose tissue by adenosine 3′,5′-phosphate, *J. Biol. Chem.*, **1964**, *239*, 392–395.

225 P. Strålfors and P. Belfrage, Phosphorylation of hormone-sensitive lipase by cyclic AMP-dependent protein kinase, *J. Biol. Chem.*, **1983**, *258*, 15146–15152.

226 J.N. Fain and J.A. Garcia-Sainz, Adrenergic regulation of adipocyte metabolism, *J. Lipid Res.*, **1983**, *24*, 945–966.

227 C. Londos, RS. Honnor, and G.S. Dhillon, cAMP-dependent protein kinase and lipolysis in rat adipocytes. III. Multiple modes of insulin regulation of lipolysis and regulation of insulin responses by adenylate cyclase regulators, *J. Biol. Chem.*, **1985**, *260*, 15139–15145.

228 J.K. Huttunen and D. Steinberg, Activation and phosphorylation of purified adipose tissue hormone-sensitive lipase by cyclic AMP-dependent protein kinase, *Biochim. Biophys. Acta*, **1971**, *239*, 411–427.

229 J.K. Huttunen, D. Steinberg, and S.E. Mayer, ATP-dependent and cyclic AMP-dependent activation of rat adipose tissue lipase by protein kinase from rabbit skeletal muscle, *Proc. Natl. Acad. Sci. USA*, **1970**, *67*, 290–295.

230 R.C. Honnor, G.S. Dillon, and C. Londos, cAMP-dependent protein kinase and lipolysis in rat adipocytes. II. Definition of steady-state relationship with lipolytic and antilipolytic modulators, *J. Biol. Chem.*, **1985**, *260*, 15130–15138.

231 A.J. Garton, D.G. Campbell, P. Cohen, and S.J. Yeaman, Primary structure of the site on bovine hormone-sensitive lipase phosphorylated by cyclic AMP-dependent protein kinase, *FEBS Lett.*, **1988**, *229*, 68–72.

232 P. Strålfors and P. Belfrage, Phosphorylation of hormone-sensitive lipase by cyclic AMP-dependent protein kinase, *J. Biol. Chem.*, **1983**, *258*, 15146–15152.

233 A.J. Garton, D.G. Campbell, P. Cohen, and S.J. Yeaman, Primary structure of the site on bovine hormone-sensitive lipase phosphorylated by cyclic AMP-dependent protein kinase, *FEBS Lett.*, **1988**, *229*, 68–72.

234 C. Holm, T.G. Kirchgessner, and K.L. Svenson, et al., Hormone-sensitive lipase: sequence, expression, and chromosomal localization to 19 cent-q13.3, *Science*, **1988**, *241*, 1503–1506.

235 P. Strålfors, P. Björgell, and P. Belfrage, Hormonal regulation of hormone-sensitive lipase in intact adipocytes: identification of phosphorylated sites and effects on the phosphorylation by lipolytic hormones and insulin, *Proc. Natl. Acad. Sci. USA*, **1984**, *81*, 3317–3321.

236 M.W. Anthonsen, L. Rönnstrand, C. Wernstedt, E. Degerman, and C. Holm, Identification of novel phosphorylation sites in hormone-sensitive lipase that are phosphorylated in response to isoproterenol and govern activation properties in vitro, *J. Biol. Chem.*, **1998**, *273*, 215–221.

237 H. Olsson, P. Strålfors, and P. Belfrage, Phosphorylation of the basal site of hormone-sensitive lipase by glycogen synthase-4, *FEBS Lett.*, **1986**, *209*, 175–180.

238 A.J. Garton, D.G. Campbell, and D. Carling, et al., Phosphorylation of bovine hormone-sensitive lipase by the AMP-activated protein kinase. A possible antilipolytic mechanism, *Eur. J. Biochem.*, **1989**, *179*, 249–254.

239 D.G. Hardie and D. Carling, The AMP-activated protein kinase – fuel gauge of the mammalian cell?, *Eur. J. Biochem.*, **1997**, *246*, 259–273.

240 J.M. Corton, J.G. Gillespie, S.A. Hawley, and D.G. Hardie, 5-Aminoimidazole-4-carboxamide ribonucleoside. A specific method for activating AMP-activating protein kinase in intact cells?, *Eur. J. Biochem.*, **1995**, *229*, 558–565.

241 J.E. Sullivan, K.J. Brocklehurst, A.E. Marley, F. Carey, and D. Carling, et al., Inhibition of lipolysis and lipogenesis in isolated rat adipocytes with AICAR, a cell-permeable activator of AMP-activated protein kinase, *FEBS Lett.*, **1994**, *353*, 33–36.

242 B.E. Kemp, K.I. Mitchelhill, and D. Stapleton, et al., Dealing with energy demand: the AMP-activated protein kinase, *Trends Biochem. Sci.*, **1999**, *24*, 22–25.

243 B.E. Kemp, K.I. Mitchelhill, D. Stapleton, B.J. Michell, Z.P. Chen, and L. Witters, Dealing with energy demand: the AMP-activated protein kinase, *Trends Biochem. Sci.*, **1999**, *24*, 22–35.

244 S.K. Moule and R.M. Denton, The activation of p38 MAPK by the β-adrenergic agonist isoproterenol in rat epididymal fat cells, *FEBS Lett.*, **1998**, *439*, 287–290.

245 W. Yin, J. Mu, and M.J. Birnbaum, Role of AMP-activated protein kinase in cyclic-AMP-dependent lipolysis in 3T3-L1 adipocytes, *J. Biol. Chem.*, **2003**, *278*, 43 074–43 080.

246 C.L. Su, C. Sztalryd, J.A. Contreras, C. Holm, A.R. Kimmel, and C. Londos, Mutational analysis of the hormone-sensitive lipase translocation reaction in adipocytes, *J. Biol. Chem.*, **2003**, *278*, 43 615–43 619.

247 H. Olsson and P. Belfrage, The regulation and basal phosphorylation sites of hormone-sensitive lipase are dephosphorylated by protein phosphatase-1, 2A and 2C but not by protein phosphatase-2B, *Eur. J. Biochem.*, **1987**, *168*, 399–405.

248 S.L. Wood, N. Emmison, A.C. Borthwick, and S.J. Yeaman, The protein phosphorylation responsible for dephosphorylation of hormone-sensitive lipase in isolated rat adipocytes, *Biochem. J.*, **1993**, *295*, 531–535.

249 P. Strålfors and R.C. Honnor, Insulin-induced dephosphorylation of hormone-sensitive lipase. Correlation with lipolysis and cAMP-dependent protein kinase activity, *Eur. J. Biochem.*, **1989**, *182*, 379–385.

250 P. Strålfors, H. Olsson, and P. Belfrage, Hormone-sensitive lipase. In *The Enzymes*, "XVIII", Academic Press, London, **1987**, pp. 147–175.

251 J.J. Egan, A.S. Greenberg, M.K. Chang, S.A. Wek, M.C. Moos, and C. Londos, Mechanism of hormone-stimulated lipolysis in adipocytes: translocation of hormone-sensitive lipase to the lipid storage droplet, *Proc. Natl. Acad. Sci. USA*, **1992**, *89*, 8537–8541.

252 D.L. Brasaemle, D.M. Levin, D.C. Adler-Wailes, and C. Londos, The lipolytic stimulation of 3T3-L1 adipocytes promotes the translocation of cytosolic hormone-sensitive lipase to the surface of lipid storage droplets, *Biochim. Biophys. Acta*, **1999**, *1483*, 251–262.

253 H. Okuda, C. Morimoto, and T. Tsujita, Relationship between cyclic AMP production and lipolysis induced by forskolin in rat fat cells, *J. Lipid Res.*, **1992**, *33*, 225–231.

254 H.A. Hirsch and O.M. Rosen, Lipolytic stimulation modulates the subcellular distribution of hormone-sensitive lipase in 3T3-L1 cells, *J. Lipid Res.*, **1984**, *25*, 665–677.

255 C. Holm, G. Fredrikson, R. Sundler, and P. Belfrage, Incorporation of hormone-sensitive lipase into phosphatidylcholine vesicles, *Lipids*, **1990**, *25*, 254–259.

256 C. Holm, J.A. Contreras, R. Verger, and M.C. Schotz, Large-scale purification and kinetic properties of recombinant hormone-sensitive lipase from baculovirus-insect cell systems, *Methods Enzymol.*, **1997**, *284*, 272–284.

257 J. Hermoso, D. Pignol, and S. Penel, et al., Neutron crystallography evidence of lipase-colipase complex activation by a micelle, *EMBO J.*, **1997**, *16*, 5531–5536.

258 S. Lucas, G. Tavernier, C. Tiraby, M. Aline, and D. Langin, Expression of hormone-sensitive lipase in white adipose tissue of transgenic mice increases lipase activity but does not enhance in vitro lipolysis, *J. Lipid Res.*, **2003**, *44*, 154–163.

259 W.J. Shen, K. Sridhar, D.A. Bernlohr, and F.B. Kraemer, Interaction of rat hormone-sensitive lipase with adipocyte lipid-binding protein, *Proc. Natl. Acad. Sci. USA*, **1999**, *96*, 5528–5532.

260 S.C. Souza, K.V. Muliro, and L. Liscum, et al., Modulation of hormone-sensitive lipase and protein kinase A-mediated lipolysis by perilipin A in an adenoviral reconstituted system, *J. Biol. Chem.*, **2002**, *277*, 8267–8272.

261 J.T. Tansey, A.M. Huml, and R. Vogt, et al., Functional studies on native and mutated forms of perilipins, *J. Biol. Chem.*, **2003**, *278*, 8401–8408.

262 W. Doerfler, K.R. Feingold, and C. Grunfeld, Cytokines induce catabolic effects in cultured adipocytes by multiple mechanisms, *Cytokine*, **1994**, *6*, 478–484.

263 K.R. Feingold, W. Doerfler, C.A. Dinarello, W. Fiers, and C. Grunfeld, Stimulation of lipolysis in cultured fat cells by tumor necrosis factor, interleukin-1, and the interferons is blocked by inhibition of prostaglandin synthesis, *Endocrinology*, **1992**, *130*, 10–16.

264 S.C. Souza, L.M. Vargas De, M.T. Yamamoto, P. Lien, M.D. Franciosa, L.G. Moss, and A.S. Greenberg, Overexpression of perilipin A and B blocks the ability of tumor necrosis factor alpha to increase lipolysis in 3T3-L1 adipocytes, *J. Biol. Chem.*, **1998**, *273*, 24665–24669.

265 G.M. Clifford, C. Londos, F.B. Kraemer, R.G. Vernon, and S.J. Yeaman, Translocation of hormone-sensitive lipase and perilipin upon lipolytic stimulation of rat adipocytes, *J. Biol. Chem.*, **2000**, *275*, 5011–5015.

266 L.-J. Syu and A.R. Saltiel, Lipotransin: A novel docking protein for hormone-sensitive lipase, *Mol. Cell*, **1999**, *4*, 109–115.

267 F.J. McNally and S. Thomas, Katanin is responsible for the M-phase microtubule-severing activity in Xenopus eggs, *Mol. Biol. Cell.*, **1998**, *9*, 1847–1861.

268 R. Erdmann, F.F. Wiebel, A. Flessau, J. Rytka, A. Beyer, K.U. Frohlich, and W.H. Kunau, PAS1, a yeast gene required for peroxisome biogenesis, encodes a member of a novel family of putative ATPases, *Cell*, **1991**, *64*, 499–510.

269 T. Rowe and W.E. Balch, Membrane fusion. Bridging the gap by AAA ATPases, *Nature*, **1997**, *388*, 20–21.

270 S. Patel and M. Latterich, The AAA team: related ATPases with diverse functions, *Trends Cell. Biol.*, **1998**, *8*, 65–71.

271 S.W. Whiteheart, K. Rossnagel, S.A. Buhrow, M. Brunner, R. Jaenicke, and J.E. Rothman, N-ethylmaleimide-sensitive fusion: a trimeric ATPase whose hydrolysis of ATP is required for membrane fusion, *J. Cell. Biol.*, **1994**, *126*, 945–954.

272 C.A. Jepson and S.J. Yeaman, Inhibition of hormone-sensitive lipase by intermediary lipid metabolites, *FEBS Lett.*, **1992**, *310*, 197–200.

273 J.N. Fain and R.E. Shepherd, Free fatty acids as feedback regulators of adenylate cyclase and cyclic 3:5-AMP accumulation in rat fat cells, *J. Biol. Chem.*, **1975**, *250*, 6586–6592.

274 T. Rahn, L. Rönnstrand, C. Wernstedt, M.-J. Leroy, H. Tornqvist, V. Manganiello, P. Belfrage, and E. Degerman, Identification of the site in the cGMP-inhibited phosphodiesterase phosphorylated in adipocytes in response to insulin and isoproterenol, *J. Biol. Chem.*, **1996**, *271*, 11575–11580.

275 G. Fredrikson, H. Tornqvist, and P. Belfrage, Hormone-sensitive lipase and monoacylglycerol lipase are both required for complete degradation of adipocyte triacylglycerol, *Biochim. Biophys. Acta*, **1986**, *876*, 288–293.

276 M.A. Lunzer, J.A. Manning, and R.K. Ockner, Inhibition of rat liver acetyl coenzyme A carboxylase by long chain acyl coenzyme A and fatty acid. Modulation by fatty acid-binding protein, *J. Biol. Chem.*, **1977**, *252*, 5483–5487.

277 G. Lehrer, S. R. Panini, D. H. Rogers, and H. Rudney, Modulation of rat liver 3-hydroxy-3-methylglutaryl coenzyme A reductase by lipid inhibitors, substrates, and cytosolic factors, *J. Biol. Chem.*, **1981**, *256*, 5612–5619.

278 D. Carling, V. A. Zammit, and D. G. Hardie, A common bicyclic protein kinase cascade inactivates the regulatory enzymes of fatty acid and cholesterol biosynthesis, *FEBS Lett.*, **1987**, *223*, 217–222.

279 W.-J. Shen, K. Sridhar, D. A. Bernlohr, and F. B. Kraemer, Interaction of rat hormone-sensitive lipase with adipocyte lipid-binding protein, *Proc. Natl. Acad. Sci. USA*, **1999**, *96*, 5528–5532.

280 B. M. Spiegelman and H. Green, Control of specific protein biosynthesis during the adipose conversion of 3T3 cells, *J. Biol. Chem.*, **1980**, *255*, 8811–8818.

281 T. G. Ramsay, Fat cells, *Endocrinol. Metab. Clin. North. Am.*, **1996**, *25*, 847–879.

281a M. G. Wootan, D. A. Bernlohr, and J. Storch, Mechanism of fluorescent fatty acid transfer from adipocyte fatty acid binding to membranes, *Biochemistry*, **1993**, *32*, 8622–8627.

282 D. A. Bernlohr, M. A. Simpson, A. Vogel Hertzel, and L. J. Banaszak, Intracellular lipid-binding proteins and their genes, *Annu. Rev. Nutr.*, **1997**, *17*, 277–303.

283 N. R. Coe, M. A. Simpson, and D. A. Bernlohr, Targeted disruption of the adipocyte lipid-binding protein (aP2 protein) gene impairs fat cell lipolysis and increases cellular fatty acid levels, *J. Lipid Res.*, **1999**, *40*, 967–972.

284 L. Scheja, L. Makowski, K. T. Uysal, S. M. Wiesbrock, D. R. Shimshek, D. S. Meyers, M. Morgans, R. A. Parker, and G. S. Hotamisligil, Altered insulin secretion associated with reduced lipolytic efficiency in aP2–/– mice, *Diabetes*, **1999**, *48*, 1987–1994.

285 G. S. Hotamisligil, R. S. Johnson, R. J. Distel, R. Ellis, V. E. Papaioannou, and B. M. Spiegelman, Uncoupling of obesity from insulin resistance through a targeted mutation in aP2, the adipocyte fatty acid binding protein, *Science*, **1996**, *274*, 1377–1379.

286 W.-J. Shen, Y. Liang, R. Hong, S. Patel, V. Natu, K. Sridhar, A. Jenkins, D. A. Bernlohr, and F. B. Kraemer, Characterization of the functional interaction of adipocyte lipid-binding protein with hormone-sensitive lipase, *J. Biol. Chem.*, **2001**, *276*, 49443–49448.

287 W.-J. Shen, S. Patel, and F. B. Kraemer, Hormone-sensitive lipase functions as an oligomer, *Biochemistry*, **2000**, *39*, 2392–2398.

288 C. Sztalryd and F. B. Kraemer, Differences in hormone-sensitive lipase expression in white adipose tissue from various anatomical locations of the rat, *Metabolism*, **1994**, *43*, 241–247.

289 E. Plee-Gautier, J. Grober, E. Duplus, D. Langin, and C. Forest, Inhibition of hormone-sensitive lipase gene expression by cAMP and phorbol esters in 3T3-F442A and BFC-1 adipocytes, *Biochem. J.*, **1996**, *318*, 1057–1063.

290 J. Antras-Ferry, P. Robin, D. Robin, and C. Forest, Fatty acids and fibrates are potent inducers of transcription of the phosphoenolpyruvate carboxykinase gene in adipocytes, *Eur. J. Biochem.*, **1995**, *234*, 390–396.

291 T. Raclot, M. Dauzats, and D. Langin, Regulation of hormone-sensitive lipase expression by glucose in 3T3-F442A adipocytes, *Biochem. Biophys. Res. Commun.*, **1998**, *245*, 510–513.

292 B. G. Slavin, J. M. Ong, and P. A. Kern, Hormonal regulation of hormone-sensitive lipase activity and mRNA levels in isolated rat adipocytes, *J. Lipid Res.*, **1994**, *35*, 1535–1541.

293 S. Reynisdottir, M. Dauzats, A. Thorne, and D. Langin, Comparison of hormone-sensitive lipase activity in visceral and subcutaneous human adipose tissue, *J. Clin. Endocrinol. Metab.*, **1997**, *82*, 4162–4166.

294 V. Large, P. Arner, S. Reynisdottir, J. Grober, V. Harmelen van, C. Holm, and D. Langin, Comparison of hormone-sensitive lipase expression and activity in relation to lipolysis in human fat cells, *J. Lipid Res.*, **1998**, *39*, 1688–1695.

295 H. Laurell, J. Grober, C. Vindis, T. Lacombe, M. Dauzats, C. Holm, and M. Langin, Species-specific alternative splicing generates a catalytically inactive form

of human hormone-sensitive lipase (LIPE), *Biochem. J.*, **1997**, *328*, 137–143.

296 V. Large, P. Arner, S. Reynisdottir, J. Grober, V. Harmelen van, C. Holm, and D. Langin, Hormone-sensitive lipase expression and activity in relation to lipolysis in human fat cells, *J. Lipid Res.*, **1998**, *39*, 1688–1695.

297 H. Ray, M. Beylot, P. Arner, D. Larrouy, D. Langin, C. Holm, and V. Large, The presence of a catalytically inactive form of hormone-sensitive lipase is associated with decreased lipolysis in abdominal subcutaneous adipose tissue of obese subjects, *Diabetes*, **2003**, *52*, 1417–1422.

298 J.E. Schaffer, Fatty acid transport: the roads taken, *Am. J. Physiol. Endocrinol. Metab.*, **2002**, *282*, 239–246.

299 K. Motojima, P. Passilly, J.M. Peters, F.J. Gonzales, and N. Latruffe, Expression of putative fatty acid transporter genes are regulated by peroxisome proliferator-activated receptor alpha and gamma activators in a tissue- and inducer-specific manner, *J. Biol. Chem.*, **1998**, *273*, 16710–16714.

300 W. Stremmel, G. Strohmeyer, F. Borchard, S. Kochwa, and P.D. Berk, Isolation and partial characterization of a fatty acid binding protein in rat liver plasma membranes, *Proc. Natl. Acad. Sci. USA*, **1985**, *82*, 4–8.

301 P.D. Berk, S.L. Zhou, C.L. Kiang, D. Stump, M. Bradbury, and L.M. Isola, Uptake of long chain free fatty acids is selectively up-regulated in adipocytes of Zucker rats with genetic obesity and non-insulin-dependent diabetes mellitus, *J. Biol. Chem.*, **1997**, *272*, 8830–8835.

302 M. Febbraio, N.A. Abumrad, and D.P. Hajjar, et al., A null mutation in CD36 reveals an important role in fatty acid and lipoprotein metabolism, *J. Biol. Chem.*, **1999**, *274*, 19055–19062.

303 A. Ibrahimi, A. Bonen, and W.D. Blinn, Muscle-specific overexpression of FAT/CD36 enhances fatty acid oxidation by contracting muscle, reduces plasma triglycerides and fatty acids, and increases plasma glucose and insulin, *J. Biol. Chem.*, **1999**, *274*, 26761–26766.

304 T.Y. Hui, B.I. Frohnert, A.J. Smith, J. E. Schaffer, and D.A. Bernlohr, Characterization of the murine fatty acid transport protein gene and its insulin response sequence, *J. Biol. Chem.*, **1998**, *273*, 27420–27429.

305 M.Z. Man, T.Y. Hui, J.E. Schaffer, H.F. Lodish, and D.A. Bernlohr, Regulation of the murine adipocyte fatty acid transporter gene by insulin, *Mol. Endocrinol.*, **1996**, *10*, 1021–1028.

306 H. Kuriyama, S. Kawamoto, and N. Ishida, et al., Molecular cloning and expression of a novel human aquaporin from adipose tissue with glycerol permeability, *Biochem. Biophys. Res. Commun.*, **1997**, *241*, 53–58.

307 K. Ishibashi, K. Yamauchi, and Y. Kageyama, et al., Molecular characterization of human aquaporin-7 gene and its chromosomal mapping, *Biochim. Biophys. Acta*, **1998**, *1399*, 62–66.

308 K. Kishida, H. Kuriyama, and T. Funahashi, et al., Aquaporin adipose, a putative glycerol channel in adipocytes, *J. Biol. Chem.*, **2000**, *275*, 20896–20902.

309 A.S. Verkman and A.K. Mitra, Structure and function of aquaporin water channels, *Am. J. Physiol. Renal Physiol.*, **2000**, *278*, 13–28.

310 J.B. Heymann and A. Engel, Structural clues in the sequences of the aquaporins, *J. Mol. Biol.*, **2000**, *295*, 1039–1053.

311 S. Nielsen, C.L. Chou, D. Marples, E.I. Christensen, B.K. Kishore, and M.A. Knepper, Vasopressin increases water permeability of kidney collecting duct by inducing translocation of aquaporin-CD water channels to plasma membrane, *Proc. Natl. Acad. Sci. USA*, **1995**, *92*, 1013–1017.

312 P.A. Jansson, A. Larsson, U. Smith, and P. Lonnroth, Glycerol production in subcutaneous adipose tissue in lean and obese humans, *J. Clin. Invest.*, **1992**, *89*, 1610–1617.

313 I. Puhakainen, V.A. Koivisto, and H. Yki-Jarvinen, Lipolysis and gluconeogenesis from glycerol are increased in patients with noninsulin-dependent diabetes mellitus, *J. Clin. Endocrinol. Metab.*, **1992**, *75*, 789–794.

314 N. Nurjhan, A. Consoli, and J. Gerich, Increased lipolysis and its consequences on gluconeogenesis in non-insu-

lin-dependent diabetes mellitus, *J. Clin. Invest.*, **1992**, *89*, 169–175.

315 K. Rebrin, G.M. Steil, S.D. Mittelman, and R.N. Bergman, Causal linkage between insulin suppression of lipolysis and suppression of liver glucose output in dogs, *J. Clin. Invest.*, **1996**, *98*, 741–749.

316 N. Nurjhan, F. Kennedy, A. Consoli, C. Martin, J. Miles, and J. Gerich, Quantification of the glycolytic origin of plasma glycerol: implications for the use of the rate of appearance of plasma glycerol as an index of lipolysis in vivo, *Metabolism*, **1988**, *37*, 386–389.

317 H. Kondo, I. Shimomura, and K. Kishida, et al., Human aquaporin adipose (AQPap) gene, *Eur. J. Biochem.*, **2002**, *269*, 1814–1826.

318 P. Arner, E. Kriegholm, P. Engfeldt, and J. Bolinder, Adrenergic regulation of lipolysis in situ at rest and during exercise, *J. Clin. Invest.*, **1990**, *85*, 893–898.

319 I. Dugail, S. Le Lay, M. Varret, X. Le Liepvre, G. Dagher, and P. Ferré, New insights into how adipocytes sense their triglyceride stores. Is cholesterol a signal, *Horm. Metab. Res.*, **2003**, *35*, 204–210.

320 J.F. Oram and R.M. Lawn, The gatekeeper for eliminating excess tissue cholesterol, *J. Lipid Res.*, **2001**, *42*, 1173–1179.

321 Y. Luo and A.R. Tail, Sterol upregulation of human CETP expression in vitro and in transgenic mice by an LXR element, *J. Clin. Invest.*, **2000**, *105*, 513–520.

322 B.A. Laffitte, J.J. Repa, S.B. Joseph, D.C. Wilpitz, H.R. Kast, D.J. Mangelsdorf, and P. Tontonoz, LXRs control lipid-inducible expression of the apolipoprotein E gene in macrophages and adipocytes, *Proc. Natl. Acad. Sci. USA*, **2001**, *98*, 507–512.

323 J.F. Oram, R.M. Lawn, M.R. Garvin, and D.P. Wade, ABCA1 is the cAMP-inducible apolipoprotein receptor that mediates cholesterol secretion from macrophages, *J. Biol. Chem.*, **2000**, *275*, 34508–34511.

324 J. Langfort, T. Ploug, J. Ihlemann, J.H. Enevoldsen, B. Stallknecht, M. Saldo, M. Kjaer, C. Holm, and H. Galbo, Hormone-sensitive lipase (HSL) expression and regulation in skeletal muscle, *Adv. Exp. Med. Biol.*, **1998**, *441*, 219–228.

325 D.J. Dyck and A. Bonen, Muscle contraction increases palmitate esterification and oxidation and triacylglycerol oxidation, *Am. J. Physiol. Endocrinol. Metab.*, **1998**, *275*, 888–896.

326 J.F. Hopp and W.K. Palmer, Effect of electrical stimulation on intracellular triacylglycerol in isolated skeletal muscle, *J. Appl. Physiol.*, **1990**, *68*, 348–354.

327 L.L. Spriet and R.A. Howlett, Metabolic control of energy production during physical activity, In *Perspectives in Exercise Science and Sports Medicine*, (D.R. Lamb, R. Murray, and I.N. Carmel, ed.), Cooper Medical Press, Tampa, FL, **1999**, pp. 1–45.

328 D.J. Dyck, G. Steinberg, and A. Bonen, Insulin increases FA uptake and esterification but reduces lipid utilization in isolated contracting muscle, *Am. J. Physiol. Endocrinol. Metab.*, **2001**, *281*, 600–607.

329 R. Mora-Rodriguez and E.F. Cole, Effects of plasma epinephrine on fat metabolism during exercise: interactions with exercise intensity, *Am. J. Physiol. Endocrinol. Metab.*, **2000**, *278*, 669–676.

330 D.O. Allen and R.R. Beck, Role of calcium ion in hormone-stimulated lipolysis, *Biochem. Pharmacol.*, **1986**, *35*, 767–772.

331 S. Efendic, B. Alm, and H. Löw, Effects of Ca^{2+} on lipolysis in human omental adipose tissue in vitro, *Horm. Metab. Res.*, **1970**, *2*, 287–291.

332 R.J. Schimmel, Characterization of a colchicine receptor protein in rat epididymal adipose tissue, *Biochim. Biophys. Acta*, **1975**, *399*, 181–190.

333 F. Tebar, M. Soley, and I. Ramirez, The antilipolytic effects of insulin and epidermal growth factor in rat adipocytes are mediated by different mechanisms, *Endocrinology*, **1996**, *137*, 4181–4188.

334 B. Xue, N. Moustaid, W.O. Wilkison, and M.B. Zemel, The agouti gene product inhibits lipolysis in human adipocytes via a Ca^{2+}-dependent mechanism, *FASEB J.*, **1998**, *12*, 1391–1396.

335 F.B. Kraemer, S. Patel, M.S. Saedi, and C. Sztalryd, Detection of hormone-sen-

sitive lipase in various tissues. I. Expression of an HSL/bacterial fusion protein and generation of anti-HSL antibodies, *J. Lipid Res.*, **1993**, *34*, 663–671.

336 S. J. Peters, D. J. Dyck, A. Bonen, and L. L. Spriet, Effects of epinephrine on lipid metabolism in resting skeletal muscle, *Am. J. Physiol. Endocrinol. Metab.*, **1998**, *275*, 300–309.

337 J. Langfort, T. Ploug, J. Ihlemann, M. Saldo, C. Holm, and H. Galbo, Expression of hormone-sensitive lipase and its regulation by adrenaline in skeletal muscle, *Biochem. J.*, **1999**, *340*, 459–465.

338 M. Z. Tucker and L. P. Turcotte, Impaired fatty acid oxidation in muscle of aging rats perfused under basal conditions, *Am. J. Physiol. Endocrinol. Metab.*, **2002**, *282*, 1102–1109.

339 J. Langfort, T. Ploug, J. Ihlemann, C. Holm, and H. Galbo, Stimulation of hormone-sensitive lipase activity by contractions in rat skeletal muscle, *Biochem. J.*, **2000**, *351*, 207–214.

340 L. Enevoldsen, B. Stallknecht, J. Langfort, L. Petersen, C. Holm, T. Ploug, and H. Galbo, The effect of exercise training on hormone-sensitive lipase in rat intra-abdominal adipose tissue and muscle, *J. Physiol.*, **2001**, *536*, 871–877.

341 J. Suzuki, W.-J. Shen, B. D. Nelson, S. Patel, J. H. Veerkamp, S. P. Selwood, G. M. Murphy, E. Reaven, and F. B. Kraemer, Absence of cardiac lipid accumulation in transgenic mice with heart-specific HSL overexpression, *Am. J. Physiol. Endocrinol. Metab.*, **2001**, *281*, 857–866.

342 J. J. Berger and R. J. Barnard, Effect of diet on fat cell size and hormone-sensitive lipase activity, *J. Appl. Physiol.*, **1990**, *87*, 227–232.

343 C. Sztalryd and F. B. Kraemer, Regulation of hormone-sensitive lipase during fasting, *Am. J. Physiol.*, **1994**, *266*, 179–185.

344 B. E. Wilson, S. Deeb, and G. I. Florant, Seasonal changes in hormone-sensitive and lipoprotein lipase messenger RNA concentrations in marmot white adipose tissue, *Am. J. Physiol.*, **1992**, *262*, 177–181.

345 L. M. Botion and A. Green, Long-term regulation of lipolysis and hormone-sensitive lipase by insulin and glucose, *Diabetes*, **1999**, *48*, 1691–1697.

346 F. Smith, P. Rouet, S. Lucas, A. Mairal, C. Sengenes, M. Lafontan, S. Vaulont, M. Casado, and D. Langin, Transcriptional regulation of adipocyte hormone-sensitive lipase by glucose, *Diabetes*, **2002**, *51*, 293–300.

347 V. C. Manganiello, T. Murata, M. Taira, P. Belfrage, and E. Degerman, Diversity in cyclic nucleotide phosphodiesterase isoenzyme families, *Arch. Biochem. Biophys.*, **1995**, *322*, 1–13.

348 E. Degerman, M.-J. Leroy, M. Taira, P. Belfrage, and V. Manganiello, A role for insulin-mediated regulation of cyclic monophosphate (cGMP)-inhibited phosphodiesterase in the antilipolytic action of insulin, In D. Le Roith, J. M. Olefsky, and S. Taylor, (ed) *Diabetes Mellitus: A Fundamental and Clinical Text*, Lippincott-Raven, Philadelphia, **1996**, pp. 197–204.

349 M. L. Elks and V. C. Manganiello, Antilipolytic action of insulin: role of cAMP phosphodiesterase activation, *Endocrinology*, **1985**, *116*, 2119–2221.

350 E. Degerman, P. Belfrage, and V. C. Manganiello, Structure, localization, and regulation of cGMP-inhibited phosphodiesterase (PDE3), *J. Biol. Chem.*, **1997**, *272*, 6823–6826.

351 H. Eriksson and H. Tornqvist, Specific inhibition of the cGMP-inhibited cAMP phosphodiesterase blocks the insulin-like antilipolytic effect of growth hormone in rat adipocytes, *Mol. Cell. Biochem.*, **1997**, *169*, 37–42.

352 S. H. Soderling, S. J. Bayuga, and J. A. Beavo, Identification and characterization of a novel family of cyclic nucleotide phosphodiesterases, *J. Biol. Chem.*, **1998**, *273*, 15553–15558.

353 Y. Shechter, Differential effects of two phosphodiesterase inhibitors on fat cell metabolism, *Endocrinology*, **1984**, *115*, 1787–1791.

354 C. Schmitz-Peiffer, M. L. Reeves, and R. M. Denton, Characterization of the cyclic nucleotide phosphodiesterase isoenzymes present in rat epididymal fat cell, *Cell. Signal.*, **1992**, *4*, 37–49.

355 H. Eriksson, M. Ridderstrale, E. Degerman, D. Ekholm, C. J. Smith, V. Manganiello, P. Belfrage, and H. Tornqvist, Evidence for the key role of

the adipocyte cGMP-inhibited cAMP phosphodiesterase in the antilipolytic action of insulin, *Biochim. Biophys. Acta*, **1995**, *1266*, 101–104.

356 E. Hagström-Toft, J. Bolinder, S. Eriksson, and P. Arner, Role of phosphodiesterase III in the antilipolytic effect of insulin in vivo, *Diabetes*, **1995**, *44*, 1170–1175.

357 S. Enoksson, E. Degerman, E. Hagström-Toft, V. Large, and P. Arner, Various phosphodiesterase subtypes mediate the in vivo antilipolytic effect of insulin on adipose tissue and skeletal muscle in man, *Diabetologia*, **1998**, *41*, 560–568.

358 E. Degerman, C.J. Smith, H. Tornqvist, V. Vasta, V. Manganiello, and P. Belfrage, Evidence that insulin and isoprenaline activate the cGMP-inhibited low-Km cAMP phosphodiesterase in rat fat cells by phosphorylation, *Proc. Natl. Acad. Sci. USA*, **1990**, *87*, 533–537.

359 C.J. Smith, V. Vasta, E. Degerman, P. Belfrage, and V. Manganiello, Hormone-sensitive cyclic GMP-inhibited cyclic AMP phosphodiesterase in rat adipocytes. Regulation of insulin- and cAMP-dependent activation by phosphorylation, *J. Biol. Chem.*, **1991**, *266*, 13385–13390.

360 N. Tennagels, D. Telting, S. Parvaresch, J.A. Maassen, and H.W. Klein, Identification of Ser(1275) and Ser(1309) as autophosphorylation sites of the human insulin receptor in intact cells, *Biochem. Biophys. Res. Commun.*, **2001**, *282*, 387–393.

361 H. Al-Hasani, B. Eisermann, and N. Tennagels, et al., Identification of Ser-1275 and Ser-1309 as autophosphorylation sites of the insulin receptor, *FEBS Lett.*, **1997**, *400*, 65–70.

362 T. Rahn, M. Ridderstrale, H. Tornqvist, V. Manganiello, G. Fredrikson, P. Belfrage, and E. Degerman, Essential role of phosphatidylinositol 3-kinase in insulin-induced activation and phosphorylation of the cGMP-inhibited cAMP phosphodiesterase in rat adipocytes studies using the selective inhibitor wortmannin, *FEBS Lett.*, **1994**, *350*, 314–317.

363 C. Holm, D. Langin, V. Manganiello, P. Belfrage, and E. Degerman, Regulation of hormone-sensitive lipase activity in adipose tissue, *Methods Enzymol.*, **1997**, *286*, 45–67.

364 J. Wijkander, L.S. Holst, T. Rahn, S. Resjö, and I. Castan, et al., Regulation of protein kinase B in rat adipocytes by insulin, vanadate, and peroxovanadate. Membrane translocation in response to peroxovanadate, *J. Biol. Chem.*, **1997**, *272*, 21520–21526.

365 J. Wijkander, T.R. Landström, V. Manganiello, P. Belfrage, and E. Degerman, Insulin-induced phosphorylation and activation of phosphodiesterase 3B in rat adipocytes: possible role of protein kinase B but not mitogen-activated protein kinase or p70 S6 kinase, *Endocrinology*, **1998**, *139*, 219–227.

366 C. Sztalryd and F.B. Kraemer, Regulation of hormone-sensitive lipase in streptozotocin-treated rats, *Metabolism*, **1995**, *44*, 1391–1396.

367 E.S. Kang, D. Betts, J.N. Fain, S.W. Bahouth, and L.K. Myers, Chronic exposure of rat fat cells to insulin enhances lipolysis and activation of partially purified hormone-sensitive lipase, *Diabetes*, **1993**, *42*, 1415–1424.

368 M. Rosenstock, A.S. Greenberg, and A. Rudich, Distinct long-term regulation of glycerol and non-esterified fatty acid release by insulin and TNF-α in 3T3-L1 adipocytes, *Diabetologia*, **2001**, *44*, 55–62.

369 S.S. Chernick, P.M. Spooner, M.M. Garrison, and R.O. Scow, Effect of epinephrine and other lipolytic agents on intracelluar lipolysis and lipoprotein lipase activity in 3T3-L1 adipocytes, *J. Lipid Res.*, **1986**, *27*, 286–294.

370 C. Londos, Hormone-sensitive lipase and the control of lipolysis in adipocytes, In D. LeRoith, J.M. Olefsky, and S.I. Taylor (ed.) Diabetes Mellitus: A Fundamental and Clinical Text, Lippincott-Raven, Philadelphia, **1996**, pp. 223–227.

371 J.R. Zierath, J.N. Livingston, and A. Thorne, et al., Regional difference in insulin inhibition of non-esterified fatty acid release from human adipocytes: relation to insulin receptor phosphorylation and intracellular signalling through the insulin receptor substrate-1 pathway, *Diabetologia*, **1998**, *41*, 1343–1354.

372 T. Okada, Y. Kawano, T. Sakakibara, O. Hazeki, and M. Ui, Essential role of PI 3-kinase in insulin-induced glucose transport and antilipolysis in rat adipocytes, *J. Biol. Chem.*, **1994**, *269*, 3568–3573.

373 M.Y. Wang, Y. Lee, and R.H. Unger, Novel form of lipolysis induced by leptin, *J. Biol. Chem.*, **1999**, *274*, 17541–17544.

374 G. Frühbeck, M. Aguardo, and J.A. Martinez, In vitro lipolytic effect of leptin on mouse adipocytes: evidence for a possible autocrine/paracrine role of leptin, *Biochem. Biophys. Res. Commun.*, **1997**, *240*, 590–594.

375 G. Frühbeck, M. Aguardo, J. Gomez-Ambrosi, and J.A. Martinez, Lipolytic effect of in vivo leptin administration on adipocytes of lean and ob/ob mice, but not db/db mice, *Biochem. Biophys. Res. Commun.*, **1998**, *250*, 99–102.

376 U. Sarmiento, B. Benson, and S. Kaufman, Morphologic and molecular changes induced by recombinant human leptin in the white and brown adipose tissues of C57BL/6 mice, *Lab. Invest.*, **1997**, *77*, 243–256.

377 G. Müller, J. Ertl, M. Gerl, and G. Preibisch, Leptin impairs metabolic actions of insulin in isolated rat adipocytes, *J. Biol. Chem.*, **1997**, *272*, 10585–10593.

378 M.B. Davidson, Effect of growth hormone on carbohydrate and lipid metabolism, *Endocrinol. Rev.*, **1987**, *8*, 115–131.

379 B.G. Slavin, J.M. Ong, and P.A. Kern, Hormonal regulation of hormone-sensitive lipase activity and mRNA levels in isolated rat adipocytes, *J. Lipid Res.*, **1994**, *35*, 1535–1541.

380 P. Strålfors and P. Belfrage, Phosphorylation of hormone-sensitive lipase by cyclic GMP-dependent protein kinase, *FEBS Lett.*, **1985**, *180*, 280–284.

381 J.C. Khoo, P.J. Sperry, G.N. Gill, and D. Steinberg, Activation of hormone-sensitive lipase and phosphorylase kinase by purified cyclic GMP-dependent protein kinase, *Proc. Natl. Acad. Sci. USA*, **1977**, *74*, 4843–4847.

382 T.M. Lincoln, C.L. Hall, C.R. Park, and J.D. Corbin, Guanosine 3′:5′-cyclic monophosphate binding proteins in rat tissues, *Proc. Natl. Acad. Sci. USA*, **1976**, *73*, 2559–2563.

383 N. Gaudiot, A.M. Jaubert, and E. Charbonnier, et al., Modulation of white adipose tissue lipolysis by nitric oxide, *J. Biol. Chem.*, **1998**, *273*, 13475–13481.

384 C.H. McIntosh, I. Bremsak, and F.C. Lynn, et al., Glucose-dependent insulinotropic polypeptide stimulation of lipolysis in differentiated 3T3-L1 cells: wortmannin-sensitive inhibition by insulin, *Endocrinology*, **1999**, *140*, 398–404.

385 C.P. Sewter, J.E. Digby, F. Blows, J. Prins, and S. O'Rahilly, Regulation of tumor necrosis factor-α release from human adipose tissue in vitro, *J. Endocrinol.*, **1999**, *163*, 33–38.

386 G. Hotamisligil, N. Shargill, and B. Spiegelman, Adipose expression of tumor necrosis factor alpha: direct role in obesity-linked insulin resistance, *Science*, **1993**, *259*, 87–91.

387 G.S. Hotamisligil, P. Arner, J.F. Caro, R.L. Atkinson, and B.M. Spiegelman, Increased adipose tissue expression of tumor necrosis factor-alpha in human obesity and insulin resistance, *J. Clin. Invest.*, **1995**, *95*, 2409–2415.

388 H. Hauner, T. Petruschka, M. Russ, K. Röhrig, and J. Eckel, Effects of tumour necrosis factor alpha (TNF alpha) on glucose transport and lipid metabolism of newly-differentiated human fat cells in cell culture, *Diabetologia*, **1995**, *38*, 764–771.

389 M. Sumida, K. Sekiya, H. Okuda, Y. Tanaka, and T. Shiosaka, Inhibitory effect of tumor necrosis factor on gene expression of hormone sensitive lipase in 3T3-L1 adipocytes, *J. Biochem. (Tokyo)*, **1990**, *107*, 1–2.

390 S.C. Souza, M.T. Yamamoto, M.D. Franciosa, P. Lien, and A.S. Greenberg, BRL 49653 blocks the lipolytic actions of tumor necrosis factor-alpha: a potential new insulin-sensitizing mechanism for thiazolidinediones, *Diabetes*, **1998**, *47*, 691–695.

391 M. Sumida, K. Sekiya, H. Okuda, Y. Tanaka, and T. Shiosaka, Inhibitory effect of tumor necrosis factor on gene expression of hormone sensitive lipase in 3T3-L1 adipocytes, *J. Biochem.*, **1990**, *107*, 1–2.

392 A. Green, S.B. Dobias, D.J. Walters, and A.R. Brassier, Tumor necrosis factor increases the rate of lipolysis in primary cultures of adipocytes without altering levels of hormone-sensitive lipase, *Endocrinology*, **1994**, *134*, 2581–2588.

393 P.A. Kern, M. Saghizadeh, J.M. Ong, R.J. Bosch, R. Deem, and R.B. Simsolo, The expression of tumor necrosis factor in human adipose tissue. Regulation by obesity weight loss, and relationship to lipoprotein lipase, *J. Clin. Invest.*, **1995**, *95*, 2111–2119.

394 R.A. Memon, J. Fuller, A.H. Moser, P.J. Smith, K.R. Feingold, and C. Grunfeld, In vivo regulation of acyl-CoA synthetase mRNA activity by endotoxin and cytokines, *Am. J. Physiol.*, **1998**, *275*, 64–72.

395 D. Ron, A.R. Brasier, R.E. McGehee, and J.F. Habener, Tumor necrosis factor-induced reversal of adipocytic phenotype of 3T3-L1 cells is preceded by a loss of nuclear CCAAT/enhancer binding protein (C/EBP), *J. Clin. Invest.*, **1992**, *89*, 223–233.

396 G.S. Hotamisligil and B.M. Spiegelman, Tumor necrosis factor alpha: a key component of the obesity-diabetes link, *Diabetes*, **1994**, *43*, 1271–1278.

397 S. Gasic, B. Tian, and A. Green, Tumor necrosis factor alpha stimulates lipolysis in adipocytes by decreasing Gi protein concentrations, *J. Biol. Chem.*, **1999**, *274*, 6770–6775.

398 S.C. Souza, L.M. de Vargas, M.T. Yamamoto, P. Lien, M.D. Franciosa, L.G. Moss, and A.S. Greenberg, Overexpression of perilipin A and B blocks the ability of tumor necrosis factor α to increase lipolysis in 3T3-L1 adipocytes, *J. Biol. Chem.*, **1998**, *273*, 24665–24669.

399 A.R. Saltiel and J.M. Olefsky, Thiazolidinediones in the treatment of insulin resistance and type II diabetes, *Diabetes*, **1996**, *45*, 1661–1669.

400 T. Rahn-Landstrom, J. Mei, M. Karlsson, V. Manganiello, and E. Degerman, Down-regulation of cyclic-nucleotide phosphodiesterase 3B in 3T3-L1 adipocytes induced by tumour necrosis factor α and cAMP, *Biochem. J.*, **2000**, *346*, 337–343.

401 M. Ryden, A. Dicker, and V. Harmelen van, et al., Mapping of early signaling events in tumor necrosis factor-α-mediated lipolysis in human fat cells, *J. Biol. Chem.*, **2002**, *277*, 1085–1091.

402 H.H. Zhang, M. Halbleib, F. Ahmad, V.C. Manganiello, and A.S. Gruenberg, Tumor necrosis factor-alpha stimulates lipolysis in differentiated human adipocytes through activation of extracellular signal-related kinase and elevation of intracellular cAMP, *Diabetes*, **2002**, *51*, 2029–2035.

403 K. Cianflone, Z. Xia, and L.Y. Chen, Critical review of acylation-stimulating protein physiology in humans and rodents, *Biochim. Biophys. Acta*, **2003**, *1609*, 127–143.

404 J.K. Sethi and G.S. Hotamisligil, The role of TNF-α in adipocyte metabolism, *Sem. Cell. Develop. Biol.*, **1999**, *10*, 19–29.

405 M.J. Tisdale, Cachexia in cancer patients, *Nat. Rev. Cancer*, **2002**, *2*, 862–871.

406 J.M. Argiles, J. Lopez-Soriano, S. Busquets, and F.J. Lopez-Soriano, Journey from cachexia to obesity by TNF-α, *FASEB J.*, **1997**, *11*, 743–751.

407 J.W. Warne, Tumour necrosis factor α: a key regulator of adipose tissue mass, *J. Endocrinol.*, **2003**, *177*, 351–355.

408 A.D. Sniderman, M. Maslowska, and K. Cianflone, Of mice and men (and women) and the acylation-stimulating protein pathway, *Curr. Opin. Lipidol*, **2000**, *11*, 291–296.

409 K. Cianflone, D.A.K. Roncari, M. Maslowska, A. Baldo, J. Forden, and A.D. Sniderman, Adipsin/acylation stimulating protein system in human adipocytes: Regulation of triacylglycerol synthesis, *Biochemistry*, **1994**, *33*, 9489–9495.

410 K. Cianflone and M. Maslowska, Differentiation-induced production of ASP in human adipocytes, *Eur. J. Clin. Invest.*, **1995**, *25*, 817–825.

411 K. Cianflone, A.D. Sniderman, D. Kalant, E.B. Marlies, and R. Gougeon, Response of plasma ASP to a prolonged fast, *Int. J. Obes.*, **1995**, *19*, 604–609.

412 M. Maslowska, H. Vu, S. Phelis, A.D. Sniderman, B. Rhodes, D. Blank, and K. Cianflone, Plasma acylation stimulating protein, adipsin and lipids in non-obese and obese populations, *Eur. J. Clin. Invest.*, **1999**, *29*, 679–686.

413 M. Maslowska, T. Scantlebury, R. Germinario, and K. Cianflone, Acute in vitro production of acylation stimulating protein in differentiated human adipocytes, *J. Lipid Res.*, **1997**, *38*, 1–11.

414 T. Scantlebury, M. Maslowska, and K. Cianflone, Chylomicron-specific enhancement of acylation stimulating protein and precursor protein C3 production in differentiated human adipocytes, *J. Biol. Chem.*, **1998**, *273*, 20903–20909.

415 J. Saleh, L. K. M. Summers, K. Cianflone, B. A. Fielding, A. D. Sniderman, and K. N. Frayn, Coordinated release of acylation stimulating protein (ASP) and triacylglycerol clearance by human adipose tissue in vivo in the postprandial period, *J. Lipid Res.*, **1998**, *39*, 884–891.

416 J. A. Charlesworth, P. W. Peake, L. V. Campbell, B. A. Pussell, R. O'Grady, and T. Tzilopoulos, *Int. J. Obes. Rel. Metab. Dis.*, **1998**, *22*, 1096–1102.

417 K. Cianflone, Acylation stimulating protein and the adipocyte, *J. Endocrinol.*, **1997**, *155*, 203–206.

418 K. Cianflone, M. Maslowska, and A. D. Sniderman, Acylation stimulating protein (ASP), an adipocyte autocrine: new directions, *Semin. Cell. Dev. Biol.*, **1999**, *10*, 31–41.

419 Z. Yasruel, K. Cianflone, A. D. Sniderman, M. Rosenbloom, M. Walsh, and M. A. Rodriguez, Effect of acylation stimulating protein on the triacylglycerol synthetic pathway of human adipose tissue, *Lipids*, **1991**, *26*, 495–499.

420 Y. Z. Tao, K. Cianflone, A. D. Sniderman, S. P. Colby-Germinario, and R. J. Germinario, Acylation-stimulating protein (ASP) regulates glucose transport in the rat L6 muscle cell line, *Biochim. Biophys. Acta*, **1997**, *1344*, 221–229.

421 R. Germinario, A. D. Sniderman, S. Manuel, S. P. Lefebre, A. Baldo, and K. Cianflone, Coordinate regulation of triacylglycerol synthesis and glucose transport by acylation-stimulating protein, *Metabolism*, **1993**, *42*, 574–580.

422 V. Harmelen, S. Reynisdottir, K. Cianflone, E. Degerman, J. Hoffstedt, K. Nilsell, A. D. Sniderman, and P. Arner, Mechanisms involved in the regulation of free fatty acid release from isolated human fat cells by acylation-stimulating protein and insulin, *J. Biol. Chem.*, **1999**, *274*, 18243–18251.

423 A. Baldo, A. D. Sniderman, S. St-Luce, X. J. Zhang, and K. Cianflone, Signal transduction pathway of acylation stimulating protein: involvement of protein kinase C, *J. Lipid Res.*, **1995**, *36*, 1415–1426.

424 P. Arner, M. Goldfinger, C. Marcus, and P. Engfeldt, Effects of lipolytic and antilipolytic agents on glucose transport in human fat cells, *Int. J. Obes.*, **1991**, *15*, 327–331.

425 P. Tornvall and G. W. Walldius, A comparison between nicotinic acid and acipimox in hypertriglyceridaemia – effects on serum lipids, lipoproteins, glucose tolerance and tolerability, *J. Int. Med.*, **1991**, *230*, 415–421.

426 C. R. Sirtori, G. Gianfranceschi, and M. Sirtori, et al., Reduced triglyceridaemia and increased high density lipoprotein cholesterol levels after treatment with acipimox, a new inhibitor of lipolysis, *Atherosclerosis*, **1981**, *38*, 267–271.

427 P. M. J. Stuyt, A. F. H. Stalenhoef, and P. N. M. Demacker, and A. van't Laar, A comparative study of the effects of acipimox and clofibrate in type III and type IV hyperlipoproteinaemia, *Atherosclerosis*, **1985**, *55*, 51–62.

428 M. J. Ball, M. Vella, and J. P. D. Reckless, et al., Acipimox in the treatment of patients with hyperlipidaemia: a double blind trial, *Eur. J. Clin. Pharmacol.*, **1986**, *31*, 201–204.

429 G. Crepaldi, P. Avogaro, and G. C. Descovich, et al., Plasma lipid lowering activity of acipimox in patients with type II and type IV hyperlipoproteinaemia: a double blind trial, *Atherosclerosis*, **1988**, *70*, 115–121.

430 S. Tavintharan and M. L. Kashyap, The benefits of niacin in atherosclerosis, *Curr. Atheroscler. Rep.*, **2001**, *3*, 74–82.

431 R. H. Stern, J. D. Spence, D. J. Freeman, and A. Parbtani, Tolerance to nicotinic acid flushing, *Clin. Pharmacol. Ther.*, **1991**, *50*, 66–70.

432 J. D. Morrow, J. A. Awad, J. A. Oates, and L. Jackson-Roberts, Identification of skin as a major site of prostaglandin

D2 release following oral administration of niacin in humans, *J. Invest. Dermatol.*, **1992**, *98*, 812–815.

433 K. Aktories, G. Schultz, and K. H. Jakobs, Action principles of hormones and neurotransmitters, *Arzneimittelforschung*, **1980**, *30*, 1981–1986.

434 K. Aktories, G. Schultz, and K. H. Jakobs, Islet-activating protein prevents nicotinic acid-induced GTPase stimulation and GTP but not GTPS-induced adenylate cyclase inhibition in rat adipocytes, *FEBS Lett.*, **1983**, *156*, 88–92.

435 C. Landau, Y. D. Chen, R. Skowronski, C. B. Hollenbeck, J. B. Japan, and G. M. Reaven, Effect of nicotinic acid on plasma glucose concentration in normal individuals, *Horm. Metab. Res.*, **1992**, *24*, 424–428.

436 O. Peroni, V. Large, and M. Beylot, Measuring gluconeogenesis with [2-^{13}C]glycerol and mass isotopomer distribution analysis of glucose, *Am. J. Physiol. Endocrinol. Metab.*, **1995**, *269*, 516–523.

437 E. O. Balasse and M. A. Neef, Influence of nicotinic acid on the rates of turnover and oxidation of plasma glucose in man, *Metabolism*, **1973**, *22*, 1193–1204.

438 J. P. Felber and V. Buber, Effect of nicotinic acid derivative on glucose levels and glucose tolerance in animal and man, In: K. F. Gey and L. A. Carlson (eds) *Metabolic Effects of Nicotinic Acid and its Derivatives*, Huber, Berne, **1970**, pp. 695–710.

439 T. A. Miettinen, M. R. Taskinen, R. Pelkonen, and E. A. Nikkila, Glucose tolerance and plasma insulin in man during acute and chronic administration of nicotinic acid, *Acta Med. Scand.*, **1969**, *186*, 247–253.

440 A. Garg and S. M. Grundy, Nicotinic acid therapy for dyslipidemia in non-insulin-dependent diabetes mellitus, *JAMA*, **1990**, *264*, 723–726.

441 L. W. Gibbons, V. Gonzales, N. Gordon, and S. Grundy, The prevalence of side effects with regular and sustained-release nicotinic acid, *Am. J. Med.*, **1995**, *99*, 378–385.

442 G. F. Tutwiler and P. Dellevigne, Action on the oral hypoglycemic agent 2-tetradecylglycidic acid on hepatic fatty acid oxidation and gluconeogenesis, *J. Biol. Chem.*, **1979**, *254*, 2935–2941.

443 A. W. Christie, D. K. T. McCormick, N. Emmison, F. B. Kraemer, K. G. M. M. Alberti, and S. J. Yeaman, Mechanism of anti-lipolytic action of acipimox in isolated rat adipocytes, *Diabetologia*, **1996**, *39*, 45–53.

444 J. N. Fain and C. C. Malbon, Regulation of adenylate cyclase by adenosine, *Mol. Cell. Biochem.*, **1979**, *25*, 143–169.

445 K. Aktories, S. Gunter, and K. H. Jakobs, Na^+ amplifies adenosine receptor-mediated inhibition of adipocyte adenylate cyclase, *Eur. J. Pharmacol.*, **1981**, *71*, 157–160.

446 L. Olansky, G. A. Myers, S. L. Pohl, and E. L. Hewlett, Promotion of lipolysis in rat adipocytes by pertussis toxin: reversal of endogenous inhibition, *Proc. Natl. Acad. Sci. USA*, **1983**, *80*, 6547–6551.

447 P. B. Wieser and J. N. Fain, Insulin, prostaglandin E1, phenylisopropyl-adenosine and nicotinic acid as regulators of fat cell metabolism, *Endocrinology*, **1975**, *96*, 1221–1225.

448 L. Musatti, E. Maggi, E. Moro, G. Valzelli, and V. Tamassia, Bioavailability and pharmacokinetics in man of acipimox, a new antilipolytic and hypolipemic agent, *J. Int. Med. Res.*, **1981**, *9*, 381–386.

449 C. Stirling, M. McAleer, and J. P. D. Reckless, et al., Effects of acipimox, a nicotinic acid derivative, on lipolysis in human adipose tissue and on cholesterol synthesis in human jejunal mucosa, *Clin. Sci.*, **1985**, *68*, 83–88.

450 F. J. Moreno, R. E. Shepherd, and J. N. Fain, Effect of nicotinamide and nicotinic acid on cyclic AMP accumulation by fat cells. *Naunyn-Schmiedebergs Arch. Pharmacol.* **1979**, *306*, 179–183.

451 K. Aktories, K. H. Jakobs, and G. Schultz, Nicotinic acid inhibits adipocyte adenylate cyclase in a hormone-like manner, *FEBS Lett.*, **1980**, *115*, 11–14.

452 K. Aktories, G. Schultz, and K. H. Jakobs, Regulation of adenylate cyclase in hamster adipocytes. Inhibition by prostaglandins, α-adrenergic agonists and nicotinic acid, *Naunyn-Schmiedebergs Arch. Pharmacol.* **1980**, *312*, 167–173.

453 H. Kather, K. Aktories, G. Schultz, and K.H. Jakobs, Islet-activating protein discriminates the antilipolytic mechanism of insulin from that of other antilipolytic compounds, *FEBS Lett.*, **1983**, *161*, 149–152.

454 C. Londos, D.M.F. Cooper, and M. Rodbell, Receptor-mediated stimulation and inhibition of adenylate cyclase: the cyclase, the fat cell as a model system, In: J.E. Dumont, P. Greenberg, and G.A. Robinson, *Advances in Cyclic Nucleotide Research*, (eds.) Raven Press, New York, **1981**, vol. 14, pp. 163–167.

455 A. Lorenzen, C. Stannek, H. Lang, V. Andrianov, I. Kalvinsh, and U. Schwabe, Characterization of a G protein-coupled receptor for nicotinic acid, *Mol. Pharmacol.*, **2001**, *59*, 349–357.

456 A. Wise, S.M. Foord, N.J. Fraser, A.A. Barnes, and N. Elshourbagy, et al., Molecular identification of high and low affinity receptors for nicotinic acid, *J. Biol. Chem.*, **2003**, *278*, 9869–9874.

457 H. Nomura, B.W. Nielsen, and K. Matsushima, *Int. Immunol.*, **1993**, *5*, 1239–1249.

458 J.L. Benovic, F. Mayor, C. Staniszewski, R.J. Lefkowitz, and M.G. Caron, Purification and characterization of the beta-adrenergic receptor kinase, *J. Biol. Chem.*, **1987**, *262*, 9026–9032.

459 R.J. Lefkowitz, W.P. Hausdorff, and M.G. Caron, Role of phosphorylation in desensitization of the β-adrenoceptor, *Trends Pharmacol. Sci.*, **1990**, *11*, 190–194.

460 A. Green and J.M. Olefsky, Evidence for insulin-induced internalization and degradation of insulin receptors in rat adipocytes, *Proc. Natl. Acad. Sci. USA*, **1982**, *79*, 427–431.

461 S. Marshall, A. Green, and J.M. Olefsky, Evidence for recycling of insulin receptors in isolated rat adipocytes, *J. Biol. Chem.*, **1981**, *256*, 11464–11470.

462 M. Bouvier, S. Collins, B.F. O'Dowd, P.T. Campbell, and A. Blasi de, et al., Two distinct pathways for cAMP-mediated down-regulation of the beta 2-adrenergic receptor. Phosphorylation of the receptor and regulation of its mRNA level, *J. Biol. Chem.*, **1989**, *264*, 16786–16792.

463 J.R. Hadcock, H. Wang, and C.C. Malbon, Agonist-induced destabilization of beta-adrenergic receptor mRNA. Attenuation of glucocorticoid-induced up-regulation of beta-adrenergic receptors, *J. Biol. Chem.*, **1989**, *264*, 19928–19933.

464 T.R. Premont and R. Iyengar, Heterologous desensitization of the liver adenylyl cyclase: analysis of the role of G-proteins, *Endocrinology*, **1989**, *125*, 1151–1160.

465 C. Reithmann, P. Gierschik, D. Sidiropoulos, K. Werdan, and K.H. Jakobs, Mechanism of noradrenaline-induced heterologous desensitization of adenylate cyclase stimulation in rat heart muscle cells: increase in the level of inhibitory G-protein alpha-subunits, *Eur. J. Pharmacol.*, **1989**, *172*, 211–221.

466 C.E. Grier, P. Nambi, N. Aiyar, and S.T. Crooke, Molecular mechanisms of homologous and heterologous desensitization mediated by vasopressin in smooth muscle cells, *J. Biol. Chem.*, **1989**, *264*, 5384–5391.

467 J.B. Millar and E. Rozengurt, Heterologous desensitization of bombesin-induced mitogenesis by prolonged exposure to vasopressin: a post-receptor signal transduction block, *Proc. Natl. Acad. Sci. USA*, **1989**, *86*, 3204–3208.

468 A. Green, Adenosine receptor down-regulation and insulin resistance following prolonged incubation of adipocytes with an A1 adenosine receptor agonist, *J. Biol. Chem.*, **1987**, *262*, 15702–15707.

469 G.L. Stiles, Adenosine receptors: structure, function and regulation, *Trends Pharmacol. Sci.*, **1986**, *7*, 486–490.

470 T. Trost and U. Schwabe, Adenosine receptors in fat cells. Identification by (–)-N6-[^{3}H]phenylisopropyladenosine binding, *Mol. Pharmacol.*, **1981**, *19*, 228–235.

471 J.A. Garcia-Sainz and M.L. Torner, Rat fat-cells have three types of adenosine receptors (Ra, Ri and P). Differential effects of pertussis toxin, *Biochem. J.*, **1985**, *232*, 439–443.

472 A. Green, J.L. Johnson, and G. Milligan, Down-regulation of Gi sub-types by prolonged incubation of adipocytes with an A1 adenosine receptor agonist, *J. Biol. Chem.*, **1990**, *265*, 5206–5210.

473 A. Green and J. L. Johnson, Evidence for impaired coupling of receptors to Gi protein in adipocytes from streptozocin-induced diabetic rats, *Diabetes*, **1991**, *40*, 88–94.
474 W. J. Parsons and G. L. Stiles, Heterologous desensitization of the inhibitory A1 adenosine receptor-adenylate cyclase system in rat adipocytes. Regulation of both Ns and Ni, *J. Biol. Chem.*, **1987**, *262*, 841–847.
475 J. P. Longabaugh, J. Didsbury, A. Spiegel, and G. L. Stiles, Modification of the rat adipocyte A1 adenosine receptor-adenylate cyclase system during chronic exposure to an A1 adenosine receptor agonist: alterations in the quantity of Gs alpha and Gi alpha are not associated with changes in their mRNAs, *Mol. Pharmacol.*, **1989**, *36*, 681–688.
476 A. Green, G. Milligan, and S. B. Dobias, Gi Down-regulation as a mechanism for heterologous desensitization in adipocytes, *J. Biol. Chem.*, **1992**, *267*, 3223–3229.
477 G. Müller, Dynamics of plasma membrane microdomains and cross-talk to the insulin signalling cascade, *FEBS Lett.*, **2002**, *531*, 81–87.
478 W. Frick, A. Bauer, J. Bauer, S. Wied, and G. Müller, Structure-activity relationship of synthetic phosphoinositolglycans mimicking metabolic insulin action, *Biochemistry*, **1998**, *37*, 13421–13426.
479 G. Müller, S. Wied, A. Crecelius, A. Kessler, and J. Eckel, Phosphoinositolglycan-peptides from yeast potently induce metabolic insulin actions in isolated rat adipocytes, cardiomyocytes and diaphragm, *Endocrinology*, **1997**, *138*, 3459–3475.
480 O. Nosjean, A. Briolay, and B. Roux, Mammalian GPI proteins: sorting, membrane residence and functions, *Biochim. Biophys. Acta*, **1997**, *1331*, 153–186.
481 G. Müller, The molecular mechanism of the insulin-mimetic/sensitizing activity of the antidiabetic sulfonylurea drug amaryl, *Mol. Med.*, **2000**, *6*, 907–933.
482 G. Müller, S. Wied, E.-M. Wetekam, A. Crecelius, A. Unkelbach, and J. Pünter, Stimulation of glucose utilization in 3T3 adipocytes and rat diaphragm in vitro by the sulfonylurea, glimepiride and glibenclamide, is correlated with modulations of the cAMP regulatory cascade, *Biochem. Pharmacol.*, **1994**, *48*, 985–996.
483 G. Müller and S. Wied, The sulfonylurea drug, glimepiride, stimulates glucose transport, glucose transporter translocation, and dephosphorylation in insulin-resistant rat adipocytes in vitro, *Diabetes*, **1993**, *42*, 1852–1867.
484 G. Müller, N. Hanekop, W. Kramer, W. Bandlow, and W. Frick, Interaction of phosphoinositolglycan(-peptides) with plasma membrane lipid rafts of rat adipocytes, *Arch. Biochem. Biophys.*, **2002**, *408*, 17–32.
485 G. Müller, C. Jung, W. Frick, W. Bandlow, and W. Kramer, Interaction of phosphoinositolglycan(-peptides) with plasma membrane lipid rafts triggers insulin-mimetic signaling in rat adipocytes, *Arch. Biochem. Biophys.*, **2002**, *408*, 7–16.
486 G. Müller and K. Geisen, Characterization of the molecular mode of action of the sulfonylurea, glimepiride, at adipocytes, *Horm. Metab. Res.*, **1996**, *28*, 469–487.
487 G. Müller and S. Welte, Lipid raft domains are the targets for the insulin-independent blood glucose-decreasing activity of the sulfonylurea glimepiride, *Rec. Res. Develop. Endocrinol.*, **2002**, *3*, 401–423.
488 A. Schlegel and M. P. Lisanti, The caveolin triad: caveolae biogenesis, cholesterol trafficking, and signal transduction, *Cytokine Growth Factor Rev.*, **2001**, *12*, 41–51.
489 T. V. Kurzchalia and R. G. Parton, Membrane microdomains and caveolae, *Curr. Opin. Cell. Biol.*, **1999**, *11*, 424–431.
490 F. Galbiati, B. Razani, and M. P. Lisanti, Emerging themes in lipid rafts and caveolae, *Cell*, **2001**, *106*, 403–411.
491 G. Müller, C. Jung, S. Wied, S. Welte, H. Jordan, and W. Frick, Redistribution of glycolipid raft domain components induces insulin-mimetic signaling in rat adipocytes, *Mol. Cell. Biol.*, **2001**, *21*, 4553–4567.
492 G. Müller, W. Frick, A. Crecelius, and S. Wied, The insulin- and sulphonylurea-inducible glycosylphosphatidylinositol-specific phospholipase C controles the localization and activity of signaling proteins in lipid rafts of adipocytes, *Signal Transduction, Mediator, Genes*, **2003**, *4*, p. 175.

493 A. Schlegel, D. Volonte, and J.A. Engelmann, Crowded little caves: structure and function of caveolae, *Cell. Signal.*, **1999**, *10*, 457–463.
494 P.E. Bickel, Lipid rafts and insulin signaling, *Am. J. Physiol. Endocrinol. Metab.*, **2002**, *282*, 1–10.
495 G. Müller, N. Hanekop, S. Wied, and W. Frick, Cholesterol depletion blocks redistribution of lipid raft components and insulin-mimetic signaling by glimepiride and phosphoinositolglycans in rat adipocytes, *Mol. Med.*, **2002**, *8*, 120–136.
496 G. Müller, E.A. Dearey, A. Korndörfer, and W. Bandlow, Stimulation of a glycosyl-phosphatidylinositol-specific phospholipase by insulin and the sulfonylurea, glimepiride, in rat adipocytes depends on increased glucose transport, *J. Cell. Biol.*, **1994**, *126*, 1267–1276.
497 S. Movahedi and N. Hooper, Insulin stimulates the release of the glycosyl phosphatidylinositol-anchored membrane dipeptidase from 3T3-L1 adipocytes through the action of a phospholipase C, *Biochem. J.*, **1997**, *326*, 531–537.
498 G. Müller, E. A. Dearey, and J. Pünter, The sulphonylurea drug, glimepiride, stimulates release of glycosylphosphatidylinositol-anchored plasma membrane proteins from 3T3 adipocytes, *Biochem. J.*, **1993**, *289*, 509–521.
499 M. Lafontan and M. Berlan, Do regional differences in adipocyte biology provide new pathophysiological insights?, *Trends Pharmacol. Sci.*, **2003**, *24*, 276–283.
500 M. Bergman, Non-esterified fatty acids and the liver: why is insulin secreted into the portal vein?, *Diabetologia*, **2000**, *43*, 946–952.
501 Z. Guo, et al., Regional postprandial fatty acid metabolism in different obesity phenotypes, *Diabetes*, **1999**, *48*, 1586–1592.
502 A. Basu, et al., Systemic and regional free fatty acid metabolism in type 2 diabetes, *Am. J. Physiol. Endocrinol. Metab.*, **2001**, *280*, 1000–1006.
503 A.-M. Levebvre, et al., Depot-specific differences in adipose tissue gene expression in lean and obese subjects, *Diabetes*, **1998**, *47*, 98–103.
504 J. Zierath, et al., Regional difference in insulin inhibition of non-esterified fatty acid release from human adipocytes: release to insulin receptor phosphorylation and intracellular signalling through the insulin receptor substrate-1 pathway, *Diabetologia*, **1998**, *41*, 1343–1354.
505 S.E. Meek, et al., Insulin regulation of regional free fatty acid metabolism, *Diabetes*, **1999**, *48*, 10–14.
506 X. Wu, et al., Depot-specific variation in protein-tyrosine phosphatase activities in human omental and subcutaneous adipose tissue: a potential contribution to differential insulin sensitivity, *J. Clin. Endocrinol. Metab.*, **2001**, *86*, 5973–5980.
507 F. Lönnqvist, et al., Catecholamine-induced lipolysis in adipose tissue of the elderly, *J. Clin. Invest.*, **1990**, *85*, 1614–1621.
508 P. Mauriège, et al., Heterogenous distribution of beta- and alpha2-adrenoceptor binding sites in human fat cells from various fat deposits: functional consequences, *Eur. J. Clin. Invest.*, **1987**, *17*, 156–165.
509 C. Garenc, et al., The alpha2-adrenergic receptor gene and body fat content and distribution: the HERITAGE family study, *Mol. Med.*, **2002**, *8*, 88–94.
510 J.-I. Osuga, S. Ishibashi, T. Oka, H. Yagyu, R. Tozawa, A. Fujimoto, F. Shionoira, N. Yahagi, F.B. Kraemer, O. Tsutsumi, and N. Yamada, Targeted disruption of hormone-sensitive lipase results in male sterility and adipocyte hypertrophy, but not in obesity, *Proc. Natl. Acad. Sci. USA*, **2000**, *97*, 787–792.
511 S.P. Wang, N. Laurin, J. Himms-Hagen, M.A. Rudnicki, E. Levy, M.-F. Robert, L. Pan, L. Oligny, and G.A. Mitchell, The adipose tissue phenotype of hormone-sensitive lipase deficiency in mice, *Obes. Res.*, **2001**, *9*, 119–128.
512 G. Haemmerle, R. Zimmermann, M. Hayn, C. Theussl, G. Waeg, E. Wagner, W. Sattler, T.M. Magin, E.F. Wagner, and R. Zechner, Hormone-sensitive lipase deficiency in mice causes diglyceride accumulation in adipose tissue, muscle, and testis, *J. Biol. Chem.*, **2002**, *277*, 4806–4815.
513 G. Haemmerle, R. Zimmermann, J.G. Strauss, D. Kratky, M. Riederer, G. Knipping, and R. Zechner, Hormone-

sensitive lipase deficiency in mice changes the plasma lipid profile by affecting the tissue-specific expression pattern of lipoprotein lipase in adipose tissue and muscle, *J. Biol. Chem.*, **2002**, *277*, 12946–12952.

514 P.J. Voshol, G. Haemmerle, and D.M. Ouwens, et al., Increased hepatic insulin sensitivity together with decreased hepatic triglyceride stores in hormone-sensitive lipase-deficient mice, *Endocrinology*, **2003**, *144*, 3456–3462.

515 W.-J. Shen, J. Wang, K. Harada, S. Patel, and F.B. Kraemer, Gene expression profile in islets of wild type and HSL–/– mice, *Diabetes*, **2003**, *52* (Suppl), A380.

516 M.S. Winzell, H. Svensson, S. Enerbäck, K. Ravnskjaer, S. Mandrup, V. Esser, P. Arner, M.-C. Alves-Guerra, B. Miroux, F. Sundler, B. Ahren, and C. Holm, Pancreatic β-cell lipotoxicity induced by overexpression of hormone-sensitive lipase, *Diabetes*, **2003**, *52*, 2057–2065.

517 M. Cnop, J.C. Hannaert, A. Hoorens, D.L. Eizirik, and D.G. Pipeleers, Inverse relation between cytotoxicity of free fatty acids in pancreatic islet cells and cellular triglyceride accumulation, *Diabetes*, **2001**, *50*, 1771–1777.

518 M.S. Winzell, C. Holm, and B. Ahren, Down-regulation of islet hormone-sensitive lipase during long-term high-fat feeding, *Biochem. Biophys. Res. Commun.*, **2003**, *304*, 273–278.

519 G.S. Hotamisligil, R.S. Johnson, R.J. Distel, R. Ellis, V.E. Papaioannou, and B.M. Spiegelman, Uncoupling of obesity from insulin resistance through a targeted mutation in aP2, the adipocyte fatty acid binding protein, *Science*, **1996**, *274*, 1377–1379.

520 N.R. Coe, M.A. Simpson, and D.A. Bernlohr, Targeted disruption of the adipocyte lipid-binding (aP2 protein) gene impairs fat cell lipolysis and increases cellular fatty acid levels, *J. Lipid Res.*, **1999**, *40*, 967–972.

521 K.T. Uysal, L. Scheja, S.M. Wiesbrock, S. Bonner-Weir, and G.S. Hotamisligil, Improved glucose and lipid metabolism in genetically obese mice lacking aP2, *Endocrinology*, **2000**, *141*, 3388–3396.

522 S. Shaughnessy, E.R. Smith, S. Kodukula, J. Storch, and S.K. Fried, Adipocyte metabolism in adipocyte fatty acid binding protein knockout (aP2–/–) mice after short-term high-fat feeding: functional compensation by keratinocyte fatty acid binding protein, *Diabetes*, **2000**, *49*, 904–911.

523 S. Mottagui-Tabar, M. Ryden, P. Löfgren, G. Faulds, J. Hoffstedt, A.J. Brookes, I. Andersson, and P. Arner, Evidence for an important role of perilipin in the regulation of human adipocyte lipolysis, *Diabetologia*, **2003**, *46*, 789–797.

524 J. Martinez-Botas, J.B. Anderson, and D. Tessier, et al., Absence of perilipin results in leanness and reverses obesity in Leprdb/db mice, *Nat. Gen.*, **2000**, *26*, 474–479.

525 J.T. Tansey, C. Sztalryd, and J. Gruia Gray, et al., Perilipin ablation results in a lean mouse with aberrant adipocyte lipolysis, enhanced leptin production, and resistance to diet-induced obesity, *Proc. Natl. Acad. Sci. USA*, **2001**, *98*, 6494–6499.

526 C. Sztalryd, G. Xu, and H. Dorward, et al., Perilipin A is essential for the translocation of hormone-sensitive lipase during lipolytic activation, *J. Cell. Biol.*, **2003**, *161*, 1093.

527 W.B. Huttner and A. Schmidt, Lipids, lipid modification and lipid-protein interaction in membrane budding and fission-insights from the roles of endophilin A1 and synaptophysin in synaptic vesicle endocytosis, *Curr. Opin. Neurobiol.*, **2000**, *10*, 543–551.

528 G.S. McKnight, D.E. Cummings, P.S. Amieux, M.A. Sikorski, E.P. Brandon, J.V. Planas, K. Motamed, and R.L. Idzerda, Cyclic AMP, PKA, and the physiological regulation of adiposity, *Rec. Prog. Horm. Res.*, **1998**, *53*, 139–161.

529 D.E. Cummings, E.P. Brandon, J.V. Planas, K. Motamed, R.L. Idzerda, and G.S. McKnight, Genetically lean mice result from targeted disruption of the RII beta subunit of protein kinase A, *Nature*, **1996**, *382*, 622–626.

530 L. Emorine, N. Blin, and A.D. Strosberg, The human beta 3-adrenoceptor: the search for a physiological function, *Trends Pharmacol. Sci.*, **1994**, *15*, 3–7.

531 M.K. Meier, L. Alig, M.E. Burgi-Saville, and M. Muller, Phenethanolamine derivatives with calorigenic and antidiabetic qualities, *Int. J. Obes.*, **1984**, *8*, 215–225.

532 E.E. Largis, M.G. Burns, H.A. Meunkel, J.A. Dolan, and T.H. Claus, Antidiabetic and antiobesity effects of a highly selective β_3-adrenoceptor agonist, *Drug. Dev. Res.*, **1994**, *32*, 69–76.

533 J. Himms-Hagen, J. Cui, E. Danforth, D.J. Tatjes, S.S. Lang, B.L. Waters, and T.H. Claus, Effect of Cl-316,243, a thermogenic beta 3-agonist, on energy balance and brown and white adipose tissues in rats, *Am. J. Physiol.*, **1994**, *266*, 1371–1382.

534 P.S. Amieux, D.E. Cummings, K. Motamed, E.P. Brandon, L.A. Wailes, K. Le, R.L. Idzerda, and G.S. Knight, Compensatory regulation of RIα protein levels in protein kinase A mutant mice, *J. Biol. Chem.*, **1997**, *272*, 3993–3998.

535 C.M. Moxham and C.C. Malbon, Insulin action impaired by deficiency of the G-protein subunit Gi alpha2, *Nature*, **1996**, *379*, 840–844.

536 S.A. Schreyer, D.E. Cummings, G.S. McKnight, and R.C. LeBoeuf, Mutations of the RIIβ subunit of protein kinase A prevents diet-induced insulin resistance and dyslipidemia in mice, *Diabetes*, **2001**, *50*, 2555–2562.

537 R.S. Surwit, C.M. Kuhn, C. Cochrane, J.A. McCubbin, and M.N. Feinglos, Diet induced type II diabetes in C57BL/6J mice, *Diabetes*, **1988**, *37*, 1163–1170.

538 R.S. Surwit, M.F. Seldin, C.M. Kuhn, C. Cochrane, and M.N. Feinglos, Control of expression of insulin resistance and hyperglycemia by different genetic factors in diabetes C57BL/6J mice, *Diabetes*, **1991**, *40*, 82–87.

539 D.O. Allen and J.T. Quesenberry, Quantitative differences in the cyclic AMP-lipolysis relationships for isoproterenol and forskolin, *J. Pharmacol. Exp. Ther.*, **1988**, *244*, 852–858.

540 D.O. Allen, B. Ahmed, and K. Naseer, Relationships between cyclic AMP levels and lipolysis in fat cells after isoproterenol and forskolin stimulation, *J. Pharmacol. Exp. Ther.*, **1986**, *238*, 659–664.

541 I. Litosch, T.H. Hudson, I. Mills, S.Y. Li, and J.N. Fain, Forskolin as an activator of cyclic AMP accumulation and lipolysis in rat adipocytes, *Mol. Pharmacol.*, **1982**, *22*, 109–115.

542 G.J. Murphy, D.M. Kirkham, M.A. Cawthorne, and P. Young, Correlation of β3-adrenoceptor-induced activation of cyclic amp-dependent protein kinase with activation of lipolysis in rat white adipocytes, *Biochem. Pharmacol.*, **1993**, *46*, 575–581.

543 C. Hollenga, F. Brouwer, and J. Zaagsma, Differences in functional cyclic AMP compartments mediating lipolysis by isoprenaline and BRL 37344 in four adipocyte types, *Eur. J. Pharmacol.*, **1991**, *200*, 325–330.

544 D. Langin, D. Ekholm, M. Ridderstrale, M. Lafontan, and P. Belfrage, cAMP-dependent protein kinase activation mediated by β3-adrenergic receptors parallels lipolysis in rat adipocytes, *Biochim. Biophys. Acta*, **1992**, *1135*, 349–352.

545 M. Shih, F. Lin, J.D. Scott, H. Wang, and C.C. Malbon, Dynamic complexes of β2-adrenergic receptors with protein kinases and phosphatases and the role of gravin, *J. Biol. Chem.*, **1999**, *274*, 1588–1595.

546 C.S. Rubin, A kinase anchor proteins and the intracellular targeting of signals carried by cyclic AMP, *Biochim. Biophys. Acta*, **1994**, *1224*, 467–479.

547 M. Colledge and J.D. Scott, AKAPs: from structure to function, *Trends Cell. Biol.*, **1999**, *9*, 216–221.

548 C. Rosenmund, D.W. Carr, S.E. Bergeson, G. Nilaver, J.D. Scott, and G.L. Westbrook, Anchoring of protein kinase A is required for modulation of AMPA/kainate receptors on hippocampal neurons, *Nature*, **1994**, *368*, 853–856.

549 R.S. Westphal, S.J. Tavalin, J.W. Lin, N.M. Alto, I.D. Fraser, L.K. Langeberg, M. Sheng, and J.D. Scott, Regulation of NMDA receptors by an associated phosphatase-kinase signaling complex, *Science*, **1999**, *285*, 93–96.

550 I. Murray, P.J. Havel, A.D. Sniderman, and K. Cianflone, Reduced body weight, adipose tissue, and leptin levels despite increased energy intake in female

mice lacking acylation-stimulating protein, *Endocrinology*, **2000**, *141*, 1041–1049.

551 I. MURRAY, A. D. SNIDERMAN, and K. CIANFLONE, Acylation stimulating protein (ASP) deficiency alters postprandial and adipose tissue metabolism in male mice, *J. Biol. Chem.*, **1999**, *274*, 36219–36225.

552 B. RAZANI, Caveolin-1 deficient mice are lean, resistant to diet-induced obesity, and show hypertriglyceridemia with adipocyte abnormalities, *J. Biol. Chem.*, **2002**, *277*, 8635–8647.

553 S. L. ZHOU, D. STUMP, D. SORRENTINO, B. J. POTTER, and P. D. BERK, Adipocyte differentiation of 3T3-L1 cells involves augmented expression of a 43-kDa plasma membrane fatty acid-binding protein, *J. Biol. Chem.*, **1992**, *267*, 14456–14461.

554 G. E. GERBER, D. MANGROO, and B. L. TRIGATTI, Identification of high affinity membrane-bound fatty acid-binding proteins using a photoreactive fatty acid, *Mol. Cell. Biochem.*, **1993**, *123*, 39–44.

555 B. L. TRIGATTI, R. G. ANDERSON, and G. E. GERBER, Identification of caveolin-1 as a fatty acid binding protein, *Biochem. Biophys. Res. Commun.*, **1999**, *255*, 34–39.

556 C. A. BAUMANN, V. RIBON, and M. KANZAKI, et al., CAP defines a second signalling pathway required for insulin-stimulated glucose transport, *Nature*, **2000**, *407*, 202–207.

557 C. HOLM, D. LANGIN, V. MANGANIELLO, P. BELFRAGE, and E. DEGERMAN, Regulation of hormone-sensitive lipase activity in adipose tissue, *Methods Enzymol.*, **1997**, *286*, 45–66.

558 C. HOLM, T. OSTERLUND, H. LAURELL, and J. A. CONTRERAS, Molecular mechanisms regulating hormone-sensitive lipase and lipolysis, *Annu. Rev. Nutr.*, **2000**, *20*, 365–393.

559 G. FRÜHBECK, J. GOMEZ-AMBROSI, F. J. MURUZABAL, and M. A. BURRELL, The adipocyte: a model for integration of endocrine and metabolic signaling in energy metabolism regulation, *Am. J. Physiol. Endocrinol. Metab.*, **2001**, *280*, 827–847.

Subject Index

a

ABCA1 277
accumulated hydrolysis products 46
acetylcholinesterase 128
acetyl-CoA carboxylase (ACC) 126, 271 f.
acid lipase 6
acipimox 286 ff.
acylation stimulating protein (ASP) 285, 293, 300 f.
acyl-CoA synthase-1 (ACS1) 266
acyl-CoA: cholesterol acyltransferase 244
acyl-CoA: diacylglycerol acyltransferase 244
adenosine 285, 287 ff.
adenylate cyclase (AC) 106, 270, 287 ff., 293, 300
adipocyte differentiation related protein (ADRP) 248 f., 257 f., 267, 298
adipocyte lipid-binding protein 272
adipocyte lipid-binding protein (ALBP) 128, 273, 293, 296, 297
adipophilin 250
adrenalin 126
ADRP 248 f.
adsorption chromatographic techniques 4
affinity chromatography 3, 6
AKAPs 300
albumin 122, 125
allosteric activator 126
alutenusin 89
AMP-activated protein kinase (AMPK) 260, 261 ff., 272
amphipathic lid 170
anticancer drug 42
anti-obesity 156
apolipoprotein CII 143
apoptosis 79 ff.
appetite suppressants 216
aqua(glycerol)porin (AQPap) 276, 277
adenosine receptor (AR) 287
Arf 63 ff., 68, 69 f.
ARNO 65
atherosclerosis 35, 81, 84, 134, 139

b

bacterial lipase 216
bile lipids 202
bile salts 156
bile salt-stimulated lipase 128
bile-lipoprotein complex 202
body-mass index 222
brefeldin A 65
brefeldin A esterase 128
butanol 67 ff.

c

C3 exoenzyme 64
Ca^{2+}/calmodulin-dependent kinase 278
Caelyx 42
cAMP 126, 259 ff., 264, 270 f., 274, 277 ff., 283 f., 287 f., 296, 299 f.
carbachol 65
carbamates 129
carbobenzyloxy-leucine-tyrosine-chloromethyl ketone 67zcatalytic domain 6, 34, 127
catalytic triad 24 ff., 159, 199, 165, 274, 298
catecholamines 126, 259, 278 f., 292 ff., 300
caveolae 83, 104, 110, 251, 298
caveolin 83, 95, 122, 247, 250 f., 252 f., 255, 257, 259, 301
Cbl-CAP 302
cGMP 279 f., 283
chiro-inositol 106 f.
chlorogentisylquinone 89
cholecystokinin 196
cholesterol 139, 277, 286, 291
cholesterol esterase 216
cholesterol esters 122

Lipases and Phospholipases in Drug Development
Edited by G. Müller and S. Petry

ISBN: 3-527-30677-3

chronic pancreatitis 215
chylomicrons 233 f., 243, 285
coatamer 35, 37, 68
colipase 156, 158, 202
complement pathway 285
computer-assisted experimental design strategy (CAED) 129
continuous assays 123
COS-1 cells 142
cyclipostine 128 f.
cytokines 101, 105, 232, 243, 267, 279, 284, 295

d
degranulation 30
desensitization 289
DFP 36
DGAT 240, 245, 258, 285
diabetes 101, 105, 107, 129, 134, 236, 237 ff., 244, 249, 283, 286, 290, 292, 296, 302,
diacylglycerol (DAG) 21, 71 f., 69, 103, 122, 196, 294
diacylglycerol lipase 6
di-capryl PC lipids (DCPC) 48
dietary lipids 195
diglycerides (DG) 21, 71 f., 69, 103, 122, 196, 294
DIGs 247, 251, 259, 290 f., 302
2,3-diphosphoglycerate 67
domains 44 ff.
Doxil 42
DPPC 47 ff.
DPPE-PEG$_{2000}$ 48
drug delivery systems 43
drug-carrying liposomes 50

e
EL knockout mice 148
electrostatic effect 46
emulsification 155, 195 f.
encapsulated drugs 43
endothelial cells 139, 142
endothelial lipase (EL) 122, 139 f., 143 ff., 148
energy homeostasis 232
enterocytes 155
EST2 esterase 128
eudesmanolide 129

f
fatty acid transport protein (FATP) 266, 275 f.
fatty acid-binding protein (FABP) 297
flip-flop 94
fluid lipid patches in solid bilayers 44
free fatty acids (FFA) 195, 196, 211
fungal lipases 128, 216
fusogenic 94
futile cycle 236

g
galenic form 224
gastric emptying 210
gastric inhibitory polypeptide 283, 296
gastric lipase 5 f., 165, 216
gastrointestinal (GI) tract 195
gastrointestinal lipolysis 206 f., 211
GDP/GTP exchange factor (GEF) 65
gel filtration 3 ff.
gel filtration chromatography 6
glimepiride 290 ff., 302
glucagon 126
glucagon-like peptide 1 296
gluconeogenesis 276
glucose homeostasis 237
glucose-6-phosphate 279
glucose-dependent insulinotropic polypeptide 283, 296
GLUT4 110
glycerol-3-phosphate acyltransferase 244
glycosyl-phosphatidylinositol-anchored (GPI) proteins 102, 251, 290 f.
Gly-X-Ser-X-Gly motif 127
Golgi 58 f.
growth hormone 283
GTPases 64 ff.
GW4869 89

h
HDL-cholesterol (HDL) 139, 143 ff., 147, 151
HDL binding 144
HDL metabolism 146
HDL phospholipids 144
heparin 140, 148
hepatic lipase (HL) 23, 25, 36, 122, 139, 143 f., 147 f.
high-throughput screening 89, 125
histamine release 30
HKD 55, 57, 58, 142
HMG-CoA reductase 271 f.
hormone-sensitive lipase (HSL) 122, 125 f., 217, 234 f., 237 f., 240, 258, 260 ff., 265 ff., 268 ff., 274 ff., 284 f., 287, 291, 293 f.
human gastric lipase (HGL) 180, 195 f., 204, 224
human pancreatic lipase (HPL) 6, 25, 122, 155 f., 170, 176, 180, 195 f., , 202, 204, 265

α/β-hydrolase fold 128, 157, 159
hydrophobic domains 122, 170
hydrophobic-interaction chromatography 4 ff.
hyperglycaemia 110
hyperlipidemia 234, 241, 286
hyperthermia 42
hypothyroidism 142

i
IL-1β 145
insulin 126zinsulin receptor 108 ff., 280, 289 ff., 295, 302
insulin receptor substrate protein (IRS) 280
insulin resistance 110, 238, 243 f., 292 f., 297
insulin secretion 295 f.
interfacial activation 122 f., 158, 167, 199
interfacial denaturation 160
interfacial quality 160, 163, 176, 180
intestinal absorption 216
ion-exchange chromatography 4 f., 6

j
JAK/STAT 282 f.

l
lag time 46
Langmuir 92
Laplace transformation 12 f.
lateral heterogeneity 44, 49
lipid droplets (LD) 235, 238, 243, 246 ff., 251 ff., 262 ff., 275, 277, 279, 287, 293, 296 ff., 301 ff.
LDL-cholesterol (LDL) 81, 83 f., 95, 139
LDL/HDL ratio 223
lecithin-cholesterol acyltransferase 148
leptin 243, 281 f., 293, 301
lid 199
lipase consensus sequence 23
lipase inhibitor 144
lipase secretions 206
lipase-colipase complex 156
lipid disorders 129
lipid/water interface 122, 167, 202
lipid-binding domain 127, 140
lipoapoptosis 238 ff.
lipolysis 108, 112, 206, 232, 234 ff., 239 f., 259 ff., 267 f., 277 ff., 284, 286 f., 291
lipoprotein 122, 139 f., 232, 234, 240, 242, 252
lipoprotein lipase (LPL) 6, 23, 25, 122, 139, 143, 145, 176, 180, 217, 232, 234, 243, 283, 293 f.
liposomal drug targeting 41, 50
liposomes 41, 43
lipotoxicity 234, 237 ff., 295 f.
lipotransin 262, 268 ff., 290, 302 f.
lipstatin 216
LXR 277
lysolipids 43, 49
lysophosphatidic acid (LPA) 67 ff.
lysophospholipase 35, 70
lysosomal acid lipase 5

m
macquarimicin A 89
α-mangostin 89
manumycin A 89
metabolic syndrome 236 ff., 302
methyl arachidonyl fluorophosphonate 36
micellization 155
Michaelis-Menten equilibrium 160
Michaelis-Menten kinetics 122
microdialysis 276
microdomain 95
mixed bile salt/orlistat micelles 176
mixed micelles 196zmobilization of NEFA 121 f.
monoacylglycerol (MAG) 122, 195 f., 211
monoacylglycerol lipase (ML) 125, 235, 270 f.
monoglycerides (MG) 122, 195 f., 211
monolayer technique 164
monomolecular film technique 169
2-monopalmitoylglycerol 270 f.
myo-inositol 106 f., 111

n
NBD 123 f.
NEFA 125, 235
neomycin 60
NF-ξB pathway 145
nicotinic acid (NA) 286, 288 ff.
NIDDM 101, 105, 107, 129, 134, 236, 237 ff., 244, 249, 283, 286, 290, 292, 296, 302
Niemann-Pick syndrome 80
nitrix oxide 283
noradrenalin 126

o
obesity 216
1-O-DPPC 50
oil drop 164
oil-water interface 199
orlistat 175, 180, 185, 216
1,2-O-stearoyl PC (di-O-SPC) lipids 47
oxadiazolone 129

β-oxidation 126
oxyanion hole 159

p
pancreatic β-cells 232, 238 ff., 295
pancreatic (HPL) lipases 6, 25, 122, 155 f., 170, 176, 180, 195 f., 202, 204, 265
pancreatic lipase-related proteins 198
pancreatic proteases 214
parasympathic nervous system 126
phosphodiesterase (PDE) 231, 270, 279 f., 284 ff., 290 ff.
(PEG)-lipids 42
perilipin 248, 250, 255 ff., 262, 266 ff., 284, 290, 293, 297 f., 302
permeability enhancers 49
phase boundaries 45
phorbol esters 61 ff., 274
phosphatases 261, 263, 270, 300
phosphatidic acid (PA) 23, 25, 33, 35 ff., 55, 67 ff., 103
phosphatidylcholine 55
phosphoinositolglycans (PIG) 290 f., 302
phospholipase 139, 253, 255, 289, 291
phospholipase A_2 41, 43, 46, 95
phospholipase B 29
phospholipase C 79, 82 ff., 88 f., 93 ff., 110, 102 f., 241
phospholipase D 33, 35, 102 f., 109 f., 242, 257 f.
phospholipid 122, 139, 142, 199
phospholipid substrates 143
phospholipid vesicles 127
phosphorylation 127
Phox homology 55
pH-stat assay 204
phosphatidylinositol-3-kinase (PI3K) 262, 280, 285, 291, 293, 295
platelets 30
pleckstrin homology (PH) domains 55 f., 58, 60
PMA 61 ff.
polycystic ovarian disease 107
polyunsaturated fatty acids 126
porcine lipase 7
preeclampsia 107
proenzymes 199
programmable fusogenetic liposomes 42
protein kinase A (PKA) 106, 126, 231, 260 ff., 267, 270, 272, 275 f. , 279 ff., 284, 290, 297 f., 299 f., 302
protein kinase B (PKB) 231, 280, 295
proteolytic cleavage 199, 214
PX domains 56, 58
pyrrolopyrazinediones 129

r
RACK1 62
rafts 33, 44, 80, 247, 251, 259, 290 f., 302
reactive oxygen species 83
re-esterification 234 ff.
regulatory domain 127
Rho 63 f., 66 f.

s
scyphostatin 89
second messenger 102
secretory phospholipase 43, 148
short-chain lipid species 46
size-exclusion chromatography 7 f., 19, 36
sn-3 position 199
stealth liposomes 46
stereoselectivity 168
steroid fatty acid 126
super-substrates 122
supramolecular aggregate 44
surface area density 46
surface dilution phenomena 160
surface loop 158, 167
surface tension 202, 204

t
tensio-active 204
test meals 205
tetrahydrolipstatin 144, 158
thiazolidinediones 284
TNF-α 145,267,283f,293
toxin B 60, 64
transgenic mouse model 141
transition state 167
triacylglycerols (TAG) 122, 139, 142, 144, 196, 199
tributyrin 204
triglyceride (TG) 122, 139, 142, 144, 196, 199
triglyceride digestion 216
triglyceride lipases 139
triglyceride-rich lipoproteins 144

u
unilamellar liposomes 46zv
VLDL 125, 232 ff., 237, 242 ff., 258, 294

w
weight loss/regain 224
white adipose tissue (WAT) 256, 275, 282, 294, 299 ff.